Advances in Crop Molecular Breeding and Genetics

Advances in Crop Molecular Breeding and Genetics

Guest Editors

Zhiyong Li
Chaolei Liu
Jiezheng Ying

Basel • Beijing • Wuhan • Barcelona • Belgrade • Novi Sad • Cluj • Manchester

Guest Editors

Zhiyong Li
State Key Lab of Rice Biology
and Breeding
China National Rice
Research Institute
Hangzhou
China

Chaolei Liu
State Key Lab of Rice Biology
and Breeding
China National Rice
Research Institute
Hangzhou
China

Jiezheng Ying
State Key Lab of Rice Biology
and Breeding
China National Rice
Research Institute
Hangzhou
China

Editorial Office
MDPI AG
Grosspeteranlage 5
4052 Basel, Switzerland

This is a reprint of the Special Issue, published open access by the journal *Agronomy* (ISSN 2073-4395), freely accessible at: www.mdpi.com/journal/agronomy/special_issues/SO96IP2443.

For citation purposes, cite each article independently as indicated on the article page online and using the guide below:

Lastname, A.A.; Lastname, B.B. Article Title. *Journal Name* **Year**, *Volume Number*, Page Range.

ISBN 978-3-7258-2910-1 (Hbk)
ISBN 978-3-7258-2909-5 (PDF)
https://doi.org/10.3390/books978-3-7258-2909-5

© 2025 by the authors. Articles in this book are Open Access and distributed under the Creative Commons Attribution (CC BY) license. The book as a whole is distributed by MDPI under the terms and conditions of the Creative Commons Attribution-NonCommercial-NoDerivs (CC BY-NC-ND) license (https://creativecommons.org/licenses/by-nc-nd/4.0/).

Contents

Yaqi Su, Zhen Cheng, Jiezheng Ying, Chaolei Liu and Zhiyong Li
New Insights into Crop Molecular Breeding and Genetics
Reprinted from: *Agronomy* **2024**, *14*, 2999, https://doi.org/10.3390/agronomy14122999 1

Hong Wang, Zhixing Nie, Tonglin Wang, Shuhuan Yang and Jirong Zheng
Comparative Transcriptome Analysis of Eggplant (*Solanum melongena* L.) Peels with Different Glossiness
Reprinted from: *Agronomy* **2024**, *14*, 3063, https://doi.org/10.3390/agronomy14123063 7

Yichen Ye, Shuting Wen, Guo Zhang, Xingzhe Yang, Dawei Xue and Yunxia Fang et al.
Quantitative Trait Locus Analysis for Panicle and Flag Leaf Traits in Barley (*Hordeum vulgare* L.) Based on a High-Density Genetic Linkage Map
Reprinted from: *Agronomy* **2024**, *14*, 2953, https://doi.org/10.3390/agronomy14122953 19

Guoqing Dong, Zihao Gui, Yi Yuan, Yun Li and Dengxiang Du
Expression Analysis of the Extensive Regulation of Mitogen-Activated Protein Kinase (MAPK) Family Genes in Buckwheat (*Fagopyrum tataricum*) During Organ Differentiation and Stress Response
Reprinted from: *Agronomy* **2024**, *14*, 2613, https://doi.org/10.3390/agronomy14112613 28

Lucia Urbanová, Jana Bilčíková, Dagmar Moravčíková and Jana Žiarovská
Natural Variability of Genomic Sequences of Mal d 1 Allergen in Apples as Revealed by Restriction Profiles and Homolog Polymorphism
Reprinted from: *Agronomy* **2024**, *14*, 2056, https://doi.org/10.3390/agronomy14092056 50

Yonghui He, Chenxi Wang, Xueyou Hu, Youle Han, Feng Lu and Huanhuan Liu et al.
Genetic Basis and Exploration of Major Expressed QTL *qLA2-3* Underlying Leaf Angle in Maize
Reprinted from: *Agronomy* **2024**, *14*, 1978, https://doi.org/10.3390/agronomy14091978 63

Jirapong Yangklang, Jirawat Sanitchon, Jonaliza L. Siangliw, Tidarat Monkham, Sompong Chankaew and Meechai Siangliw et al.
Yield Performance of RD6 Glutinous Rice near Isogenic Lines Evaluated under Field Disease Infection at Northeastern Thailand
Reprinted from: *Agronomy* **2024**, *14*, 1871, https://doi.org/10.3390/agronomy14081871 76

Yong Han, Alexander Abair, Julian van der Zanden, Madhugiri Nageswara-Rao, Saipriyaa Purushotham Vasan and Roopali Bhoite et al.
Transcriptome-Wide Genetic Variations in the Legume Genus *Leucaena* for Fingerprinting and Breeding
Reprinted from: *Agronomy* **2024**, *14*, 1519, https://doi.org/10.3390/agronomy14071519 91

Panpan Jia, Shenghui Liu, Wenqiu Lin, Honglin Yu, Xiumei Zhang and Xiou Xiao et al.
Authenticity Identification of F_1 Hybrid Offspring and Analysis of Genetic Diversity in Pineapple
Reprinted from: *Agronomy* **2024**, *14*, 1490, https://doi.org/10.3390/agronomy14071490 105

Mu Peng, Zhiyan Wang, Zhibiao He, Guorui Li, Jianjun Di and Rui Luo et al.
Combining Ability, Heritability, and Heterosis for Seed Weight and Oil Content Traits of Castor Bean (*Ricinus communis* L.)
Reprinted from: *Agronomy* **2024**, *14*, 1115, https://doi.org/10.3390/agronomy14061115 118

Yiwang Zhu, Xiaohuai Yang, Peirun Luo, Jingwan Yan, Xinglan Cao and Hongge Qian et al.
Creation of Bacterial Blight Resistant Rice by Targeting Homologous Sequences of *Xa13* and *Xa25* Genes
Reprinted from: *Agronomy* **2024**, *14*, 800, https://doi.org/10.3390/agronomy14040800 . . . . . . **132**

Xinghua Qi, Ying Zhao, Ningning Cai, Jian Guan, Zeji Liu and Zhiyong Liu et al.
Characterization and Transcriptome Analysis Reveal Exogenous GA_3 Inhibited Rosette Branching via Altering Auxin Approach in Flowering Chinese Cabbage
Reprinted from: *Agronomy* **2024**, *14*, 762, https://doi.org/10.3390/agronomy14040762 . . . . . . **141**

Ruisi Yang, Fei Wang, Ping Luo, Zhennan Xu, Houwen Wang and Runze Zhang et al.
The Role of the ADF Gene Family in Maize Response to Abiotic Stresses
Reprinted from: *Agronomy* **2024**, *14*, 717, https://doi.org/10.3390/agronomy14040717 . . . . . . **159**

Cuiping Chen, Xuebing Zhu, Zhi Zhao, Dezhi Du and Kaixiang Li
Fine Mapping and Functional Verification of the *Brdt1* Gene Controlling Determinate Inflorescence in *Brassica rapa* L.
Reprinted from: *Agronomy* **2024**, *14*, 281, https://doi.org/10.3390/agronomy14020281 . . . . . . **175**

Demba Dramé, Amy Bodian, Daniel Fonceka, Hodo-Abalo Tossim, Mouhamadou Moussa Diangar and Joel Romaric Nguepjop et al.
Agro-Morphological Variability of Wild *Vigna* Species Collected in Senegal
Reprinted from: *Agronomy* **2023**, *13*, 2761, https://doi.org/10.3390/agronomy13112761 . . . . . **190**

Junying Zhang, Jifeng Zhu, Liyong Yang, Yanli Li, Weirong Wang and Xirong Zhou et al.
Mapping of the Waxy Gene in *Brassica napus* L. via Bulked Segregant Analysis (BSA) and Whole-Genome Resequencing
Reprinted from: *Agronomy* **2023**, *13*, 2611, https://doi.org/10.3390/agronomy13102611 . . . . . **206**

Yang Xu, Fan Yan, Zhengwei Liang, Ying Wang, Jingwen Li and Lei Zhao et al.
Overexpression of the Peanut *AhDGAT3* Gene Increases the Oil Content in Soybean
Reprinted from: *Agronomy* **2023**, *13*, 2333, https://doi.org/10.3390/agronomy13092333 . . . . . **222**

Wanning Liu, Guan Li, Jiezheng Ying and Zhiyong Li
Advances in Crop Molecular Breeding and Genetics
Reprinted from: *Agronomy* **2023**, *13*, 2311, https://doi.org/10.3390/agronomy13092311 . . . . . **235**

Zanping Han, Yunqian Jin, Bin Wang and Yiyang Guo
Multi-Omics Revealed the Molecular Mechanism of Maize (*Zea mays* L.) Seed Germination Regulated by GA3
Reprinted from: *Agronomy* **2023**, *13*, 1929, https://doi.org/10.3390/agronomy13071929 . . . . . **239**

Diankai Gong, Xue Zhang, Fei He, Ying Chen, Rui Li and Jipan Yao et al.
Genetic Improvements in Rice Grain Quality: A Review of Elite Genes and Their Applications in Molecular Breeding
Reprinted from: *Agronomy* **2023**, *13*, 1375, https://doi.org/10.3390/agronomy13051375 . . . . . **252**

Lian Yin, Yudong Sun, Xuehao Chen, Jiexia Liu, Kai Feng and Dexu Luo et al.
Genome-Wide Analysis of the HD-Zip Gene Family in Chinese Cabbage (*Brassica rapa* subsp. *pekinensis*) and the Expression Pattern at High Temperatures and in Carotenoids Regulation
Reprinted from: *Agronomy* **2023**, *13*, 1324, https://doi.org/10.3390/agronomy13051324 . . . . . **278**

Qian Jiang, Yu Wang, Aisheng Xiong, Hui Zhao, Ruizong Jia and Mengyao Li et al.
Genome-wide Analysis of NBS-type Genes in Papaya Gives an Insight to Simplified Disease Resistance Genes of Eudicots
Reprinted from: *Agronomy* **2023**, *13*, 970, https://doi.org/10.3390/agronomy13040970 . . . . . . **296**

So-Myeong Lee, Nkulu Rolly Kabange, Ju-Won Kang, Youngho Kwon, Jin-Kyung Cha and Hyeonjin Park et al.
Identifying QTLs Related to Grain Filling Using a Doubled Haploid Rice (*Oryza sativa* L.) Population
Reprinted from: *Agronomy* **2023**, *13*, 912, https://doi.org/10.3390/agronomy13030912 . . . . . . **314**

Editorial

New Insights into Crop Molecular Breeding and Genetics

Yaqi Su [1], Zhen Cheng [1,2], Jiezheng Ying [1], Chaolei Liu [1] and Zhiyong Li [1,*]

[1] State Key Laboratory of Rice Biology and Breeding, China National Rice Research Institute, Hangzhou 311400, China; zhenc0311@gmail.com (Z.C.); yingjiezheng@caas.cn (J.Y.); liuchaolei@caas.cn (C.L.)
[2] School of Life Sciences, Hubei University, Wuhan 430061, China
* Correspondence: lizhiyong@caas.cn

1. Introduction

As the global population continues to grow, the need to increase agricultural productivity is becoming increasingly urgent. Cultivation of crops are the basis of agricultural production, but traditional planting patterns have shortcomings, such as low yield, poor resistance to adversity, and weak resistance to diseases and pests. In order to solve these problems, researchers have begun to use biotechnologies such as gene editing, transgenic technology, molecular marker-assisted selection (MAS), genomics, and transcriptomics to optimize crops. This paper focuses on the latest discoveries of different crops published in *agronomy* regarding the regulation of plant developmental processes or agronomic traits, especially methods to increase crop yield, such as searching for elite genetic resources, adaptation to adversity stress, improvement of crop resistance to pests and diseases, and the potential utilization of biotechnology in the genetic improvement of crops.

2. Research in Food Crops

2.1. Research in Rice

Rice is one of the world's major food crops and is used as a major source of energy by more than 3 billion people worldwide, accounting for 25% of the world's total caloric intake [1]. In recent years, rice yield has increased dramatically with the improvement of hybridization technology and cultivation management, but it is still difficult to meet the growing demand of the population [2]. Therefore, most of the current rice breeders concentrate their efforts on seeking various methods, such as gene editing, MAS, and genomics, to improve rice yield. Rice yield is associated with a variety of traits, among which the defining trait is grain filling, and the development of rice grains during filling directly affects rice fruiting rate and grain weight. Lee et al. [3] identified three quantitative trait loci (QTLs), *qFG3*, *qFG5-1*, and *qFG5-2*, significantly associated with the control of grain filling by a doubled haploid rice population, of which *qFG3* is a novel and stable QTL that was detected in both early and normal cultivation seasons. In addition, genes harbored by *qFG3* are related to cell division and differentiation, photosynthesis, and starch synthesis and can also mediate the transition between seed filling and abiotic stress response mechanisms. Rice is susceptible to a variety of diseases throughout its reproductive stage, which is one of the main reasons affecting rice yield. Bacterial blight (BB) is one of the most serious bacterial diseases in rice caused by *Xanthomonas oryzae pv. oryzae* (*Xoo*) [4]. Currently, pesticide control is the primary method to control BB. Long-term use of pesticides will result in environmental pollution and the development of pathogen resistance, so the most effective and eco-friendly approach is to breed resistant rice cultivars. Zhu et al. [5] created a single Cas9/gRNA within a 30 bp homologous sequence of the *Xa13* and *Xa25* genes that targets both, and successfully obtained *xa13*, *xa25*, and *xa13/xa25* double mutants. The *xa13*, *xa25*, and *xa13/xa25* double mutants exhibit greater resistance to BB than the wild type (WT) in all five rice varieties. These results provide an effective idea for creating rice cultivars with multiple resistance genes. RD6 is the famous Thailand glutinous rice cultivar, which is popular for its high flavor quality. While it is susceptible to BB and rice blast disease infection. Jirapong et al. [6] pyramided multiple resistance genes by MAS and generated eight near isogenic lines (NILs) of RD6. Nine genotypes, including RD6, were evaluated at three different locations in two consecutive years, and it was found that most of the genotypes (G1, G3, G4, G5, G6, and G7) had higher yield stability than RD6, whereas the stability of yield was poorer in G2 and G8. In case of severe BB disease infection, G2 was able to maintain greater yield while the yield of other genotypes strikingly reduced.

2.2. Research in Maize

Maize is a high-quality food crop with high nutritional value. It is an essential raw material for human food, cultured feed, and industrial production [7]. During the growth of maize, it is susceptible to various kinds of adversity stresses, among which drought is one of the most serious stresses that seriously affects maize yield. The hazard of drought is becoming more severe with climate change and population pressure [8,9]. Yang et al. [10] identified 15 actin depolymerizing factor (ADF) genes in the form of tandem duplication or segmental duplication. Among which, *ZmADF5* has been confirmed to be a drought tolerance gene. The analysis of candidate gene association revealed that its promoter region contains three mutation sites associated with drought resistance. In addition, the ADF gene family with a high degree of conservation shows structural similarity. The presence of abundant abiotic stress regulatory elements in the promoter region of the ADF gene family suggests that it has great potential to participate in plant growth and adversity response. With the continuous development of advanced technologies, there is an increasing abundance of methods to analyze the regulatory mechanisms of maize life activities. Maize seed germination, which is directly related to yield and quality, is a complex process involving multiple genes and regulatory networks. Han et al. [11] decoded the molecular mechanism of gibberellin (GA) in the regulation of maize germination by using multi-omics analysis, including transcriptomics, miRNA, and degradome. In the study, differentially expressed genes (DEGs) were identified by RNA-seq, whose number revealed that the regulation of early stages was more complicated since more DEGs were needed than in the later stages. Moreover, miRNA sequencing revealed that the number of differentially expressed miRNAs and their target genes was increased after treatment with GA3, suggesting that GA3 induces more differentially expressed miRNAs to regulate genes associated with maize seed germination and participates in the regulation of maize seed germination. Interestingly, a gene named *ZmSLP* was found in the glycerophospholipid metabolic pathway, which negatively regulated maize seed germination and inhibited growth at the seedling stage. Maize leaf angle is the angle between the leaf blade and the main stem, and its size is closely related to the photosynthetic efficiency and planting density [12]. He et al. [13] detected a new QTL named *qLA2-3* controlling leaf angle, which was finely localized at a physical distance of about 338.46 kb and contained 16 genes in total. Further sequencing and transcriptomics analyses of the NILs identified five candidate genes that may be involved in the regulation of leaf angle.

3. Research in Oilseed Crops

Soybean is one of the widely grown oilseed crops in the world, but its total oil content is much lower than that of peanut. In order to increase the total oil content of soybean, Xu et al. [14] transformed *DGAT3*, a gene only found in peanut and *Arabidopsis*, into soybean. It was found that the contents of oleic acid composition and total fatty acid content in transgenic soybean plants overexpressing *AhDGAT3* were significantly higher than those of WT. Meanwhile, transgenic soybean plants that overexpressed *AhDGAT3* exhibited better agronomic traits compared with WT, including size of pods at different periods, number of effective branches, and weight of seeds at harvest. It was hypothesized that the gene might be related to soybean growth and development. Castor is a special industrial oilseed crop; seed weight and fatty acid content are two main indexes for evaluating its quality. Recently, Peng et al. [15] explored the effect of castor parental combinations on seed weight and fatty acid content of offspring for the first time. It showed that the levels of the indexes in the F1 generation were closely correlated to their parents, with the additive general combining ability and non-additive specific combining ability genetic effects significantly contributing to the changes in fatty acid composition. Moreover, two lines, CSR181 and 20111149, were recommended for use in the castor hybridization program based on the analysis of the comprehensive advantages of combining ability and heritability. *Brassica napus* L. is a major oilseed crop covered with a layer of wax powder to protect it against external stresses and ensure normal plant growth and development. Zhang et al. [16] found that the formation of wax powder is controlled by two pairs of genes. Through block segregation analysis (BSA) and whole-wide resequencing, the wax gene was localized in a region of 590,663–1,657,546 bp on chromosome A08, where 16 candidate genes were analyzed by quantitative reverse transcription polymerase chain reaction (qRT-PCR). It was revealed that three genes showed significant differences in the leaves of waxed and unwaxed parents, and they may inextricably link to wax formation. Brassica oil crops contain a wide range of species and all have indeterminate growth habits, among which *Brassica rapa* L. is famous for its indeterminate inflorescence [17]. Since indeterminate inflorescences can lead to the appearance of prolonged plant growth and maturity and susceptibility to collapse, which ultimately affects yield, it is crucial to investigate the mode of inheritance and molecular mechanisms that determine inflorescence traits. Chen et al. [18] found that inflorescences are controlled by *Brdt1* and *Brdt2*, and *Brdt1* was successfully localized within an interval of approximately 72.7 kb. Furthermore, there is a gene, *Bra009508 (BraA10.TFL1)*, which is homologous to *TFL1* in *Arabidopsis* and may play an important role in controlling the determinant inflorescence trait.

4. Research in Vegetable and Fruits

Flowering Chinese cabbage is prevalent in China due to its exceptional nutritional composition and delightful flavor, whose main edible part is the stem, and yield is directly connected with branching. Gibberellin (GA) is a phytohormone, which is considered an inhibitor of branching that regulates the growth and development of branch buds [19]. Qi et al. [20] investigated the effect of gibberellin on the branching of flowering Chinese cabbage by exogenous spraying of GA3, which showed that GA3 could effectively inhibit the primary rosette branching of it and had a direct impact on the whole plant yield of flowering Chinese cabbage. In addition, auxin-related genes can interact with GA to negatively regulate the number of branches on the rosette stems in flowering Chinese cabbage. Chinese cabbage is another kind of Chinese specialty vegetable that prefers cool and humid environments. Heat stress has an influence on its growth and development and in turn affects its production and quality [21]. The homeodomain leucine zipper (HD-zip) gene family plays an important role in the regulation of carotenoid content, which is closely associated with the color of the inner leaf blades of Chinese cabbage [22,23]. Yin et al. [24] divided the HD-Zip gene family into four subfamilies with three common motifs through research of three Brassicaceae plants. In this study, 14 HD-Zip genes were highly expressed under heat stress treatment in Chinese cabbage, 11 of which were from the HD-Zip I subfamily. Moreover, three genes related to carotenoid content were successfully identified, which also belonged to HD-Zip I. Among them, the expression of the HD-Zip I gene, *BraA09g011460.3C*, was up-regulated after heat stress treatment but significantly reduced in the carotenoid-rich cultivars, which showed the potential to tolerate heat stress and regulate carotenoid content.

Papaya is a tropical fruit of great commercial and nutritional value. In its growing areas, however, PRSV virus infestation has severely reduced its yield and commercial value. The papaya genome has fewer disease resistance genes, but there are nucleotide-binding site leucine-rich repeat receptor (NLR) family genes in its genome that exert a unique disease resistance role [25,26]. In order to provide a new perspective on disease resistance breeding in papaya, Jiang et al. [27] conducted a comprehensive analysis of the 59 NLR gene in the papaya genome. It was shown that papaya retains all NLR subclasses, and the dominated subclass is coiled-coil(CC)-NBS-LRR(CNL). Meanwhile, they speculate that relatively conserved resistance to powdery mildew8(RPW8)-NBS-LRs(RNLs) and CNLs may help relatively abundant toll/interleukin-1 receptor (TIR)-NBS-LRRs(TNLs) and CNLs to recognize variable pathogen effectors triggering immune responses in papaya. In addition, an insertion cluster of five duplicated CNLs was identified in the papaya genome, where dosage effects and expression divergence of disease resistance genes during evolution were observed. Three genes of this cluster were also strongly correlated to fungal infection. Pineapple is a tropical fruit with great economic value, but its self-incompatibility limits industrialization progress. Selection and breeding of new pineapple cultivars is a vital measure to improve pineapple yield and promote industrialization. Single nucleotide polymorphism (SNP) markers, as a new generation of molecular markers, have verified hybrid authenticity in several crops but have not yet been applied to the study of pineapple F_1 generation. Jia et al. [28] successfully identified the F_1 hybrid generation of pineapple, which was constructed with male parent 'MD2' and female parent 'Josapine' with a true hybridization rate of 87.58% based on SNP molecular markers. In addition, clustering analysis involving the parents and 313 hybrid offspring revealed that 68.5% of offspring aggregated with 'MD2', while only 31.95% were grouped with 'Josapine'. This study facilitates rapid and accurate identification of target progeny at the seedling stage, which will improve selection efficiency and ensure superior traits in planted cultivars.

5. Research in Other Crops

Leucaena leucocephala (Lam.) de Wit is a mimosoid legume genus plant with high protein content, which has been popularized for livestock feeding in some regions. Currently, hybrid triploids have been successfully created among *Leucaena* species [29]. Han et al. [30] identified transcriptome-based genetic variation in 21 *Leucaena* taxa for the first time and established an efficient and high-throughput platform for the rapid identification of triploids in *Leucaena*. Tartary buckwheat is a flavonoid-rich medicinal plant with anti-inflammatory, antiviral, and anti-allergic properties. However, it often encounters abiotic stresses, including drought and high salt, due to the constraints of the growing environment, which affects its medicinal quality and yield. The mitogen-activated protein kinase (MAPK) signaling cascade is one of the most common signaling pathways in plants, which casts a significant function in the process of abiotic stress. Dong et al. [31] identified 16 MAPK family genes based on the conserved structural domains of MAPK in Tartary buckwheat and analyzed their expression level. The results showed that 15 genes were significantly expressed variably in different organs, indicating that *MAPKs* are pivotal in Tartary buckwheat growth and development. Furthermore, the expression of *FtMAPKs* was also analyzed under drought and salt stress, in which 12 and 14 genes, respectively, showed significant expression variations with drought and salt treatment. Meanwhile, three candidate genes, *FtMAPK3*, *FtMAPK4*, and *FtMAPK8*, demonstrated significant expression changes across both abiotic stress treatments.

6. Conclusions and Perspective

In conclusion, we reviewed studies of different crops on yield, quality, resistance, and stress tolerance, which provide new ideas for genetic improvement of crops and the progression of agricultural production. With the emergence of new technologies, genetic improvement means of plant biotechnology will be more abundant; the transformation and application of research results will progress to the realization of sustainable development in agriculture.

Author Contributions: All the authors participated in the editing of this research topic. Y.S. wrote the draft, and all the other authors provided suggestive comments on the editorial. All authors have read and agreed to the published version of the manuscript.

Funding: This project was funded by the National Natural Science Foundation of China (32201805) and the Zhejiang Provincial Natural Science Foundation of China (LD24C130001).

Conflicts of Interest: The authors declare that the research was conducted in the absence of any commercial or financial relationships that could be construed as a potential conflict of interest.

References

1. Kusano, M.; Yang, Z.; Okazaki, Y.; Nakabayashi, R.; Fukushima, A.; Saito, K. Using metabolomic approaches to explore chemical diversity in rice. *Mol. Plant* **2015**, *8*, 58–67. [CrossRef] [PubMed]
2. Birla, D.S.; Malik, K.; Sainger, M.; Chaudhary, D.; Jaiwal, R.; Jaiwal, P.K. Progress and challenges in improving the nutritional quality of rice (*Oryza sativa* L.). *Crit. Rev. Food Sci. Nutr.* **2017**, *57*, 2455–2481. [CrossRef]
3. Lee, S.-M.; Kabange, N.R.; Kang, J.-W.; Kwon, Y.; Cha, J.-K.; Park, H.; Oh, K.-W.; Seo, J.; Koh, H.-J.; Lee, J.-H. Identifying QTLs related to grain filling using a doubled haploid rice (*Oryza sativa* L.) population. *Agronomy* **2023**, *13*, 912. [CrossRef]
4. Ji, Z.; Wang, C.; Zhao, K. Rice routes of countering *Xanthomonas oryzae*. *Int. J. Mol. Sci.* **2018**, *19*, 3008. [CrossRef]
5. Zhu, Y.; Yang, X.; Luo, P.; Yan, J.; Cao, X.; Qian, H.; Zhu, X.; Fan, Y.; Mei, F.; Fan, M.; et al. Creation of Bacterial Blight Resistant Rice by Targeting Homologous Sequences of *Xa13* and *Xa25* Genes. *Agronomy* **2024**, *14*, 800. [CrossRef]
6. Yangklang, J.; Sanitchon, J.; Siangliw, J.L.; Monkham, T.; Chankaew, S.; Siangliw, M.; Sirithunya, K.; Toojinda, T. Yield performance of RD6 glutinous rice near isogenic lines evaluated under field disease infection at northeastern Thailand. *Agronomy* **2024**, *14*, 1871. [CrossRef]
7. Liu, J.; Guo, X.; Zhai, T.; Shu, A.; Zhao, L.; Liu, Z.; Zhang, S. Genome-wide identification and characterization of microRNAs responding to ABA and GA in maize embryos during seed germination. *Plant Biol.* **2020**, *22*, 949–957. [CrossRef]
8. Kopecká, R.; Kameniarová, M.; Černý, M.; Brzobohatý, B.; Novák, J. Abiotic stress in crop production. *Int. J. Mol. Sci.* **2023**, *24*, 6603. [CrossRef] [PubMed]
9. Wang, B.; Liu, C.; Zhang, D.; He, C.; Zhang, J.; Li, Z. Effects of maize organ-specific drought stress response on yields from transcriptome analysis. *BMC Plant Biol.* **2019**, *19*, 335. [CrossRef]
10. Yang, R.; Wang, F.; Luo, P.; Xu, Z.; Wang, H.; Zhang, R.; Li, W.; Yang, K.; Hao, Z.; Gao, W. The Role of the ADF Gene Family in Maize Response to Abiotic Stresses. *Agronomy* **2024**, *14*, 717. [CrossRef]
11. Han, Z.; Jin, Y.; Wang, B.; Guo, Y. Multi-Omics Revealed the Molecular Mechanism of Maize (*Zea mays* L.) Seed Germination Regulated by GA3. *Agronomy* **2023**, *13*, 1929. [CrossRef]
12. Mantilla-Perez, M.B.; Fernandez, M.G.S. Differential manipulation of leaf angle throughout the canopy: Current status and prospects. *J. Exp. Bot.* **2017**, *68*, 5699–5717. [CrossRef] [PubMed]
13. He, Y.; Wang, C.; Hu, X.; Han, Y.; Lu, F.; Liu, H.; Zhang, X.; Yin, Z. Genetic Basis and Exploration of Major Expressed QTL *qLA2-3* Underlying Leaf Angle in Maize. *Agronomy* **2024**, *14*, 1978. [CrossRef]
14. Xu, Y.; Yan, F.; Liang, Z.; Wang, Y.; Li, J.; Zhao, L.; Yang, X.; Wang, Q.; Liu, J. Overexpression of the peanut *AhDGAT3* gene increases the oil content in soybean. *Agronomy* **2023**, *13*, 2333. [CrossRef]
15. Peng, M.; Wang, Z.; He, Z.; Li, G.; Di, J.; Luo, R.; Wang, C.; Huang, F. Combining Ability, Heritability, and Heterosis for Seed Weight and Oil Content Traits of Castor Bean (*Ricinus communis* L.). *Agronomy* **2024**, *14*, 1115. [CrossRef]
16. Zhang, J.; Zhu, J.; Yang, L.; Li, Y.; Wang, W.; Zhou, X.; Jiang, J. Mapping of the Waxy Gene in *Brassica napus* L. via Bulked Segregant Analysis (BSA) and Whole-Genome Resequencing. *Agronomy* **2023**, *13*, 2611. [CrossRef]
17. Kaur, H.; Banga, S.S. Discovery and mapping of *Brassica juncea Sdt$_1$* gene associated with determinate plant growth habit. *Theor. Appl. Genet.* **2015**, *128*, 235–245. [CrossRef] [PubMed]
18. Chen, C.; Zhu, X.; Zhao, Z.; Du, D.; Li, K. Fine Mapping and Functional Verification of the *Brdt1* Gene Controlling Determinate Inflorescence in *Brassica rapa* L. *Agronomy* **2024**, *14*, 281. [CrossRef]
19. Tan, M.; Li, G.; Liu, X.; Cheng, F.; Ma, J.; Zhao, C.; Zhang, D.; Han, M. Exogenous application of GA$_3$ inactively regulates axillary bud outgrowth by influencing of branching-inhibitors and bud-regulating hormones in apple (*Malus domestica* Borkh.). *Mol. Genet. Genom.* **2018**, *293*, 1547–1563. [CrossRef]
20. Qi, X.; Zhao, Y.; Cai, N.; Guan, J.; Liu, Z.; Liu, Z.; Feng, H.; Zhang, Y. Characterization and Transcriptome Analysis Reveal Exogenous GA$_3$ Inhibited Rosette Branching via Altering Auxin Approach in Flowering Chinese Cabbage. *Agronomy* **2024**, *14*, 762. [CrossRef]
21. Sun, X.; Feng, D.; Liu, M.; Qin, R.; Li, Y.; Lu, Y.; Zhang, X.; Wang, Y.; Shen, S.; Ma, W.; et al. Single-cell transcriptome reveals dominant subgenome expression and transcriptional response to heat stress in Chinese cabbage. *Genome Biol.* **2022**, *23*, 262. [CrossRef] [PubMed]
22. Sharif, R.; Raza, A.; Chen, P.; Li, Y.; El-Ballat, E.M.; Rauf, A.; Hano, C.; El-Esawi, M.A. HD-ZIP gene family: Potential roles in improving plant growth and regulating stress-responsive mechanisms in plants. *Genes* **2021**, *12*, 1256. [CrossRef] [PubMed]
23. Yuan, P.; Umer, M.J.; He, N.; Zhao, S.; Lu, X.; Zhu, H.; Gong, C.; Diao, W.; Gebremeskel, H.; Kuang, H.; et al. Transcriptome regulation of carotenoids in five flesh-colored watermelons (*Citrullus lanatus*). *BMC Plant Biol.* **2021**, *21*, 203. [CrossRef] [PubMed]
24. Yin, L.; Sun, Y.; Chen, X.; Liu, J.; Feng, K.; Luo, D.; Sun, M.; Wang, L.; Xu, W.; Liu, L.; et al. Genome-Wide Analysis of the HD-Zip Gene Family in Chinese Cabbage (*Brassica rapa* subsp. *pekinensis*) and the Expression Pattern at High Temperatures and in Carotenoids Regulation. *Agronomy* **2023**, *13*, 1324. [CrossRef]
25. Yue, J.; VanBuren, R.; Liu, J.; Fang, J.; Zhang, X.; Liao, Z.; Wai, C.M.; Xu, X.; Chen, S.; Zhang, S.; et al. SunUp and sunset genomes revealed impact of particle bombardment mediated transformation and domestication history in papaya. *Nat. Genet.* **2022**, *54*, 715–724. [CrossRef] [PubMed]

26. Ming, R.; Hou, S.; Feng, Y.; Yu, Q.; Dionne-Laporte, A.; Saw, J.H.; Senin, P.; Wang, W.; Ly, B.V.; Lewis, K.L.T.; et al. The draft genome of the transgenic tropical fruit tree papaya (*Carica papaya* Linnaeus). *Nature* **2008**, *452*, 991–996. [CrossRef]
27. Jiang, Q.; Wang, Y.; Xiong, A.; Zhao, H.; Jia, R.; Li, M.; An, H.; Ji, C.; Guo, A. Phylogenetic Analyses and Transcriptional Survey Reveal the Characteristics, Evolution, and Expression Profile of NBS-Type Resistance Genes in Papaya. *Agronomy* **2023**, *13*, 970. [CrossRef]
28. Jia, P.; Liu, S.; Lin, W.; Yu, H.; Zhang, X.; Xiao, X.; Sun, W.; Lu, X.; Wu, Q. Authenticity Identification of F_1 Hybrid Offspring and Analysis of Genetic Diversity in Pineapple. *Agronomy* **2024**, *14*, 1490. [CrossRef]
29. Real, D.; Revell, C.; Han, Y.; Li, C.; Castello, M.; Bailey, C.D. Successful creation of seedless (sterile) leucaena germplasm developed from interspecific hybridisation for use as forage. *Crop. Pasture Sci.* **2022**, *74*, 783–796. [CrossRef]
30. Han, Y.; Abair, A.; van der Zanden, J.; Nageswara-Rao, M.; Vasan, S.P.; Bhoite, R.; Castello, M.; Bailey, D.; Revell, C.; Li, C.; et al. Transcriptome-Wide Genetic Variations in the Legume Genus *Leucaena* for Fingerprinting and Breeding. *Agronomy* **2024**, *14*, 1519. [CrossRef]
31. Dong, G.; Gui, Z.; Yuan, Y.; Li, Y.; Du, D. Expression Analysis of the Extensive Regulation of Mitogen-Activated Protein Kinase (MAPK) Family Genes in Buckwheat (*Fagopyrum tataricum*) During Organ Differentiation and Stress Response. *Agronomy* **2024**, *14*, 2613. [CrossRef]

Disclaimer/Publisher's Note: The statements, opinions and data contained in all publications are solely those of the individual author(s) and contributor(s) and not of MDPI and/or the editor(s). MDPI and/or the editor(s) disclaim responsibility for any injury to people or property resulting from any ideas, methods, instructions or products referred to in the content.

Article

Comparative Transcriptome Analysis of Eggplant (*Solanum melongena* L.) Peels with Different Glossiness

Hong Wang [1,†], Zhixing Nie [1,*,†], Tonglin Wang [1], Shuhuan Yang [2] and Jirong Zheng [1,*]

1. Vegetable Research Institute, Hangzhou Academy of Agricultural Sciences, Hangzhou 310024, China; hongwang201010@163.com (H.W.); tlwang@zju.edu.cn (T.W.)
2. College of Plant Science and Technology, Huazhong Agricultural University, Wuhan 430070, China; luoyeguigen2@163.com
* Correspondence: niezhixing@126.com (Z.N.); topzheng2003@163.com (J.Z.)
† These authors contributed equally to this work.

Abstract: Peel glossiness is an important commercial trait of eggplant (*Solanum melongena* L.). In this study, two eggplant-inbred lines with different levels of peel glossiness were used to identify genes related to peel glossiness. Paraffin section analysis showed that increased wax thickness and wrinkles on the wax surface of eggplant peels decreased glossiness. Differential gene expression related to eggplant peel glossiness was analyzed by comparing the transcriptomes of eggplant peels with different gloss levels and at different developmental stages. The results identified 996 differentially expressed genes (DEGs), including 502 upregulated and 494 downregulated genes, possibly related to eggplant peel glossiness. GO enrichment and KEGG enrichment analyses revealed that the DNA replication pathway (GO:0003688, GO:0006270) and the photosynthesis pathway (map00195) were downregulated and thus may be associated with reduced eggplant peel glossiness. Expression level analysis of eggplant peels with different glossiness levels revealed that a C2H2 transcription factor gene, two ERF transcription factor genes, one long-chain acyl-CoA synthetase gene, and four wax- or cutin-related genes may be associated with the glossiness of eggplant fruit peels. These findings will help guide future genetic improvements in eggplant peel glossiness.

Keywords: eggplant (*Solanum melongena* L.); peel glossiness; RNA-Seq; wax

Citation: Wang, H.; Nie, Z.; Wang, T.; Yang, S.; Zheng, J. Comparative Transcriptome Analysis of Eggplant (*Solanum melongena* L.) Peels with Different Glossiness. *Agronomy* 2024, 14, 3063. https://doi.org/10.3390/agronomy14123063

Academic Editor: Xingwang Liu

Received: 20 November 2024
Revised: 18 December 2024
Accepted: 20 December 2024
Published: 22 December 2024

Copyright: © 2024 by the authors. Licensee MDPI, Basel, Switzerland. This article is an open access article distributed under the terms and conditions of the Creative Commons Attribution (CC BY) license (https://creativecommons.org/licenses/by/4.0/).

1. Introduction

Eggplant (*Solanum melongena* L.) is an important vegetable crop worldwide and Asia's third most important vegetable crop, with extensive cultivation in both China and the Indian subcontinent [1]. According to the Food and Agriculture Organization (FAO), approximately 818,100 ha of eggplant were cultivated in China in 2022, yielding 3.83 million tons (https://www.fao.org/faostat/, accessed on 25 October 2024). The glossiness of eggplant peel is one of its most significant commercial traits. Consumers favor eggplants with higher glossiness because it is an important indicator of eggplant fruit freshness [2]. Consequently, research on eggplant peel glossiness is both theoretically and practically significant. The glossiness of eggplant skin can be affected by cultivation measures such as grafting and water management [3]. Nine quantitative trait locus (QTLs) associated with eggplant luster have been identified [4]. However, compared to other crops, such as cucumber (*Cucumis sativus* L.), there has been little research on the glossiness of eggplant peel.

Plant glossiness refers to the ability of plant surfaces to reflect light, which not only affects their appearance but also significantly affects their physiological functions, environmental adaptability, and resistance to pests and diseases [5,6]. Glossiness is also directly related to the market value of crops, as it is a consumer preference, particularly for vegetable crops such as peppers, tomatoes, and eggplants. Consequently, elucidating plant glossiness's formation and regulation mechanisms could help improve crop quality significantly.

Studies have shown that plant glossiness is closely related to the composition of the epidermal wax and cuticle [7,8]. Plant epidermal wax is a complex mixture of very-long-chain fatty acids and their derivatives; however, the predominant constituents are very long-chain alkanes [9]. The cuticle is primarily composed of cutin and wax [10], and changes in these components can affect vegetable glossiness. For example, wax crystal-deficient mutants of Chinese cabbage (*Brassica rapa* L. ssp. *pekinensis*), such as wdm4 and wdm8, have almost no wax crystals and exhibit a glossy phenotype. In contrast, wild-type plants have obvious wax crystals and do not exhibit a glossy phenotype [11]. Liu et al. [12] found that the peel surface of control navel oranges was rough and covered with small flaky wax crystals, whereas mutant navel oranges had a shiny peel surface with almost no wax crystals. Cucumber *Csgp* mutants exhibit increased epidermal glossiness, lower wax content, and a thinner cuticle structure than wild-type plants [13]. The wax content of the cucumber fruit epidermis is negatively correlated with fruit glossiness [14]. However, increased wax content can improve the glossiness of tomato (*Solanum lycopersicum*) fruits [15]. Decreased cutin content and thinner cuticles also enhance tomato glossiness [16,17]. Wax content may also be unrelated to fruit glossiness; for example, in citrus, fruit surface glossiness is determined by the wax structure rather than the wax content [18].

The basic pathway for wax biosynthesis consists of three steps: de novo fatty acid synthesis, very-long-chain fatty acid (VLCFA) synthesis, and the synthesis of various wax components. The reported biosynthetic pathways of plant cuticular wax indicate that acetyl-CoA in the endoplasmic reticulum undergoes a tricarboxylic acid (TCA) cycle followed by de novo fatty acid synthesis to produce C16 and C18 fatty acids. These fatty acids are then elongated into VLCFAs, precursors for wax synthesis catalyzed by acyl reductases. The primary components of wax are generated via acyl reduction and decarboxylation pathways [19]. Numerous studies have identified genes related to plant wax and cutin. For example, the *Arabidopsis* gene *CER4* encodes an alcohol-forming fatty acyl-CoA reductase involved in cuticular wax production [10]. Furthermore, the *AP2/ERF* transcription factor *WRINKLED4* regulates cuticular wax biosynthesis [20,21]. He et al. [22] identified a key gene, *TaCER1-6A*, for cuticular alkane biosynthesis in wheat leaves, specifically catalyzing the biosynthesis of C27–C33 alkanes and promoting drought tolerance. In Chinese cabbage, the glossiness trait of line Y1211-1 is controlled by a single recessive locus, *WAX2* (*BrWAX2*). The absence of *BrWAX2* in the alkane formation pathway and the expression of other genes reduced the wax content, resulting in a glossy phenotype [8]. Cucumber transcription factor *CsGLF1* is a locus that determines the glossiness trait. It encodes a homolog of the Cys2His2-like fold group (C2H2) zinc finger protein 6 (ZFP6), whose deletion leads to reduced cuticular wax accumulation and a glossy fruit peel [23]. Another gene, *CsDULL*, encodes a C2H2-type zinc finger transcription factor that regulates the biosynthesis and transport of cutin and wax by targeting two genes, *CsGPAT4* and *CsLTPG1*, thereby affecting the glossiness of cucumber fruit peel [24]. The wax biosynthesis gene *CsCER6* (*ECERIFERUM6*) and the regulatory gene *CsCER7* may affect wax accumulation in cucumber fruits. *CsCER6* and *CsCER7* positively regulate epidermal wax accumulation in fruits, negatively affecting fruit glossiness [14]. Mutations in the *CsSEC23* gene in cucumber lead to increased permeability and thinning of the fruit cuticle, thus impairing or altering the original vesicle transport function. Simultaneously, some substances that affect fruit glossiness, such as waxes and cutins, also undergo changes [13]. *SlSHN1* acts as a transcriptional activator of wax synthesis in tomatoes. The overexpression of this gene significantly upregulates the expression of the GDSL lipase and acyl-CoA synthetase genes, thereby increasing the wax content of tomato peels [25]. The long-chain acyl-CoA synthetase (LACS) gene *SlLACS1* is a key factor in epidermal wax synthesis, which enhances drought resistance in tomatoes and prolongs their shelf life [26]. In *Brassica napus* L., *BnUC1* affects the formation of epidermal wax by regulating the expression of LTP and genes related to VLCFA biosynthesis, significantly influencing leaf glossiness [27].

This study aimed to identify genes related to eggplant peel glossiness using a comparative transcriptome analysis of eggplant peels with different levels of glossiness. The results will provide a theoretical basis and technical support for the molecular genetic improvement of eggplant quality.

2. Materials and Methods

2.1. Plant Samples and Preparation

Fruit peels from inbred eggplant lines A21 and A32 were used for transcriptome sequencing. Three glossy peel eggplant inbred lines (GZ1, ZE11, and ZSQ) and three non-glossy peel eggplant inbred lines (B209, B203, and HNQQ) were used to determine the expression of genes related to eggplant peel glossiness. The inbred eggplant lines used in this study were obtained from the Hangzhou Academy of Agricultural Sciences. All inbred lines were produced after at least eight generations of inbreeding. All materials were grown in a greenhouse at the Zhijiang Base of the Hangzhou Academy of Agricultural Sciences, Hangzhou, China (30°15' N, 120°09' E). During the fruiting period, the fruits of A21 were purple-black and those of A32 were purple-red.

The fruit peels of inbred lines A21 and A32 exhibited glossiness 7 d after flowering. However, 21 d after flowering, the fruit peel of inbred line A21 retained its glossiness, whereas that of inbred line A32 lost its glossiness. Fruit peels from inbred lines A21 and A32 were collected 7 and 21 d after flowering, respectively. These peels were cut into small pieces, approximately 0.5×0.5 cm, and quickly placed in liquid nitrogen for preservation until transcriptome sequencing. The peels collected 7 d after flowering from the two inbred lines were named A21a and A32a, whereas those collected 21 days after flowering were named A21b and A32b. Three biological replicates were used at each time point. None of the plants in this study were grafted.

2.2. Pericarps Paraffin Sectioning

Pericarps of different sizes from A21 and A32 were collected in Carnoy's solution (alcohol). Paraffin sections (8-μm) of the pericarps were prepared using a microtome (RM2135, Leica, Wetzlar, Germany) and stained with 1% saffron solution and 1% fast green solution, according to Dai et al. [28]. The chromatic CIEL*a*b* parameters of the eggplant peel were registered using a colorimeter (CM-600D, MINOLTA, Osaka, Japan). In the leaf chromatic analysis, L* represents lightness, a* indicates the red/green coordinate, and b* signifies the yellow/blue coordinates.

2.3. Total RNA Extraction, Library Construction, and Sequencing

Total RNA was extracted from tissues using an RNAprep Pure Plant Plus Kit (DP441; Tiangen Biotech, Beijing, China) following the manufacturer's instructions. RNA quality was assessed using a 5300 Bioanalyzer (Agilent, Santa Clara, CA, USA) and quantified using an ND-2000 spectrophotometer (NanoDrop Technologies, Wilmington, DE, USA). According to the manufacturer's instructions, RNA purification, reverse transcription, library construction, and sequencing were performed at Shanghai Majorbio Bio-Pharm Biotechnology Co., Ltd. (Shanghai, China). The RNA-seq transcriptome library was prepared using an Illumina® Stranded mRNA Prep, Ligation Kit (Illumina, San Diego, CA, USA) with 1 μg of total RNA following the manufacturer's instructions. After completing 10–15 cycles of cDNA PCR using the LongAmp Taq PCR Kit (New England Biolabs, Ipswich, MA, USA), as per the manufacturer's guidelines, PCR adapters were attached directly to both termini of the first-strand cDNA. The sequencing library was constructed on an Illumina NovaSeq 6000 using a NovaSeq Reagent Kit (Illumina, San Diego, CA, USA) following the manufacturer's instructions.

2.4. Bioinformatic Analysis of the Sequencing Data and Differential Expression Analysis

Raw reads obtained from sample sequencing were subjected to quality control using Fastp v0.23.4 [29]. Reads with adapters containing >10% N bases, consisting entirely of A bases, or classified as low-quality were excluded. The obtained clean reads were aligned with the eggplant reference genome HQ-1315 [30] using TopHat v2.1.1 [31].

To identify differentially expressed genes (DEGs) between the two samples, the expression level of each transcript was calculated according to the transcripts per million reads (TPM) method. RSEM v1.1.11 [32] was used to quantify gene abundance. Differential expression analysis was performed using DESeq2 v3.20 [33]. DEGs were defined by $|\log2FC| \geq 1$ and p adjust < 0.01.

2.5. Functional Enrichment and Transcription Factor Identification

Functional enrichment analyses were performed using gene ontology (GO) and the Kyoto Encyclopedia of Genes and Genomes (KEGG). DEGs were identified that were significantly enriched in GO terms and metabolic pathways at a Bonferroni-corrected p adjust < 0.05 and then compared with the whole-transcriptome background. Functional annotation of the DEGs was performed according to the GO analysis [34], and enrichment analysis was conducted using the GOseq v3.20, which uses the Wallenius non-central hypergeometric distribution [35] to account for potential gene length bias in the DEGs. KEGG pathway annotation was performed using BLAST against the KEGG database (accessible at http://www.genome.jp/kegg/, accessed on 3 September 2024). Subsequently, pathway enrichment analysis of the DEGs was performed using a KEGG orthology-based annotation system [36]. GO terms and KEGG pathways were significantly enriched if they exhibited a corrected p adjust < 0.05.

Transcription factors (TFs) were identified using iTAK (Plant Transcription Factor and Protein Kinase Identifier and Classifier) [37] to predict the DEGs, with reference to the PlnTFDB and PlantTFDB databases.

2.6. Quantitative Real-Time PCR

All primers used in this study were designed using Primer-Blast (https://www.ncbi.nlm.nih.gov/tools/primer-blast/, accessed on 3 September 2024) from NCBI (Bethesda, MD, USA) and synthesized by Sangon Biotech (Shanghai, China). Pepper β-actin gene [38] was used as an internal control for standardizing the expression data. The primer pairs used in this study are listed in Table S1. Quantitative real-time PCR (qRT-PCR) was performed following the guidelines provided for the 2× Hieff UNICON® Universal Blue qPCR SYBR Green Master Mix by YEASEN (Shanghai, China), and the results were analyzed using the Quant Studio 5 real-time PCR system manufactured by Thermo Fisher Scientific, Inc. (Waltham, MA, USA). The relative mRNA expression levels were calculated using the $2^{-\Delta\Delta Ct}$ method. Each experiment included three biological and three technical replicates.

3. Results

3.1. Morphological Characteristics of Eggplant Peel Glossiness

The fruits of the inbred eggplant lines A21 and A32 are purple-black and purple-red, respectively. Seven days post-flowering (dpf), the fruits of A21 were approximately 5 cm and 12 g, and those of A32 were approximately 10 cm and 20 g (Figure 1a,c). By 21 dpf, the A21 and A32 fruits were approximately 20 cm and 85 g (Figure 1b,d). On day 7, the pericarps of both inbred lines had a glossy appearance; the CIEL*a*b* of A21a was L* = 27.17, a* = 11.27, and b* = −7.19, whereas the CIEL*a*b* of A32a was L* = 43.04, a* = 26.06, and b* = −10.74. However, 21 dpf, the pericarp of A21 remained glossy, and the CIEL*a*b* was L* = 24.63, a* = 4.68, and b* = −5.90, whereas that of A32b, which was no longer glossy, was L* = 46.20, a* = 19.90, and b* = −8.26. Paraffin section analysis revealed that at 7 dpf, there was little wax on the pericarp surface of either variety, with only thin and nearly absent layers (Figure 1e,g). By 21 dpf, the wax content on the pericarp surface increased, but the thickness of the epidermal wax in A32 (8.75 μm) was significantly greater than that in A21 (7.79 μm). The epidermal wax of A21 fruits appeared smoother, whereas the surface of A32 epidermal wax exhibited severe wrinkling (Figure 1f,h). We hypothesized that an increase in wax thickness and wrinkling on the epidermal surface of eggplant fruits would reduce the glossiness of the pericarps.

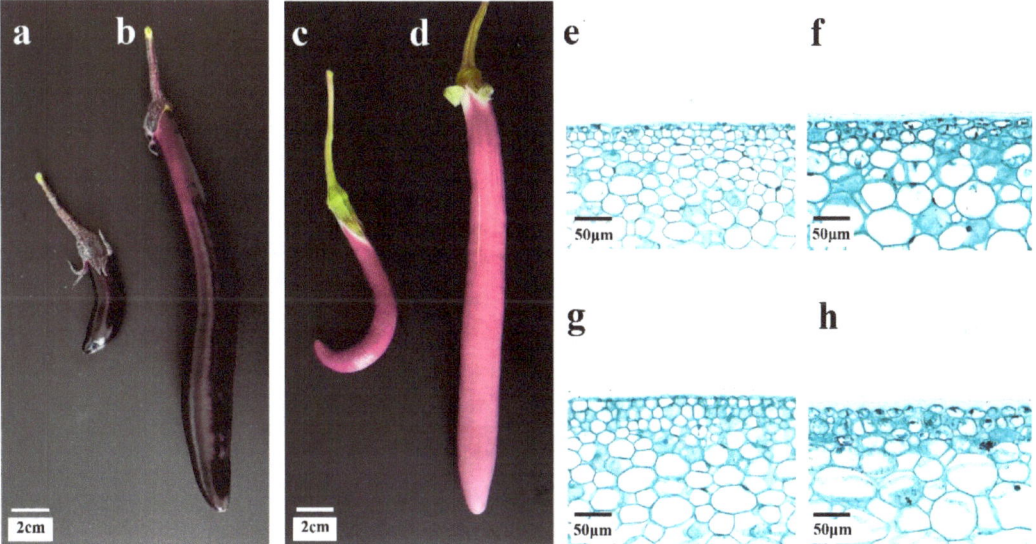

Figure 1. Morphological characteristics related to fruit peel glossiness in the inbred eggplant lines A21 and A32. Fruits of A21 7 d (**a**) and 21 d (**b**) and A32 7 d (**c**) and 21 d (**d**) after flowing. Paraffin sections of the fruit epidermis of A21 7 d (**e**) and 21 d (**f**) and A32 7 d (**g**) and 21 d (**h**) after flowing.

3.2. Quality Results of the Eggplant Peel Transcriptome Sequencing

Transcriptome sequencing was conducted on eggplant fruit peels (Figure 1) at different stages with three biological replicates per group. The raw sequencing data of the 12 samples ranged from 9.77 to 12.94 Gb. After statistical and quality assessments of the raw sequencing data and the removal of sequences contaminated with adapters, the proportion of high-quality sequence bases obtained was >98% for all samples (Table 1). This indicated that sequencing quality control was good and the data were reliable. The degree of matching between the transcriptome data and reference genome was >96% for all samples, indicating that the alignment results were satisfactory (Table 1) and could be used for further analysis.

Table 1. Statistical analysis of RNA-Seq read quality in eggplant inbred lines A21 and A32.

Sample	Raw Reads	Raw Data (Gb)	Clean Reads	Clean Reads Rates Ratio	Clean Data (Gb)	Q20	GC Content	Mapped Reads	Mapped Rates Ratio
A32a_1	85,723,096	12.94	84,601,196	98.69%	11.96	96.91%	43.68%	81,966,641.00	96.89%
A32a_2	72,334,058	10.92	71,496,558	98.84%	10.09	96.8%	43.6%	69,182,538.00	96.76%
A32a_3	66,204,714	10.00	65,298,078	98.63%	9.18	96.77%	43.67%	63,217,896.00	96.81%
A32b_1	66,688,094	10.07	65,590,548	98.35%	9.26	96.74%	43.61%	63,501,960.00	96.82%
A32b_2	80,983,980	12.23	79,614,942	98.31%	11.23	96.79%	43.55%	77,083,011.00	96.82%
A32b_3	70,561,868	10.65	69,218,874	98.10%	9.76	96.8%	43.62%	67,102,207.00	96.94%
A21a_1	68,753,352	10.38	67,582,288	98.30%	9.54	96.82%	43.66%	65,240,362.00	96.53%
A21a_2	68,370,232	10.32	67,442,464	98.64%	9.46	96.8%	43.57%	65,221,053.00	96.71%
A21a_3	75,566,314	11.41	74,195,388	98.19%	10.41	96.82%	43.55%	71,929,609.00	96.95%
A21b_1	74,872,704	11.31	73,454,914	98.11%	10.35	96.89%	43.74%	71,100,144.00	96.79%
A21b_2	71,217,114	10.75	70,154,346	98.51%	9.90	96.77%	43.73%	67,851,652.00	96.72%
A21b_3	64,693,436	9.77	63,827,850	98.66%	8.76	96.6%	43.84%	61,653,848.00	96.59%

3.3. Gene Expression in Eggplant Peels During Development

Compared with the reference genome, 28,693 expressed genes were detected in this transcriptome analysis, including 25,924 known and 2769 novel genes. Principal component analysis (PCA) revealed that the expressed genes in the peels of the two inbred eggplant lines were significantly separated at different developmental stages, whereas those within the same group clustered to varying degrees (Figure 2a). The number of differentially expressed genes (DEGs) between peels 7 and 21 dpf was generally consistent in both A21 and A32 (Figure 2b–d). In A21, there were 2670 DEGs, of which 1174 were upregulated in A21b and 1496 were downregulated. In A32, there were 2960 DEGs, with 1475 upregulated in A32b and 1485 downregulated. Because there was no difference in peel glossiness between 7 and 21 DPA in A21, the DEGs in A21 were unrelated to peel glossiness. Therefore, after excluding common DEGs in A32 and A21, 1993 DEGs were identified. Due to differences in peel glossiness between A32b and A21b, 4234 DEGs that may be related to glossiness were identified. At the intersection of the 1993 and 4234 DEGs, 996 DEGs (Table S2) were most likely associated with eggplant peel glossiness. These 996 DEGs were used for further analyses. A32b contained 502 upregulated DEGs and 494 downregulated DEGs compared with A32a.

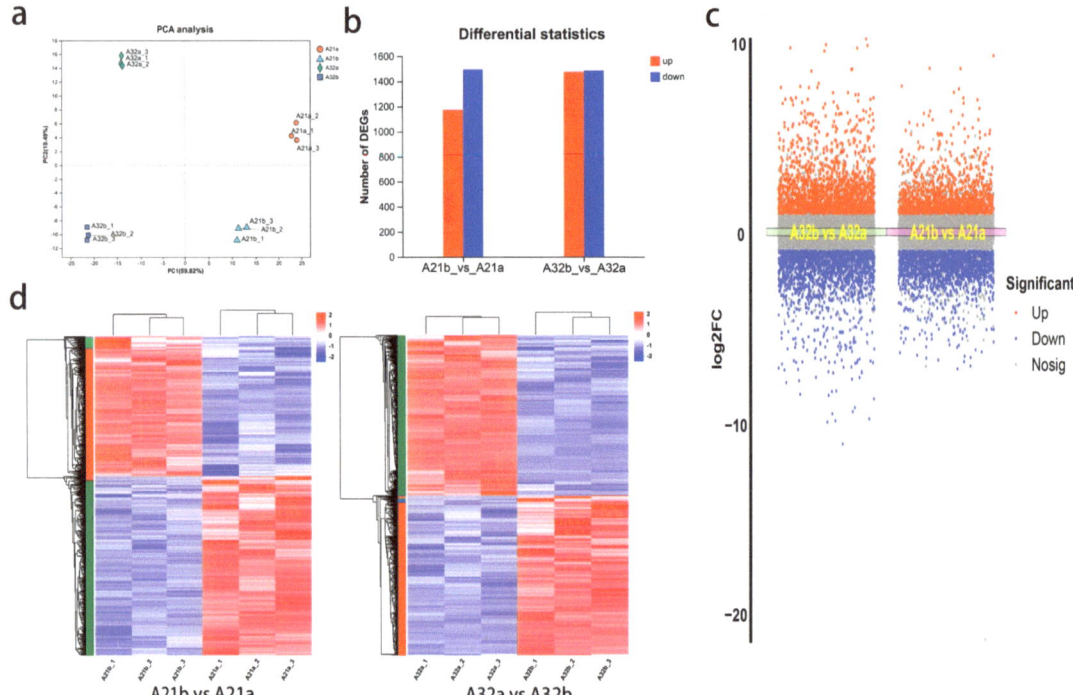

Figure 2. Gene expression in eggplant peels from inbred lines A21 and A32. (**a**) Principal component analysis (PCA) of expressed genes in eggplant inbred lines A21 and A32. (**b**) Histogram of the differentially expressed genes (DEGs) in inbred eggplant lines A21 and A32. (**c**) Volcano plots of the DEGs in inbred eggplant lines A21 and A32. (**d**) Clustering heatmap of the DEGs in inbred eggplant lines A21 and A32. The horizontal coordinate represents the names of the samples and their clustering results, while the vertical coordinate indicates the differential genes and the clustering results of these genes. The different colors on the right represent different subclusters.

3.4. GO Functional Annotation and KEGG Pathway Analysis of DEGs Related to Glossiness

GO enrichment analysis was conducted on the 996 DEGs, resulting in 303 enriched GO pathways, including 17 significantly related (p adjust < 0.05) pathways (Figure 3a, Table S3). Cellular components were primarily classified into the MCM complex, plasma membrane, endoplasmic reticulum lumen, chloroplast thylakoid membrane, plastid thylakoid membrane, and extracellular region. Molecular functions were primarily categorized as DNA replication-origin binding. Biological processes are primarily grouped into double-strand break repair, DNA replication initiation, recombination repair, double-strand break repair via homologous recombination, response to stimulus, cell cycle process, cellular response to stress, and cellular response to stimuli. DNA replication-related pathways were enriched in both molecular functions and biological processes, encompassing nine DEGs. Compared with A32a cells, all nine DEGs were downregulated in A32b cells. These results indicate that downregulation of the DNA replication pathway (GO:0003688, GO:0006270) may be associated with reduced glossiness in eggplant peels.

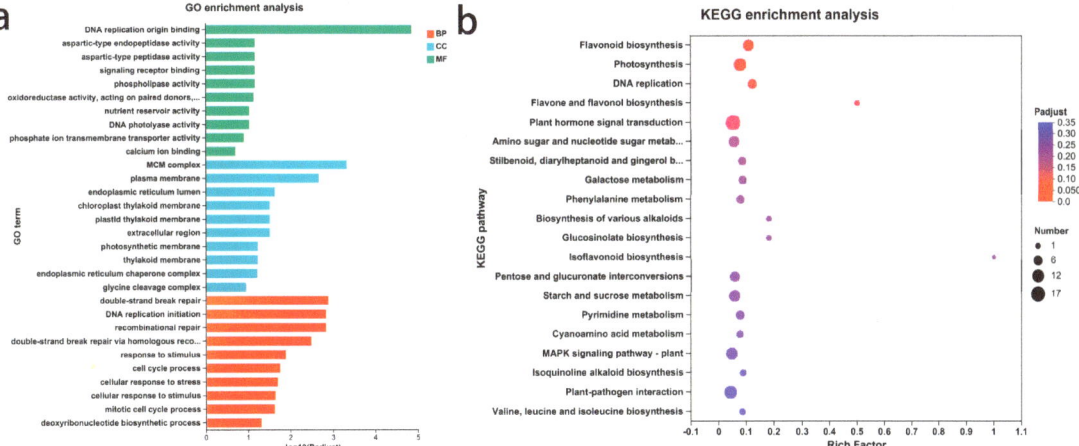

Figure 3. Gene Ontology (GO) and Kyoto Encyclopedia of Genes and Genomes (KEGG) pathway enrichment analysis results for the differentially expressed genes in eggplant peel from inbred line A32. (**a**) GO enrichment analysis results. (**b**) Scatterplot of KEGG pathway enrichment analysis results.

KEGG enrichment analysis was conducted on the 996 differentially expressed genes (DEGs), and 93 pathways were identified (Figure 3b). Among these, two pathways (Table S4), flavonoid biosynthesis and photosynthesis, were significantly enriched (p adjust < 0.05). According to the KEGG enrichment analysis network diagram (Figure S1, Network plot of top 20 KEGG enrichment pathways), the two significantly enriched pathways mentioned above do not exhibit correlation with other pathways. The NR and Swiss-Prot descriptions indicate that the flavonoid biosynthesis pathway (map00941) contains nine genes. Two of these genes were related to acylsugar acyltransferase, two were flavonoid-related, and the remaining five were associated with agmatine hydroxycinnamoyl transferase, flavanone 3-hydroxylase, chalcone-flavonone isomerase, acetyl-CoA-benzylalcohol acetyltransferase, and anthocyanin synthase. The photosynthesis pathway (map00195) comprised 12 genes: four related to photosystem II, two to photosystem I, four to ferredoxin, and the remaining two to plastocyanin A and PsbP-like proteins. It is noteworthy that compared with glossy A21b, the expression of all 12 genes in the photosynthetic pathway was significantly downregulated in A32b. This suggests that the reduction in eggplant peel glossiness may be correlated with the downregulation of the photosynthetic pathway.

3.5. Identification of Differentially Expressed TFs

Analysis of the 996 DEGs identified 71 transcription factors corresponding to 66 DEGs (Table S5). These included 25 categories of transcription factors, with 44 upregulated and 27 downregulated genes. Among them, bHLH and ERF transcription factors were the most abundant, with 12 and 8 transcription factors, respectively, followed by bZIP (5), MYB (5), Dof (4), SBP (4), WRKY (4), and 34 belonging to 20 other categories. It is of note that compared with A32a, all eight ERF transcription factors were upregulated in A32b.

3.6. Expression of Candidate Genes Related to Peel Glossiness

According to the literature, the formation of plant wax may be related to ECERIFERUM [14], wax biosynthetic processes [13], C2H2 transcription factors [24], AP2/ERF transcription factors [20,21], and long-chain acyl-CoA synthetase [26] genes. By searching for related genes among the DEGs, we identified two C2H2 transcription factors (*Smechr0104003, Smechr0902056*), eight AP2/ERF transcription factors (*Smechr0602290, Smechr0301939, Smechr0500152, Smechr0902337, Smechr0800086, Smechr0402179, Smechr1100683, Smechr1102045*), one long-chain acyl-CoA synthetase gene (*Smechr0102162*), and one ECERIFERUM gene (*Smechr0800895*) that only showed significant differences between A21b and A32b. The expression levels of these 12 genes were examined in three inbred cultivars (GZ1, ZE11, and ZSQ) with high glossiness and three inbred cultivars (B209, B203, and HNQQ) with no glossiness on their fruit peels using qRT-PCR (Figure 4). The results indicated that one C2H2 transcription factor gene (*Smechr0104003*), two ERF transcription factor genes (*Smechr0402179* and *Smechr0902337*), and the long-chain acyl-CoA synthetase gene (*Smechr0102162*) were expressed at lower levels in eggplant fruit peels with high glossiness than in those with no glossiness. These four genes may be associated with eggplant glossiness.

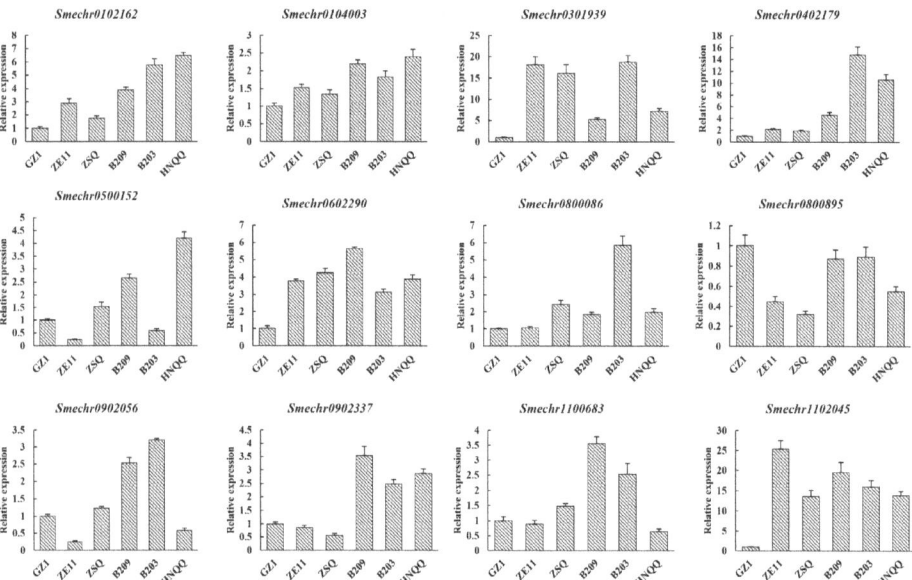

Figure 4. Expression levels for 12 candidate genes related to eggplant peel glossiness in three inbred lines with high glossiness and three inbreds without glossiness on their fruit peel by qRT-PCR. The data of columns are the mean + standard deviation. The bars mean standard deviation.

4. Discussion

Glossiness in plants is a complex trait influenced by multiple factors and closely related to various physiological and biochemical processes [39–41]. In this study, we used paraffin sections to examine the structures of eggplant peels with different levels of glossiness. We found that peel glossiness may be closely associated with the thickness of the wax layer and the degree of surface wrinkling. This finding is consistent with that of Liu et al. [14], who found that thicker wax layers resulted in lower glossiness. However, other studies have suggested that wax content may not directly affect plant glossiness; rather, it is the arrangement of wax that plays a key role. For example, Zhang et al. [18] investigated the glossiness of sweet orange (*Citrus sinensis* L. Osbeck) fruit surfaces. They found that the structure of the wax, rather than the total wax content, was the primary determinant of glossiness. In our study, we also noted significant surface wrinkling of the wax on eggplant peels with lower glossiness. These wrinkles may result from differences in the arrangement of the wax molecules [18], which further affects peel glossiness. To explore the relationship between the wax thickness, arrangement, and glossiness in greater detail, we plan to use scanning electron microscopy (SEM) to examine the microstructures of the eggplant peel surfaces. This will help us better understand the physiological basis of glossiness formation in eggplant and provide theoretical data that could help to improve its appearance.

Cultivation measures often significantly impact crop quality in agricultural production [42,43]. The nutritional status of a plant can significantly influences its overall health and performance, which in turn may manifest in observable traits such as fruit glossiness. For instance, adequate levels of essential nutrients like nitrogen, phosphorus, and potassium are vital for optimal plant growth and development [44]. These nutrients support basic physiological processes and contribute to the formation and quality of fruits. Therefore, variations in nutrient availability could impact the glossiness of eggplant fruits. Researchers can gain insights into how plant nutrition influences fruit quality and appearance by systematically examining these traits and their relationship with eggplant glossiness. The fruit peel glossiness of the inbred lines in this study exhibited consistency under the cultivation conditions across two consecutive seasons (spring and fall of 2023). It is speculated that the glossiness of the inbred line fruit peels is largely determined by genetic factors. Of course, we did not cultivate the inbred lines in extreme conditions such as drought or nutrient deficiency. Whether environmental factors would impact the fruit peel glossiness of the inbred lines in this study must be further investigated.

Wax significantly influences plant glossiness, and has multiple processing stages, such as wax synthesis and transport. Wax synthesis occurs via various pathways, including the production of C16 to C18 fatty acids, synthesis of C20 to C34 very-long-chain fatty acids (VLCFAs), and further conversion of these VLCFA derivatives [45]. The wax components, synthesized and modified by various enzymes in the endoplasmic reticulum, are first transported to the cell membrane, then undergo transmembrane transport via transporter proteins, and are finally transported across the cell wall to the cuticle layer via lipid transfer proteins [39,45,46]. The synthesis and transport of wax involves numerous genes and can alter gene expression. This study identified many DEGs between any two samples, highlighting the complex and multifaceted nature of the factors determining eggplant peel glossiness. Screening for genes directly related to eggplant glossiness from a large number of DEGs is a challenging task. In future studies, we will measure physiological and biochemical indicators in eggplant peels with different gloss levels to identify the physiological and biochemical pathways potentially involved in regulating eggplant glossiness, thereby providing data to aid in the screening of related genes.

Wax and cutin are important factors that affect the glossiness of plant surfaces, and significantly affect the appearance and adaptability of plants [7,8]. Among the four eggplant peel samples, A21a, A21b, A32a, and A32b, the first three samples (A21a, A21b, and A32a) displayed high peel glossiness, whereas A32b was the only peel to exhibit noticeably reduced glossiness. By comparing the differences in gene expression between A32b and the three other high-gloss samples, seven genes potentially related to wax and cutin synthesis were identified (see Table S6). Among these, four genes were highly expressed in A32b, while the remaining three showed relatively lower expression levels. Because the results suggest that an increase in wax content may reduce peel glossiness, wax- and cutin-related genes that are highly expressed in A32b may play inhibitory roles. These highly expressed genes included *Smechr0100391*, *Smechr0103283*, *Smechr0500624*, and *Smechr0900645*, which may be candidate genes regulating the glossiness of eggplant peel. However, further experimental studies are required to verify the specific roles of these genes in determining peel glossiness in eggplant.

In this study, GO enrichment and KEGG enrichment analysis revealed that the downregulation of the DNA replication and photosynthesis pathways may be associated with reduced eggplant peel glossiness. The glossiness of plants is related to wax biosynthesis, which is a relatively independent process from DNA replication. There is no direct biological connection between them. However, studies on the regulation of cuticular wax in the awns of barley (*Hordeum vulgare* L.) [47], specifically the characterization of glossy spike mutants in barley and the identification of candidate genes regulating epidermal wax synthesis, suggest that cuticular wax synthesis might also be linked to DNA replication. The specific mechanism connecting these two pathways requires further analysis. According to previous research, increased wax content on the leaves can prevent water from entering pores and increase resistance to gas diffusion [48]. When wax partially covers stomata, the cross-sectional area for gas diffusion is reduced. Therefore, stomatal conductance should also decrease when leaf wax is high. In addition, the wax on leaf surfaces considerably impacts light reflection. An increase in wax coverage on the leaves of Japanese yew (*Taxus cuspidata* L.) reduces stomatal conductance, thereby decreasing the photosynthetic rate of the seedlings' leaves [49]. In this study, the thickening and folding of the waxy layer reduced the glossiness of eggplant peels, potentially leading to increased light reflection and decreased stomatal conductance, subsequently decreasing photosynthesis in the eggplant peel. This finding aligns with the results of the KEGG enrichment analysis.

5. Conclusions

The glossiness of eggplant peels is an important commercial trait. Paraffin section analysis revealed that increased wax thickness and wrinkles on the wax surface of eggplant peels decreased glossiness. Differential gene expression related to eggplant peel glossiness was analyzed by comparing transcriptomes of eggplant peels with different gloss levels and at different developmental stages. In total, 996 DEGs were identified, including 502 upregulated and 494 downregulated genes, which may be related to eggplant peel glossiness. GO and KEGG enrichment analyses revealed that downregulation of the DNA replication pathway (GO:0003688, GO:0006270) and the photosynthesis pathway (map00195) may be associated with reduced eggplant peel glossiness. Expression level analysis identified eight genes associated with eggplant fruit peel glossiness. The results will provide a theoretical basis and technical support for the molecular genetic improvement of eggplant quality and, ultimately, help farmers obtain high-quality agricultural products.

Supplementary Materials: The following supporting information can be downloaded at: https://www.mdpi.com/article/10.3390/agronomy14123063/s1, Table S1: primers for qRT-PCR used in the present study; Table S2: DEGs between A32b and A32a; Table S3: the significant GO terms in in A32b vs. A32a; Table S4: the significant metabolic pathways in A32b vs. A32a; Table S5: the transcription factors from the DEGs; Table S6: the expression of Seven genes potentially related to wax and cutin synthesis in the transcriptome; Figure S1: network plot of top 20 KEGG enrichment pathways.

Author Contributions: Conceptualization, Z.N. and J.Z.; methodology, H.W. and Z.N.; investigation, H.W., Z.N., T.W., and S.Y.; resources, H.W. and J.Z.; data curation, Z.N., S.Y., and T.W.; writing—original draft preparation, Z.N.; writing—review and editing, H.W. and Z.N.; supervision, J.Z.; funding acquisition, H.W. and J.Z. All authors have read and agreed to the published version of the manuscript.

Funding: This research was funded by the Technology Project of the Hangzhou Academy of Agricultural Sciences, grant number 2022HNCT-02, the Grand Science and Technology Special Project of Zhejiang Province, grant number 2021C02065 and the Hangzhou Agricultural Science and Technology Cooperation and Innovation Project, grant number Hangnong 202158, 20241203A03.

Data Availability Statement: The original contributions presented in the study are included in the article and Supplementary Materials, further inquiries can be directed to the corresponding author.

Conflicts of Interest: The authors declare no conflicts of interest.

References

1. Alam, I.; Salimullah, M. Genetic engineering of eggplant (*Solanum melongena* L.): Progress, controversy and potential. *Horticulturae* **2021**, *7*, 78. [CrossRef]
2. Jha, S.N.; Matsuoka, T. Objective Method of estimation of post-harvest freshness of eggplant fruits. *IFAC Proc. Vol.* **2001**, *34*, 67–70. [CrossRef]
3. Argento, S.; Treccarichi, S.; Arena, D.; Rizzo, G.F.; Branca, F. Exploitation of a grafting technique for improving the water use efficiency of eggplant (*Solanum melongena* L.) grown in a cold greenhouse in mediterranean climatic conditions. *Agronomy* **2023**, *13*, 2705. [CrossRef]
4. Gaccione, L.; Martina, M.; Barchi, L.; Portis, E. A compendium for novel marker-based breeding strategies in eggplant. *Plants* **2023**, *12*, 1016. [CrossRef]
5. Jetter, R.; Kunst, L.; Samuels, A.L. Composition of Plant Cuticular Waxes. *Ann. Plant Rev. Biol. Plant Cuticle* **2007**, *23*, 145–181. [CrossRef]
6. Wang, W.; Liu, X.; Gai, X.; Ren, J.; Liu, X.; Cai, Y.; Wang, Q.; Ren, H. *Cucumis sativus* L. WAX2 plays a pivotal role in wax biosynthesis, influencing pollen fertility and plant biotic and abiotic stress responses. *Plant Cell Physiol.* **2015**, *56*, 1339–1354. [CrossRef] [PubMed]
7. Zheng, J.; He, C.; Qin, Y.; Lin, G.; Park, W.D.; Sun, M.; Li, J.; Lu, X.; Zhang, C.; Yeh, C.T.; et al. Co-expression analysis aids in the identification of genes in the cuticular wax pathway in maize. *Plant J.* **2019**, *97*, 530–542. [CrossRef]
8. Yang, S.; Liu, H.; Wei, X.; Zhao, Y.; Wang, Z.; Su, H.; Zhao, X.; Tian, B.; Zhang, X.W.; Yuan, Y. BrWAX2 plays an essential role in cuticular wax biosynthesis in Chinese cabbage (*Brassica rapa* L. ssp. *pekinensis*). *Theor. Appl. Genet.* **2022**, *135*, 693–707. [CrossRef]
9. Samuels, L.; Kunst, L.; Jetter, R. Sealing plant surfaces: Cuticular wax formation by epidermal cells. *Annu. Rev. Plant Biol.* **2008**, *59*, 683–707. [CrossRef] [PubMed]
10. Rowland, O.; Zheng, H.; Hepworth, S.R.; Lam, P.; Jetter, R.; Kunst, L. CER4 encodes an alcohol-forming fatty acyl-coenzyme A reductase involved in cuticular wax production in Arabidopsis. *Plant Physiol.* **2006**, *142*, 866–877. [CrossRef]
11. Tang, X.; Song, G.; Zou, J.; Ren, J.; Feng, H. BrBCAT1 mutation resulted in deficiency of epicuticular wax crystal in Chinese cabbage. *Theor. Appl. Genet.* **2024**, *137*, 123. [CrossRef] [PubMed]
12. Liu, D.C.; Zeng, Q.; Ji, Q.X.; Liu, C.F.; Liu, S.B.; Liu, Y. A comparison of the ultrastructure and composition of fruits' cuticular wax from the wild-type 'Newhall' navel orange (*Citrus sinensis* [L.] Osbeck cv. Newhall) and its glossy mutant. *Plant Cell Rep.* **2012**, *31*, 2239–2246. [CrossRef] [PubMed]
13. Gao, L.; Cao, J.; Gong, S.; Hao, N.; Du, Y.; Wang, C.; Wu, T. The COPII subunit CsSEC23 mediates fruit glossiness in cucumber. *Plant J.* **2023**, *116*, 524–540. [CrossRef] [PubMed]
14. Liu, X.; Ge, X.; An, J.; Liu, X.; Ren, H. CsCER6 and CsCER7 Influence Fruit Glossiness by Regulating Fruit Cuticular Wax Accumulation in Cucumber. *Int. J. Mol. Sci.* **2023**, *24*, 1135. [CrossRef]
15. Xiong, C.; Xie, Q.; Yang, Q.; Sun, P.; Gao, S.; Li, H.; Zhang, J.; Wang, T.; Ye, Z.; Yang, C. WOOLLY, interacting with MYB transcription factor MYB31, regulates cuticular wax biosynthesis by modulating CER6 expression in tomato. *Plant J.* **2020**, *103*, 323–337. [CrossRef] [PubMed]
16. Petit, J.; Bres, C.; Just, D.; Garcia, V.; Mauxion, J.P.; Marion, D.; Bakan, B.; Joubès, J.; Domergue, F.; Rothan, C. Analyses of tomato fruit brightness mutants uncover both cutin-deficient and cutin-abundant mutants and a new hypomorphic allele of GDSL lipase. *Plant Physiol* **2014**, *164*, 888–906. [CrossRef]
17. Petit, J.; Bres, C.; Mauxion, J.P.; Tai, F.W.; Martin, L.B.; Fich, E.A.; Joubès, J.; Rose, J.K.; Domergue, F.; Rothan, C. The Glycerol-3-phosphate acyltransferase GPAT6 from tomato plays a central role in fruit cutin biosynthesis. *Plant Physiol* **2016**, *171*, 894–913. [CrossRef] [PubMed]
18. Zhang, J. Establishment of an Identification Method for Citrus Fruit Surface Glossiness. Master's Thesis, Huazhong Agricultural University, Wuhan, China, 2022.
19. Li, L.; Zhao, M.; Wang, J.; Liu, S.; Wang, G.; Schnable, S.P. Research progress on genetic mechanisms of plant epidermal wax synthesis, transport and regulation. *J. China Agric. Univ.* **2023**, *28*, 1–19. [CrossRef]
20. Park, C.S.; Go, Y.S.; Suh, M.C. Cuticular wax biosynthesis is positively regulated by WRINKLED4, an AP2/ERF-type transcription factor, in Arabidopsis stems. *Plant J.* **2016**, *88*, 257–270. [CrossRef]
21. Kim, H.; Go, Y.S.; Suh, M.C. DEWAX2 transcription factor negatively regulates cuticular wax biosynthesis in *Arabidopsis* leaves. *Plant Cell Physiol.* **2018**, *59*, 966–977. [CrossRef] [PubMed]
22. He, J.; Li, C.; Hu, N.; Zhu, Y.; He, Z.; Sun, Y.; Wang, Z.; Wang, Y. ECERIFERUM1-6A is required for the synthesis of cuticular wax alkanes and promotes drought tolerance in wheat. *Plant Physiol.* **2022**, *190*, 1640–1657. [CrossRef]
23. Yang, Y.; Cai, C.; Wang, Y.; Wang, Y.; Ju, H.; Chen, X. Cucumber *glossy fruit 1* (*CsGLF1*) encodes the zinc finger protein 6 that regulates fruit glossiness by enhancing cuticular wax biosynthesis. *Hortic. Res.* **2022**, *10*, 237. [CrossRef] [PubMed]
24. Zhai, X.; Wu, H.; Wang, Y.; Zhang, Z.; Shan, L.; Zhao, X.; Wang, R.; Liu, C.; Weng, Y.; Wang, Y.; et al. The fruit glossiness locus, dull fruit (D), encodes a C(2)H(2)-type zinc finger transcription factor, CsDULL, in cucumber (*Cucumis sativus* L.). *Hortic. Res.* **2022**, *9*, 146. [CrossRef] [PubMed]
25. Al-Abdallat, A.M.; Al-Debei, H.S.; Ayad, J.Y.; Hasan, S. Over-expression of SlSHN1 gene improves drought tolerance by increasing cuticular wax accumulation in tomato. *Int. J. Mol. Sci.* **2014**, *15*, 19499–19515. [CrossRef]

26. Wu, P.; Li, S.; Yu, X.; Guo, S.; Gao, L. Identification of long-chain acyl-CoA synthetase gene family reveals that *SlLACS1* is essential for cuticular wax biosynthesis in tomato. *Int. J. Biol Macromol.* **2024**, *277*, 134438. [CrossRef] [PubMed]
27. Ni, F.; Yang, M.; Chen, J.; Guo, Y.; Wan, S.; Zhao, Z.; Yang, S.; Kong, L.; Chu, P.; Guan, R. *BnUC1* is a key regulator of epidermal wax biosynthesis and lipid transport in *Brassica napus*. *Int. J. Mol. Sci.* **2024**, *25*, 9533. [CrossRef] [PubMed]
28. Dai, J.; Zhang, R.; Wei, B.; Nie, Z.; Xing, G.; Zhao, T.; Yang, S.; Gai, J. Key biological factors related to outcrossing-productivity of cytoplasmic-nuclear male-sterile lines in soybean [*Glycine max* (L.) Merr.]. *Euphytica* **2017**, *213*, 266. [CrossRef]
29. Chen, S.; Zhou, Y.; Chen, Y.; Gu, J. fastp: An ultra-fast all-in-one FASTQ preprocessor. *Bioinformatics* **2018**, *34*, i884–i890. [CrossRef] [PubMed]
30. Wei, Q.; Wang, J.; Wang, W.; Hu, T.; Hu, H.; Bao, C. A high-quality chromosome-level genome assembly reveals genetics for important traits in eggplant. *Hortic. Res.* **2020**, *7*, 153. [CrossRef]
31. Kim, D.; Pertea, G.; Trapnell, C.; Pimentel, H.; Kelley, R.; Salzberg, S.L. TopHat2: Accurate alignment of transcriptomes in the presence of insertions, deletions and gene fusions. *Genome Biol.* **2013**, *14*, R36. [CrossRef]
32. Li, B.; Dewey, C.N. RSEM: Accurate transcript quantification from RNA-Seq data with or without a reference genome. *BMC Bioinform.* **2011**, *12*, 323. [CrossRef] [PubMed]
33. Love, M.I.; Huber, W.; Anders, S. Moderated estimation of fold change and dispersion for RNA-seq data with DESeq2. *Genome Biol.* **2014**, *15*, 550. [CrossRef] [PubMed]
34. Conesa, A.; Götz, S.; García-Gómez, J.M.; Terol, J.; Talón, M.; Robles, M. Blast2GO: A universal tool for annotation, visualization and analysis in functional genomics research. *Bioinformatics* **2005**, *21*, 3674–3676. [CrossRef] [PubMed]
35. Young, M.D.; Wakefield, M.J.; Smyth, G.K.; Oshlack, A. goseq: Gene ontology testing for RNA-seq datasets. *R. Bioconduct.* **2012**, *8*, 1–25.
36. Mao, X.; Cai, T.; Olyarchuk, J.G.; Wei, L. Automated genome annotation and pathway identification using the KEGG Orthology (KO) as a controlled vocabulary. *Bioinformatics* **2005**, *21*, 3787–3793. [CrossRef]
37. Zheng, Y.; Jiao, C.; Sun, H.; Rosli, H.G.; Pombo, M.A.; Zhang, P.; Banf, M.; Dai, X.; Martin, G.B.; Giovannoni, J.J.; et al. iTAK: A program for genome-wide prediction and classification of plant transcription factors, transcriptional regulators, and protein kinases. *Mol. Plant* **2016**, *9*, 1667–1670. [CrossRef] [PubMed]
38. Lv, J.; Liu, Z.; Liu, Y.; Ou, L.; Deng, M.-H.; Wang, J.; Song, J.; Ma, Y.; Chen, W.; Zhang, Z.; et al. Comparative transcriptome analysis between cytoplasmic male-sterile line and its maintainer during the floral bud development of pepper. *Hortic. Plant J.* **2020**, *6*, 89–98. [CrossRef]
39. Hao, Y.; Luo, H.; Wang, Z.; Lu, C.; Ye, X.; Wang, H.; Miao, L. Research progress on the mechanisms of fruit glossiness in cucumber. *Gene* **2024**, *927*, 148626. [CrossRef] [PubMed]
40. Hassanzadeh-Khayyat, M.; Akaberi, M.; Moalemzadeh Haghighi, H.; Sahebkar, A.; Emami, S.A. Distribution and variability of n-alkanes in waxes of conifers. *J. Forestry. Res.* **2019**, *30*, 429–433. [CrossRef]
41. Ezer, R.; Manasherova, E.; Gur, A.; Schaffer, A.A.; Tadmor, Y.; Cohen, H. The dominant white color trait of the melon fruit rind is associated with epicuticular wax accumulation. *Planta* **2024**, *260*, 97. [CrossRef]
42. Singh, Y.V.; Singh, K.K.; Sharma, S.K. Influence of crop nutrition on grain yield, seed quality and water productivity under two rice cultivation systems. *Rice Sci.* **2013**, *20*, 129–138. [CrossRef]
43. Vidalis, N.; Kourkouvela, M.; Argyris, D.-C.; Liakopoulos, G.; Alexopoulos, A.; Petropoulos, S.A.; Karapanos, I. The impact of salinity on growth, physio-biochemical characteristics, and quality of *Urospermum picroides* and *Reichardia picroides* plants in varied cultivation regimes. *Agriculture* **2023**, *13*, 1852. [CrossRef]
44. Besford, R.T.; Maw, G.A. Effect of potassium nutrition on tomato plant growth and fruit development. *Plant Soil* **1975**, *42*, 395–412. [CrossRef]
45. Lu, W.; Zheng, W.; Wu, Y.; Zang, Y. Research review on features and molecular mechanism of wax formation in Brassicaceae. *J. Zhejiang A&F Univ.* **2021**, *38*, 205–213.
46. Debono, A.; Yeats, T.H.; Rose, J.K.; Bird, D.; Jetter, R.; Kunst, L.; Samuels, L. Arabidopsis LTPG is a glycosylphosphatidylinositol-anchored lipid transfer protein required for export of lipids to the plant surface. *Plant Cell* **2009**, *21*, 1230–1238. [CrossRef] [PubMed]
47. Bian, X.; Yao, L.; Si, E.; Meng, Y.; Li, B.; Ma, X.; Yang, K.; Lai, Y.; Shang, X.; Li, C.; et al. Characterization of glossy spike mutants and identification of candidate genes regulating cuticular wax synthesis in barley (*Hordeum vulgare* L.). *Int. J. Mol. Sci.* **2022**, *23*, 13025. [CrossRef] [PubMed]
48. Mohammadian, M.A.; Watling, J.R.; Hill, R.S. The impact of epicuticular wax on gas-exchange and photoinhibition in *Leucadendron lanigerum* (Proteaceae). *Acta Oecologica* **2007**, *31*, 93–101. [CrossRef]
49. Li, W.; Li, J.; Wei, J.; Niu, C.; Yang, D.; Jiang, B. Response of photosynthesis, the xanthophyll cycle, and wax in Japanese yew (*Taxus cuspidata* L.) seedlings and saplings under high light conditions. *Peer J.* **2023**, *11*, e14757. [CrossRef]

Disclaimer/Publisher's Note: The statements, opinions and data contained in all publications are solely those of the individual author(s) and contributor(s) and not of MDPI and/or the editor(s). MDPI and/or the editor(s) disclaim responsibility for any injury to people or property resulting from any ideas, methods, instructions or products referred to in the content.

Article

Quantitative Trait Locus Analysis for Panicle and Flag Leaf Traits in Barley (*Hordeum vulgare* L.) Based on a High-Density Genetic Linkage Map

Yichen Ye, Shuting Wen, Guo Zhang, Xingzhe Yang, Dawei Xue ⬤, Yunxia Fang and Xiaoqin Zhang *

College of Life and Environmental Sciences, Hangzhou Normal University, Hangzhou 311121, China; 2022111010031@stu.hznu.edu.cn (Y.Y.); 2022111010026@stu.hznu.edu.cn (S.W.); zhangguo@stu.hznu.edu.cn (G.Z.); 2023111010016@stu.hznu.edu.cn (X.Y.); dwxue@hznu.edu.cn (D.X.); yxfang12@hznu.edu.cn (Y.F.)
* Correspondence: zxq@hznu.edu.cn

Abstract: The yield of barley (*Hordeum vulgare* L.) is determined by many factors, which have always been research hotspots for agronomists and molecular scientists. In this study, five important agronomic traits related to panicle and flag leaf, including awn length (AL), panicle length (PL), panicle neck length (NL), flag leaf length (LL) and flag leaf width (LW), were investigated and quantitative trait locus (QTL) analyses were carried out. Using a high-density genetic map of 134 recombinant inbred lines based on specific-locus amplified fragment sequencing (SLAF-seq) technology, a total of 32 QTLs were identified, which explained 12.4% to 50% of the phenotypic variation. Among them, *qAL5*, *qNL2*, *qNL3*, *qNL6*, *qPL2*, and *qLW2* were detected in 3 consecutive years and all of the contribution rates were more than 13.8%, revealing that these QTLs were stable major QTLs and were less affected by environmental factors. Furthermore, LL and LW exhibited significant positive correlations and the localization intervals of *qLL2* and *qLL3* were highly overlapped with those of *qLW2* and *qLW3*, respectively, indicating that *qLL2* and *qLW2*, *qLL3* and *qLW3* may be regulated by the same genes.

Keywords: barley; recombinant inbred line; panicle; flag leaf; QTL

Citation: Ye, Y.; Wen, S.; Zhang, G.; Yang, X.; Xue, D.; Fang, Y.; Zhang, X. Quantitative Trait Locus Analysis for Panicle and Flag Leaf Traits in Barley (*Hordeum vulgare* L.) Based on a High-Density Genetic Linkage Map. *Agronomy* **2024**, *14*, 2953. https://doi.org/10.3390/agronomy14122953

Academic Editor: Matthew Hegarty

Received: 25 October 2024
Revised: 1 December 2024
Accepted: 9 December 2024
Published: 11 December 2024

Copyright: © 2024 by the authors. Licensee MDPI, Basel, Switzerland. This article is an open access article distributed under the terms and conditions of the Creative Commons Attribution (CC BY) license (https://creativecommons.org/licenses/by/4.0/).

1. Introduction

In recent years, there has been a notable increase in the global demand for food, which has led to an intensified focus on enhancing crop yields. Barley (*Hordeum vulgare* L.), one of the earliest domesticated food crops, has been widely cultivated worldwide due to its high yield potential and important roles in various industries, including animal feeding, brewing and food production [1]. Improving the yield by systematically optimizing agronomic traits to better meet the growing food demand in barley breeding has become a consensus [2]. Therefore, the quantitative trait locus (QTL) analysis of barley yield-related traits can not only understand the complex genetic mechanism, but also lay the foundation for molecular breeding.

Barley yield is a complex quantitative trait, which is directly or indirectly affected by other traits. Among them, the panicle shape and grain size directly determine crop yield [3], while flag leaf length and width also affect yield by influencing the photosynthesis rate [4,5]. Although the awn is not an important organ for the growth and reproduction of triticeae crops, it can act as a photosynthetic organ and is one of the potential pathways to increase yield by significantly increasing photosynthetic products [6]. In addition, the awn also plays a role in resisting insect pests and bird pecking, which indirectly affect yield [7]. It is reported that the length of the panicle neck is also an important agronomic trait in barley breeding. A longer panicle neck increases the light and air exchange of organs above the flag leaf, reduces the incidence of fusarium head blight in the panicle and thus increases grain filling and thousand-grain weight [8]. However, too long a panicle neck length will lead to the excessive elongation of plant height, resulting in lodging [9]. Therefore, the appropriate awn length and panicle neck length are beneficial to the increase in plant yield [10].

Using a double haploid (DH) population, Wang et al. [11] constructed a high-density genetic map of barley and performed QTL analysis for 10 traits such as main panicle length, grain number per panicle and thousand-grain weight. A total of three main panicle length QTLs were detected on chromosomes 2H and 7H, explaining the contribution rates of 1.93 to 52.72. The genome wide association (GWA) analysis of yield-related traits was also carried out by using barley germplasm resources consisting of 185 cultivated and 38 wild genotypes, and seven and one QTLs related to panicle length were identified on chromosomes 2H, 3H, 4H and 5H in wet and dry environments [12]. Four flag leaf traits of barley (flag leaf thickness, length, width and area) were investigated, and five QTLs associated with flag leaf length were localized on chromosomes 1H, 2H, 3H, 4H and 6H, and two QTLs associated with flag leaf width were localized on chromosome 2H, explaining 5.8–17.9% of the phenotypic variation [13]. Du et al. [14] detected the QTLs for the length and area of flag, and the second, third and fourth leaves in barley. A total of 57 QTLs were identified, of which 6 relating to flag leaf length were located on chromosomes 2H and 7H, with variances ranging from 5.30 to 37.11%. Moreover, 12 QTLs for awn length distributed on all the 7 barley chromosomes were detected by a multiparent mapping population, in which the QTL located on chromosome 7HL explained the highest contribution rate [15]. These results provide a genetic basis for the study of panicle and flag leaf traits in barley.

Although barley has always been an important food crop, its large genome renders it difficult to construct high-density genetic maps using conventional methods [16]. Therefore, it is more feasible to construct genetic maps by SLAF-seq technology [17]. In this study, the 134 recombinant inbred lines (RILs) derived from the cross between Golden Promise (GP) and wild variety H602 were used to determine five agronomic traits, including panicle length (PL), awn length (AL), panicle neck length (NL), flag leaf length (LL) and flag leaf width (LW). The QTL analysis was conducted by a high-density genetic map based on SLAF markers, and a total of 32 QTLs were detected in consecutive three years. In this study, a scientific basis is provided for the genetic improvement of panicle and flag leaf traits in barley.

2. Materials and Methods

2.1. Plant Materials

A RILs population of F8 generation was created via single seed transmission method, which were derived from the cross of barley varieties GP and H602. The developed 134 RILs and parents were planted in the experimental field of Hangzhou Normal University, Xiasha Campus (Hangzhou, Zhejiang province, China) in the order of their numbers, with conventional field cultivation. Each line was sown one row with proximately 30 seeds, 130 cm length and 20 cm row spacing. From 2017 to 2019, the agronomic traits of 10 individuals of each line were measured in March or April.

2.2. Trait Measurement

Using a tape measure or ruler, the PL, AL, NL, LL and LW were measured. Among them, the flag leaf of the main stem was selected to examine the LL and LW, the longest awn was used to quantify AL, and the distance between the flag leaf's petiole and the base of the rachis was used to determine NL. Each trait was measured in 10 plants.

2.3. High-Density Genetic Linkage Map Construction

The specific-locus-amplified fragment sequencing (SLAF-seq) was performed following the description by Fang et al. [17]. All the reads were mapped to the barley reference genome (https://plants.ensembl.org/Hordeum_vulgare/Info/Index, accessed on 20 March 2019) and SLAFs were developed from the RILs and parents. The polymorphic SLAFs were classified into seven linkage groups (LGs). Due to the massive SNP data, a high density genetic map was constructed by HighMap mapping software (https://mybiosoftw are.com/highmap-construction-and-analysis-of-high-density-linkage-map.html#google_ vignette, accessed on 20 March 2019), and the genetic distances between two adjacent markers were computed in each LG.

2.4. Data and QTL Analysis

The descriptive statistics and correlative analysis of five traits were performed using SPSS 20.0 software (International Business Machines Corp, Armonk, NY, USA), and the frequency distributions of traits were analyzed using GraphPad Prism 9.5 (http://ww w.graphpad-prism.cn, accessed on 4 June 2024). Based on the high-density genetic map, QTL analysis was performed using the QGene 4.4.0 software [18]. During the process, the data of each trait were uploaded, and scan interval was set to one milli Morgan. Then, the composite interval mapping model was adopted for scanning seven chromosomes. The logarithm of odds (LOD) threshold greater than 3.5 was considered for significant differences ($p = 0.05$), and each QTL interval was tested by 1000 permutations. Finally, the QTLs were named obeying the rules presented by Mccouch et al. [19], and then QTLs, QTL parameters of chromosomes, marker names and intervals were obtained for drawing localization mapping using MapChart [20].

3. Results

3.1. Statistical Analysis of Measurement Data

Five traits of 134 RILs and two parents were measured in 3 consecutive years, and the data were statistically analyzed. The results demonstrated that the parent GP showed decreased AL, NL, LL and LW than H602. However, compared to H602, the PL of GP was decreased in 2017 and 2018, and increased in 2019. Except for PL, the average values of the AL, NL, LL and LW of 134 RILs were all between the parents. In addition, continuous segregation and transgressive segregation were observed for all assayed traits in 134 RILs (Table 1 and Figure 1), indicating that QTLs affecting individual's phenotype might come from two parents. Furthermore, the absolute values of skewness and kurtosis of each trait were less than 1, and the plots of the normal distribution revealed a single peak, indicating that the data fit the normal distribution model and were suitable for QTL mapping analysis.

Table 1. Descriptive statistics of parents' and RIL population's agronomic traits.

Year	Traits	GP	H602	RILs		Skewness	Kurtosis
		Mean ± SD (cm)	Mean±SD (cm)	Range (cm)	Mean ± SD (cm)		
2017	PL	7.17 ± 0.15	10.10 ± 0.75 *	5.23–12.97	8.61 ± 1.58	0.36	−0.17
	AL	4.97 ± 0.12	8.93 ± 0.52 **	1.97–111.5	6.91 ± 2.00	0.13	−0.79
	NL	3.03 ± 0.06	20.60 ± 0.51 **	1.80–28.8	12.00 ± 5.79	0.56	0.20
	LL	9.92 ± 0.79	13.60 ± 0.68 **	5.68–20.12	11.32 ± 3.10	0.51	0.21
	LW	0.88 ± 0.05	1.14 ± 0.05 **	0.58–1.58	0.99 ± 0.18	0.42	0.58
2018	PL	8.10 ± 0.21	8.47 ± 0.27	5.43–12.33	8.51 ± 1.35	0.21	−0.37
	AL	6.23 ± 0.34	9.97 ± 0.03 **	5.03–12.87	8.64 ± 1.70	0.01	−0.61
	NL	0.00	15.57 ± 0.25 **	0.00–21.83	9.41 ± 5.20	0.19	−0.82
	LL	6.17 ± 0.47	7.87 ± 0.41 *	3.1–12.43	7.27 ± 2.07	0.56	−0.09
	LW	0.62 ± 0.04	0.78 ± 0.04	0.30–1.43	0.72 ± 0.20	0.82	0.79
2019	PL	8.58 ± 0.23	8.14 ± 0.21	5.32–11.86	8.94 ± 1.23	0.01	−0.11
	AL	6.14 ± 0.15	10.16 ± 0.32 **	5.14–13.58	9.58 ± 1.97	−0.38	−0.59
	NL	0.00	20.5 ± 0.73 **	0.00–23.44	11.70 ± 5.75	0.10	−0.68
	LL	7.16 ± 0.51	9.96 ± 0.63 **	4.72–13.56	8.84 ± 2.12	0.10	−0.91
	LW	0.58 ± 0.09	0.9 ± 0.17 **	0.42–1.32	0.79 ± 0.20	0.52	0.08

GP, Golden Promise; RILs, recombinant inbred lines; PL, panicle length; AL, awn length; NL, neck length; LL, flag leaf length; LW, flag leaf width. Mean ± SD indicates the means and the standard deviation; "*" indicates significant difference between parents; "**" indicates a highly significant difference between parents.

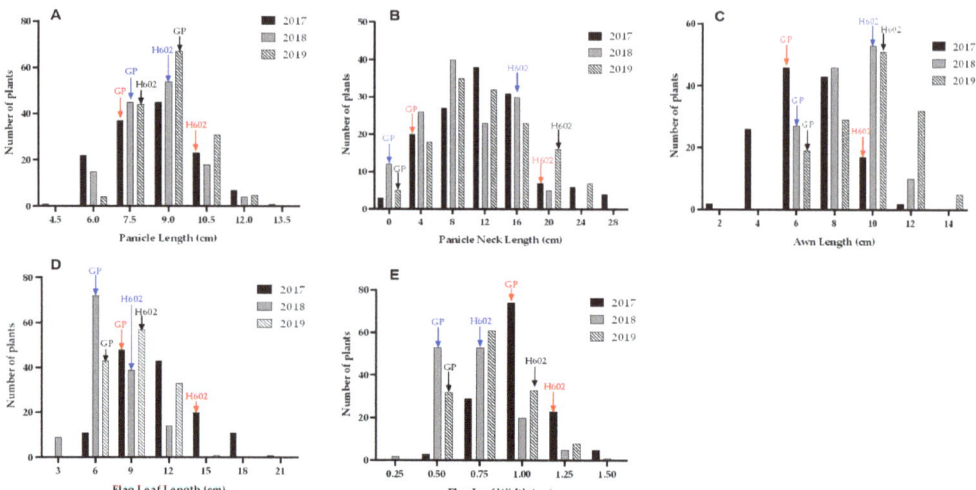

Figure 1. Frequency distribution of agronomic traits in the RIL population. (**A–E**) are the frequency distribution of panicle lengthe, panicle neck length, awn length, flag leaf length and flag leaf width, respectively. The *X*-axis represents the length in centimeters, and the *Y*-axis represents the number of plants. Red, blue and black arrows represent the location of the parental GP and H602 in 2017, 2018 and 2019, respectively. RIL, recombinant inbred line; GP, golden promise.

3.2. Correlation Analysis

The Spearman correlation coefficients between the PL, AL, NL, LL and LW of RILs in 3 consecutive years were analyzed by SPSS 20.0. It was found that LL versus LW exhibited extremely significant positive correlations in 3 consecutive years, and the most correlation coefficients of LL versus PL, LL versus AL, LL versus NL and LW versus NL showed significant or extremely significant positive correlations, indicating that correlation traits may be affected by the same genes or may have a strong linkage effect. The correlation data are presented in Table 2.

Table 2. Spearman correlation analysis among different traits in the RIL population.

	PL17	PL18	PL19	AL17	AL18	AL19	NL17	NL18	NL19	LL17	LL18	LL19	LW17	LW18	LW19
PL17	1														
PL18	0.72**	1													
PL19	0.66**	0.74**	1												
AL17	0.35**	0.09	0.05	1											
AL18	0.27**	0.17	0.08	0.65**	1										
AL19	0.20*	0.10	0.09	0.57**	0.70**	1									
NL17	0.24**	−0.11	−0.05	0.29**	0.17*	0.18*	1								
NL18	0.06	0.04	0.02	−0.09	0.04	−0.02	0.59**	1							
NL19	0.09	0.03	0.16	−0.03	0.06	0.20*	0.52**	0.62**	1						
LL17	0.32**	0.08	0.02	0.32**	0.19*	0.09	0.31**	0.14	0.14	1					
LL18	0.22*	0.30**	0.19*	0.28**	0.19*	0.18*	0.21*	0.39**	0.33**	0.54**	1				
LL19	0.27**	0.22*	0.22*	0.19*	0.06	0.22**	0.26**	0.28**	0.46**	0.45**	0.56**	1			
LW17	0.20*	0.133	0.04	0.01	−0.14	−0.18*	0.15	0.21*	0.14	0.76**	0.48**	0.42**	1		
LW18	0.09	0.21*	0.08	0.07	−0.10	−0.06	0.18*	0.41**	0.30**	0.41**	0.80**	0.45**	0.64**	1	
LW19	0.10	0.16	0.06	−0.02	−0.22*	−0.07	0.18*	0.34**	0.37**	0.27**	0.42**	0.77**	0.53**	0.62**	1

17, 18 and 19 were the abbreviations of 2017, 2018 and 2019, respectively; PL, panicle length; AL, awn length; NL, neck length; LL, flag leaf length; LW, flag leaf width. "*" indicates significant difference between traits; "**" indicates a highly significant difference between traits.

3.3. QTL Analysis

Through analysis, a total of 32 QTLs were detected in the 3 years (Table 3 and Figure 2), which were localized on chromosomes 2H, 3H, 4H, 5H, 6H and 7H. Their LOD values ranged from 3.85 to 20.16, and contribution to phenotypic variation ranged from 12.4 to 50%. Among them, the QTL qAL5 detected in 2017 exhibited the highest LOD value of 20.16 and a contribution rate of 50%. The additive effects of QTLs were positive, indicating that the favorable alleles were derived from the parental H602. Among the 32 QTLs, qAL5, qNL2, qNL3, qNL6, qNL7, qPL2, qLL3, qLW2 and qLW3 were detected in three consecutive years, and showed high contribution rates, revealing that these QTLs are stably inherited and are less affected by the external environment and other factors.

Table 3. QTL analysis of agronomic traits in the RIL population.

Year	QTLs	Chr	Interval (cM)	Peak LOD	R^2 (%)	Effect	Flank Markers	Physical Pos. (Mb)
2017	qAL5	5	58.9–65.6	20.16	50	1.406	Marker5580080–Marker6842284	467.44–488.28
	qNL2	2	0–8.4	9.01	26.6	2.978	Marker24990312–Marker23545719	63.53–64.08
	qNL3	3	41.8–48.2	7.4	22.4	2.736	Marker19746494–Marker21929437	374.11–477.74
	qNL4	4	21.3–23.5	4.4	14	2.162	Marker14966870–Marker11037594	338.34–401.70
	qNL5	5	35.1–41.5	7.95	23.9	2.849	Marker5710486–Marker6018261	354.49–416.31
	qNL6	6	38.8–47.4	13.79	37.7	3.655	Marker1551125–Marker2570397	303.29–480.82
	qNL7	7	69.1–75.9	8.57	25.5	2.94	Marker8576789–Marker8377405	153.50–211.47
	qPL2	2	103.9–107.3	7.05	21.5	0.736	Marker24884231–Marker24834355	730.26–743.57
	qPL5	5	61.9–65.5	6.41	19.8	0.701	Marker6998274–Marker6842284	481.62–488.28
	qLL5	5	53.6–58.1	4.48	14.3	1.159	Marker3862529–Marker4756954	446.24–462.05
	qLW2	2	0–3.8	4.94	15.6	0.072	Marker24990312–Marker26781447	46.72–236.45
	qLW3	3	38–44.8	7.21	21.9	0.086	Marker21428815–Marker18993774	173.69–408.378
2018	qAL5	5	61.5–67.2	11.39	33.30	0.978	Marker4410393–Marker5466564	481.70–488.67
	qNL2	2	0–4.9	10.87	31.2	2.906	Marker24990312–Marker25394564	46.72–236.45
	qNL3	3	41.4–47	11.87	33.5	3.019	Marker19120814–Marker21365319	287.83–458.84
	qNL6	6	38.1–43.9	4.67	14.9	2.009	Marker1438212–Marker2973766	303.29–430.94
	qNL7	7	59.4–63.3	7.79	23.5	2.527	Marker8501384–Marker10200503	54.37–127.93
	qPL2	2	103.9–108.1	5.48	17.2	0.559	Marker24884231–Marker23702882	730.26–747.92
	qLL3	3	38.7–42.5	3.86	12.4	0.726	Marker20887162–Marker18512482	253.66–416.32
	qLW2	2	0–2.2	3.85	12.4	0.071	Marker24990312–Marker24565213	46.72–236.45
	qLW3	3	38.7–43.3	6.32	19.5	0.089	Marker20887162–Marker21996421	173.69–416.32
2019	qAL5	5	61.5–67.2	6.45	19.9	0.877	Marker4410393–Marker5466564	481.70–488.67
	qNL2	2	0–3.8	11.1	31.7	3.237	Marker24990312–Marker26781447	46.72–236.45
	qNL3	3	41.4–47.8	11.05	31.6	3.235	Marker19120814–Marker18774403	374.11–479.05
	qNL6	6	37.3–40.7	4.32	13.8	2.129	Marker1066730–Marker1300046	295.15–390.93
	qNL7	7	59.4–63.3	7.01	21.4	2.662	Marker8501384–Marker10200503	54.37–127.93
	qPL2	2	103.9–107.3	6.38	19.7	0.545	Marker24884231–Marker24834355	730.26–743.57
	qLL2	2	0–3.8	5.16	16.2	0.856	Marker24990312–Marker26781447	46.72–236.45
	qLL3	3	40.6–46.3	5.26	16.5	0.857	Marker18862742–Marker18763097	287.83–444.09
	qLL7	7	61.4–65	4.22	12.3	0.746	Marker10083213–Marker9339689	59.29–127.93
	qLW2	2	0–2.2	6.11	18.9	0.085	Marker24990312–Marker24565213	46.72–236.45
	qLW3	3	39.1–43.6	5.19	16.3	0.079	Marker21868285–Marker19438077	253.66–416.32

Chr represents the chromosome number; peak LOD represents the maximum likelihood LOD value; R2 (%) represents the contribution rate of phenotypic variation; effect indicates additive effect, and the additive effect is positive, indicating that the favorable allele comes from H602, on the contrary, it comes from GP.

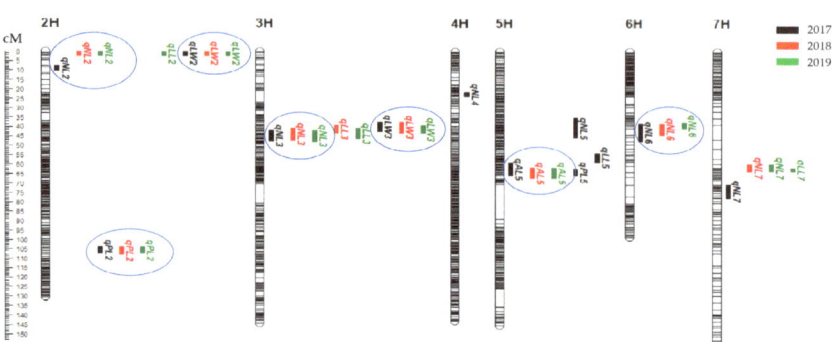

Figure 2. Localization of 32 QTLs for barley yield on the genetic linkage map. Black, red and green boxes represent QTLs in 2017, 2018 and 2019, respectively. Circles represent QTLs detected in 3 consecutive years. Black lines indicate SLAF markers.

3.3.1. QTL Analysis for Awn Length

From 2017 to 2019, the awn length-related QTL *qAL5* on chromosome 5H was detected in 3 consecutive years, which showed LOD values from 6.45 to 20.16 and phenotypic variance from 11% to 50%. Among them, the maximum LOD value of *qAL5* detected in 2017 was 20.16, and the phenotypic variance was as high as 50%. This result indicated that *qAL5* from parental H602 was a major QTL with a high contribution rate and stable inheritance, rendering it suitable for genetic improvement.

3.3.2. QTL Analysis for Panicle Neck Length

A total of 14 QTLs of panicle neck length were identified on chromosomes 2H, 3H, 4H, 5H, 6H and 7H. The LOD values of these QTLs ranged from 4.32 to 13.79, and the contribution to phenotypic variance ranged from 13.8% to 37.7%. Four QTLs named *qNL2*, *qNL3*, *qNL6* and *qNL7* were detected in 3 consecutive years. In 2017, the maximum LOD value of *qNL6* was detected to be 13.79, and its contribution to phenotypic variation reached 37.7%. In 2018 and 2019, the contribution rates of *qNL2* and *qNL3* were detected to be greater than 30%. However, the *qNL4* and *qNL5* were only detected in 2017, and the LOD values and phenotypic variation rate were relatively small, indicating that these two QTLs should be minor genes and are susceptible to environmental factors.

3.3.3. QTL Analysis for Panicle Length

A total of four QTLs for panicle length trait were detected on chromosomes 2H and 5H. The LOD values ranged from 5.48 to 7.05 and the contribution to phenotypic variation ranged from 17.2% to 21.5%. The *qPL2* was detected in all 3 years and the localization intervals were highly overlapped, suggesting that *qPL2* is a major and stable genetic QTL locus for panicle length. Moreover, the *qPL5* was only detected in 2017 and explained the phenotypic variance of 17.2%, indicating that it was highly susceptible to environmental impacts.

3.3.4. QTL Analysis for Flag Leaf Length and Width

According to the result of QTL analysis for flag leaf length, five QTLs were identified, which were localized on chromosomes 2H, 3H, 5H and 7H with LOD values ranging from 3.86 to 5.26 and contribution to phenotypic variation ranging from 12.4 to 16.5%. Among them, the *qLL3* was detected in 2018 and 2019, *qLL2* and *qLL7* were detected in 2019 and *qLL5* was only detected in 2017. None of the flag leaf length QTLs could be mapped in 3 consecutive years, indicating that they were greatly affected by environmental factors. In flag leaf width, a total of six QTLs were detected, which were localized on chromosomes 2H and 3H. Among them, the *qLW2* and *qLW3* were all detected in 3 consecutive years with the LOD values ranged from 3.85 to 6.32 and a phenotypic variation of 12.4% to 19.5%, revealing that leaf width can be stably inherited.

4. Discussion

The yield of gramineous crops is determined by both genetic and environmental factors. Therefore, the key to increasing yield is by selecting high heritability gene loci. The panicle length can directly affect the yield, whereas the panicle neck length and awn length indirectly affect the yield by resisting diseases and pests [21]. The flag leaf is the topmost leaf, and its size is determined by length and width, which influences photosynthesis rate to a certain extent, thereby affecting crop yield [22]. Although yield-related QTLs have been extensively studied in various crops, such as rice (*Oryza sativa* L.) [23–25], wheat (*Triticum aestivum* L.) [26,27] and maize (*Zea mays* L.) [28,29], the related research progress is relatively limited in barley.

There is a strong correlation between panicle length and yield. Longer panicles usually produce more seeds, which will be beneficial for improving the harvest index. Previous QTL studies of panicle length in barley have mainly focused on chromosome 2H. Chutimanitsakun et al. [30] identified a QTL on chromosome 2H that significantly affects various traits, including panicle length, plant height and panicle kernels. Similarly, Xue et al. [31] identified a QTL with close location for panicle length between markers bPb-6088 and bPb-5440. Moreover, Rozanova et al. [32] also identified three related loci on chromosome 2H. Islamovic et al. [33] positioned a QTL related to panicle length at 102.8 cm on chromosome 2H, which is very close to the localization interval of *qPL2* in the present study, suggesting that they may be controlled by the same QTL.

Barley awn is essentially degraded leaves, which play an important role in protecting plant from exogenous aggression, efficiently dispersing seeds and promoting photosynthetic efficiency [34]. Using a multiparent mapping population, Liller et al. [15] identified 12 QTLs related to awn length, which were distributed on all seven chromosomes of barley, respectively. In our study, a QTL related to awn length, *qAL5*, was mapped to an interval of 58.9–67.2 cM on chromosome 5H, which was close to the position of *qAL5.1* detected by Liller et al. [15], suggesting that the two may be the same QTL. However, only 1 QTL for awn length was detected in 3 consecutive years in this study, indicating that it may be related to the parental material used to construct the population. Considering the high contribution and stable genetic ability of *qAL5*, it can be used for breeding applications.

In this study, no QTL related to panicle leaf length was repeatedly detected in 3 consecutive years, and the contribution to phenotypic variance were also lower that of other trait QTLs, which may be related to the mapped population and environmental factors. However, two QTLs related to panicle leaf width, *qLW2* and *qLW3*, were detected on chromosomes 2 and 3H in 3 consecutive years. In addition, *qLL2* and *qLW2* were mapped to the same interval, and *qLL3* and *qLW3* were also highly overlapped, suggesting that there may be genetic pleiotropy that can control both LL and LW. It has been reported that a QTL on the 2H chromosome can simultaneously influence the length, width, area and chlorophyll content of flag leaves in barley [35]. Similarly, another QTL on chromosome 2H has also been revealed as affecting both the length and width of flag leaves [36]. These studies validate our conclusion that LL and LW are governed by the same gene.

In gramineous plants, there is also a relationship between the panicle neck length and yield. A longer panicle neck length helps to efficiently transport nutrients produced by photosynthesis to the grain, thereby increasing grain filling and yield [37]. However, excessively long panicle neck length directly leads to the increase in plant height, which in turn induces lodging and has a negative effect on yields [38]. To date, there are still only a few reports on QTLs related to panicle neck length in barley. In this study, we identified 14 QTLs associated with panicle neck length. Among them, the *qNL2*, *qNL3* and *qNL6*, located on chromosomes 2H, 3H and 6H, respectively, were detected in 3 consecutive years, while the others were only identified for 1 or 2 years. A previous report revealed that the panicle neck length was easily affected by environmental factors in wheat [39]. By analyzing the phenotypic data of panicle neck length from 2017 to 2019, we found that the panicle neck length of parent GP and some RIL lines were zero or close to zero in 2018 and 2019, but were significantly elongated in 2017, suggesting that barley panicle neck length may be also easily influenced by the environment. Combined with the high overlap between qLL, qLW and qNL on 2H and 3H chromosomes, we speculate that there may be a locus that regulates leaf morphology and panicle neck length at the same time.

5. Conclusions

In this study, the QTL analysis related to panicle and flag leaf traits was carried out by a high-density genetic map based on 134 RILs. A total of 32 QTLs related to five yield-related traits, panicle length, awn length, panicle neck length and flag leaf length and width were detected, which were distributed on chromosomes 2H to 7H. Among them, qPL2, qLW2, qLW3, qNL2, qNL3, qNL6 and qAL5 were detected in 3 consecutive years and their mapping positions were overlapped, indicating that these QTLs were genetically stable and less affect by environmental factors. These results lay a foundation for molecular breeding of these traits.

Author Contributions: Conceptualization: Y.F., X.Z. and D.X.; methodology: Y.F., X.Z. and D.X.; investigation: Y.Y., S.W., X.Y. and G.Z.; writing—original draft preparation: Y.Y. and S.W.; writing—review and editing: Y.F., X.Z. and D.X. All authors have read and agreed to the published version of the manuscript.

Funding: This research received no external funding.

Data Availability Statement: Specific data from this study can be obtained by contacting the authors.

Conflicts of Interest: The authors declare no conflicts of interest.

References

1. Abou-Elwafa, S.F. Association mapping for drought tolerance in barley at the reproductive stage. *Comptes Rendus Biol.* **2016**, *339*, 51–59. [CrossRef] [PubMed]
2. Cossani, C.M.; Palta, J.; Sadras, V.O. Genetic yield gain between 1942 and 2013 and associated changes in phenology, yield components and root traits of Australian barley. *Plant Soil* **2022**, *480*, 151–163. [CrossRef]
3. Du, B.; Wu, J.; Wang, Q.; Sun, C.; Sun, G.; Zhou, J.; Zhang, L.; Xiong, Q.; Ren, X.; Lu, B. Genome-wide screening of meta-QTL and candidate genes controlling yield and yield-related traits in barley (*Hordeum vulgare* L.). *PLoS ONE* **2024**, *19*, e0303751. [CrossRef] [PubMed]
4. Alqudah, A.M.; Schnurbusch, T. Barley leaf area and leaf growth rates are maximized during the pre-anthesis phase. *Agronomy* **2015**, *5*, 107–129. [CrossRef]
5. Ma, J.; Tu, Y.; Zhu, J.; Luo, W.; Liu, H.; Li, C.; Li, S.; Liu, J.; Ding, P.; Habib, A.; et al. Flag leaf size and posture of bread wheat: Genetic dissection, QTL validation and their relationships with yield-related traits. *Theor. Appl. Genet.* **2020**, *133*, 297–315. [CrossRef]
6. Abebe, T.; Wise, R.P.; Skadsen, R.W. Comparative transcriptional profiling established the awn as the major photosynthetic organ of the barley spike while the lemma and the palea primarily protect the seed. *Plant Genome* **2009**, *2*, 247–259. [CrossRef]
7. Abebe, T.; Skadsen, R.W.; Kaeppler, H.F. Cloning and Identification of Highly Expressed Genes in Barley Lemma and Palea. *Crop Sci.* **2004**, *44*, 942–950. [CrossRef]
8. Jian, M.A.; Min, S.; Puyang, D.; Wei, L.; Xiaohong, Z.; Congcong, Y.; Zhang, H.; Qin, N.; Yang, Y.; Lan, X. Genetic Identification of QTL for Neck Length of Spike in Wheat. *J. Triticeae Crops* **2017**, *37*, 319–324.
9. Bridgemohan, P.; Bridgemohan, R.S. Evaluation of anti-lodging plant growth regulators on the growth and development of rice (*Oryza sativa* L.). *J. Clin. Oncol.* **2014**, *5*, 12–16.
10. Elbaum, R.; Zaltzman, L.; Burgert, I.; Fratzl, P. The Role of Wheat Awns in the Seed Dispersal Unit. *Science* **2007**, *316*, 884–886. [CrossRef]
11. Wang, J.; Sun, G.; Ren, X.; Li, C.; Liu, L.; Wang, Q.; Du, B.; Sun, D. QTL underlying some agronomic traits in barley detected by SNP markers. *BMC Genet.* **2016**, *17*, 103. [CrossRef] [PubMed]
12. Varshney, R.K.; Paulo, M.J.; Grando, S.; Eeuwijk, F.V.; Keizer, L.C.; Guo, P.; Ceccarelli, S.; Kilian, A.; Baum, M.; Graner, A. Genome wide association analyses for drought tolerance related traits in barley (*Hordeum vulgare* L.). *Field Crops Res.* **2012**, *126*, 171–180. [CrossRef]
13. Niu, Y.; Chen, T.; Zheng, Z.; Zhao, C.; Liu, C.; Jia, J.; Zhou, M. A new major QTL for flag leaf thickness in barley (*Hordeum vulgare* L.). *BMC Plant Biol.* **2022**, *22*, 305. [CrossRef]

14. Du, B.; Liu, L.; Wang, Q.; Sun, G.; Ren, X.; Li, C.; Sun, D. Identification of QTL underlying the leaf length and area of different leaves in barley. *Sci. Rep.* **2019**, *9*, 4331. [CrossRef] [PubMed]
15. Liller, C.B.; Walla, A.; Boer, M.P.; Hedley, P.; Macaulay, M.; Effgen, S.; von Korff, M.; van Esse, G.W.; Koornneef, M. Fine mapping of a major QTL for awn length in barley using a multiparent mapping population. *Theor. Appl. Genet.* **2017**, *130*, 269–281. [CrossRef] [PubMed]
16. Sato, K. History and future perspectives of barley genomics. *DNA Res.* **2020**, *27*, dsaa023. [CrossRef]
17. Fang, Y.; Zhang, X.; Zhang, X.; Tong, T.; Zhang, Z.; Wu, G.; Hou, L.; Zheng, J.; Niu, C.; Li, J.; et al. A High-Density Genetic Linkage Map of SLAFs and QTL Analysis of Grain Size and Weight in Barley (*Hordeum vulgare* L.). *Front. Plant Sci.* **2020**, *11*, 620922. [CrossRef]
18. Joehanes, R.; Nelson, J.C. Qgene 4.0, an extensible java QTL-analysis platform. *Bioinformatics* **2008**, *24*, 2788–2789. [CrossRef]
19. McCouch, S.R.; Cho, Y.G.; Yano, M.; Paul, E.; Blinstrub, M.; Morishima, H.; Kinoshita, T. Report on QTL nomenclature. *Rice Genet. Newsl.* **1997**, *14*, 11–13.
20. Voorrips, R.E. MapChart: Software for the graphical presentation of linkage maps and QTLs. *J. Hered.* **2002**, *93*, 77–78. [CrossRef]
21. Hua, L.; Wang, D.R.; Tan, L.; Fu, Y.; Liu, F.; Xiao, L.; Zhu, Z.; Fu, Q.; Sun, X.; Gu, P.; et al. *LABA1*, a Domestication Gene Associated with Long, Barbed Awns in Wild Rice. *Plant Cell* **2015**, *27*, 1875–1888. [CrossRef] [PubMed]
22. Xue, D.W.; Chen, M.C.; Zhou, M.X.; Chen, S.; Mao, Y.; Zhang, G.P. QTL analysis of flag leaf in barley (*Hordeum vulgare* L.) for morphological traits and chlorophyll content. *J. Zhejiang Univ. Sci. B* **2008**, *9*, 938–943. [CrossRef] [PubMed]
23. Xu, Z.; Li, M.; Du, Y.; Li, X.; Wang, R.; Chen, Z.; Tang, S.; Liu, Q.; Zhang, H. Characterization of *qPL5*: A novel quantitative trait locus (QTL) that controls panicle length in rice (*Oryza sativa* L.). *Mol. Breed.* **2022**, *42*, 70. [CrossRef]
24. Zhao, D.D.; Son, J.H.; Farooq, M.; Kim, K.M. Identification of Candidate Gene for Internode Length in Rice to Enhance Resistance to Lodging Using QTL Analysis. *Plants* **2021**, *10*, 1369. [CrossRef] [PubMed]
25. Wang, P.; Zhou, G.; Yu, H.; Yu, S. Fine mapping a major QTL for flag leaf size and yield-related traits in rice. *Theor. Appl. Genet.* **2011**, *123*, 1319–1330. [CrossRef]
26. Kuang, C.H.; Zhao, X.F.; Yang, K.; Zhang, Z.P.; Ding, L.; Pu, Z.E.; Ma, J.; Jiang, Q.T.; Chen, G.Y.; Wang, J.R.; et al. Mapping and characterization of major QTL for spike traits in common wheat. *Physiol. Mol. Biol. Plants* **2020**, *26*, 1295–1307. [CrossRef]
27. Liu, K.; Xu, H.; Liu, G.; Guan, P.; Zhou, X.; Peng, H.; Yao, Y.; Ni, Z.; Sun, Q.; Du, J. QTL mapping of flag leaf-related traits in wheat (*Triticum aestivum* L.). *Theor. Appl. Genet.* **2018**, *131*, 839–849. [CrossRef]
28. Yang, J.; Liu, Z.; Chen, Q.; Qu, Y.; Tang, J.; Lübberstedt, T.; Li, H. Mapping of QTL for Grain Yield Components Based on a DH Population in Maize. *Sci. Rep.* **2020**, *10*, 7086. [CrossRef]
29. Zhang, J.; Fengler, K.A.; Van Hemert, J.L.; Gupta, R.; Mongar, N.; Sun, J.; Allen, W.B.; Wang, Y.; Weers, B.; Mo, H.; et al. Identification and characterization of a novel stay-green QTL that increases yield in maize. *Plant Biotechnol.* **2019**, *17*, 2272–2285. [CrossRef]
30. Chutimanitsakun, Y.; Nipper, R.W.; Cuesta-Marcos, A.; Cistué, L.; Corey, A.; Filichkina, T.; Johnson, E.A.; Hayes, P.M. Construction and application for QTL analysis of a Restriction Site Associated DNA (RAD) linkage map in barley. *BMC Genom.* **2011**, *12*, 4. [CrossRef]
31. Xue, D.W.; Zhou, M.X.; Zhang, X.Q.; Chen, S.; Wei, K.; Zeng, F.R.; Mao, Y.; Wu, F.B.; Zhang, G.P. Identification of QTLs for yield and yield components of barley under different growth conditions. *J. Zhejiang Univ. Sci. B* **2010**, *11*, 169–176. [CrossRef] [PubMed]
32. Rozanova, I.V.; Grigoriev, Y.N.; Efimov, V.M.; Igoshin, A.V.; Khlestkina, E.K. Genetic Dissection of Spike Productivity Traits in the Siberian Collection of Spring Barley. *Biomolecules* **2023**, *13*, 909. [CrossRef] [PubMed]
33. Islamovica, E.; Obertc, D.E.; Oliverk, R.E.; Marshall, J.; Miclausd, K.J.; Hanga, A.; Chaof, S.; Lazog, G.R.; Harrisonh, S.A.; Ibrahimi, A.; et al. A new genetic linkage map of barley (*Hordeum vulgare* L.) facilitates genetic dissection of height and spike length and angle. *Field Crops Res.* **2013**, *154*, 91–99. [CrossRef]
34. Huang, B.; Huang, D.; Hong, Z.; Owie, S.O.; Wu, W. Genetic analysis reveals four interacting loci underlying awn trait diversity in barley (*Hordeum vulgare*). *Sci. Rep.* **2020**, *10*, 12535. [CrossRef] [PubMed]
35. Liu, L.; Sun, G.; Ren, X.; Li, C.; Sun, D. Identification of QTL underlying physiological and morphological traits of flag leaf in barley. *BMC Genet.* **2015**, *16*, 29. [CrossRef] [PubMed]
36. Gyenis, L.; Yun, S.J.; Smith, K.P.; Steffenson, B.J.; Bossolini, E.; Sanguineti, M.C.; Muehlbauer, G.J. Genetic architecture of quantitative trait loci associated with morphological and agronomic trait differences in a wild by cultivated barley cross. *Genome* **2007**, *50*, 714–723. [CrossRef]
37. Li, C.; Tang, H.; Luo, W.; Zhang, X.; Mu, Y.; Deng, M.; Liu, Y.; Jiang, Q.; Chen, G.; Wang, J.; et al. A novel, validated, and plant height-independent QTL for spike extension length is associated with yield-related traits in wheat. *Theor. Appl. Genet.* **2020**, *133*, 3381–3393. [CrossRef]
38. Berry, P.M.; Sylvester-Bradley, R.; Berry, S. Ideotype design for lodging-resistant wheat. *Euphytica* **2007**, *154*, 165–179. [CrossRef]
39. Jia, Y.; Zhang, Y.; Sun, Y.; Ma, C.; Bai, Y.; Zhang, H.; Hou, J.; Wang, Y.; Ji, W.; Bai, H.; et al. QTL Mapping of Yield-Related Traits in Tetraploid Wheat Based on Wheat55K SNP Array. *Plants* **2024**, *13*, 1285. [CrossRef]

Disclaimer/Publisher's Note: The statements, opinions and data contained in all publications are solely those of the individual author(s) and contributor(s) and not of MDPI and/or the editor(s). MDPI and/or the editor(s) disclaim responsibility for any injury to people or property resulting from any ideas, methods, instructions or products referred to in the content.

Article

Expression Analysis of the Extensive Regulation of Mitogen-Activated Protein Kinase (MAPK) Family Genes in Buckwheat (*Fagopyrum tataricum*) During Organ Differentiation and Stress Response

Guoqing Dong [†], Zihao Gui [†], Yi Yuan, Yun Li and Dengxiang Du *

School of Biological and Pharmaceutical Engineering, Wuhan Polytechnic University, Wuhan 430023, China
* Correspondence: ddx@whpu.edu.cn
[†] These authors contributed equally to this work.

Citation: Dong, G.; Gui, Z.; Yuan, Y.; Li, Y.; Du, D. Expression Analysis of the Extensive Regulation of Mitogen-Activated Protein Kinase (MAPK) Family Genes in Buckwheat (*Fagopyrum tataricum*) During Organ Differentiation and Stress Response. *Agronomy* 2024, 14, 2613. https://doi.org/10.3390/agronomy14112613

Academic Editors: Francis Drummond and Fengjie Sun

Received: 16 July 2024
Revised: 21 October 2024
Accepted: 31 October 2024
Published: 6 November 2024

Copyright: © 2024 by the authors. Licensee MDPI, Basel, Switzerland. This article is an open access article distributed under the terms and conditions of the Creative Commons Attribution (CC BY) license (https://creativecommons.org/licenses/by/4.0/).

Abstract: The mitogen-activated protein kinase (MAPK) signaling cascade is a unique and relatively conserved signaling pathway in eukaryotes, transmitting extracellular signals into cells through successive phosphorylation and eliciting appropriate responses from the organism. While its mechanism in plant immune response has been partially elucidated in *Arabidopsis*, it has been rarely examined in Tartary buckwheat (*Fagopyrum tataricum*). Based on the conserved MAPK domain, we identified 16 MAPK family genes in Tartary buckwheat. The FtMAPKs have similar structures and motif compositions, indicating that this gene family is conserved yet functionally diverse. Using quantitative reverse transcription polymerase chain reaction (qRT-PCR) analysis, we observed significant expression variation in 15 genes across different organs, except for *FtMAPK12*. *FtMAPK9* showed specific expression in vegetative organs, *FtMAPK4* in reproductive organs, and *FtMAPK1* and *FtMAPK10* in leaves and flowers, respectively, indicating their regulatory roles in Tartary buckwheat development. Following drought and salt stress treatments, 12 and 14 *FtMAPKs*, respectively, showed significantly altered expression in leaves exhibiting notable biological oxidation. Among these, *FtMAPK3*, *FtMAPK4*, and *FtMAPK8* demonstrated highly significant changes across both treatments. Transcriptome analysis confirmed these findings, suggesting that these three genes play pivotal roles in Tartary buckwheat's response to abiotic stress and hold potential for molecular breeding improvements.

Keywords: gene expression; mitogen-activated protein kinase; tartary buckwheat; organ differentiation; salt stress; abiotic stress response; stress signaling pathways

1. Introduction

Plants, as sessile organisms, are continuously exposed to a range of external environmental stimuli throughout their entire growth. These stimuli include abiotic stresses such as drought, salt, heavy metals, and low temperatures, as well as biotic stresses like bacterial and viral infections, all of which can significantly affect plant growth and development [1,2]. To ensure normal growth, plants have evolved sophisticated protective mechanisms that enable them to sense and respond to environmental changes by modulating metabolic and physiological processes, including gene expression [3,4]. Initially, environmental stimuli or intercellular signaling molecules bind to cell surface receptors. The signal is then transmitted intracellularly via secondary messengers, and through a signaling pathway, where reversible protein phosphorylation regulates cellular responses and induces gene expression [5]. Protein kinases, enzymes that transfer phosphate groups from adenosine triphosphate (ATP) to specific amino acid residues, mediate this phosphorylation process [6,7]. Based on the type of amino acid phosphorylated, protein kinases are categorized into five types: serine/threonine kinases, tyrosine kinases, histidine kinases, tryptophan kinases, and aspartyl/glutamyl kinases [8]. Among these, serine/threonine kinases include mitogen-activated protein kinase (MAPK), which is part of a highly conserved signaling pathway across eukaryotes, playing a crucial role in various biological processes [9].

1.1. Structure and Classification of MAPK-Type Protein Kinases

The MAPK cascade plays critical roles in processes such as plant development, cell differentiation, and responses to both biotic and abiotic stresses [10–13]. A complete MAPK signaling cascade comprises three core components: mitogen-activated protein kinase kinase kinase (MAPKKK), mitogen-activated protein kinase kinase (MAPKK), and MAPK [14–16]. This cascade receives upstream stimulus signals, amplifies them, and relays them to downstream proteins, thereby triggering specific cellular responses [9,17]. MAPKs are characterized by their sequence similarity and the presence of a threonine-x-tyrosine (TXY) activation motif located between kinase domains VII and VIII. Activation of MAPKs is triggered by the dual phosphorylation of threonine (Thr) and tyrosine (Tyr) residues by upstream MAPKKs. Additionally, plant MAPKs are further categorized into two subtypes based on their TXY motif: the TEY subtype, associated with the yeast and animal extracellular signal-regulated kinase (ERK) subfamilies, and the TDY subtype, which is specific to plants [18,19].

The *Arabidopsis* genome contains 20 MAPKs, 10 MAPKKs, and around 21 MEKK1-type MAPKKKs [20,21]. Additionally, *Arabidopsis* harbors about 48 Raf-class MAPKKKs, and analogous gene repertoires encoding MAPK cascade components identified in various other plant species [16,22]. MAPK family genes have been progressively discovered across multiple plant species, with reports indicating 17 MAPK genes in rice [23,24], 19 in corn [25,26], 16 in tomato [27], and 14 in cucumber [28]. Recent research has also identified 19 MAPK genes in chickpeas [29]. These findings underscore the ubiquity of MAPK family genes in plants, where they serve distinct functional roles.

1.2. Function of MAPK and Its Cascade Pathway in Plant Development

Plant growth and development necessitate constant communication and coordination among diverse cells, tissues, and organs. Research has demonstrated that the MAPK cascade serves as a pivotal signaling module downstream of cell surface receptors, facilitating the transduction of ligand-receptor interaction signals into the cell interior through the phosphorylation of MAPK substrates [11,19]. When Bush et al. studied *Arabidopsis* MAPK, they discovered that the *mapk6* mutant exhibited defects in flower development. Further phenotypic observations revealed that *MAPK6* in *Arabidopsis* plays a crucial role in biological processes encompassing anther, inflorescence, and embryo development [30]. Similarly, Hord et al., in their study of MAPK functionality in *Arabidopsis thaliana*, observed that the *mpk3/mpk6* double mutant failed to undergo anther dehiscence, corroborating the involvement of *MAPK6* in regulating anther cell division and differentiation [31,32]. Furthermore, *TaMPK4* in *Arabidopsis thaliana* and found abnormalities in pollen development among *mpk4* mutants. Their subsequent research revealed that the pollen of *mpk4* mutants was incapable of undergoing normal male meiosis [33]. Recent studies have also demonstrated that MAPKs are intricately involved in regulating traits such as grain size, thereby significantly influencing crop yield [34]. For instance, Xu et al., in their investigation of the MAPK cascade in rice, found that downregulation of *OsMKKK10* led to a reduction in both grain size and 1000-grain weight, whereas overexpression of *OsMKKK10* resulted in enlarged grains [35]. They further elucidated that *OsMKKK10* phosphorylates *OsMAPKK4*, which in turn phosphorylates the downstream *OsMAPK6*, establishing that the *OsMKKK10-OsMAPKK4-OsMAPK6* pathway regulates rice grain development. Similarly, Guo et al. also discovered the regulatory role of the *OsMKKK10-OsMAPKK4-OsMAPK6* pathway in determining rice grain size [36–38]. In their study of rice *OsMAPK12-1*, Xiao et al. discovered that the expression of *OsMAPK12-1* is induced by jasmonic acid, salicylic acid, melatonin, and bacterial pathogens. Their further investigation revealed that while overexpression of *OsMAPK12-1* suppresses seed germination and seedling growth, it significantly enhances the resistance of rice against rice bacterial leaf blight strain PX099 and bacterial leaf streak disease strain Rs105. This finding suggests that *OsMAPK12-1* negatively regulates rice growth and development while positively modulating its resistance to bacterial diseases [39]. Additionally, Shao et al. reported that the absence of the *MKK4/MKK5-MPK3/MPK6* regulatory pathway results in shorter primary roots in Arabidopsis thaliana [40,41].

1.3. Function of MAPK and Its Cascade Pathway in Plant Response to Abiotic Stress

When plants encounter adversity, they undergo a series of morphological, physiological, and molecular biological changes to mitigate stress damage and adapt to environmental fluctuations. The MAPK cascade system is one of the most prevalent signaling pathways in plants, facilitating their response to both biotic and abiotic stresses. MAPKK, a crucial component of this system, plays a key role in mediating responses to various abiotic stresses, including salinity, drought, extreme temperatures, and physical damage.

Studies have shown that, in *Arabidopsis thaliana*, *AtMAPK7* is induced and activated by H_2O_2 and *Pseudomonas syringae*, participating in the regulation of downstream pathogen defense gene expression [42]. The *AtMAPKKK17/18-AtMKK3-AtMAPK7* cascade has also been reported to intersect with the abscisic acid (ABA) signaling pathway, contributing to the ABA response process. In line with the function of *AtMAPK7* in *Arabidopsis thaliana* [43,44], Zong et al. discovered that the *ZmMAPK7* gene in maize is also induced by ABA and H_2O_2, positively regulating the ABA and reactive oxygen species (ROS)-mediated osmotic stress response [45]. Additionally, the *GhMAPK7* gene in cotton is reported to respond to abiotic stresses, including H_2O_2, salt, physical damage, salicylic acid (SA), and methyl jasmonate (MeJA), as well as to defense responses against *Rhizoctonia solani*, *Colletotrichum gossypii*, and *Fusarium oxysporum* f. sp. vasinfectum [46,47]. Furthermore, the *GhMKK3-GhMAPK7* cascade pathway has been shown to be induced by ABA and drought, playing a significant role in drought resistance by regulating the growth and water loss rate of cotton roots [48,49].

MAPK typically functions via its three-tiered cascade signaling pathway. During the investigation of cold tolerance in rice, it was discovered that the *bHLH002* (basic helix-loop-helix protein) transcription factor is a pivotal factor in regulating cold tolerance. Further analysis revealed that *OsbHLH002* in rice is activated through phosphorylation by *OsMAPK3*, which subsequently targets and activates the downstream *OsTPP1* (trehalose-6-phosphate phosphatase 1). This activation leads to trehalose accumulation, thereby enhancing the cold resistance in rice [50–52]. In their study of rice *OsMPK3*, Singh et al. observed that SUB1A1 (SUBMERGENCE1) can bind to the promoter of *OsMPK3*, jointly regulating rice's tolerance to waterlogging stress [53,54]. Similarly, Zhang et al. demonstrated that *OsMAPK3* interacts with *OsZFP213* to synergistically modulate salt tolerance in rice [55]. Additionally, Wang et al. observed that salt stress triggers the expression of *OsMKK1* in rice. They further demonstrated that *OsMKK1* mutants exhibit increased sensitivity to salt stress, while the interaction between *OsMKK1* and *OsMPK4* enhances rice's salt tolerance [56]. Alternatively, research on maize identified the MAPK-like protein (MPKL), a kinase that possesses a TXY motif highly similar to the MAPK sequence. Overexpression of *ZmMPKL1* in maize seedlings resulted in increased stomatal aperture and accelerated dehydration, indicating heightened sensitivity to drought and confirming that *ZmMPKL1* regulates drought sensitivity in maize seedlings [26,57].

1.4. MAPK and Its Cascade Pathway in Fagopyrum tataricum

Fagopyrum tataricum (Tartary buckwheat) is a dicotyledonous plant and an important crop valued for both its medicinal and nutritional properties [58,59]. Tartary buckwheat is rich in starch, protein, dietary fiber, vitamins, mineral elements, flavonoids, and trace elements. Its high flavonoid content classifies it as a "triple-lowering" food, known for its ability to effectively reduce blood pressure, blood sugar, and blood lipids, thus providing both nutritional and medicinal benefits [60,61]. Among the flavonoids in Tartary buckwheat, rutin stands out due to its diverse pharmacological properties, including reducing capillary permeability, as well as its anti-inflammatory, antioxidant, anti-allergic, anti-tumor, antiviral, and liver-protecting effects. The rutin present in Tartary buckwheat grains holds immense nutritional, healthcare, and medicinal value [62–64].

Due to its cultivation in predominantly remote and infertile environments, buckwheat frequently encounters various abiotic stress conditions, such as drought, extreme temperatures, and high salinity, which negatively impact both yield and quality. With the rapid advancements in sequencing technology, transcriptome and metabolome analyses of Tartary buckwheat under abiotic stress conditions—including drought [65,66], salt stress [67,68], aluminum stress [69–71], cadmium stress [72,73], and lead stress [74,75]—have revealed significant variations in the expression of the MAPK signaling pathway in response to environmental stresses. Additionally, the MAPK signaling pathway was observed to be enriched during tissue differentiation and grain development in Tartary buckwheat [58,76,77], indicating the potential of genes associated with this pathway for breeding high-quality, high-yield, and multi-resistant buckwheat varieties.

Concurrently, research on the reference genome of Tartary buckwheat has facilitated the identification and analysis of various gene families. Several gene families and transposable elements, including auxin response factor (ARF) [78], NAC (NAM: no apical meristem, ATAF: Arabidopsis thaliana transcription activator factor, and CUC: cup-shaped cotyledon) [79], and auxin/indole-3-acetic acid (Aux/IAA) [80], have been identified in bulk. The preliminary identification of MAPK family members in Tartary buckwheat by Yao et al. has laid a foundation for further functional exploration of this important pathway family [81]. Building on this groundwork, this study conducted expression analysis of MAPK family members in Tartary buckwheat during tissue differentiation and under abiotic stress conditions, focusing on expression regulation and gene variants within MAPK family members. This analysis provides valuable insights into the effective identification of genes involved in stress responses.

2. Materials and Methods

2.1. Plant Materials and Growth Conditions

The Tartary buckwheat cultivar Jinqiao 2 was sourced from the College of Biological Sciences and Technology, Taiyuan Normal University (Taiyuan, China). The experiment was conducted in a greenhouse at Wuhan Polytechnic University (Wuhan, China) using standard planting methods [72]. The cultivation conditions were meticulously controlled, with air humidity maintained at 80%, light intensity at 400 $\mu mol \cdot m^{-2} \cdot s^{-1}$, and a photoperiod of 16 h of light followed by 8 h of darkness. During the light phase, the temperature was maintained at 28 °C, while during the dark phase, it was lowered to 23 °C. Fresh samples of roots, stems, and leaves were harvested at the seedling stage, while flowers and seeds were collected at maturity. These samples were immediately frozen in liquid nitrogen and stored at -80 °C for subsequent expression analysis.

Under standard greenhouse conditions, plants were cultivated in a soil-vermiculite mixture (1:1) and subjected to stress treatments at the seven-leaf stage [82], with plants grown under normal growth conditions serving as the control (CK). For the drought treatment, identical buckwheat seedlings were exposed to 100 mmol·L^{-1} polyethylene glycol (PEG) for seven consecutive days, with ten plants per treatment group. Similarly, for the salt stress treatment, identical buckwheat seedlings were treated with 100 mmol·L^{-1} NaCl for seven days. Fresh leaf samples from both treatments were frozen in liquid nitrogen and stored at -80 °C for further analysis. In accordance with standard procedures, morphological observations, chlorophyll content measurements, proline content determinations malondialdehyde (MDA) content, and activity of catalase (CAT) were conducted on leaves exhibiting similar growth patterns. Subsequently, fresh leaf samples were frozen in liquid nitrogen and stored at -80 °C for future gene expression verification.

2.2. Identification and Structural Analysis of MAPK Genes in Tartary Buckwheat

Utilizing the *Arabidopsis* MAPK cascade protein obtained from the *Arabidopsis* Information Resource (TAIR) database (from https://www.arabidopsis.org/) as a reference, the corresponding MAPK protein in Tartary buckwheat was identified by searching the Molecular Breeding Knowledgebase (MBKbase) (http://mbkbase.org/Pinku1/) using the BLASTp program. The BLASTp parameters were set with a score value ≥ 100 and an e-value $\leq 1 \times e^{-10}$ [83]. Subsequently, HMMER3.0 (with default parameters) and a cutoff value of 0.01 were employed, along with the SMART tool (http://smart.embl-heidelberg.de/) to confirm the presence of MAPK domains [84,85]. The sequence length, molecular weight, isoelectric point (pI), and subcellular localization of the FtMAPK proteins were determined using tools available on the ExPASy website (https://web.expasy.org/compute/) [81].

The full-length amino acid sequences of the FtMAPK proteins served as the basis for the phylogenetic analysis. A phylogenetic tree was constructed using the neighbor-joining (NJ) method in MEGA 7.0 through the Geneious R11 platform, with the following parameters: the Jukes-Cantor model, global alignment with free-end gaps, and a bootstrap value of 1000 [86,87]. Furthermore, the exon-intron structure of all *FtMAPK* genes was analyzed using a Gene Structure Display Server, considering both the coding sequence (CDS) length and the corresponding full-length sequences [80]. Additionally, the online MEME program was utilized to analyze protein sequences, with parameters set for an optimal motif width of 6 to 200 amino acids and a maximum of 10 motifs [88].

2.3. Total RNA Extraction, cDNA Reverse Transcription, and qRT-PCR Analysis

Total RNA was extracted from plant samples using the RNAprep Pure Polysaccharide Polyphenol Plant Total RNA Extraction Kit (DP441, TIANGEN, Beijing, China) and subsequently used to produce cDNA from a 1 mg RNA sample. The reverse transcription was carried out using the FastKing One-Step Reverse Transcription Fluorescence Quantitative Kit (FP314, vazyme, Wuhan, China) [89]. Sequencing was performed using a Real-Time Fluorescent Quantitative PCR 7500 System (Thermo Fisher Scientific, Waltham, MA, USA). Gene expression of the selected genes was analyzed by quantitative real-time PCR (qRT-PCR), with primer designed using Primer 5.0, and each reaction was repeated at least three times (Table 1).

Table 1. Primer sequences of FtMAPK genes for qRT-PCR.

Gene	Primer Sequene (5'-3')
FtMAPK1	GCTGAACTTCTTGGGCGA//ACAACGGGATGCCTAGT
FtMAPK2	TTGTCACTCGTTGGTACC//TCCGGTGAACCTATGAGC
FtMAPK3	CTGCTCGTCCTTGAATTC//GGTGCAGATCTGTGTCCA
FtMAPK4	AGCAACTTCCTCATGTCC//TGGGCTCCTCGTTGATTT
FtMAPK5	ATCAAGATGGTCGGCAAT//GTTTGGTTGGAGCGGATT
FtMAPK6	ACTTCCTCGTGTGCCAAA//TCAGTCAAGCTCACTCGC
FtMAPK7	ACAAATGCCACCACAACC//TCTCGGGGCAAATAGGT
FtMAPK8	GAAGGGAAGCTATGGTGT//ATCTGGGTGGTGTAGCA
FtMAPK9	ACTTGACCCCTGAGCATT//TGTACCACCTTGTTGCAA
FtMAPK10	GCCTGCTACTACTACACA//AGCGTACGGGTTAGTGTT
FtMAPK11	ACCAAGAAAACGCCAGTG//TTCCCAAGTTTGTGCGG
FtMAPK12	AAGCCGATATCAAATCCG//TGGTGGAGAAGACGAAA
FtMAPK13	TACCTCCCAAACCACATC//TCTGCCTTTGTTGTCATG
FtMAPK14	CAGCCAGAAAACGAACCA//AACCGGTCTACTCCACT
FtMAPK15	TTCGATTGCTCCATCACC//TTGGCTTCAAATCACGGT
FtMAPK16	ATCCATCACCAAAGCCAG//TTCGGTATCATGGTAGGC

The *FtH3* gene was selected as an internal control, as it is stably expressed across most tissues at each growth stage [90]. Gene expression levels were calculated using the $2^{-(\Delta\Delta CT)}$ method [91]. Each qRT-PCR reaction was performed in a total volume of 20 µL, comprising

2 µL diluted cDNA, 1 µL each of the forward and reverse primers, 10 µL SYBR Premix Ex Taq, and 6 µL ddH$_2$O. The qPCR program was as follows: 95 °C for 3 min, followed by 30 cycles of 95 °C for 15 s, 60 °C for 30 s, and 72 °C for 20 s.

2.4. Oxidative Biomarker and Antioxidant Enzyme Activity Under Drought/Salt Treatment

Chlorophyll content was determined using an ultraviolet spectrophotometer. Briefly, fresh leaves were ground in liquid nitrogen, and the cells were resuspended in an 80% acetone solution. The mixture was then extracted and agitated for 48 h (80 g, 25 °C). The supernatant was collected, and the chlorophyll content was measured using spectrophotometry at a wavelength of 625 nm [82].

Proline content was determined using an ultraviolet spectrophotometer (Multiskan SkyHigh, Waltham, MA, USA). Briefly, a standard solution was prepared by increasing the concentration of the proline standard solution by a proportional gradient of 0.2. Next, using a blank control tube, colorimetric analysis at a wavelength of 520 nm was performed, and its absorbance value was measured. Based on this measurement, a standard curve was drawn with proline mass on the x-axis and the absorbance value on the y-axis, and the linear regression equation was obtained. Next, 0.3 g of fresh leaf sample was accurately weighed and ground into powder using liquid nitrogen, and the sample solution was prepared according to the experimental procedure [92]. For testing, 2 mL of appropriately diluted supernatant was combined with 2 mL of glacial acetic acid and 3 mL of ninhydrin developer, and the mixture was heated in a boiling water bath for 40 min. The sample absorbance was then measured using the same method as the standard curve preparation, with distilled water serving as the blank control. Lastly, proline content was calculated by inputting the absorbance value into a linear regression equation.

The content of malondialdehyde (MDA) was determined using the thiobarbituric acid (TBA) method [93]. A 0.5 g sample of fresh leaves was ground into powder with liquid nitrogen, followed by the addition of 6 mL of 10% trichloroacetic acid (TCA) to form a homogenate. The homogenate was centrifuged at $4000 \times g$ for 10 min, and the supernatant was diluted to 10 mL for testing. 2 mL of distilled water was used as a blank control. For the assay, 2 mL of supernatant was mixed with 2 mL of 0.6% TBA solution, thoroughly mixed, and boiled in a boiling water bath for 15 min. After cooling to room temperature, the mixture was centrifuged at $2000 \times g$ for 10 min. The absorbance of the supernatant was measured at 450 nm, 532 nm, and 600 nm.

CAT activity was determined using the TBA-TCA method [94]. A 0.5 g of fresh leaves was ground into powder in liquid nitrogen, and 5 mL of precooled phosphate-buffer solution (200 mM, containing 1% PVP) with a pH of 7.8 was added until a homogenate was formed. The homogenate was centrifuged at $7000 \times g$ for 15 min, and the supernatant was diluted to 10 mL and stored at 4 °C. The enzyme assay was carried out in a 10 mL centrifuge tube, mixing 0.2 mL of crude enzyme solution, 1.5 mL of phosphate buffer solution, and 1 mL of distilled water. After preheating at 26 °C for 10 min, 0.3 mL of 0.1 mol L^{-1} hydrogen peroxide was added to the main tube. Absorbance was measured at 240 nm at 1 min intervals for 4 min. After measuring all three tubes, the enzyme activity was determined. CAT activity was calculated based on the change in absorbance, with CAT activity defined as a decrease of 0.1 activity unit (min^{-1} g^{-1}) in the optical density (OD) value per minute.

2.5. Transcriptome Analysis of Tartary Buckwheat After Drought/Salt Treatment

After drought/salt stress treatment, transcriptome analysis was conducted on Tartary buckwheat organs exhibiting significant physiological changes, with untreated samples as controls. Total RNA was extracted from fresh leaves using a total RNA extraction kit (Sangon, SK1321, Shanghai, China), and genomic DNA was removed through RNase-free DNase-I treatment [95]. A cDNA library was constructed using a one-step reverse transcription kit (FP314, vazyme, Wuhan, China), and the sequencing library was constructed using the TruSeq Stranded mRNA LTSample Prep Kit (Illumina, San Diego, CA, USA) according to the manufacturer's instructions. Sequencing was performed on the Illumina Hi-Seq platform (San Diego, CA, USA, Illumina®), yielding high-throughput transcriptome sequencing data. Raw reads were processed using Trimmomatic to filter and trim the raw data, removing low-quality reads, adapters, and primer sequences to obtain clean reads [96]. The filtered reads were then aligned to the genome using HISAT2 software, and gene expression levels were calculated using HTseq. Additionally, TPM (Transcripts Per Million mapped reads) was utilized to normalize gene expression levels. Principal component analysis (PCA) was performed to assess the gene expression profiles of the experimental groups, with the screening criteria set to |Log2 fold change| > 1 and p-value < 0.05. Differentially expressed genes (DEGs) were identified through volcano plot analysis [97,98]. Cluster Profiler software was employed to conduct gene ontology (GO) functional enrichment analysis on the DEGs. Significantly enriched GO terms were identified using hypergeometric distribution to determine their primary biological functions [83,99]. Additionally, KEGG pathway enrichment analysis was conducted to identify the main enriched metabolic pathways involved. Data visualization analysis was performed using R packages (R version 4.3.3).

3. Results

3.1. Identification of MAPK Family Genes in Tartary Buckwheat

In this study, we identified 16 *FtMAPK* genes, named *FtMAPK1* to *FtMAPK16* based on conserved domains. The distribution of these genes across the chromosomes was analyzed, revealing that they are distributed across all chromosomes except chromosome 6, with a notable enrichment on chromosome *Ft7*. Additionally, we examined the basic characteristics of the *FtMAPKs* in Tartary buckwheat samples, including protein length, molecular weight, pI, and predicted subcellular localization (Table 2). Most of the identified proteins range from 380 to 580 amino acids in length. FtMAPK6 is the smallest, with 371 amino acids, while FtMAPK16 is the largest with 609 amino acids. The molecular weights of the proteins vary from 42.6 kDa (FtMAPK6) to 69.4 kDa (FtMAPK16), and the pI values range from 5.39 (FtMAPK4) to 9.47 (FtMAPK16). Predicted subcellular localization results indicated that four FtMAPKs (25%) are located in the nucleus, six (50%) in the cytoskeleton, three (18.75%) in the microbody, and one (6.25%) in the chloroplast matrix. The subcellular localization patterns of the FtMAPKs align with their glycosylation function.

Table 2. The characteristics of FtMAPK genes in buckwheat.

Gene	Gene ID	Chromosomal	CDS[1]	aa	Introns	pI	MV	Location
FtMAPK1	FtPinG0006545900	Ft1	1158	385	2	6.86	44.2	M
FtMAPK2	FtPinG0004152000	Ft7	1167	388	5	6.19	44.2	N
FtMAPK3	FtPinG0005197700	Ft8	1224	407	5	5.43	46.7	C
FtMAPK4	FtPinG0000626000	Ft5	1119	372	5	5.39	43.0	N
FtMAPK5	FtPinG0004095200	Ft3	1221	406	6	6.90	46.4	CS
FtMAPK6	FtPinG0001815800	Ft7	1116	371	5	5.41	42.6	C
FtMAPK7	FtPinG0006454600	Ft2	1146	381	5	5.66	44.3	C
FtMAPK8	FtPinG0008104600	Ft4	1359	452	10	8.34	51.2	C
FtMAPK9	FtPinG0005315200	Ft7	1704	567	9	8.84	64.7	N
FtMAPK10	FtPinG0009258900	Ft8	1827	608	9	9.40	68.9	C
FtMAPK11	FtPinG0005630300	Ft1	1770	589	9	9.25	66.9	C
FtMAPK12	FtPinG0005324100	Ft7	1218	405	6	8.93	47.1	M
FtMAPK13	FtPinG0003584300	Ft7	1701	570	9	9.33	64.6	M
FtMAPK14	FtPinG0009311600	Ft5	1632	543	10	6.51	61.8	C
FtMAPK15	FtPinG0009346800	Ft4	1716	571	10	7.41	65.5	C
FtMAPK16	FtPinG0001873100	Ft1	1830	609	9	9.47	69.4	N

[1] CDS, coding sequence, bp; aa, protein length; pI, isoelectric points; MW, molecular weight, kDa; Location, Subcellular location; M, microbody; N, nucleus; C, cytoplasm; CS, chloroplast stroma.

3.2. Phylogenetic Structure and Conserved Motifs of FtMAPKs

Based on the amino acid sequences of the 16 identified FtMAPKs proteins, we constructed a phylogenetic tree (Figure 1a). The results indicate that the FtMAPK proteins are divided into four distinct groups, with sequence identities greater than 40% within each group. In each of the four FtMAPK gene groups, the number of FtMAPKs varies, with the largest group containing nine MAPK members and the smallest group having only one member (FtMAPK1).

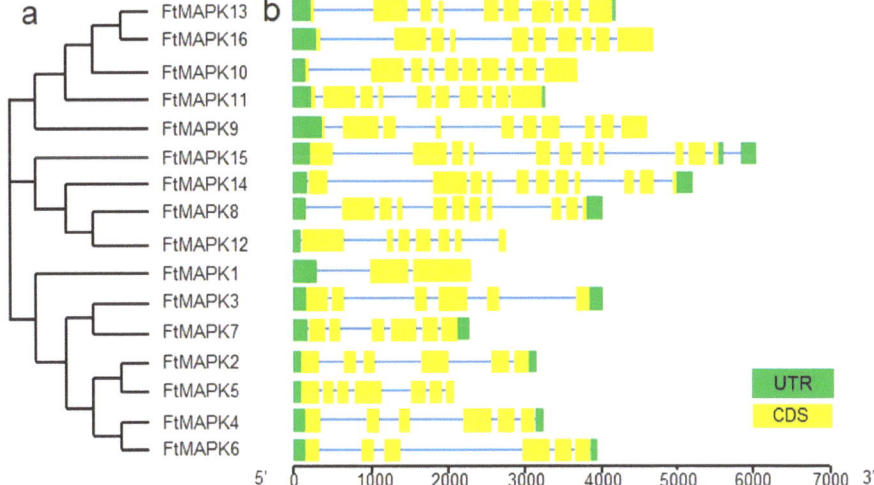

Figure 1. Conserved motifs and gene structure analysis of FtMAPKs in buckwheat. (**a**) Phylogenetic tree was constructed by the NJ method. (**b**) Gene structure of FtMAPK exons and introns are indicated by yellow rectangles and gray lines, respectively.

To further understand the structural diversity of different *FtMAPK* genes, we analyzed their exon and intron structures. The results showed high levels of structural conservation within each group, although there were differences across groups (Figure 1b). For example, *FtMAPK3* and *FtMAPK7* in group A shared the same exon-intron distribution, while the two most closely related members in group B, *FtMAPK4* and *FtMAPK6*, also had the same exon-intron distribution pattern. *FtMAPK1*, which is the sole member of group C, contained only two exons and two introns. In group D, the structures were more complex, with varying numbers of exons. *FtMAPK12* had 7 exons, while four genes (*FtMAPK8*, *FtMAPK9*, *FtMAPK14*, and *FtMAPK15*) had 11 exons, and four other genes (*FtMAPK10*, *FtMAPK11*, *FtMAPK13*, and *FtMAPK16*) had 13 exons.

We analyzed the conserved motifs of the identified MAPK proteins and found 15 distinct motifs (Figure 2). All 16 FtMAPK proteins contained the TDY signature motif (motif 2) or the TEY signature motif (motif 8), confirming their classification within the MAPK gene family. Notably, members within the same subfamily shared similar conserved motifs, indicating a high degree of conservation. However, the group with nine *FtMAPK* genes contains the TDY/TEY motif, including specific motif 9 at the N-terminus and specific motifs 7, 13, 14, and 15 at the C-terminus.

Groups A, B, and C all featured TEY characteristic motifs, whereas group D featured TDY characteristic motifs. Although group D exhibited less overall conservation compared to the other three groups, it possessed additional motif features, including TDY/TEY motifs at the N-terminal sequence 9 and specific motifs 7, 13, 14, and 15 at the C-terminus. These variations in motifs suggest structural and potential functional differences among different proteins, which may result in diverse biological functions. Further research is necessary to verify and explore these functional differences.

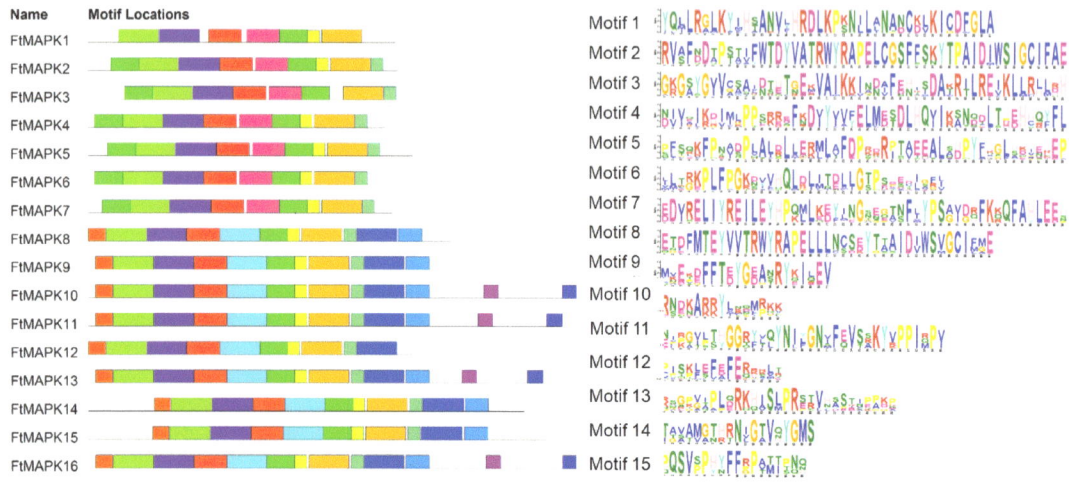

Figure 2. Sequence logo of the MAPK proteins motifs. The height of each amino acid represents the relative frequency of the amino acid at that position.

3.3. Expression Analysis of FtMAPKs in Different Organs

We conducted an expression analysis of the 16 *FtMAPKs* genes in different organs of Tartary buckwheat, including roots, stems, leaves, flowers, and seeds, using qRT-PCR (Figure 3). The results indicated that all genes, except for *FtMAPK12*, were expressed, albeit with significant differences in expression levels across organs. Specifically, the expression range of the *FTMAPK1* gene spanned from 0.490 ± 0.095 (flower) to 1.596 ± 0.064 (leaf). Notably, there was a significant disparity in gene expression between leaves and flowers, whereas no notable differences were observed in expression levels among other tissues. *FTMAPK2* gene expression ranged from 3.952 ± 7.067 (seed) to 19.480 ± 0.382 (stem), with significantly lower expression in seeds compared to other organs. *FTMAPK3* exhibited expression levels ranging from 30.612 ± 0.911 (stem) to 52.570 ± 0.269 (flower), with no significant differences in expression among organs. *FTMAPK4* had an expression range from 1.442 ± 0.116 (leaf) to 6.928 ± 0.071 (seed), showing significant differences in expression between reproductive organs, such as flowers and seeds, and vegetative organs. *FTMAPK5* expression spanned from 3.352 ± 0.161 (seed) to 20.050 ± 0.428 (leaf), with significantly higher expression in leaves and roots compared to other organs. *FTMAPK6* expression spanned from 0.084 ± 0.131 (flowers) to 9.710 ± 0.032 (stems), with notably higher expression levels in roots and stems compared to other organs. *FTMAPK7* expression ranged from 19.906 ± 0.431 (leaves) to 42.622 ± 0.562 (stems), showing significantly higher expression in stems compared to leaves, while the differences in expression among other organs were insignificant. *FTMAPK8* expression ranged from 5.614 ± 0.069 (seeds) to 15.658 ± 0.323 (flowers), with flowers exhibiting significantly higher expression compared to seeds, while the differences in expression among other organs are not significant. *FTMAPK9* expression spanned from 0.690 ± 0.059 (seeds) to 93.990 ± 0.342 (leaves), revealing significant differences between vegetative and reproductive organs, with roots, stems, and leaves showing significantly higher expression compared to flowers and seeds. *FTMAPK10* expression ranged from 49.522 ± 0.812 (leaves) to 102.980 ± 0.902 (flowers), with flowers exhibiting significantly higher expression levels compared to other organs. *FTMAPK11* expression range spanned from 5.998 ± 0.316 (stems) to 23.072 ± 0.343 (seeds), with stems showing significantly lower expression compared to other organs. *FTMAPK13* expression ranged from 5.314 ± 0.125 (root) to 7.162 ± 0.139 (leaf), with no significant differences across the organs. *FTMAPK14* ranged from 5.520 ± 0.180 (stem) to 35.348 ± 1.003 (root), with significantly lower expression in stems compared to leaves. *FTMAPK15* expression spanned from 3.258 ± 0.063 (seed) to 8.816 ± 0.214 (leaf), with higher expression in roots and leaves compared to seeds. Lastly, the expression range *FTMAPK16* was 15.316 ± 0.216 (stem) to 49.658 ± 0.293 (seed), with significantly lower expression in rhizomes, leaves, and roots compared to flowers and seeds.

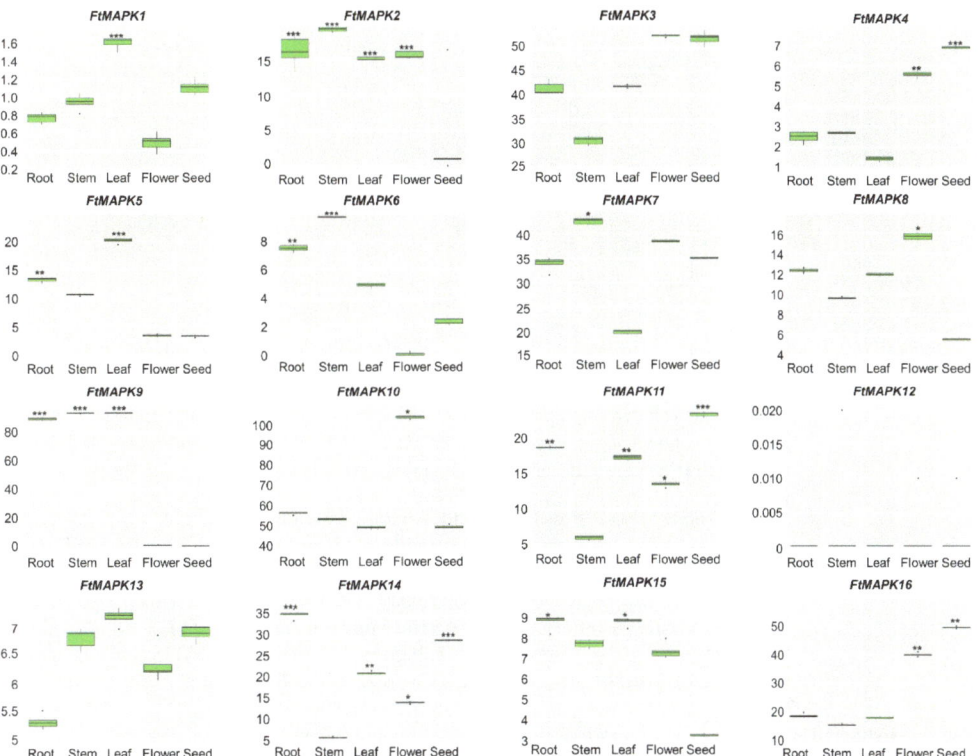

Figure 3. Expression analysis of FtMAPKs in different tissues. Expression patterns of FtMAPKs in roots, stems, leaves, flowers, and seeds were examined by qRT-PCR. Different letters above columns indicate statistically significant differences between tissues (LSD test, $p < 0.05$). Statistically significance was denoted as *** $p \leq 0.001$, ** $p \leq 0.01$, and * $p \leq 0.05$. Error bars represent SE (n = 10).

Two genes, *FtMAPK6* and *FtMAPK12*, were not detected in flowers, and *FtMAPK12* was not expressed in seeds. Notably, *FtMAPK14* and *FtMAPK15* exhibited the highest expression levels in roots, while *FtMAPK2*, *FtMAPK6*, and *FtMAPK7* had the highest expression in stems. In leaves, the highest expression levels were observed for five genes *FtMAPK1*, *FtMAPK5*, *FtMAPK9*, *FtMAPK13*, and *FtMAPK15*. Notably, *FtMAPK12* had the highest expression level in flowers, whereas four genes—*FtMAPK3*, *FtMAPK4*, *FtMAPK11*, and *FtMAPK16*—had the highest expression levels in seeds. These results demonstrate distinct expression patterns for *FtMAPKs*, suggesting that these genes may play different roles in plant development. The expression patterns within each group were significantly correlated, indicating that the genes within the same tissue may share similar expression and functional roles.

3.4. Physiological Analysis of Tartary Buckwheat Seedlings Under Drought/Salt Treatment

Under the drought treatment, the edges of the leaves turned visibly yellow and wilted, with chlorophyll content significantly lower than in the control group (CK). On day 7, the chlorophyll content of the leaves measured 4.54 ± 0.34 mg/g in CK and 2.63 ± 0.29 mg/g in the treatment group (7D) (Figure 4a). In contrast, proline content significantly increased, from 8.08 ± 0.31 ug/g in the CK to 19.99 ± 0.59 ug/g in 7D (Figure 4b). The MDA content also increased significantly, reaching 38.65 ± 1.99 nmol/g in 7D (Figure 4c). Additionally, CAT activity demonstrated a significant downward trend during the drought treatment period (Figure 4d).

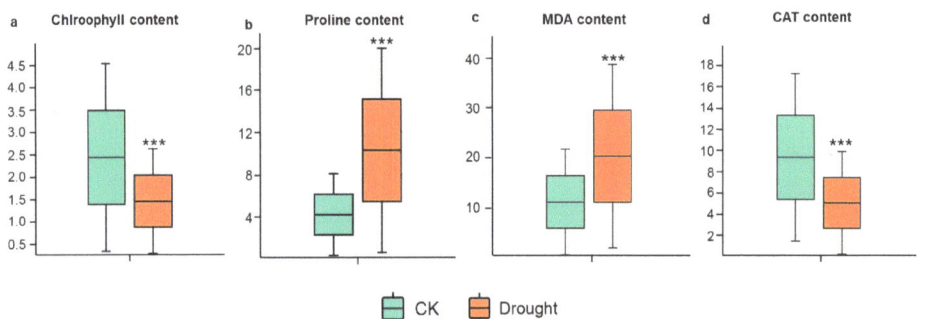

Figure 4. Phenotypic and physiological analysis of buckwheat seedling under drought treatment. (**a**) The diversity in chlorophyll contents, on the basis of units mg/g. (**b**) Proline content in leaves, on the basis of units ug/g. (**c**) MDA content in leaves, on the basis of units nmol/g. (**d**) CAT content in leaves, on the basis units $min^{-1} g^{-1}$. *** means $p \leq 0.001$.

Under the salt stress treatment, similar physiological and phenotypic responses were observed when compared to normal water management (CK). On day 7, the chlorophyll content dropped to 1.99 ± 0.09 mg/g in the treatment group (7D) compared to 4.35 ± 0.17 mg/g in the CK (Figure 5a). Proline content increased significantly, from 7.91 ± 0.11 µg/g in the CK to 36.54 ± 0.29 µg/g in 7D (Figure 5b). Similarly, MDA content rose to 38.65 ± 3.11 nmol/g in 7D (Figure 5c). However, CAT activity decreased from 15.77 ± 0.68 units $min^{-1} g^{-1}$ in the CK to 7.34 ± 0.15 units $min^{-1} g^{-1}$ in 7D (Figure 5d).

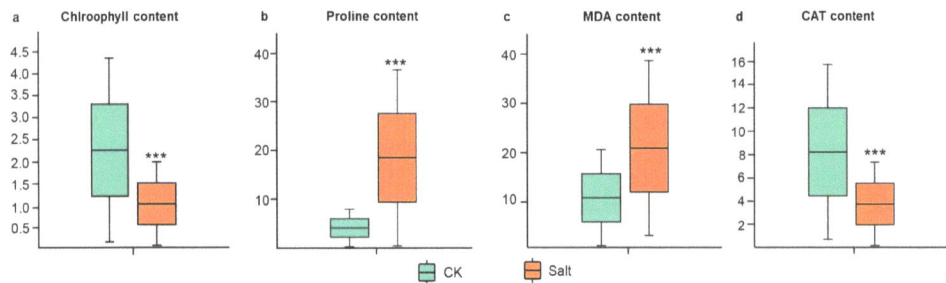

Figure 5. Phenotypic and physiological analysis of buckwheat seedling under salt treatment. (**a**) The diversity in chlorophyll contents, on the basis of units mg/g. (**b**) Proline content in leaves, on the basis of units ug/g. (**c**) MDA content in leaves, on the basis of units nmol/g. (**d**) CAT content in leaves, on the basis units $min^{-1} g^{-1}$. *** means $p \leq 0.001$.

3.5. Expression Analysis of FtMAPKs Under Drought and Salt Stress

Using untreated materials as a control, fresh leaves were collected from plants subjected to drought and salt treatments for seven days. RNA was extracted to detect the expression changes in the *FtMAPKs* family genes after stress treatment. qRT-PCR analysis was performed, and the results are presented in Figure 6. Following seven days of drought treatment, six genes (*FtMAPK1*, *FtMAPK2*, *FtMAPK4*, *FtMAPK7*, *FtMAPK13*, and *FtMAPK15*) were significantly upregulated, whereas seven genes (*FtMAPK3*, *FtMAPK5*,

FtMAPK8, *FtMAPK9*, *FtMAPK10*, *FtMAPK14*, and *FtMAPK16*) were significantly downregulated. However, *FtMAPK6* and *FtMAPK11* demonstrated no significant changes in expression. Following seven days of salt treatment, four genes (*FtMAPK1*, *FtMAPK4*, *FtMAPK12*, and *FtMAPK15*) were significantly upregulated, and four genes (*FtMAPK7*, *FtMAPK11*, *FtMAPK13*, and *FtMAPK16*) remained significantly upregulated. Conversely, five genes (*FtMAPK3*, *FtMAPK5*, *FtMAPK8*, *FtMAPK9*, *FtMAPK10*, and *FtMAPK14*) were significantly downregulated. However, *FtMAPK2*, *FtMAPK6*, and *FtMAPK7* demonstrated no significant changes in expression. Comparing the expression of *FtMAPKs* under drought and salt treatments revealed abundant variations in the response of this gene family in Tartary buckwheat. A total of 14 genes showed significant changes, with 12 genes demonstrating consistent changes under both stresses and two genes (*FtMAPK11* and *FtMAPK16*) showing inconsistent expression patterns.

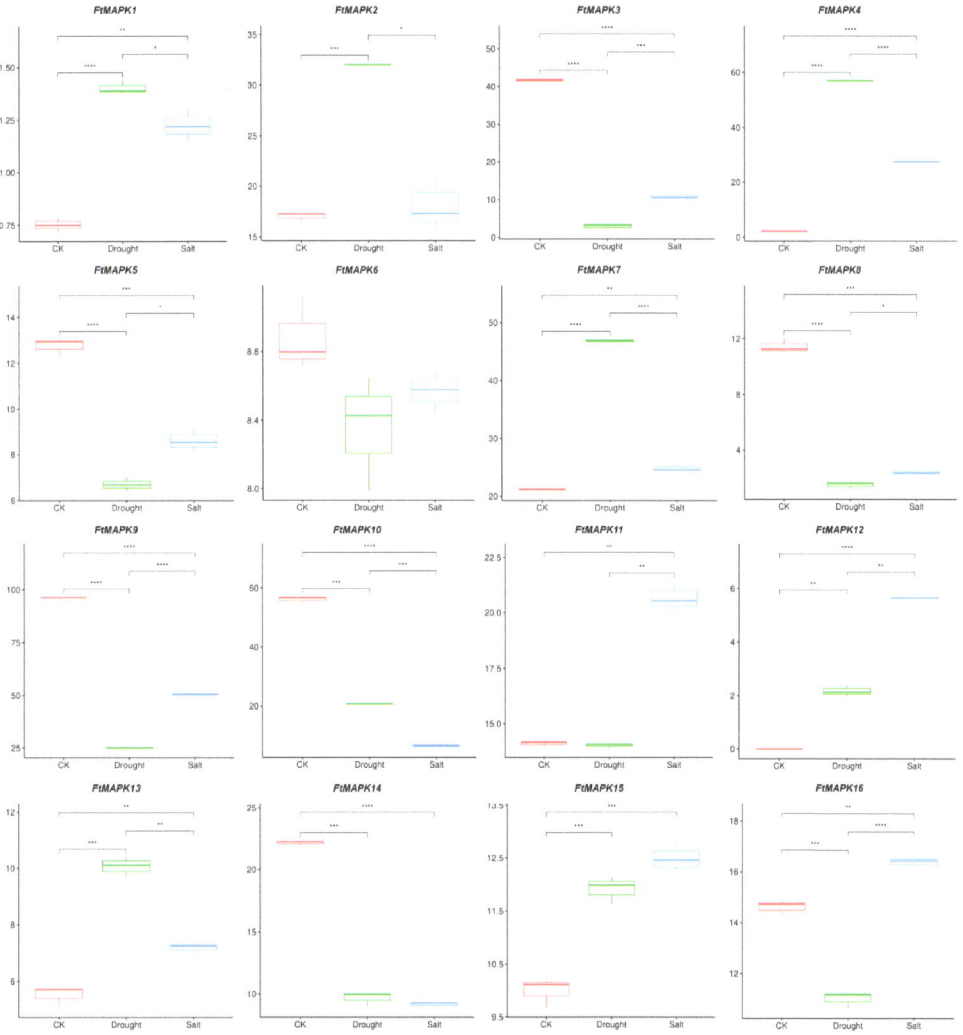

Figure 6. Expression analysis of FtMAPKs in buckwheat under drought and salt. * represented a *p*-value less than 0.05, ** represented a *p*-value less than 0.01, *** represented a *p*-value less than 0.001, and **** represented a *p*-value less than 0.0001.

Among the 12 genes with similar responses, 10 demonstrated significant changes. Specifically, the expression level of *FtMAPK3* was 41.71 in the CK, which decreased to 3.25 ± 0.45 after drought treatment and 10.45 ± 1.11 after salt treatment, representing more than a 3.99-fold change. Notably, the expression level dropped by 12.83-fold after drought treatment. Similarly, *FtMAPK8* was significantly downregulated after stress treatment, with its expression decreasing from 11.04 ± 0.84 in the CK to 1.64 ± 0.05 after drought and 2.29 ± 0.03 salt treatment, representing 6.73- and 4.83-fold decreases, respectively. In contrast, *FtMAPK4* was significantly upregulated after stress treatment, increasing from 2.32 ± 0.12 to 56.83 ± 2.73 (24.50-fold change) after drought treatment and 27.54 ± 2.24 (11.87-fold change) after salt treatment. Due to these extremely significant expression changes, *FtMAPK3*, *FtMAPK8*, and *FtMAPK4* can be considered candidate genes for drought and salt stress resistance in the *FtMAPK* family, as determined by expression analysis.

3.6. Transcriptomic Analysis of Tartary Buckwheat Seedlings Under Drought/Salt Treatment

A comprehensive expression analysis was conducted to investigate the changes in Tartary buckwheat following stress treatment using transcriptome analysis. High-throughput sequencing yielded a total of 81.84 GB of clean data, with an average of 6.82 GB per sample. Among the total clean reads from the twelve libraries, approximately 90.49–97.18% uniquely when aligned with the buckwheat genome, with alignment rates ranging from 89.93 to 99.96%.

After seven days of drought treatment, 194 upregulated genes and 179 downregulated genes were detected (Figure 7a). KEGG clustering analysis of all DEGs revealed 20 top-level pathways (Figure 7b), with the most significant pathways including MAPK signaling, β-alanine metabolism, glycerophospholipid metabolism, glycerolipid metabolism, and fatty acid biosynthesis. Similarly, after seven days of salt treatment, 273 upregulated genes and 196 downregulated genes were detected (Figure 7c). DEGs were significantly enriched in 20 pathways, including plant-pathogen interaction, endocytosis, plant hormone signal transduction, MAPK signaling, and spliceosome (Figure 7d).

Both transcriptome analyses revealed a significant enrichment of the MAPK signaling pathway, indicating that MAPK family genes are extensively involved in Tartary buckwheat's response to abiotic stress. In this study, five MAPK family genes (*FtMAPK1*, *FtMAPK3*, *FtMAPK4*, *FtMAPK8*, and *FtMAPK16*) showed significant differential expression after drought treatment, with *FtMAPK3* and *FtMAPK4* exhibiting the most pronounced changes. Following drought treatment, the expression level of *FtMAPK3* dropped to 3.45 ± 0.11 (compared to 47.68 ± 3.49 in the CK), while *FtMAPK4* showed a 25.06-fold upregulation, reaching 55.45 ± 3.29, compared to the CK. After salt treatment, seven differentially expressed genes (*FtMAPK3*, *FtMAPK5*, *FtMAPK8*, *FtMAPK9*, *FtMAPK10*, *FtMAPK12*, and *FtMAPK15*). Among these, *FtMAPK3*, *FtMAPK4*, and *FtMAPK8* exhibited consistent expression changes under both drought and salt treatments, with increases of 4.3, 13.5, and 5.4 times, respectively. These findings align with the overall expression patterns of *FtMAPK* genes in response to both drought and salt stress.

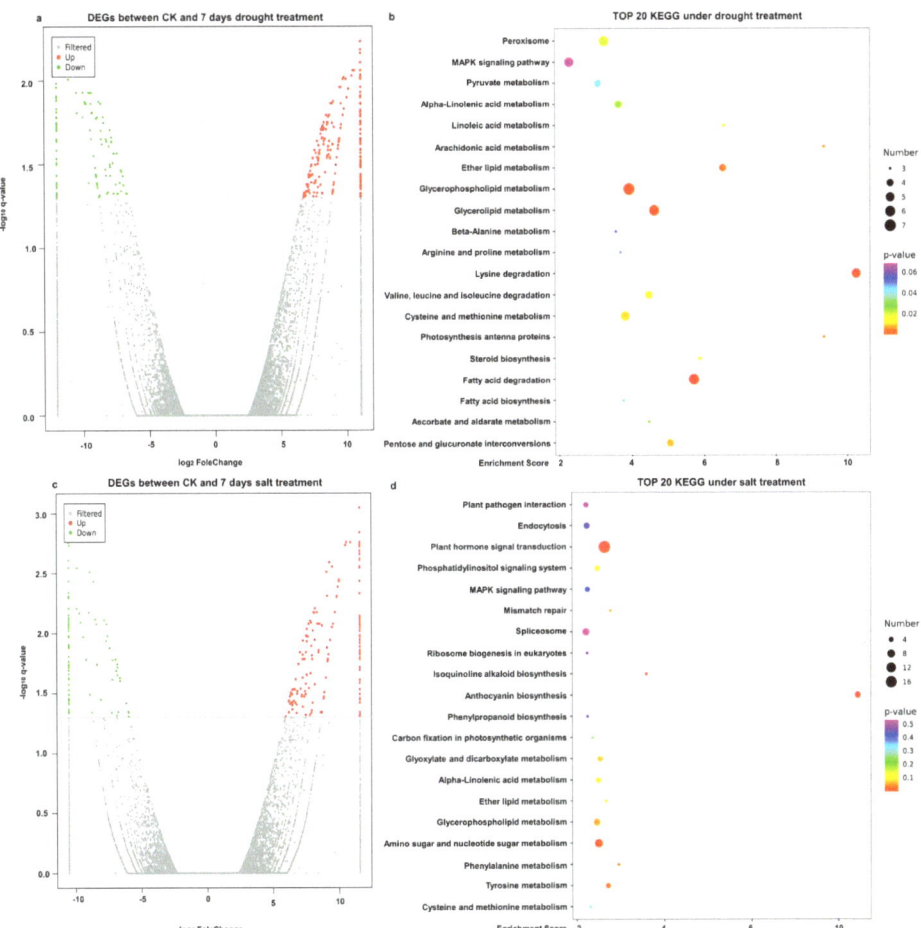

Figure 7. Identification and annotation of differentially expressed genes (DEGs) under drought/salt treatment. Data generated using high-throughput deep sequencing technology. (**a**) Identification of DEGs between CK and 7 days of drought treatment. (**b**) KEGG analysis of DEGs between CK and drought treatment. (**c**) Identification of DEGs between CK and 7 days salt treatment. (**d**) KEGG analysis of DEGs between CK and salt treatment. The volcano plot presented the expression of the DEGs, with red dots representing up-regulated DEGs and green dots representing down-regulated DEGs. The bubble map showed the KEGG-enriched pathways. The larger the bubble, the more the genes; the darker the bubble color, the higher the Q-value of the differentially expressed genes.

4. Discussion

4.1. Identification and Expression Analysis of MAPK Genes in Tartary Buckwheat

MAPKs are highly conserved protein kinases found in plants and other eukaryotic organisms [100]. In plants, the MAPK cascade pathway is a key cell signaling pathway, activated by signals related to normal growth and development signals or stress signals [101,102], playing a role in regulating plant growth and environmental adaptation [21,103]. Systematic identification and characterization of the MAPK gene family have been conducted in several plant species. For instance, 20, 17, 16, 19, 20, and 54 MAPK genes have been identified in model plants such as *Arabidopsis thaliana* [104], rice (*Oryza sativa*) [105], *Brachypodium distachyon* [106], maize (*Zea mays*) [26], barley (*Hordeum vul-*

gare) [107], and wheat (*Triticum aestivum*) [108], respectively. In buckwheat, 16 MAPK family genes were identified through gene family analysis [81]. This study further explores the structure and function of these MAPK genes in Tartary buckwheat. This gene family shows high conservation, with most genes ranging from 380 to 580 amino acids in length. Gene structure analysis reveals that most members of the Tartary buckwheat MAPK superfamily possess a straightforward gene structure, which may contribute to their high activity and rapid transcription.

The expression pattern of genes reflects their functions to a certain extent. In this study, we analyzed the expression patterns of *FtMAPKs* using qRT-PCR. All genes, except for *FtMAPK12*, showed significant expression, indicating that *FtMAPKs* play a crucial regulatory role in the development of Tartary buckwheat, similar to findings in other plant species. Members of the MAPK cascade pathway, which consists of multiple components, are involved in the growth and development of plants at various stages. For instance, in tobacco (*Nicotiana tabacum*), the *NACK1/2, NPK1-MQK1-NRK1-MAP65-1* signaling pathway regulates cytoplasmic division [109]. In *Arabidopsis thaliana*, *MPK8* promotes seed germination, and recent studies have further shown that this cascade signaling pathway also regulates the development of the hypocotyls and early embryos [110,111].

In addition, the *FtMAPK* family exhibits specific expression patterns across different organs. For instance, *FtMAPK9* exhibits high expression in nutritional organs, whereas *FtMAPK4* and *FtMAPK16* are highly expressed in reproductive organs. Moreover, *FtMAPK14* is specifically expressed in roots, aligning with studies in other plants where the *MKK4/5-MPK3/6* pathway regulates lateral root formation by modulating the *IDA HAE/HSL2* interaction [112]. Similarly, *FtMAPK6* and *FtMAPK7* are highly expressed in stems, reflecting findings in *Arabidopsis*, where this family of genes plays a regulatory role in stem elongation [111]. *FtMAPK1* and *FtMAPK5* were specifically overexpressed in leaves, and studies in *Arabidopsis* have found that the *YDA-MKK4/5-MPK3/6*-level connectivity pathway regulates stomatal development [113]. *FtMAPK8* is highly expressed in flowers, indicating its potential role in the development of bitter buckwheat flowers. Similar studies in rice have demonstrated that the *OsMKKK10-OsMKK4-OsMPK6* signaling pathway regulates inflorescence structure [36,37]. Recent studies have also shown that *MPK3/6* can phosphorylate squamosa promoter-binding protein-like (SPL) genes in vitro, which plays an important role in the development of *Arabidopsis* anthers [96].

4.2. FtMAPK Cascade Pathway and Its Role in Abiotic Stress

When plants are subjected to environmental stress signals, the MAPK cascade is induced, enhancing the activity of MAPK and its substrates [58,114]. Specifically, when plants face abiotic stress, MAPK cascade signaling molecules play an important role through interactions with hormones and facilitating the clearance of reactive oxygen species (ROS) [115,116]. Despite advances in understanding this pathway, a significant challenge remains in identifying the downstream effector proteins or substrates of MAPK that mediate these stress responses.

Drought stress, a common abiotic stress, triggers various physiological processes, molecular mechanisms, and morphological changes in plants. To combat drought stress, plants activate the ABA signaling pathway and regulate ROS production via the MAPK signaling pathway [117–119]. The primary mechanism for ROS clearance involves enhancing the enzymatic activity of peroxidase (POD), CAT, and superoxide dismutase (SOD). For example, *PtMKK4* in poplar enhances drought resilience by elevating the activity of SOD and POD [120], while *OsDSM1* in rice boosts drought tolerance by augmenting POD activity [121]. Similarly, signaling pathways like *GhMAP3K14-GhMKK11-GhMPK31* in cotton [122] and *ZmMPK6-ZmWRKY104* in maize [123] can also amplify plant drought tolerance. In addition to ROS regulation, plants employ other molecular mechanisms to cope with drought stress. In *Arabidopsis*, the *AtMAPKKK18-AtMAPKK3* cascade pathway reacts to drought stress via ABA-mediated stomatal closure [124], and in rice, *OsMKK10.2-OsMPK3* similarly reacts to drought stress through the ABA signaling pathway [125].

Additionally, cotton *MPK2* also copes with drought stress through ABA-mediated stomatal closure [126]. In this study, Tartary buckwheat exhibited significant redox changes after drought treatment, such as an increase in proline content from 8.08 ± 0.31 ug/g to 19.99 ± 0.59 ug/g and an increase in MDA content to 38.65 ± 1.99 nmol/g, alongside a significant decrease in CAT activity. These results align with similar findings in other plant species. Furthermore, expression analysis of 16 *FtMAPKs* after drought treatment revealed significant upregulation of seven genes and significant downregulation of seven genes, consistent with reports that *MAPK* genes are widely involved in the plant response to drought stress. This confirms that the *MAPK* gene family plays an important role in the drought response of Tartary buckwheat.

Salt stress is a complex abiotic stress that primarily affects plant growth and development by disrupting ion homeostasis within the plant body. Imbalances in ion homeostasis can cause membrane damage, severe dehydration, and other physiological issues. The MAPK cascade helps maintain water balance in plants by regulating the expression of genes related to osmotic stress and salt tolerance, thereby enhancing plant tolerance to salt stress [127,128]. In *Arabidopsis thaliana*, salt stress resistance occurs through three pathways. The first involves phosphatidic acid (PA), which activates the downstream *MKK7/9*, which then phosphorylates *MPK6*, leading to the activation of SOS [129,130]. The second pathway involves *MEKK1*, which subsequently activates *MKK2*, which further phosphorylates *MPK4* and *MPK6* [131]. The third pathway involves *MKK1*, *MKK3*, and *MKK4*. *MKK3* activates *MPK6*, *MKK4* activates *MPK3*, and *MKK1* activates *MPK4* [42]. Other plants, such as maize *MKK1* [25], rice *OsMKKK63* [132], and cotton *GhRaf19* [48,133] are all capable of responding to salt stress. Overexpression of *GhMPK2* in tobacco plants increases tolerance to NaCl treatment during growth, and improves higher germination and survival rates after ABA treatment, indicating that *GhMPK2* actively regulates salt stress in an ABA-dependent manner [134]. Similar to drought stress, the MAPK cascade can enhance plant salt tolerance by altering the activity of antioxidant enzymes. For instance, overexpression of the *ZmMPK5* gene in tobacco increases the activity of CAT, POD, and SOD enzymes, rendering transgenic plants more salt-tolerant [135]. In our study, Tartary buckwheat exhibited significant physiological changes after salt treatment, including changes in chlorophyll content, proline content, MDA content, and CAT enzyme activity. Additionally, 13 *FtMAPK* genes exhibited significant expression changes, suggesting that this gene family plays a crucial role in Tartary buckwheat's response to salt stress, aligning with similar findings in other plant species.

Furthermore, transcriptome analysis of plant samples subjected to drought and salt stress revealed that DEGs were significantly enriched in the MAPK signaling pathway. Through various linkage analyses, we identified three key candidate genes—*FtMAPK3*, *FtMAPK4*, and *FtMAPK8*—which play a crucial role in Tartary buckwheat's adaptation to abiotic stress, holding potential for further validation and breeding applications.

5. Conclusions

The MAPK signaling cascade in plants is extensively involved in physiological activities such as growth, development, stress response, and immune response. In this study, we identified 16 *FtMAPKs* in the Tartary buckwheat genome based on the conserved MAPK domain. Phylogenetic analysis grouped these genes into three groups, with *FtMAPKs* from the same group generally sharing similar structures and motif compositions, highlighting the conservation and diverse functional potential of the gene family in Tartary buckwheat. Through qRT-PCR analysis, we detected 15 genes that exhibit abundant expression variation across different organs of Tartary buckwheat, except for *FtMAPK12*. The specific expression of *FtMAPK9* in vegetative organs, high expression of *FtMAPK4* in reproductive organs, and high expression of *FtMAPK1* and *FtMAPK10* in leaves and flowers, respectively, indicate that *FtMAPKs* have regulatory effects on the development and tissue differentiation of Tartary buckwheat. After undergoing drought and salt stress treatments, 12 and 14 *FtMAPKs*, respectively, showed significantly altered expressions in leaves exhibiting notable biological oxidation. Among these, *FtMAPK3*, *FtMAPK4*, and *FtMAPK8* demonstrated the most pronounced expression changes across both treatments. Transcriptome analysis further verified these findings, suggesting that *FtMAPK3*, *FtMAPK4*, and *FtMAPK8* play a pivotal role in Tartary buckwheat's response to abiotic stress, thereby exhibiting potential as candidate genes for molecular breeding improvements.

Author Contributions: Conceptualization, D.D. and G.D.; methodology, Y.Y. and Y.L.; software, Z.G.; writing—original draft preparation, D.D.; writing—review and editing, D.D. and G.D. All authors have read and agreed to the published version of the manuscript.

Funding: This research received no external funding.

Data Availability Statement: The data are contained within the article; further inquiries may be directed to the corresponding author.

Conflicts of Interest: The authors declare no conflicts of interest.

References

1. Zandalinas, S.I.; Balfagón, D.; Gómez-Cadenas, A.; Mittler, R. Plant responses to climate change: Metabolic changes under combined abiotic stresses. *J. Exp. Bot.* **2022**, *73*, 3339–3354. [CrossRef] [PubMed]
2. Napier, J.D.; Heckman, R.W.; Juenger, T.E. Gene-by-environment interactions in plants: Molecular mechanisms, environmental drivers, and adaptive plasticity. *Plant Cell* **2023**, *35*, 109–124. [CrossRef] [PubMed]
3. Zhou, Y.; Wang, B.; Yuan, F. The role of transmembrane proteins in plant growth, development, and stress responses. *Int. J. Mol. Sci.* **2022**, *23*, 13627. [CrossRef] [PubMed]
4. Nawaz, M.; Sun, J.; Shabbir, S.; Khattak, W.A.; Ren, G.; Nie, X.; Sonne, C. A review of plants strategies to resist biotic and abiotic environmental stressors. *Sci. Total Environ.* **2023**, *900*, 165832. [CrossRef]
5. Riccardi, B. Specificity of cellular communication, from signal to functional response. *World J. Adv. Res. Rev.* **2023**, *18*, 549–560. [CrossRef]
6. Morris, P.C. MAP kinase signal transduction pathways in plants. *New Phytol.* **2001**, *151*, 67–89. [CrossRef] [PubMed]
7. Zhang, W.J.; Zhou, Y.; Zhang, Y.; Su, Y.H.; Xu, T. Protein phosphorylation: A molecular switch in plant signaling. *Cell Rep.* **2023**, *42*, 112729. [CrossRef]
8. Ma, R.; Li, S.; Li, W.; Yao, L.; Huang, H.D.; Lee, T.Y. KinasePhos 3.0: Redesign and expansion of the prediction on kinase-specific phosphorylation sites. *Genom. Proteom. Bioinform.* **2023**, *21*, 228–241. [CrossRef]
9. Shi, G.; Song, C.; Torres Robles, J.; Salichos, L.; Lou, H.J.; Lam, T.T.; Turk, B.E. Proteome-wide screening for mitogen-activated protein kinase docking motifs and interactors. *Sci. Signal.* **2023**, *16*, eabm5518. [CrossRef]
10. Plotnikov, A.; Zehorai, E.; Procaccia, S.; Seger, R. The MAPK cascades: Signaling components, nuclear roles and mechanisms of nuclear translocation. *Biochim. Biophys. Acta Mol. Cell Res.* **2011**, *1813*, 1619–1633. [CrossRef]
11. Yue, J.; López, J.M. Understanding MAPK signaling pathways in apoptosis. *Int. J. Mol. Sci.* **2020**, *21*, 2346. [CrossRef] [PubMed]
12. Maik-Rachline, G.; Wortzel, I.; Seger, R. Alternative splicing of MAPKs in the regulation of signaling specificity. *Cells* **2021**, *10*, 3466. [CrossRef] [PubMed]
13. Ma, H.; Gao, Y.; Wang, Y.; Dai, Y.; Ma, H. Regulatory mechanisms of mitogen-activated protein kinase cascades in plants: More than sequential phosphorylation. *Int. J. Mol. Sci.* **2022**, *23*, 3572. [CrossRef] [PubMed]
14. Guo, Y.J.; Pan, W.W.; Liu, S.B.; Shen, Z.F.; Xu, Y.; Hu, L.L. ERK/MAPK signalling pathway and tumorigenesis. *Exp. Ther. Med.* **2020**, *19*, 1997–2007. [CrossRef]
15. Kciuk, M.; Gielecińska, A.; Budzinska, A.; Mojzych, M.; Kontek, R. Metastasis and MAPK pathways. *Int. J. Mol. Sci.* **2022**, *23*, 3847. [CrossRef] [PubMed]
16. Manna, M.; Rengasamy, B.; Sinha, A.K. Revisiting the role of MAPK signalling pathway in plants and its manipulation for crop improvement. *Plant Cell Environ.* **2023**, *46*, 2277–2295. [CrossRef]
17. Zhang, H.J.; Liao, H.Y.; Bai, D.Y.; Wang, Z.Q.; Xie, X.W. MAPK/ERK signaling pathway: A potential target for the treatment of intervertebral disc degeneration. *Biomed. Pharmacother.* **2021**, *143*, 112170. [CrossRef] [PubMed]
18. González-Rubio, G.; Sellers-Moya, Á.; Martín, H.; Molina, M. A walk-through MAPK structure and functionality with the 30-year-old yeast MAPK Slt2. *Int. Microbiol.* **2021**, *24*, 531–543. [CrossRef]
19. Zhang, M.; Zhang, S. Mitogen-activated protein kinase cascades in plant signaling. *J. Integr. Plant Biol.* **2022**, *64*, 301–341. [CrossRef] [PubMed]
20. Colcombet, J.; Hirt, H. Arabidopsis MAPKs: A complex signalling network involved in multiple biological processes. *Biochem. J.* **2008**, *413*, 217–226. [CrossRef]
21. Kumar, K.; Raina, S.K.; Sultan, S.M. Arabidopsis MAPK signaling pathways and their cross talks in abiotic stress response. *J. Plant Biochem. Biotechnol.* **2020**, *29*, 700–714. [CrossRef]
22. Hamel, L.P.; Nicole, M.C.; Sritubtim, S.; Morency, M.J.; Ellis, M.; Ehlting, J.; Ellis, B.E. Ancient signals: Comparative genomics of plant MAPK and MAPKK gene families. *Trends Plant Sci.* **2006**, *11*, 192–198. [CrossRef] [PubMed]

23. Yamada, K.; Yamaguchi, K.; Yoshimura, S.; Terauchi, A.; Kawasaki, T. Conservation of chitin-induced MAPK signaling pathways in rice and Arabidopsis. *Plant Cell Physiol.* **2017**, *58*, 993–1002. [CrossRef] [PubMed]
24. Chen, J.; Wang, L.; Yuan, M. Update on the roles of rice MAPK cascades. *Int. J. Mol. Sci.* **2021**, *22*, 1679. [CrossRef] [PubMed]
25. Kong, X.; Pan, J.; Zhang, D.; Jiang, S.; Cai, G.; Wang, L.; Li, D. Identification of mitogen-activated protein kinase kinase gene family and MKK–MAPK interaction network in maize. *Biochem. Biophys. Res. Commun.* **2013**, *441*, 964–969. [CrossRef] [PubMed]
26. Zhu, D.; Chang, Y.; Pei, T.; Zhang, X.; Liu, L.; Li, Y.; Zhuang, J.; Yang, H.; Qin, F.; Song, C.; et al. MAPK-like protein 1 positively regulates maize seedling drought sensitivity by suppressing ABA biosynthesis. *Plant J.* **2020**, *102*, 747–760. [CrossRef] [PubMed]
27. Pedley, K.F.; Martin, G.B. Identification of MAPKs and their possible MAPK kinase activators involved in the Pto-mediated defense response of tomato. *J. Biol. Chem.* **2004**, *279*, 49229–49235. [CrossRef]
28. Wang, J.; Pan, C.; Wang, Y.; Ye, L.; Wu, J.; Chen, L.; Lu, G. Genome-wide identification of MAPK, MAPKK, and MAPKKK gene families and transcriptional profiling analysis during development and stress response in cucumber. *BMC Genom.* **2015**, *16*, 386. [CrossRef]
29. Singh, A.; Nath, O.; Singh, S.; Kumar, S.; Singh, I.K. Genome-wide identification of the MAPK gene family in chickpea and expression analysis during development and stress response. *Plant Gene* **2018**, *13*, 25–35. [CrossRef]
30. Bush, S.M.; Krysan, P.J. Mutational evidence that the Arabidopsis MAP kinase MPK6 is involved in anther, inflorescence, and embryo development. *J. Exp. Bot.* **2007**, *58*, 2181–2191. [CrossRef]
31. Hord, C.L.; Sun, Y.J.; Pillitteri, L.J.; Torii, K.U.; Wang, H.; Zhang, S.; Ma, H. Regulation of Arabidopsis early anther development by the mitogen-activated protein kinases, MPK3 and MPK6, and the ERECTA and related receptor-like kinases. *Mol. Plant* **2008**, *1*, 645–658. [CrossRef]
32. Hann, C.T.; Ramage, S.F.; Negi, H.; Bequette, C.J.; Vasquez, P.A.; Stratmann, J.W. Dephosphorylation of the MAP kinases MPK6 and MPK3 fine-tunes responses to wounding and herbivory in Arabidopsis. *Plant Sci.* **2024**, *339*, 111962. [CrossRef] [PubMed]
33. Yao, S.F.; Wang, Y.X.; Yang, T.R.; Lin, H.; Lu, W.J.; Kai, X. Expression and functional analyses of the mitogen-activated protein kinase (MPK) cascade genes in response to phytohormones in wheat (*Triticum aestivum* L.). *J. Integr. Agric.* **2017**, *16*, 27–35. [CrossRef]
34. Wu, C.J.; Shan, W.; Liu, X.C.; Zhu, L.S.; Wei, W.; Yang, Y.Y.; Kuang, J.F. Phosphorylation of transcription factor bZIP21 by MAP kinase MPK6-3 enhances banana fruit ripening. *Plant Physiol.* **2022**, *188*, 1665–1685. [CrossRef] [PubMed]
35. Xu, R.; Duan, P.; Yu, H.; Zhou, Z.; Zhang, B.; Wang, R.; Li, Y. Control of grain size and weight by the OsMKKK10-OsMKK4-OsMAPK6 signaling pathway in rice. *Mol. Plant* **2018**, *11*, 860–873. [CrossRef]
36. Guo, T.; Chen, K.; Dong, N.Q.; Shi, C.L.; Ye, W.W.; Gao, J.P.; Lin, H.X. GRAIN SIZE AND NUMBER1 negatively regulates the OsMKKK10-OsMKK4-OsMPK6 cascade to coordinate the trade-off between grain number per panicle and grain size in rice. *Plant Cell* **2018**, *30*, 871–888. [CrossRef]
37. Guo, T.; Lu, Z.Q.; Shan, J.X.; Ye, W.W.; Dong, N.Q.; Lin, H.X. ERECTA1 acts upstream of the OsMKKK10-OsMKK4-OsMPK6 cascade to control spikelet number by regulating cytokinin metabolism in rice. *Plant Cell* **2020**, *32*, 2763–2779. [CrossRef]
38. Liu, Z.; Mei, E.; Tian, X.; He, M.; Tang, J.; Xu, M.; Bu, Q. OsMKKK70 regulates grain size and leaf angle in rice through the OsMKK4-OsMAPK6-OsWRKY53 signaling pathway. *J. Integr. Plant Biol.* **2021**, *63*, 2043–2057. [CrossRef] [PubMed]
39. Xiao, X.; Tang, Z.; Li, X.; Hong, Y.; Li, B.; Xiao, W.; Chen, Y. Overexpressing OsMAPK12-1 inhibits plant growth and enhances resistance to bacterial disease in rice. *Funct. Plant Biol.* **2017**, *44*, 694–704. [CrossRef]
40. Shao, Y.; Yu, X.; Xu, X.; Li, Y.; Yuan, W.; Xu, Y.; Xu, J. The YDA-MKK4/MKK5-MPK3/MPK6 cascade functions downstream of the RGF1-RGI ligand–receptor pair in regulating mitotic activity in root apical meristem. *Mol. Plant* **2020**, *13*, 1608–1623. [CrossRef]
41. Xie, C.; Yang, L.; Gai, Y. MAPKKKs in plants: Multidimensional regulators of plant growth and stress responses. *Int. J. Mol. Sci.* **2023**, *24*, 4117. [CrossRef] [PubMed]
42. Doczi, R.; Brader, G.; Pettko-Szandtner, A.; Rajh, I.; Djamei, A.; Pitzschke, A.; Hirt, H. The Arabidopsis mitogen-activated protein kinase kinase MKK3 is upstream of group C mitogen-activated protein kinases and participates in pathogen signaling. *Plant Cell* **2007**, *19*, 3266–3279. [CrossRef]
43. Danquah, A.; de Zélicourt, A.; Boudsocq, M.; Neubauer, J.; Frei dit Frey, N.; Leonhardt, N.; Colcombet, J. Identification and characterization of an ABA-activated MAP kinase cascade in Arabidopsis thaliana. *Plant J.* **2015**, *82*, 232–244. [CrossRef] [PubMed]
44. Wu, P.; Wang, W.; Li, Y.; Hou, X. Divergent evolutionary patterns of the MAPK cascade genes in Brassica rapa and plant phylogenetics. *Hortic. Res.* **2017**, *4*, 17079. [CrossRef] [PubMed]
45. Zong, X.J.; Li, D.P.; Gu, L.K.; Li, D.Q.; Liu, L.X.; Hu, X.L. Abscisic acid and hydrogen peroxide induce a novel maize group C MAP kinase gene, ZmMPK7, which is responsible for the removal of reactive oxygen species. *Planta* **2009**, *229*, 485–495. [CrossRef] [PubMed]
46. Sun, J.; An, H.; Shi, W.; Guo, X.; Li, H. Molecular cloning and characterization of GhWRKY11, a gene implicated in pathogen responses from cotton. *South Afr. J. Bot.* **2012**, *81*, 113–123. [CrossRef]
47. Wang, N.N.; Zhao, L.L.; Lu, R.; Li, Y.; Li, X.B. Cotton mitogen-activated protein kinase4 (GhMPK4) confers the transgenic Arabidopsis hypersensitivity to salt and osmotic stresses. *Plant Cell Tissue Organ Cult. (PCTOC)* **2015**, *123*, 619–632. [CrossRef]
48. Yin, Z.; Zhu, W.; Zhang, X.; Chen, X.; Wang, W.; Lin, H.; Ye, W. Molecular characterization, expression and interaction of MAPK, MAPKK and MAPKKK genes in upland cotton. *Genomics* **2021**, *113*, 1071–1086. [CrossRef]
49. Zhang, X.M.; Wu, G.Q.; Wei, M. The role of MAPK in plant response to abiotic stress. *Acta Prataculturae Sin.* **2024**, *33*, 182.

50. Zhang, Z.; Li, J.; Li, F.; Liu, H.; Yang, W.; Chong, K.; Xu, Y. OsMAPK3 phosphorylates OsbHLH002/OsICE1 and inhibits its ubiquitination to activate OsTPP1 and enhances rice chilling tolerance. *Dev. Cell* **2017**, *43*, 731–743. [CrossRef]
51. Xia, C.; Gong, Y.; Chong, K.; Xu, Y. Phosphatase OsPP2C27 directly dephosphorylates OsMAPK3 and OsbHLH002 to negatively regulate cold tolerance in rice. *Plant Cell Environ.* **2021**, *44*, 491–505. [CrossRef]
52. Chen, Y.; Qi, H.; Yang, L.; Xu, L.; Wang, J.; Guo, J.; Song, S. The OsbHLH002/OsICE1-OSH1 module orchestrates secondary cell wall formation in rice. *Cell Rep.* **2023**, *42*, 112702. [CrossRef] [PubMed]
53. Singh, P.; Sinha, A.K. A positive feedback loop governed by SUB1A1 interaction with MITOGEN-ACTIVATED PROTEIN KINASE3 imparts submergence tolerance in rice. *Plant Cell* **2016**, *28*, 1127–1143. [CrossRef]
54. Singh, A.; Singh, Y.; Mahato, A.K.; Jayaswal, P.K.; Singh, S.; Singh, R.; Rai, V. Allelic sequence variation in the Sub1A, Sub1B and Sub1C genes among diverse rice cultivars and its association with submergence tolerance. *Sci. Rep.* **2020**, *10*, 8621. [CrossRef] [PubMed]
55. Zhang, Z.; Liu, H.; Sun, C.; Ma, Q.; Bu, H.; Chong, K.; Xu, Y. A C_2H_2 zinc-finger protein OsZFP213 interacts with OsMAPK3 to enhance salt tolerance in rice. *J. Plant Physiol.* **2018**, *229*, 100–110. [CrossRef] [PubMed]
56. Wang, F.; Jing, W.; Zhang, W. The mitogen-activated protein kinase cascade MKK1–MPK4 mediates salt signaling in rice. *Plant Sci.* **2014**, *227*, 181–189. [CrossRef] [PubMed]
57. Li, Y.; Su, Z.; Lin, Y.; Xu, Z.; Bao, H.; Wang, F.; Gao, J. Utilizing transcriptomics and metabolomics to unravel key genes and metabolites of maize seedlings in response to drought stress. *BMC Plant Biol.* **2024**, *24*, 34. [CrossRef]
58. Li, H.; Lv, Q.; Liu, A.; Wang, J.; Sun, X.; Deng, J.; Wu, Q. Comparative metabolomics study of Tartary (*Fagopyrum tataricum* (L.) Gaertn) and common (*Fagopyrum esculentum* Moench) buckwheat seeds. *Food Chem.* **2022**, *371*, 131125. [CrossRef] [PubMed]
59. Sofi, S.A.; Ahmed, N.; Farooq, A.; Rafiq, S.; Zargar, S.M.; Kamran, F.; Mousavi Khaneghah, A. Nutritional and bioactive characteristics of buckwheat, and its potential for develop gluten-free products: An updated overview. *Food Sci. Nutr.* **2023**, *11*, 2256–2276. [CrossRef]
60. Zhu, F. Chemical composition and health effects of Tartary buckwheat. *Food Chem.* **2016**, *203*, 231–245. [CrossRef]
61. Zou, L.; Wu, D.; Ren, G.; Hu, Y.; Peng, L.; Zhao, J.; Xiao, J. Bioactive compounds, health benefits, and industrial applications of Tartary buckwheat (*Fagopyrum tataricum*). *Crit. Rev. Food Sci. Nutr.* **2023**, *63*, 657–673. [CrossRef]
62. Fabjan, N.; Rode, J.; Košir, I.J.; Wang, Z.; Zhang, Z.; Kreft, I. Tartary buckwheat (*Fagopyrum tataricum* Gaertn.) as a source of dietary rutin and quercitrin. *J. Agric. Food Chem.* **2003**, *51*, 6452–6455. [CrossRef] [PubMed]
63. Ruan, J.; Zhou, Y.; Yan, J.; Zhou, M.; Woo, S.H.; Weng, W.; Zhang, K. Tartary buckwheat: An under-utilized edible and medicinal herb for food and nutritional security. *Food Rev. Int.* **2022**, *38*, 440–454. [CrossRef]
64. Zhang, L.; Li, X.; Ma, B.; Gao, Q.; Du, H.; Han, Y.; Qiao, Z. The tartary buckwheat genome provides insights into rutin biosynthesis and abiotic stress tolerance. *Mol. Plant* **2017**, *10*, 1224–1237. [CrossRef] [PubMed]
65. Aubert, L.; Konrádová, D.; Barris, S.; Quinet, M. Different drought resistance mechanisms between two buckwheat species Fagopyrum esculentum and Fagopyrum tataricum. *Physiol. Plant.* **2021**, *172*, 577–586. [CrossRef] [PubMed]
66. Hossain, M.S.; Li, J.; Sikdar, A.; Hasanuzzaman, M.; Uzizerimana, F.; Muhammad, I.; Feng, B. Exogenous melatonin modulates the physiological and biochemical mechanisms of drought tolerance in tartary buckwheat (*Fagopyrum tataricum* (L.) Gaertn). *Molecules* **2020**, *25*, 2828. [CrossRef]
67. Zhang, X.; He, P.; Guo, R.; Huang, K.; Huang, X. Effects of salt stress on root morphology, carbon and nitrogen metabolism, and yield of Tartary buckwheat. *Sci. Rep.* **2023**, *13*, 12483. [CrossRef]
68. Kim, N.S.; Kwon, S.J.; Cuong, D.M.; Jeon, J.; Park, J.S.; Park, S.U. Accumulation of phenylpropanoids in tartary buckwheat (*Fagopyrum tataricum*) under salt stress. *Agronomy* **2019**, *9*, 739. [CrossRef]
69. Qi, A.; Yan, X.; Liu, Y.; Zeng, Q.; Yuan, H.; Huang, H.; Wan, Y. Silicon Mitigates Aluminum Toxicity of Tartary Buckwheat by Regulating Antioxidant Systems. *Phyton* **2024**, *93*, 1–13. [CrossRef]
70. Wang, H.; Chen, R.F.; Iwashita, T.; Shen, R.F.; Ma, J.F. Physiological characterization of aluminum tolerance and accumulation in tartary and wild buckwheat. *New Phytol.* **2015**, *205*, 273–279. [CrossRef]
71. Zhu, H.; Wang, H.; Zhu, Y.; Zou, J.; Zhao, F.J.; Huang, C.F. Genome-wide transcriptomic and phylogenetic analyses reveal distinct aluminum-tolerance mechanisms in the aluminum-accumulating species buckwheat (*Fagopyrum tataricum*). *BMC Plant Biol.* **2015**, *15*, 16. [CrossRef] [PubMed]
72. Du, D.; Xiong, H.; Xu, C.; Zeng, W.; Li, J.; Dong, G. Nutrient Metabolism Pathways Analysis and Key Candidate Genes Identification Corresponding to Cadmium Stress in Buckwheat through Multiomics Analysis. *Genes* **2023**, *14*, 1462. [CrossRef] [PubMed]
73. Huo, D.; Hao, Y.; Zou, J.; Qin, L.; Wang, C.; Du, D. Integrated transcriptome and metabonomic analysis of key metabolic pathways in response to cadmium stress in novel buckwheat and cultivated species. *Front. Plant Sci.* **2023**, *14*, 1142814. [CrossRef] [PubMed]
74. Pirzadah, T.B.; Malik, B.; Tahir, I.; Hakeem, K.R.; Alharby, H.F.; Rehman, R.U. Lead toxicity alters the antioxidant defense machinery and modulate the biomarkers in Tartary buckwheat plants. *Int. Biodeterior. Biodegrad.* **2020**, *151*, 104992. [CrossRef]
75. Domańska, J.; Leszczyńska, D.; Badora, A. The possibilities of using common buckwheat in phytoremediation of mineral and organic soils contaminated with Cd or Pb. *Agriculture* **2021**, *11*, 562. [CrossRef]
76. Li, J.; Hossain, M.S.; Ma, H.; Yang, Q.; Gong, X.; Yang, P.; Feng, B. Comparative metabolomics reveals differences in flavonoid metabolites among different coloured buckwheat flowers. *J. Food Compos. Anal.* **2020**, *85*, 103335. [CrossRef]

77. Song, C.; Xiang, D.B.; Yan, L.; Song, Y.; Zhao, G.; Wang, Y.H.; Zhang, B.L. Changes in seed growth, levels and distribution of flavonoids during tartary buckwheat seed development. *Plant Prod. Sci.* **2016**, *19*, 518–527. [CrossRef]
78. Hao, Y.R.; Du, W.; Hou, S.Y.; Wang, D.H.; Feng, H.M.; Han, Y.H.; Sun, Z.X. Identification of ARF gene family and expression pattern induced by Auxin in *Fagopyrum tataricum*. *Sci. Agric. Sin.* **2020**, *53*, 4738–4749.
79. Huang, J.; Ren, R.; Rong, Y.; Tang, B.; Deng, J.; Chen, Q.; Shi, T. Identification, expression, and functional study of seven NAC transcription factor genes involved in stress response in Tartary buckwheat (*Fagopyrum tataricum* (L.) Gaertn.). *Agronomy* **2022**, *12*, 849. [CrossRef]
80. Yang, F.; Zhang, X.; Tian, R.; Zhu, L.; Liu, F.; Chen, Q.; Huo, D. Genome-Wide Analysis of the Auxin/Indoleacetic Acid Gene Family and Response to Indole-3-Acetic Acid Stress in Tartary Buckwheat (*Fagopyrum tataricum*). *Int. J. Genom.* **2021**, *2021*, 3102399. [CrossRef]
81. Yao, Y.; Zhao, H.; Sun, L.; Wu, W.; Li, C.; Wu, Q. Genome-wide identification of MAPK gene family members in *Fagopyrum tataricum* and their expression during development and stress responses. *BMC Genom.* **2022**, *23*, 96. [CrossRef] [PubMed]
82. Zhang, J.; Zhang, C.; Huang, S.; Chang, L.; Li, J.; Tang, H.; Zhao, L. Key cannabis salt-responsive genes and pathways revealed by comparative transcriptome and physiological analyses of contrasting varieties. *Agronomy* **2021**, *11*, 2338. [CrossRef]
83. Huang, X.; Bai, X.; Guo, T.; Xie, Z.; Laimer, M.; Du, D.; Yi, K. Genome-wide analysis of the PIN auxin efflux carrier gene family in coffee. *Plants* **2020**, *9*, 1061. [CrossRef] [PubMed]
84. Finn, R.D.; Clements, J.; Eddy, S.R. HMMER web server: Interactive sequence similarity searching. *Nucleic Acids Res.* **2011**, *39*, W29–W37. [CrossRef] [PubMed]
85. Bateman, A.; Birney, E.; Durbin, R.; Eddy, S.R.; Howe, K.L.; Sonnhammer, E.L. The Pfam protein families database. *Nucleic Acids Res.* **2000**, *32*, 263–266. [CrossRef]
86. Stecher, G.; Tamura, K.; Kumar, S. Molecular evolutionary genetics analysis (MEGA) for macOS. *Mol. Biol. Evol.* **2020**, *37*, 1237–1239. [CrossRef] [PubMed]
87. Liu, M.; Ma, Z.; Sun, W.; Huang, L.; Wu, Q.; Tang, Z.; Bu, T.; Li, C.; Chen, H. Genome-wide analysis of the NAC transcription factor family in Tartary buckwheat (*Fagopyrum tataricum*). *BMC Genom.* **2019**, *20*, 113. [CrossRef]
88. Xue, G.; He, A.; Yang, H.; Song, L.; Li, H.; Wu, C.; Ruan, J. Genome-wide identification, abiotic stress, and expression analysis of PYL family in Tartary buckwheat (*Fagopyrum tataricum* (L.) Gaertn.) during grain development. *BMC Plant Biol.* **2024**, *24*, 725. [CrossRef] [PubMed]
89. Dong, G.; Xiong, H.; Zeng, W.; Li, J.; Du, D. Ectopic Expression of the Rice Grain-Size-Affecting Gene GS5 in Maize Affects Kernel Size by Regulating Endosperm Starch Synthesis. *Genes* **2022**, *13*, 1542. [CrossRef]
90. Liu, M.; Sun, W.; Ma, Z.; Zheng, T.; Huang, L.; Wu, Q.; Chen, H. Genome-wide investigation of the AP2/ERF gene family in tartary buckwheat (*Fagopyum tataricum*). *BMC Plant Biol.* **2019**, *19*, 84. [CrossRef]
91. Sun, W.; Jin, X.; Ma, Z.; Chen, H.; Liu, M. Basic helix-loop-helix (bHLH) gene family in Tartary buckwheat (*Fagopyrum tatari-cum*): Genome-wide identification, phylogeny, evolutionary expansion and expression analyses. *Int. J. Mol. Sci.* **2020**, *155*, 1478–1490.
92. Shahbaz, M.; Akram, A.; Raja, N.I.; Mukhtar, T.; Mehak, A.; Fatima, N.; Abasi, F. Antifungal activity of green synthesized selenium nanoparticles and their effect on physiological, biochemical, and antioxidant defense system of mango under mango malformation disease. *PLoS ONE* **2023**, *18*, e0274669. [CrossRef] [PubMed]
93. Wang, W.; Zhang, Z.; Liu, X.; Cao, X.; Wang, L.; Ding, Y.; Zhou, X. An improved GC-MS method for malondialdehyde (MDA) detection: Avoiding the effects of nitrite in foods. *Foods* **2022**, *11*, 1176. [CrossRef] [PubMed]
94. Zhao, F.; Zheng, Y.F.; Zeng, T.; Sun, R.; Yang, J.Y.; Li, Y.; Bai, S.N. Phosphorylation of SPOROCYTELESS/NOZZLE by the MPK3/6 kinase is required for anther development. *Plant Physiol.* **2017**, *173*, 2265–2277. [CrossRef] [PubMed]
95. Du, D.; Jin, R.; Guo, J.; Zhang, F. Infection of embryonic callus with Agrobacterium enables high-speed transformation of maize. *Int. J. Mol. Sci.* **2019**, *20*, 279. [CrossRef]
96. Bolger, A.M.; Lohse, M.; Usadel, B. Trimmomatic: A flexible trimmer for Illumina sequence data. *Bioinformatics* **2014**, *30*, 2114. [CrossRef] [PubMed]
97. Kim, D.; Langmead, B.; Salzberg, S.L. HISAT: A fast spliced aligner with low memory requirements. *Nat. Methods* **2015**, *12*, 357–360. [CrossRef] [PubMed]
98. Trapnell, C.; Williams, B.A.; Pertea, G.; Mortazavi, A.; Kwan, G.; Van Baren, M.J.; Pachter, L. Transcript assembly and quantification by RNA-Seq reveals unannotated transcripts and isoform switching during cell differentiation. *Nat. Biotechnol.* **2010**, *28*, 511–515. [CrossRef] [PubMed]
99. Anders, S.; Huber, W. *Differential Expression of RNA-Seq Data at the Gene Level the DESeq Package*; European Molecular Biology Laboratory (EMBL): Heidelberg, Germany, 2012; Volume 10, p. f1000research.
100. Beltrao, P.; Bork, P.; Krogan, N.J.; van Noort, V. Evolution and functional cross-talk of protein post-translational modifications. *Mol. Syst. Biol.* **2013**, *9*, 714. [CrossRef]
101. Taj, G.; Agarwal, P.; Grant, M.; Kumar, A. MAPK machinery in plants: Recognition and response to different stresses through multiple signal transduction pathways. *Plant Signal. Behav.* **2010**, *5*, 1370–1378. [CrossRef] [PubMed]
102. Li, S.; Han, X.; Lu, Z.; Qiu, W.; Yu, M.; Li, H.; Zhuo, R. MAPK cascades and transcriptional factors: Regulation of heavy met-al tolerance in plants. *Int. J. Mol. Sci.* **2022**, *23*, 4463. [CrossRef] [PubMed]
103. Behl, T.; Rana, T.; Alotaibi, G.H.; Shamsuzzaman, M.; Naqvi, M.; Sehgal, A.; Bungau, S. Polyphenols inhibiting MAPK signalling pathway mediated oxidative stress and inflammation in depression. *Biomed. Pharmacother.* **2022**, *146*, 112545. [CrossRef] [PubMed]

104. Popescu, S.C.; Popescu, G.V.; Bachan, S.; Zhang, Z.; Gerstein, M.; Snyder, M.; Dinesh-Kumar, S.P. MAPK target networks in Arabidopsis thaliana revealed using functional protein microarrays. *Genes Dev.* 2009, *23*, 80–92. [CrossRef] [PubMed]
105. Raghuram, B.; Sheikh, A.H.; Sinha, A.K. Regulation of MAP kinase signaling cascade by microRNAs in Oryza sativa. *Plant Signal. Behav.* 2014, *9*, e972130. [CrossRef]
106. Jiang, M.; Li, P.; Wang, W. Comparative analysis of MAPK and MKK gene families reveals differential evolutionary patterns in Brachypodium distachyon inbred lines. *PeerJ* 2021, *9*, e11238. [CrossRef]
107. Cui, L.; Yang, G.; Yan, J.; Pan, Y.; Nie, X. Genome-wide identification, expression profiles and regulatory network of MAPK cascade gene family in barley. *BMC Genom.* 2019, *20*, 750. [CrossRef] [PubMed]
108. Kumar, R.R.; Arora, K.; Goswami, S.; Sakhare, A.; Singh, B.; Chinnusamy, V.; Praveen, S. MAPK enzymes: A ROS activated signaling sensors involved in modulating heat stress response, tolerance and grain stability of wheat under heat stress. *3 Bio-Tech* 2020, *10*, 380. [CrossRef] [PubMed]
109. Sun, T.; Zhang, Y. MAP kinase cascades in plant development and immune signaling. *EMBO Rep.* 2022, *23*, e53817. [CrossRef]
110. Zhang, W.; Cochet, F.; Ponnaiah, M.; Lebreton, S.; Matheron, L.; Pionneau, C.; Baudouin, E. The MPK 8-TCP 14 pathway promotes seed germination in Arabidopsis. *Plant J.* 2019, *100*, 677–692. [CrossRef]
111. Lu, X.; Shi, H.; Ou, Y.; Cui, Y.; Chang, J.; Peng, L.; Li, J. RGF1-RGI1, a peptide-receptor complex, regulates Arabidopsis root meristem development via a MAPK signaling cascade. *Mol. Plant* 2020, *13*, 1594–1607. [CrossRef] [PubMed]
112. Liu, T.; Cao, L.; Cheng, Y.; Ji, J.; Wei, Y.; Wang, C.; Duan, K. MKK4/5-MPK3/6 cascade regulates Agrobacterium-mediated transformation by modulating plant immunity in Arabidopsis. *Front. Plant Sci.* 2021, *12*, 731690. [CrossRef]
113. Khan, M.; Rozhon, W.; Bigeard, J.; Pflieger, D.; Husar, S.; Pitzschke, A.; Poppenberger, B. Brassinosteroid-regulated GSK3/Shaggy-like kinases phosphorylate mitogen-activated protein (MAP) kinase kinases, which control stomata development in Arabidopsis thaliana. *J. Biol. Chem.* 2013, *288*, 7519–7527. [CrossRef] [PubMed]
114. Moustafa, K.; AbuQamar, S.; Jarrar, M.; Al-Rajab, A.J.; Trémouillaux-Guiller, J. MAPK cascades and major abiotic stresses. *Plant Cell Rep.* 2014, *33*, 1217–1225. [CrossRef]
115. Checa, J.; Aran, J.M. Reactive oxygen species: Drivers of physiological and pathological processes. *J. Inflamm. Res.* 2020, *ume 13*, 1057–1073. [CrossRef]
116. Kapoor, B.; Kumar, P.; Sharma, R.; Kumar, A. Regulatory interactions in phytohormone stress signaling implying plants resistance and resilience mechanisms. *J. Plant Biochem. Biotechnol.* 2021, *30*, 813–828. [CrossRef]
117. Danquah, A.; De Zélicourt, A.; Colcombet, J.; Hirt, H. The role of ABA and MAPK signaling pathways in plant abiotic stress responses. *Biotechnol. Adv.* 2014, *32*, 40–52. [CrossRef]
118. Zhu, J.K. Abiotic stress signaling and responses in plants. *Cell* 2016, *167*, 313–324. [CrossRef] [PubMed]
119. De Zélicourt, A.; Colcombet, J.; Hirt, H. The role of MAPK modules and ABA during abiotic stress signaling. *Trends Plant Sci.* 2016, *21*, 677–685. [CrossRef] [PubMed]
120. Wang, L.; Su, H.; Han, L.; Wang, C.; Sun, Y.; Liu, F. Differential expression profiles of poplar MAP kinase kinases in response to abiotic stresses and plant hormones, and overexpression of PtMKK4 improves the drought tolerance of poplar. *Gene* 2014, *545*, 141–148. [CrossRef]
121. Ahmad, H.; Zafar, S.A.; Naeem, M.K.; Shokat, S.; Inam, S.; Rehman, M.A.U.; Khan, M.R. Impact of pre-anthesis drought stress on physiology, yield-related traits, and drought-responsive genes in green super rice. *Front. Genet.* 2022, *256*, 832542. [CrossRef] [PubMed]
122. Chen, L.; Sun, H.; Wang, F.; Yue, D.; Shen, X.; Sun, W.; Yang, X. Genome-wide identification of MAPK cascade genes reveals the GhMAP3K14–GhMKK11–GhMPK31 pathway is involved in the drought response in cotton. *Plant Mol. Biol.* 2020, *103*, 211–223. [CrossRef] [PubMed]
123. Jiang, M.; Zhang, Y.; Li, P.; Jian, J.; Zhao, C.; Wen, G. Mitogen-activated protein kinase and substrate identification in plant growth and development. *Int. J. Mol. Sci.* 2022, *23*, 2744. [CrossRef] [PubMed]
124. Lin, L.; Wu, J.; Jiang, M.; Wang, Y. Plant mitogen-activated protein kinase cascades in environmental stresses. *Int. J. Mol. Sci.* 2021, *22*, 1543. [CrossRef] [PubMed]
125. Wang, S.; Han, S.; Zhou, X.; Zhao, C.; Guo, L.; Zhang, J.; Chen, X. Phosphorylation and ubiquitination of OsWRKY31 are integral to OsMKK10-2-mediated defense responses in rice. *Plant Cell* 2023, *35*, 2391–2412. [CrossRef]
126. Shi, J.; Zhang, L.; An, H.; Wu, C.; Guo, X. GhMPK16, a novel stress-responsive group D MAPK gene from cotton, is involved in disease resistance and drought sensitivity. *BMC Mol. Biol.* 2011, *12*, 22. [CrossRef]
127. Hao, S.; Wang, Y.; Yan, Y.; Liu, Y.; Wang, J.; Chen, S. A review on plant responses to salt stress and their mechanisms of salt resistance. *Horticulturae* 2021, *7*, 132. [CrossRef]
128. Zhao, S.; Zhang, Q.; Liu, M.; Zhou, H.; Ma, C.; Wang, P. Regulation of plant responses to salt stress. *Int. J. Mol. Sci.* 2021, *22*, 4609. [CrossRef] [PubMed]
129. Fang, Y.; Vilella-Bach, M.; Bachmann, R.; Flanigan, A.; Chen, J. Phosphatidic acid-mediated mitogenic activation of mTOR signaling. *Science* 2001, *294*, 1942–1945. [CrossRef]
130. Zhou, Y.; Zhou, D.M.; Yu, W.W.; Shi, L.L.; Zhang, J.; Lai, Y.X.; Xiao, S. Phosphatidic acid modulates MPK3-and MPK6-mediated hypoxia signaling in Arabidopsis. *Plant Cell* 2022, *34*, 889–909. [CrossRef] [PubMed]

131. Kong, Q.; Qu, N.; Gao, M.; Zhang, Z.; Ding, X.; Yang, F.; Zhang, Y. The MEKK1-MKK1/MKK2-MPK4 kinase cascade negatively regulates immunity mediated by a mitogen-activated protein kinase kinase kinase in Arabidopsis. *Plant Cell* **2012**, *24*, 2225–2236. [CrossRef] [PubMed]
132. Hu, Y.; Bai, J.; Xia, Y.; Lin, Y.; Ma, L.; Xu, X.; Chen, L. Increasing SnRK1 activity with the AMPK activator A-769662 accelerates seed germination in rice. *Plant Physiol. Biochem.* **2022**, *185*, 155–166. [CrossRef] [PubMed]
133. Jia, H.; Hao, L.; Guo, X.; Liu, S.; Yan, Y.; Guo, X. A Raf-like MAPKKK gene, GhRaf19, negatively regulates tolerance to drought and salt and positively regulates resistance to cold stress by modulating reactive oxygen species in cotton. *Plant Sci.* **2016**, *252*, 267–281. [CrossRef] [PubMed]
134. Zhang, L.; Xi, D.; Li, S.; Gao, Z.; Zhao, S.; Shi, J.; Wu, C.; Guo, X. A cotton group C MAP kinase gene, GhMPK2, positively regulates salt and drought tolerance in tobacco. *Plant Mol. Biol.* **2011**, *77*, 17–31. [CrossRef] [PubMed]
135. Zhang, A.; Zhang, J.; Ye, N.; Cao, J.; Tan, M.; Zhang, J.; Jiang, M. ZmMPK5 is required for the NADPH oxidase-mediated self-propagation of apoplastic H_2O_2 in brassinosteroid-induced antioxidant defence in leaves of maize. *J. Exp. Bot.* **2010**, *61*, 4399–4411. [CrossRef] [PubMed]

Disclaimer/Publisher's Note: The statements, opinions and data contained in all publications are solely those of the individual author(s) and contributor(s) and not of MDPI and/or the editor(s). MDPI and/or the editor(s) disclaim responsibility for any injury to people or property resulting from any ideas, methods, instructions or products referred to in the content.

Article

Natural Variability of Genomic Sequences of Mal d 1 Allergen in Apples as Revealed by Restriction Profiles and Homolog Polymorphism

Lucia Urbanová [1], Jana Bilčíková [1], Dagmar Moravčíková [2] and Jana Žiarovská [2,*]

[1] Research Centre AgroBioTech, Slovak University of Agriculture in Nitra, Tr. A. Hlinku 2, 949 76 Nitra, Slovakia; lucia.urbanova@uniag.sk (L.U.); jana.bilcikova@uniag.sk (J.B.)

[2] Institute of Plant and Environmental Sciences, Faculty of Agrobiology and Food Resources, Slovak University of Agriculture in Nitra, Tr. A. Hlinku 2, 949 76 Nitra, Slovakia; xmoravcikova@uniag.sk

* Correspondence: jana.ziarovska@uniag.sk; Tel.: +421-37-6414244

Abstract: Apples are a popular fruit worldwide, with many health and nutritional benefits. However, this fruit is also among those that, particularly in Central and Northern Europe, are allergenic due to the Mal d 1 allergen. Mal d 1 is a homologous allergen to Bet v 1—the main pollen allergen of birch. In this study, two different approaches were used to identify the natural length polymorphism of Bet v 1 homologs in apple varieties, with the aim of characterizing their effectiveness. BBAP (Bet v 1 based amplified polymorphism) and RFLP (restriction fragments length polymorphism) profiles were characterized and compared. RFLP analysis recognizes the genetic diversity of *M. domestica* Mal d 1 sequences at a relatively low level. In BBAP profiles, the genetic dissimilarity was up to 50%, which appears suitable for intraspecific fingerprinting and serves as an additional method for RFLP analysis. RFLP analysis was able to distinguish some varieties that BBAP could not, such as Sonet.

Keywords: Bet v 1; Mal d 1; polymorphism; *Malus domestica* Borkh

Citation: Urbanová, L.; Bilčíková, J.; Moravčíková, D.; Žiarovská, J. Natural Variability of Genomic Sequences of Mal d 1 Allergen in Apples as Revealed by Restriction Profiles and Homolog Polymorphism. *Agronomy* 2024, 14, 2056. https://doi.org/10.3390/agronomy14092056

Academic Editors: Jiezheng Ying, Zhiyong Li and Chaolei Liu

Received: 12 July 2024
Revised: 4 September 2024
Accepted: 6 September 2024
Published: 9 September 2024

Copyright: © 2024 by the authors. Licensee MDPI, Basel, Switzerland. This article is an open access article distributed under the terms and conditions of the Creative Commons Attribution (CC BY) license (https://creativecommons.org/licenses/by/4.0/).

1. Introduction

Different DNA-marker-based techniques have been developed to effectively analyze the genomic variability of plants. In principle, these techniques amplify coding or non-coding regions in a polymerase chain reaction (PCR) to generate variable length polymorphism, enabling the distinction or characterization of plant germplasm. For these techniques, where coding regions serve as markers, methods based on the variability of homolog genes for plant allergens have been developed [1–3]. One of them, Bet v 1-based amplified polymorphism (BBAP), amplifies the different lengths of the homologs of Bet v 1, a main birch allergen, and is anchored in the linear epitope sequence through degenerate primers [4]. Birch pollen (*Betulaceae*) is considered the most allergenic in Northern, Central, and Eastern Europe [5] and it is the main tree pollen allergen in Northern Europe [6]. Its major allergen, Bet v 1, is thought to trigger cross-reactions with some food allergens, which form the PR-10 protein family [7]. The percentage of people suffering from birch pollen allergy and who also respond to at least one of the PR-10 group is 50–93%, what is known as PFS (pollen-food syndrome).

In 2015, the EAACI reported that 7 million patients suffer from food allergies (URL4), with the pediatric population being the most vulnerable. Current studies suggest that food allergy is becoming a common population disease and affects 5–10% of the population [8]. One of the most significant Bet v 1 cross-reactions is connected to apple consumption, which affects >70% of patients allergic to birch pollen [9].

M. domestica Borkh. belongs to the *Rosaceae* family, genus *Malus*. Cultivation is widespread, especially in temperate areas, with more than 10,000 varieties being grown [10]. Apple are valued for their sweet taste and juice with a high content of minerals, vitamins, or various bioactive substances, such as polyphenols, polysaccharides, flavonoids, or various health-promoting acids [11–13]. The content of all substances is influenced by the type of cultivar or breeding purpose. The main apple allergen is Mal d 1, a 17.5 kDA protein that belongs to the PR-10 protein family.

Mal d 1 consists of a complex gene family composed of 31 different loci, each encoding a different isoallergen [14]. Moreover, the variability of the major apple allergen is amplified by the existence of isoallergen alleles with small differences, that encode isoallergenic variants/isoforms [14–16]. Differences in the amino acid sequence of isoforms can alter ligand binding sites, influencing the allergenic potential of the allergen. Reduced or increased affinity to human IgE has been confirmed between Mal d 1 isoforms from different apple varieties [17] or by amino acid substitutions in the epitopes [18,19]. Based on different isoforms, a restriction fragment length polymorphism (RFLP) can be applied to analyze genomic variability, as was previously proven for part of the *Ypr 10* gene promotor in the apple varieties, Santana, Cripps Pink, Jonagold, and Gala [20].

Despite the relatively variable residual nucleotide sequence of *ypr10* genes (genomic similarity 50–90%) [21], proteins of the PR-10 family share a very similar to identical protein structure due to highly conserved sequences [22]. The conserved sequences are probably specific IgE-binding sites, known as epitopes. A probable IgE epitope could be located between the 42nd and 52nd amino acids [23], with Glu45 considered essential. Its essentiality has been confirmed by several mutations, demonstrating the variability with which the mutants bind specifically [24–26].

In this study, two different approaches were used to identify the natural length polymorphism of Bet v 1 homologs in apple varieties, with the aim to characterizing their effectiveness. BBAP- and RFLP-generated polymorphic profiles were characterized and compared.

2. Materials and Methods

2.1. Biological Material and DNA Extraction

The biological material consisted of a total of 70 apple varieties, as listed in Table 1. All samples were collected from a private orchard (Brodno, Žilina, Slovakia) at the stage of physiological maturity and were frozen at −20 °C. Genomic DNA was isolated using GeneJet™ Plant Genomic DNA Purification Mini Kit (Thermo Scientific®, Shanghai, China). The quality and quantity of DNAs were verified three times: using a Nanophotometer™ (IMPLEN, Munich, Germany), 1% agarose gel and PCR with ITS primers [27].

Table 1. List of the apple varieties with relevant sample numbers used in all figures.

Sample Number	Apple Variety	Sample Number	Apple Variety	Sample Number	Apple Variety	Sample Number	Apple Variety
1	Santana	19	Tabor	37	Rozela	55	Hael 616
2	Spencer	20	Sonet	38	Melrose	56	Orion
3	Primadela	21	Parkerovo	39	Fiesta	57	Maj Gold
4	Ecolette	22	Jantár	40	HL 782	58	Sirius
5	Akame	23	Rubigold	41	Delor	59	Paula Red
6	Karneval	24	Topaz	42	Spigold	60	Dulcit
7	Waltz	25	Ligol	43	Rucla	61	Harmony
8	Winesap	26	Pocomoke	44	Heliodor	62	Pinova
9	Čistecké lahôdkové	27	Shalimar	45	Stela	63	Sentima
10	Melrose 24628	28	Florina	46	RubinStep	64	Mutsu
11	Selena	29	Rezistent	47	Aneta	65	Viktory
12	Lotos	30	Bolero	48	Jonagold Decosta	66	Lipno
13	Produkta	31	Ligol	49	Delbarestivale	67	Pikant
14	Jonalord	32	Alkmene	50	Goldstar	68	Fanny
15	Angold	33	Bohemia	51	Blanik	69	Admiral
16	Rezistent Opal	34	Dalila	52	HL 189	70	Freyberg

Table 1. Cont.

Sample Number	Apple Variety	Sample Number	Apple Variety	Sample Number	Apple Variety	Sample Number	Apple Variety
17	Kamzi	35	Linda	53	Rajka	71	Kristian
18	Meteor	36	Alkmene	54	Biogolden		

2.2. BBAP

Non-degenerated forward and degenerated reverse primers for the BBAP technique were designed for Bet v 1 sequences with NCBI accession numbers AJ289770.1 and AJ28977.1 [4]. The sequence of the forward primer without degeneration, matching the conservative region, was F: 5′ CCT GGA ACC ATC AAG AAG 3′. The sequence of the reverse degenerated primer (at the 12th and 14th positions), matching the variable region, was R: 5′ TTG GTG TGG TAS TKG CTG 3′, while S = G/C and K = T/G. Individual variants of the reverse primer combinations were used in the analysis separately with the codes R1, R2, R3 and R4.

For the amplification reactions, EliZyme HS Robust MIX was used with 10 ng of DNA and 400 nM of each primer. The PCR conditions were as follows: initial denaturation at 95 °C for 5 min; 40 cycles of denaturation at 95 °C, 45 s; annealing at 54 °C, 45 s; polymerization 72 °C, 35 s; and a final elongation at 72 °C for 10 min.

2.3. RFLP

The Mal d 1 sequence of the apple Bet v 1 homolog (NCBI AF020542.1) was divided into three parts for amplification, with primer sequences listed in Table 2.

Table 2. Primers used to amplify Mal d 1 sequence for RFLP.

Amplicon	Primers	Position	Length
Mal d 1–part 1	F: 5′GCTCGATCACGATAAACTAAGG 3′ R: 5′ATGAGGATGGGGTGTTGAAG 3′	nt 3–522	520 bp
Mal d 1–part 2	F: 5′ACATCCAGTACCGGGGATGA 3′ R: 5′GGGTGCAATCTTGGGGATGA 3′	nt 833–1470	638 bp
Mal d 1–part 3	F: 5′GATGCTTTGACAGACACCATTG 3′ R: 5′TTTCAAACAAATACATAAAGGGCAAC3′	nt 1801–2190	390 bp

In silico DNA restriction analysis was used to find eligible restriction enzymes by the NEBcutter V2.0 website (www.nc2.neb.com/NEBcutter2; accessed on 20 March 2024). Two restriction enzymes were selected to each sequence part: a specific enzyme for the amplified part and a common enzyme for all of the amplified parts (Table 3). Restriction cleavage conditions followed the manufacturer's instructions.

Table 3. Restriction enzymes chosen for RFLP analysis.

Restriction Enzyme	Cutting Site	Part of Mal d 1 Sequence
AseI	AT↓TA↑AT	part 1, 2, 3
NcoI	C↓CATG↑G	part 3
NlaIII	↑CATG↓	part 2
SpeI	A↓CTAG↑T	part 1

The cleaved products were separated in 10% PAGE and stained by GelRedTM (Biotium). Arrows stand for restriction site.

2.4. Data Processing

BBAP as well as RFLP profiles were analyzed using the freely available GelAnalyzer software (www.gelanalyzer.com; accessed on 15 January 2024) and converted into binary matrices. The matrices from both analyses were processed using the UPGMA method, employing the Nei-Li coefficient of genetic distances, to create dendrograms. The Mantel test, cophenetic coefficients, *phi* coefficient, effective multiplex ratio, marker index, resolving power, as well as heatmap design and dendrogram comparison were all carried out in the R studio environment [28].

3. Results

3.1. BBAP Variability

The analysis of a length-based polymorphism for one of the nucleotide sequences of the Bet v 1 epitope and its homologs is a universal technique, with different results obtained based on the individual reverse primers used in their non-degenerated forms.

Primers F + R1 (Bet v 1 isoforms with His-119) amplified a total of 13 amplicons with the following lengths: 940 bp, 910 bp, 850 bp, 690 bp, 665 bp, 600 bp, 560 bp, 385 bp, 355 bp, 340 bp, 285 bp, 200 bp, and 140 bp. A dendrogram could not be constructed due to very similar or identical profiles among the individual apple varieties analyzed. However, all the analyzed apple varieties could be grouped into four groups (Table 4) based on their F + R1 amplicon profiles. Notably, no PCR product was amplified in the Alkmene variety using this primer combination, even after repetition. Amplicons of 385 bp and 200 bp were found in all samples with a stronger signal, indicating a higher number of these loci in the analyzed genotypes and a preference for their amplification. The control amplicon of 385 bp, which matched the in silico prediction of PCR results, was amplified in all other apple varieties.

Table 4. Groups of analyzed apple varieties based on the differences in BBAP F + R1 primer combination fingerprints profiles.

Group	Variety
I.	Santana, Spencer, Primadela, Ecolette, Produkta, Rezistent Opal, Meteor, Sonet, Jantar, Rezistent, Bolero, Fiesta, Rucla, Heliodor, Rajka, Dulcit, Harmony, Sentima, Admiral
II.	Akame, Karneval, Waltz, Winesap, Melrose 24628, Selena, Lotos, Tabor, Parkerovo, Rubigold, Topaz, Pocomoke, Shalimar, Florina, Ligol, Alkmene, Bohemia, Dalila, Linda, Rozela, Melrose, HL 782, Delor, Spigold, Stela, Delbarestivale, Orion, Maigold, Sirius, Paula Red, Lipno, Pikant, Fanny, Freyberg, Kristian
III.	Čistecké lahôdkové, Kamzi, Rubinstep, Aneta, Jonagold Decosta, Goldstar, Blanik, HL 189, Biogolden, Hael 616, Pinova
IV.	Jonalord, Angold, Mutsu, Viktory

Primers F + R2 (Bet v 1 isoforms with Asp-119) amplified two lengths of 210 bp and 385 bp in all analyzed apple varieties. Only in the case of ten apple varieties—Winesap, Melrose 24628, Rubinstep, Jonagold Decosta, Goldstar, Paula Red, Dulcit, Harmony, Pinova and Sentima—were additional amplicons of 180 bp and 150 bp obtained.

Primers F + R3 (Bet v 1 isoforms with Glu-119) generated fingerprint profiles that were grouped the varieties into six different clusters, with five samples showing unique profiles (Figure 1).

The most distinct profile was observed in the Primadela variety. The greatest interspecific genetic distance was up to 50%, while the most similar groups, the fifth and sixth, shared approx. 80% genetic similarity. A total of eight PCR products were amplified: 385 bp, 220 bp, 210 bp, 150 bp, 140 bp, 125 bp, and two shorter than 50 bp, with the 385 bp and 210 bp amplicons being present in all varieties except for Primadela and Admiral.

The most polymorphic results were generated by the F + R4 primer pair annealing to the *Ypr10* gene, which has an isoform with 119-Lys in its amino acid sequences. A total of 20 PCR products were amplified, with following lengths: 1300 bp, 1200 bp, 1100 bp, 970 bp, 690 bp, 585 bp, 385 bp, 340 bp, 290 bp, 260 bp, 235 bp, 210 bp, 185 bp, 170 bp, 150 bp, 140 bp, 120 bp, 110 bp, 95 bp, and 75 bp. Despite the polymorphic BBAP profiles, identical profiles were obtained among several pairs of varieties: Maigold–Sirius; HL 782–Orion;

Winesap–Delor; Waltz–Fiesta; Spencer–Bohemia; Resistent Opal–Florina; Meteor–Bolero–Ligol. All varieties were grouped into seven clusters in the dendrogram (Figure 2), with the Fanny variety showing a unique profile. The first three groups in the dendrogram shared more than 45% similarity and were the most represented. The Jantar, Heliodor, and Aneta varieties were distinguished from all other varieties by almost 95%, with their genetic distance between 50 and 70%.

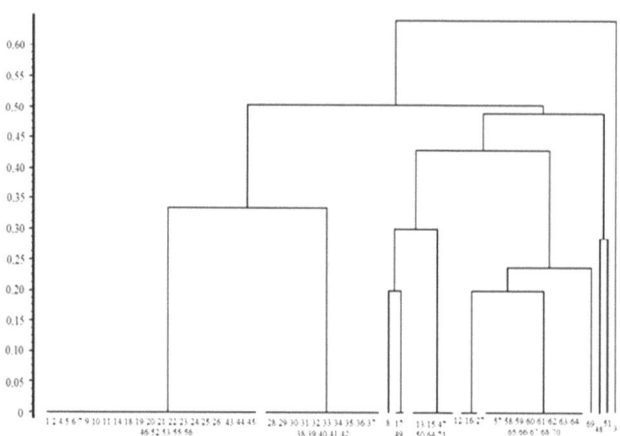

Figure 1. UPGMA dendrogram of Nei-Li coefficient of genetic similarity values among analyzed apple varieties for BBAP F + R3 fingerprints. Number codes of varieties are listed in Table 1.

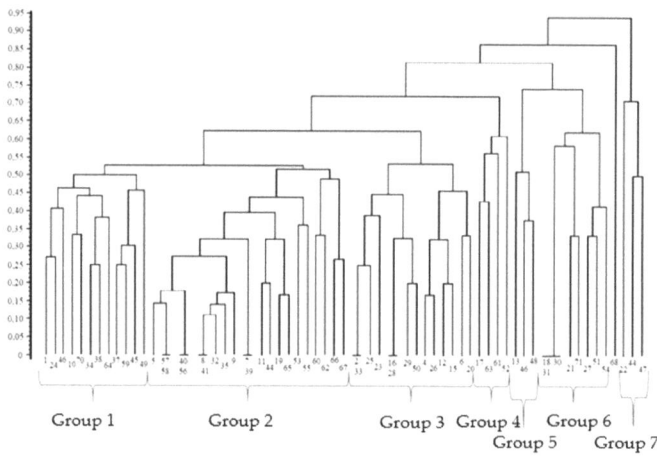

Figure 2. UPGMA dendrogram of Nei-Li coefficient of genetic similarity values among analyzed apple varieties for BBAP F + R4 fingerprints. Number codes of varieties are as listed in Table 1.

Combining the results from the individual reverse primers into one binary matrix, groups A–F were identified based on the similarity of some varieties (Table 5). The highest genetic similarities were found in group F, varying from 70 to 90%, in contrast to group K, which had the lowest similarity of all groups (only 30% between varieties Jantar and Heliodor. Two of the analyzed varieties, Maigold and Sirius, generated identical profiles for all four BBAP primers. The second-highest similarity in all BBAP profiles (almost 90%) was between Alkmene and Linda.

Table 5. Groups of varieties that share the highest similarities of BBAP profiles.

Group	Variety	Similarity in %
Group A	Topaz, Stela	~50
Group B	Dalila, Rozela, Melrose	60–70
Group C	Akame, Orion, Waltz, Selena, Tabor	60–80
Group D	Čistecké lahôdkové, Hael 616	~60
Group E	Maigold, Sirius, Lipno, Pikant	50–65
Group F	HL 782, Linda	~75
Group G	Bohemia, Florina	~55
Group H	Spencer, Ecolette, Sonet	50–55
Group I	Rubigold, Pocomoke, Karneval	50–55
Group J	Harmony, Sentima	~45
Group K	Jantar, Heliodor	~30

The results indicate a good ability of the BBAP technique to generate an intraspecific polymorphism that reflects the sequence variability of *M. domestica* Borkh. *ypr10* genes. Primers F + R2 point to intraspecific stability, while primers F + R4 demonstrate high variability of *ypr10* genes within the *M. domestica* Borkh. genome. The most stable amplicon lengths amplified across all primer combinations were 210 bp and 388 bp, what corresponding with predicted in silico data [4]. Only two completely identical profiles were obtained for varieties Maigold and Sirius; all other varieties differed by at least 10% in their BBAP profiles.

3.2. RFPL Variability

In RFLP, a total of 20 to 36 fragments were obtained, with a high degree of polymorphism. The lowest number of fragments (20) was observed in the Sonet variety, while the most abundant profile (38 fragments) was found in the Kamzi variety. Identical cleavage profiles were identified for the varieties, Blanik-HL 189-Dulcit, Jonagold-Maigold, and Alkemene-Florina.

In the first amplified part of the Mal d 1 gene, the Sonet variety profile exhibited the lowest number of cleavage fragments ($n = 3$), while the Rozela variety showed the highest (n = 12) when combining restriction fragments from both restriction enzymes used. The highest number of restriction fragments for Ase 1 was 11 and for Spe I it was 4 (Figure 3). The cleavage of the part 1 region generated 39 different cleavage profiles, 24 of which were unique. Unique cleavage profiles were found in some low-allergenic varieties, such as Santana and Pink Lady, as well as the highly allergenic variety, Golden Delicious.

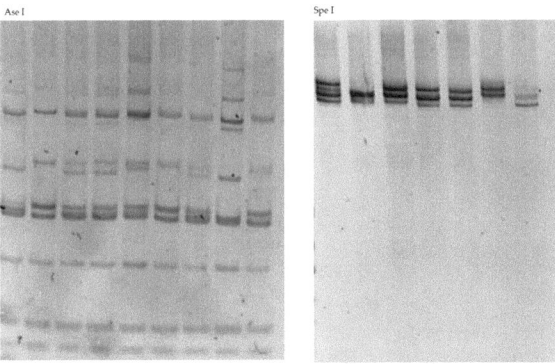

Figure 3. Selection of restriction pattern generated by Ase I and Spe I in the first amplified part of Mal d 1 gene.

In the second amplified part of the Mal d 1 gene, the AseI enzyme cut amplicon to 4–8 fragments, forming 16 restriction profiles, of which three were unique (Gala, HL 782 and Sonet). Again, the Sonet variety had the lowest number of fragments (n = 4). In the Golden Delicious variety, eight restriction fragments were obtained. The most numerous restriction profile was found in 22 varieties and consisted of five cleavage fragments. A restriction fragment of 250 bp was present in all varieties except for Sonet. Using the NlaIII enzyme, 6 to 11 fragments were obtained, resulting in 22 different restriction profiles. Unique profiles were observed in the varieties Dalila, Golden Delicious, Kamzi, Pink Lady, Produkta, and Spencer. Up to five restriction fragments were present in the restriction profiles of all 71 varieties, with lengths of 110 bp, 400 bp, 510 bp, 580 bp, and 630 bp.

In the third amplified part of the Mal d 1 gene, the cleavage profiles showed the highest degree of monomorphism with the selected enzymes AseI and NcoI. The sequence was cleaved into 2 to 5 fragments by AseI, forming five cleavage profiles, and 2 to 5 fragments by NcoI forming three restriction profiles. No variety showed a unique profile. The NcoI fragments were 83 bp and 37 bp in size, corresponding to the in silico cleavage profile fragments.

The enzymes used in the RFLP analysis cleaved all three amplified parts of the Mal d 1 sequence along the entire gene length. SpeI and NlaIII cleaved the Mal d 1/*ypr10* gene sequence upstream of the region amplified in the BBAP, NcoI cleaved downstream of this region, while AseI cleaved both upstream and downstream. The constructed dendrogram, based on the binary matrix of restriction profiles, showed that the RFLP analysis was able to differentiate between most varieties with a dissimilarity coefficient of 0.18 (Figure 4). The most unique varieties according to RFLP was Sonet, the most dissimilar to all of the other analyzed varieties was Victory. The most similar restriction profiles were obtained for the varieties Čistecké lahôdkové, Lotos, Florina, Linda, and Lipno.

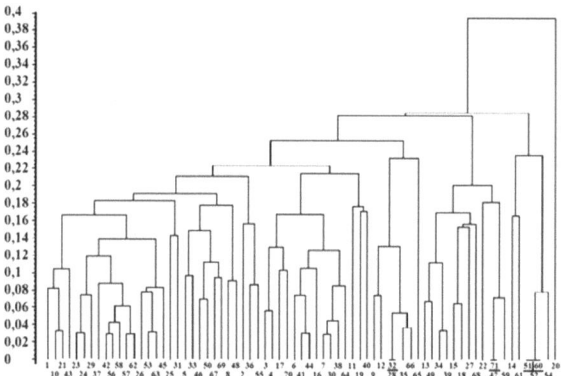

Figure 4. UPGMA dendrogram of Nei-Li coefficient of genetic similarity values among analyzed apple varieties for RFLP restriction profiles. Number codes of varieties are as listed in Table 1.

3.3. Comparative Potential of BBAP and RFLP in Mal d 1 Length Variability Analysis in the Apple Germplasm

Comparing the techniques used in the study, BBAP was able to generate fully different fingerprints, and all the analyzed apple varieties were grouped in a dendrogram without any similar pattern (Figure 5). In contrast, the RFLP restriction pattern failed to distinguish seven of the analyzed varieties in the final dendrogram. In the case of Jantár, Jonagold and Rubin Step, all of these varieties were grouped in a very different pattern, as was in differences in BBAP F + R1 primer combination fingerprint profiles.

The individual binary matrices with the Nei-Li coefficient showed no significant correlation in the Mantel test based on Pearson's product–moment correlation as well as the phi coefficient for binary matrices. Cophenetic coefficients for individual dendrograms were 0.672 for the RFLP data and 0.669 for the BBAP data.

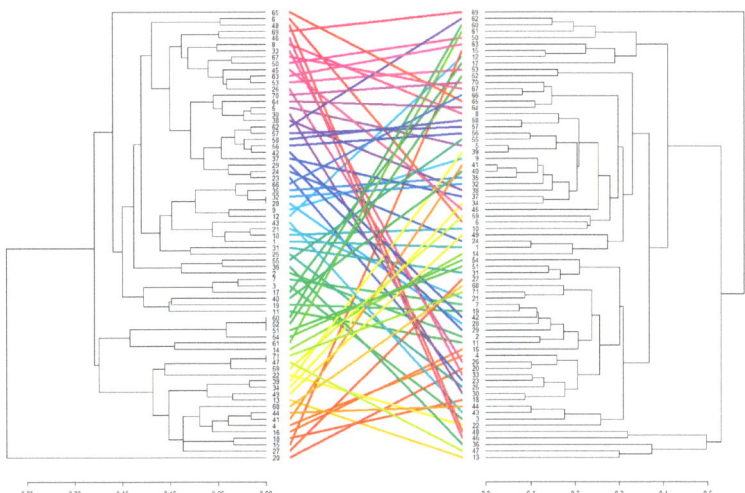

Figure 5. The relationship of the location of individual apple varieties in summary dendrograms for both of the methods used in the study. Number codes of varieties are listed in Table 1. Individual colours of lines represent individual varieties and their positions in dendrograms.

The BBAP is comparable in effectiveness and the ability to analyze genetic dissimilarities between individual varieties with the RFLP (Figures 6 and 7). The coefficient values were closely aligned when marking the universal length polymorphism of Bet v 1 homologs compared to those generated by restriction fragment length polymorphism.

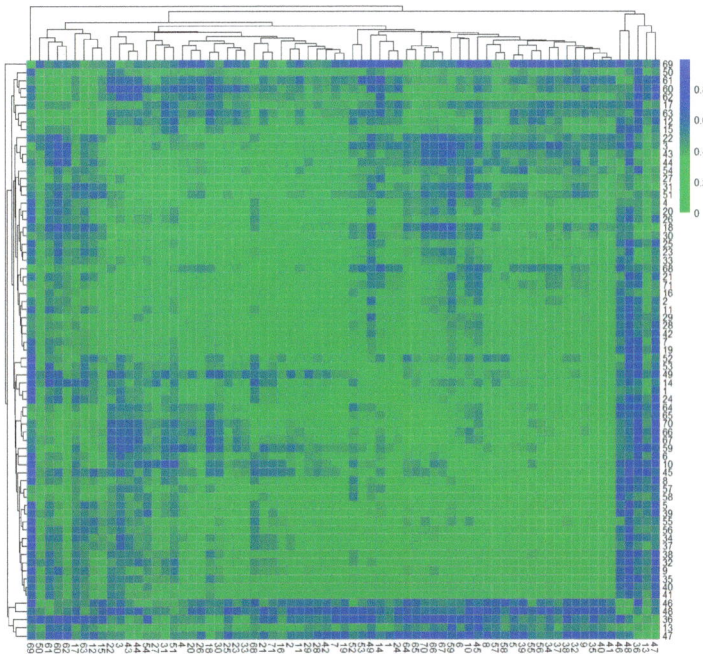

Figure 6. Heatmap of dissimilarity coefficient distribution among apple varieties analyzed by BBAP. Number codes of varieties are listed in Table 1.

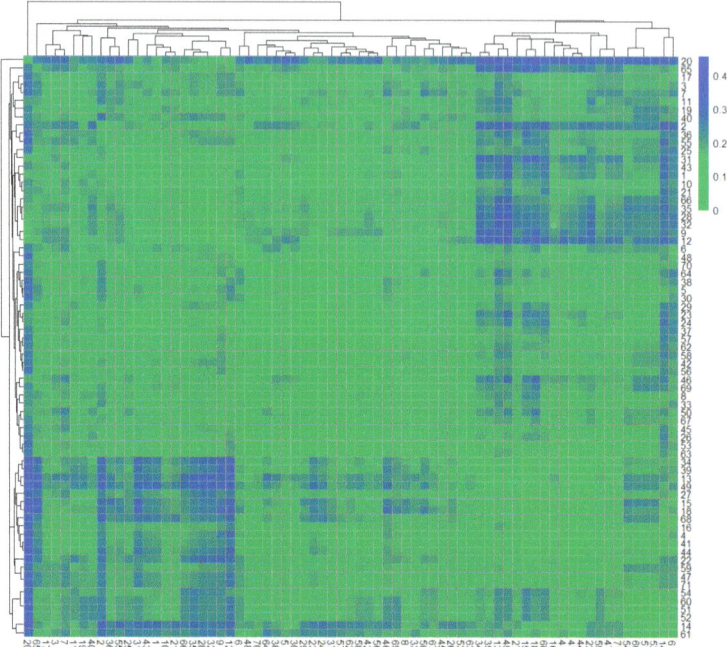

Figure 7. Heatmap of dissimilarity coefficient distribution among apple varieties analyzed by RFLP. The number codes of varieties are listed in Table 1.

The probability of detecting polymorphism with BBAP in the analysis of apple germplasm was 0.67, compared to 0.59 for RFLP. The effectiveness of both marker techniques in multiplex testing, as calculated by the effective multiplex ratio, was 32.67 for BBAP and 40.91 for RFLP. The Marker index, which estimates the total utility of the marker technique, was 18.96 for BBAP and 24.17 for RFLP. The resolving power of differences detection was 0.59 for BBAP and 0.66 for RFLP.

4. Discussion

Thus far, various DNA-based markers have been utilized in apple germplasm studies, each offering different characteristics. According to the available literature, the simple sequence repeat (SSR) technique is suitable for studying a polymorphism of *M. domestica* Borkh. due to its high level of polymorphism, co-dominant inheritance, and reproducibility [29,30]. A high level of genetic diversity was demonstrated using SSR across the species (He > 0.7) and no significant reduction in the diversity over the last eight centuries. Although, less diversity has been confirmed for commercially used varieties [31]. The conserved DNA-derived polymorphism technique (CDDP) has also been used to produce polymorphic amplification patterns in a set of 15 apple varieties from this study. In some primer combinations, identical amplification profiles were observed in a few varieties, similar to the RFPL results in this study. Identical CDDP profiles were found in the pairs of varieties: Red Delicious–Granny Smith; Maigold–Paula Red; Selena–Melodie and Maigold–Paula [32]. Gene-specific markers are reported to determine fruit storage life (ethylene production), fruit skin color, *Alternaria* resistance, and scab resistance in apples [33].

An in silico comparison of amino acid sequences of known Bet v 1 allergen isoforms typically shows significant variability among individual plant species [34]. Such variability is a good background for using these regions as DNA-based marker techniques to reveal the variability of Bet v 1 homologs in flowering plants and to be universal in their application. A high homology exists in the amino acid sequences in the region of the forward primer for the BBAP strategy, which includes a confirmed epitope for IgE [2,35]. The reverse primers amplify a variable region of the *ypr10* gene, matching the amino acid variability at position 119 of the Bet v 1 protein (P15494) [34]. This study analyzed different apple varieties to prove the efficiency of the *in silico*-generated markers for Bet v 1 homologs compared to the restriction cleavage of species-specific Bet v 1 homologs in the apple genome–Mal d 1 gene, to provide polymorphic fingerprints among them. The universality of this DNA-based technique has been reported for different groups of flowering plants relevant to food allergies, such as fruits, vegetables, nuts, or legumes [2,36,37]. For vegetables, genetic distance analysis using the Nei-Li coefficient showed a narrow range (the difference was only 0.36), but the genetic diversity was relatively high, corresponding to the findings of variability among plant Bet v 1 homolog proteins [35,38].

Several studies have shown a high degree of natural genetic diversity manifested in Mal d 1/*ypr10* genes in *M. domestica* Borkh. At the genetic level, Mal d 1 is a complex gene family with 31 loci, each encoding a different isoallergen [14], while each isoallergen can be divided into isoallergenic variants due to single-allele substitutions [14–16]. The similarity level among individual coding sequences of Mal d 1/*ypr10* genes ranges from 53.1% to 97.7% and among isovariants from 95% to 99.8%. At the protein level, the sequences of isoallergens are identical between 37% (e.g., Mal d 1.08 and Mal d 1.11 A with 102 amino acid substitutions) and 96% (Mal d 1.06 A and Mal d 1.06 D with 7 substitutions [12]. Considering the limited number of analyzed varieties [16] and the existence of more than 7000 varieties, the variability in *ypr10* genes across the entire apple genome is likely several times greater. This study focused on monitoring the natural variability in the Mal d 1 allergen across 71 varieties, including old varieties, such as Akane, Alkmene, Spencer, Spigold or Winesap, as well as newer and commercially important varieties, such as Blanik, Kamzi or Rozela. For primary screening, fingerprinting via BBAP and RFLP techniques was chosen due to its specificity and reported effectiveness [36,39].

The results of BBAP profiles confirm the ability of the primers to create intraspecific polymorphisms, reflecting the variability of the *ypr10* genes in *M. domestica* Borkh. and complementing the data obtained *in silico*. Primer pair F + R2 amplifies isoforms that are stable across varieties, while primer pair F + R4 captures the high variability of *ypr10* genes within the *M. domestica* Borkh. genome. Previous research used degenerated reverse BBAP primers that anneal to variable and conserved parts of Bet v 1 homologues genes in the BBAP technique to analyze the intraspecific variability in *M. domestica* Borkh. varieties [40]. The generated amplicons formed relatively monomorphic profiles, indicating the stability of the Bet v 1 isoforms within the selected apple varieties, corresponding to the results of F + R1 and F + R2 combinations in this study. Bet v 1 homologs in the apple genome possess a high-protein structure homology due to short amino acid sequences, which are highly conserved among plant species, resulting in a very similar or identical protein structure [22], despite the relatively variable nucleotide sequences of the *ypr10* genes (genomic similarity from 50 to more than 90%) [22].

Data mining and experimental verification were previously applied for Mal d 1 RFLP analysis [20]. Virtual cleavage maps were prepared for available sequences of the apple variety McIntosh, which has been referred to as hypoallergenic, as well as the variety Santana [41–43]. The RFLP technique has been reported to be useful for *M. domestica* when differentiating between closely related apple varieties [39] or when pedigree information is missing [44]. Phylogenetic relationships are well illustrated for apple germplasms when using RFLP [45]. The inter-apple and intra-apple variability concerning the Mal d 1 gene and allergenicity is well documented [46]. The amount of Mal d 1 in apples classified as containing low concentrations of allergen may be sufficient to induce both clinical symptoms and skin reactivity in birch-pollen-allergic patients. Differences in the concentration of Mal d 1 proteins year-on-year or even between two growing sites have also been confirmed [47]. Different aspects in the assessment of allergen identification in plant food sources, variability of allergenic molecules, and genetic relationships is important not only for the scientific community but also for the consumers as the number of allergy sufferers increases every year, which results in an impact on the quality of their life.

5. Conclusions

The results of the BBAP profiles confirm the ability of the primers to create an intraspecific polymorphism that reflects the variability of the Mal d 1/*ypr10* genes in *M. domestica* Borkh. Primer pair F + R2 captures isoforms stably occurring across the entire species, while primer pair F + R4 captures the high variability of *ypr10* genes within the *M. domestica* Borkh. genome. RFLP recognizes the genetic diversity of *M. domestica ypr10* genes at a relatively low level (with a dissimilarity coefficient up to 0,18). Compared to BBAP analysis, genetic dissimilarity was up to 50% (for F/R4), making the gene suitable for intraspecific fingerprinting and an additional method for the Bet v 1 protein family. However, RFLP analysis was able to distinguish some varieties which BBAP was not, such as Sonet.

Author Contributions: Conceptualization, J.Ž.; methodology, J.Ž.; software, L.U., D.M. and J.B.; validation, J.Ž., L.U. and J.B.; formal analysis, L.U. and D.M.; investigation, L.U. and J.B.; resources, J.Ž.; data curation, J.Ž.; writing—original draft preparation, L.U. and J.Ž.; writing—review and editing, L.U., J.Ž. and D.M.; visualization, D.M.; supervision, J.Ž. All authors have read and agreed to the published version of the manuscript.

Funding: This research was funded by project APVV-20-0058—The potential of the essential oils from aromatic plants for medical use.

Data Availability Statement: The original contributions presented in the study are included in the article, further inquiries can be directed to the corresponding author.

Acknowledgments: The authors would like to thank the G-Team for an inspirative working atmosphere and consultations related to methodological context.

Conflicts of Interest: The authors declare no conflicts of interest.

References

1. Klongová, L.; Kováčik, A.; Urbanová, L.; Kyseľ, M.; Ivanišová, E.; Žiarovská, J. Utilization of specific primers in legume allergens based polymorphism screening. *Sci. Technol. Innov.* **2021**, *13*, 12–21. [CrossRef]
2. Žiarovská, J.; Urbanová, L. Utilization of Bet v 1 homologs based amplified profile (BBAP) variability in allergenic plants fingerprinting. *Biologia* **2022**, *77*, 517–523. [CrossRef]
3. Čerteková, S.; Kováčik, A.; Klongová, L.; Žiarovská, J. Utilization of plant profilins as DNA markers. *Acta Fytotech. Zootech.* **2023**, *26*, 324–331. [CrossRef]
4. Žiarovská, J.; Zeleňáková, L. Application of Genomic Data for PCR Screening of Bet v 1 Conserved Sequence in Clinically Relevant Plant Species. In *Systems Biology*; IntechOpen: London, UK, 2019; pp. 65–82.
5. D'Amato, G.; Cecchi, L.; Bonini, S.; Nunes, C.; Annesi-Maesano, I.; Behrendt, H.; Liccardi, G.; Popov, T.; van Cauwenberge, P. Allergenic pollen and pollen allergy in Europe. *Allergy* **2007**, *62*, 976–990. [CrossRef] [PubMed]
6. Eriksson, N.E.; Holmen, A. Skin prick tests with standardized extracts of inhalant allergens in 7099 adult patients with asthma or rhinitis: Cross-sensitizations and relationships to age, sex, month of birth and year of testing. *J. Investig. Allergol. Clin. Immunol.* **1996**, *6*, 36–46. [PubMed]
7. Geroldinger-Simic, M.; Zelniker, T.; Aberer, W.; Ebner, C.; Egger, C.; Greiderer, A.; Prem, N.; Lidholm, J.; Ballmer-Weber, B.K.; Vieths, S.; et al. Birch pollen-related food allergy: Clinical aspects and the role of allergen-specific IgE and IgG4 antibodies. *J. Allergy Clin. Immunol.* **2011**, *127*, 616–622. [CrossRef]
8. Loh, W.; Tang, M.L.K. The Epidemiology of Food Allergy in the Global Context. *Int. J. Environ. Res. Public Health* **2018**, *15*, 2043. [CrossRef]
9. Vieths, S.; Scheurer, S.; Ballmer-Weber, B. Current understanding of crossreactivity of food allergens and pollen. *Ann. N. Y. Acad. Sci.* **2002**, *964*, 47–68. [CrossRef]
10. Janick, J.; Moore, J.N. (Eds.) *Fruit Breeding, Tree and Tropical Fruits*; John Wiley & Sons: New York, NY, USA, 1996; pp. 1–77. ISBN 978-0-471-31014-3.
11. Eberhardt, M.V.; Lee, C.Y.; Liu, R.H. Antioxidant activity of fresh apples. *Nature* **2000**, *405*, 903–904. [CrossRef]
12. Wolfe, K.L.; Liu, R.H. Apple peel as a value-added food ingredient. *J. Agr. Food Chem.* **2003**, *51*, 1676–1683. [CrossRef]
13. Can, Z.; Dincer, B.; Sahin, H.; Baltas, N.; Yildiz, O.; Kolayli, S. Polyphenol oxidase activity and antioxidant properties of Yomra apple (*Malus communis* L.) from Turkey. *J. Enzyme. Inhib. Med. Chem.* **2014**, *29*, 829–835. [CrossRef] [PubMed]
14. Pagliarani, G.; Paris, R.; Iorio, A.R.; Tartarini, S.; Del Duca, S.; Arens, P.; Peters, S.; Van der Weg, E. Genomic organisation of the Mal d 1 gene cluster on linkage group 16 in apple. *Mol. Breed.* **2012**, *29*, 759–778. [CrossRef]

15. Gao, Z.S.; van de Weg, W.E.; Schaart, J.G.; Schouten, H.J.; Tran, D.H.; Kodde, L.P.; van der Meer, I.M.; van der Geest, A.H.M.; Kodde, J.; Breiteneder, H.; et al. Genomic cloning and linkage mapping of the Mal d 1 (PR-10) gene family in apple (*Malus domestica*). *Theor. Appl. Genet.* **2005**, *111*, 171–183. [CrossRef]
16. Gao, Z.S.; Van de Weg, W.E.; Matos, C.I.; Arens, P.; Bolhaar, S.T.H.P.; Knulst, A.C.; Li, Y.; Hoffmann-Sommergruber, K.; Gilissen, L.J. Assessment of allelic diversity in intron-containing Mal d 1 genes and their association to apple allergenicity. *BMC Plant Biol.* **2008**, *8*, 128. [CrossRef] [PubMed]
17. Son, D.Y.; Scheurer, S.; Hoffman, A.; Haustein, D.; Vieths, S. Pollen-related food allergy: Cloning and immunological analysis of isoforms and mutants of Mal d 1, the major apple allergen, and Bet v 1, the major birch pollen allergen. *Eur. J. Nutr.* **1999**, *38*, 201–215. [CrossRef]
18. Ma, Y.; Gadermaier, G.; Bohle, B.; Bolhaar, S.; Knulst, A.; Markovic-Housley, Z.; Breiteneder, H.; Briza, P.; Hoffmann-Sommergruber, K.; Ferreira, F. Mutational analysis of amino acid positions crucial for IgE binding epitopes of the major apple (*Malus domestica*) allergen, Mal d 1. *Int. Arch. Allergy Immunol.* **2006**, *139*, 53–62. [CrossRef]
19. Holm, J.; Ferreras, M.; Ipsen, H.; Würtzen, P.A.; Gajhede, M.; Larsen, J.N.; Lund, K.; Spangfort, M.D. Epitope grafting, re-creating a conformational Bet v 1 antibody epitope on the surface of the homologous apple allergen Mal d 1. *J. Biol. Chem.* **2011**, *286*, 17569–17578. [CrossRef]
20. Žiarovská, J.; Bilčíková, J.; Fialková, V.; Zeleňáková, L.; Zamiešková, L. Restriction polymorphism of Mal d 1 allergen promotor in apple varieties. *J. Microbiol. Biotechnol. Food Sci.* **2019**, *8*, 1217–1219. [CrossRef]
21. Fernandes, H.; Michalska, K.; Sikorski, M.; Jaskolski, M. Structural and functional aspects of PR-10 proteins. *FEBS J.* **2013**, *280*, 1169–1199. [CrossRef]
22. Seutter von Loetzen, C.; Hoffmann, T.; Hartl, M.J.; Schweimer, K.; Schwab, W.; Rösch, P.; Hartl-Spiegelhauer, O. Secret of the major birch pollen allergen Bet v 1: Identification of the physiological ligand. *Biochem. J.* **2012**, *457*, 379–390. [CrossRef]
23. Mirza, O.; Henriksen, A.; Ipsen, H.; Larsen, J.N.; Wissenbach, M.; Spangfort, M.D.; Gajhede, M. Dominant epitopes and allergic cross-reactivity: Complex formation between a Fab fragment of a monoclonal murine IgG antibody and the major allergen from birch pollen Bet v 1. *J. Immunol.* **2000**, *165*, 331–338. [CrossRef]
24. Spangfort, M.D.; Mirza, O.; Ipsen, H.; van Neerven, R.J.J.; Gajhede, M.; Larsen, J.N. Dominating IgE-binding epitope of Bet v 1, the major allergen of birch pollen, characterized by X-ray crystallography and site-directed mutagenesis. *J. Immunol.* **2003**, *171*, 3084–3090. [CrossRef] [PubMed]
25. Neudecker, P.; Schweimer, K.; Nerkamp, J.; Scheurer, S.; Vieths, S.; Sticht, H.; Rösch, P. Allergic cross-reactivity made visible. Solution structure of the major cherry allergen Pru av 1. *J. Biol. Chem.* **2001**, *276*, 22756–22763. [CrossRef] [PubMed]
26. Neudecker, P.; Lehmann, K.; Nerkamp, J.; Haase, T.; Wangorsch, A.; Fötisch, K.; Hoffmann, S.; Rösch, P.; Vieths, S.; Scheurer, S. Mutational epitope analysis of Pru av 1 and Api g 1, the major allergens of cherry (*Prunus avium*) and celery (*Apium graveolens*): Correlating IgE reactivity with three-dimensional structure. *Biochem. J.* **2003**, *376*, 97–107. [CrossRef] [PubMed]
27. White, T.J.; Bruns, T.D.; Lee, S.B.; Taylor, J.W. Amplification and Direct Sequencing of Fungal Ribosomal RNA Genes for Phylogenetics. In *PCR Protocols: A Guide to Methods and Applications*; Innis, M.A., Gelfand, D.H., Sninsky, J.J., White, T.J., Eds.; Academic Press: New York, NY, USA, 1990; pp. 315–322.
28. R Core Team. *R: A Language and Environment for Statistical Computing*; R Foundation for Statistical Computing: Vienna, Austria, 2021. Available online: https://www.R-project.org/ (accessed on 12 June 2024).
29. Marconi, G.; Ferradini, N.; Russi, L.; Concezzi, L.; Veronesi, F.; Albertini, E. Genetic Characterization of the Apple Germplasm Collection in Central Italy: The Value of Local Varieties. *Front. Plant Sci.* **2018**, *9*, 1460. [CrossRef]
30. Omasheva, M.Y.; Flachowsky, H.; Ryabushkina, N.A.; Pozharskyi, A.S.; Galiakparov, N.N.; Hanke, M.V. To what extent do wild apples in Kazakhstan retain their genetic integrity? *Tree Genet Genom* **2017**, *13*, 52. [CrossRef]
31. Gross, B.L.; Henk, A.D.; Richards, C.M.; Fazio, G.; Volk, G.M. Genetic Diversity in *Malus* ×*domestica* (Rosaceae) through Time in Response to Domestication. *Am. J. Bot.* **2014**, *101*, 1770–1779. [CrossRef] [PubMed]
32. Bilčíková, J.; Farkasová, S.; Žiarovská, J. Genetic variability of commercially important apple varieties (*Malus* × *domestica* Borkh.) assessed by CDDP markers. *Acta Fytotechn. Zootech.* **2021**, *24*, 21–26.
33. Kikuchi, T.; Kasajima, I.; Morita, M.; Yoshikawa, N. Practical DNA markers to Estimate Apple (*Malus* × *domestica* Borkh.) Skin Color, Ethylene Production and Pathogen Resistance. *J. Hortic.* **2017**, *4*, 1000211. [CrossRef]
34. Breiteneder, H.; Ebner, C. Molecular and biochemical classification of plant-derived food allergens. *J. Allergy Clin. Immunol.* **2000**, *106*, 27–36. [CrossRef]
35. Uehara, M.; Sato, K.; Abe, Y.; Katagiri, M. Sequential IgE epitope analysis of a birch pollen allergen (Bet v1) and an apple allergen (Mal d1). *Allergol. Int.* **2001**, *50*, 57–62. [CrossRef]
36. Urbanová, L.; Žiarovská, J. Variability of DNA based amplicon profiles generated by Bet v 1 homologous among different vegetable species. *Acta Fytotechn. Zootech.* **2021**, *24*, 1–5.
37. Moravčíková, D.; Žiarovská, J. Variability of Genomic Profile of *ypr-10* gene in *Citrus sinensis* L. Osbeck. *Biol. Life Sci. Forum* **2024**, *30*, 2.
38. Freitas, L.B.; Bonatto, S.L.; Salzano, F.M. Evolutionary implications of infra-and interspecific molecular variability of pathogenesis-related proteins. *Braz. J. Biol.* **2003**, *63*, 437–448. [CrossRef]
39. Tignon, M.; Lateur, M.; Kettmann, R.; Watillon, B. Distinction between closely related apple cultivars of the Belle-fleur family using RFLP and AFLP markers. *Acta Hortic.* **2001**, *546*, 509–513. [CrossRef]

40. Speváková, I.; Urbanová, L.; Kyseľ, M.; Bilčíková, J.; Farkasová, S.; Žiarovská, J. BBAP amplification profiles of apple varieties. *Sci. Technol. Innov.* **2021**, *13*, 1–6. [CrossRef]
41. Pühringer, H.M.; Zinoecker, I.; Marzban, G.; Katinger, H.; Laimer, M. MdAP, a novel protein in apple, is associated with the major allergen Mal d 1. *Gene* **2003**, *321*, 173–183. [CrossRef]
42. Bolhaar, S.T.; Van de Weg, W.E.; Van Re, R.; Gonzalez-Mancebo, E.; Zuidmeer, L.; Bruijnzeel-Koomen, C.A.; Fernandez-Rivas, M.; Jansen, J.; Hoffmann-Sommergruber, K.; Knulst, A.C.; et al. In vivo assessment with prick-to-prick testing and double-blind, placebo-controlled food challenge of allergenicity of apple cultivars. *J. Allergy Clin. Immunol.* **2005**, *116*, 1080–1086. [CrossRef] [PubMed]
43. Koostra, H.S.; Vlieg-Boerstra, B.J.; Dubois, A.E.J. Assessment of the reduced properties of the Santana apple. *Ann. Allergy Asthma Immunol.* **2007**, *99*, 522–525. [CrossRef]
44. Gardiner, S.E.; Bassett, H.C.M.; Madie, C.; Noiton, D.A.M. Isozyme, Randomly Amplified Polymorphic DNA (RAPD), and Restriction Fragment-length Polymorphism (RFLP) Markers Used to Deduce a Putative Parent for the 'Braeburn' Apple. *JASHS* **1996**, *121*, 996–1001. [CrossRef]
45. Forte, A.V.; Ignatov, A.N.; Ponomarenko, V.V.; Dorokhov, D.B.; Savelyev, N.I. Phylogeny of the Malus (Apple Tree) Species, Inferred from the Morphological Traits and Molecular DNA Analysis. *Russ. J. Genet.* **2002**, *38*, 1150–1160. [CrossRef]
46. Asero, R.; Marzban, G.; Martinelli, A.; Zaccarini, M.; Laimer da Camara Machado, M. Search for low allergenic apple cultivars for birch pollen allergic patients: Is there a correlation between in vitro assays and patient response? *Eur. Ann. Allergy Clin. Immunol.* **2006**, *38*, 94–98. [PubMed]
47. Sancho, A.I.; Foxall, R.; Browne, T.; Dey, R.; Zuidmeer, L.; Marzban, G.; Waldron, K.W.; van Ree, R.; Hoffmann-Sommergruber, K.; Laimer, M.; et al. Effect of postharvest storage on the expression of the apple allergen Mal d 1. *J. Agricul. Food Chem.* **2006**, *54*, 5917–5923. [CrossRef] [PubMed]

Disclaimer/Publisher's Note: The statements, opinions and data contained in all publications are solely those of the individual author(s) and contributor(s) and not of MDPI and/or the editor(s). MDPI and/or the editor(s) disclaim responsibility for any injury to people or property resulting from any ideas, methods, instructions or products referred to in the content.

Article

Genetic Basis and Exploration of Major Expressed QTL *qLA2-3* Underlying Leaf Angle in Maize

Yonghui He [1,2,†], Chenxi Wang [1,†], Xueyou Hu [1,3], Youle Han [1], Feng Lu [1], Huanhuan Liu [1,2], Xuecai Zhang [4,*] and Zhitong Yin [1,2,*]

1 Jiangsu Key Laboratory of Crop Genomics and Molecular Breeding/Key Laboratory of Plant Functional Genomics of the Ministry of Education/Jiangsu Key Laboratory of Crop Genetics and Physiology/Joint International Research Laboratory of Agriculture and Agri-Product Safety of the Ministry of Education, Agricultural College, Yangzhou University, Yangzhou 225009, China
2 Jiangsu Co-Innovation Center for Modern Production Technology of Grain Crops, Yangzhou University, Yangzhou 225009, China
3 Crop Breeding and Cultivation Research Institute, Shanghai Academy of Agricultural Sciences, Shanghai 201403, China
4 International Maize and Wheat Improvement Center (CIMMYT), Mexico City 06600, Mexico
* Correspondence: xc.zhang@cgiar.org (X.Z.); ztyin@yzu.edu.cn (Z.Y.)
† These authors contributed equally to this work.

Abstract: Leaf angle (LA) is closely related to plant architecture, photosynthesis and density tolerance in maize. In the current study, we used a recombinant inbred line population constructed by two maize-inbred lines to detect quantitative trait loci (QTLs) controlling LA. Based on the average LA in three environments, 13 QTLs were detected, with the logarithm of odds ranging from 2.7 to 7.21, and the phenotypic variation explained by a single QTL ranged from 3.93% to 12.64%. A stable QTL, *qLA2-3*, on chromosome 2 was detected and was considered to be the major QTL controlling the LA. On the basis of verifying the genetic effect of *qLA2-3*, a fine map was used to narrow the candidate interval, and finally, the target segment was located at a physical distance of approximately 338.46 kb (B73 RefGen_v4 version), containing 16 genes. Re-sequencing and transcriptome results revealed that five candidate genes may be involved in the regulation of LA. The results enrich the information for molecular marker-assisted selection of maize LA and provide genetic resources for the breeding of dense planting varieties.

Keywords: leaf angle; maize; quantitative trait loci; QTL mapping

1. Introduction

Appropriately reducing leaf angle (LA) is one of the most effective measures to increase maize planting density and yield [1,2]. The complete maize leaf is composed of a sheath, blade and ligular region. The ligular region is located at the junction of the sheath and leaf and is divided into two parts: the ligule and the auricle [3]. The ligule is close to the maize stem and can be regarded as an extension of the leaf sheath. By contrast, the auricle is located on either side of the mid-vein and has a wedge-shaped structure. The ligule and auricle engage each other and act as hinges to keep the leaf and stem at a certain angle called the LA [4].

The identification of genes regulating maize LA is mainly involved in transcriptional regulation and hormone signalling conducted using mutants [4–6]. Transcriptional regulation is one of the main regulatory pathways affecting angle size. *lg1, lg2, lg3* and *lg4* mutants have defective or absent ligules and auricles, resulting in minimal LA [7–9]. The SQUAMOSA promoter binding transcription factor LG1 regulates LA by affecting cell autonomy [9]. The bZIP transcription factor *LG2* gene is expressed earlier than *LG1* and may be involved in the initiation of the early leaf ligular region [8]. The function defects

of *LG3* and *LG4* encoding KNOX transcription factors lead to abnormal development of the leaf primordia meristem and abnormal leaf corner morphology [7]. In addition, defective transcription factors lead to cell ectopic expression of ligule and auricle, resulting in distorted and variable LA morphology, such as *wab*, *drooping leaf1* (*drl1*) and *drl2* [10,11]. Hormone-related genes are involved in the regulation of LA. Mutations in *Brd1*, the enzyme that catalyses the last step of brassinolide (BR) synthesis, cause leaf twist and LA deformation [12]. The mutation of the *NANA PLANT2* gene related to BR biosynthesis resulted in decreased BR levels, extreme dwarfing and increased LA [13]. A natural mutation of the BR C-22 hydroxylase gene exhibits upright upper leaves and smart-canopy architecture, which is a genetic resource for breeding high-density maize varieties [6].

LA is regulated by multiple quantitative trait loci (QTL). A large number of QTLs controlling LA in maize have been detected in different genetic populations [3,4]. Nine QTLs for LA were detected using 180 recombinant inbred lines (RILs) constructed by B73 and Mo17 [14]. Using genotyping-by-sequencing analysis, 17 QTL controlling LA were detected [15]. Fourteen QTLs controlling LA were identified by a four-way strategy using a four-way cross-population [16]. A total of 30 LA QTLs were detected by association and linkage analysis using a nested association mapping population [17].

Several maize LA-related genes, including *ZmTAC1*, *ZmCLA4*, *ZmILI1*, *ZmIBH1-1*, *ZmRAVL1* and *ZmBrd1*, were cloned by fine-mapping strategies [2,18–21]. *ZmTAC1* encodes a Poaceae protein with an unknown function and is homologous to rice *OsTAC1*. Sequence variation in the 5′-UTR of *ZmTAC1* affects the expression of *ZmTAC1*, which in turn affects LA size [21,22]. *ZmCLA4* encodes a gramineous protein that negatively regulates auxin transport and is homologous to the gene *LAZY1* that controls tiller angle in rice [20]. *ZmILI1*, *ZmIBH1-1* and *ZmRAVL1* encode transcription factors that affect LA by regulating the expression of genes, including hormone responses, cell differentiation and cell wall formation [2,18–20].

It is a feasible way to improve maize plant density by introducing superior alleles that control LA into maize varieties. After the introduction of the rare allele *UPA2* in teosinte, the LA of the maize cultivar Nongda108 was significantly reduced, resulting in a higher yield under dense planting conditions [2]. In the study, we identified a novel major QTL for LA in maize, *qLA2-3*, using genetic mapping of the population. *qLA2-3* was detected in all three environments and explained up to 12.31% of the phenotypic variation. Furthermore, fine mapping was used to narrow the mapping interval to 338.46 kb. Based on the maize reference genome, this region contains 16 genes. Five candidate genes were identified based on re-sequencing and transcriptome data. The results of this study should provide genetic resources for maize plant architecture improvement.

2. Materials and Methods

2.1. Plant Materials

The inbred lines LDC-1 and YS501 are the two parent inbred lines of a commercial maize hybrid Tianyu 88 which was released by our lab. LDC-1 and YS501 carry tropical and temperate germplasm, respectively. These two inbred lines exhibit a significant difference in leaf angle. LDC-1 shows a larger leaf angle than YS501. A single-seed descent method was used to develop RILs from the cross between LDC-1 and YS501. Briefly, F2 kernels were obtained by selfing F1 plants in the spring of 2016. Approximately 200 F2 plants were grown and selfed to obtain F3 ears in the winter of 2016. One kernel from each F3 ear was used to generate the F4 plant. Then, a similar procedure was employed for F4, F5, F6, F7 and F8 ears. Finally, 186 F9 RILs were obtained in the spring of 2019 [23]. The plants of the two parents and RIL populations were grown in three environments, respectively: the Ledong experimental field of Hainan (N: 18.73°, E: 109.17°) in the winter of 2019 (E1); an experimental field on the Wenhui Road campus of Yangzhou University (N: 32.40°, E: 119.40°) in the spring of 2020 (E2); and an experimental field on the Yangzijin Campus of Yangzhou University (N: 32.40°, E: 119.40°) in the summer of 2020 (E3). The soil at the Yangzhou experimental site was sandy loam, with the duration of sunshine ranging from

12 to 14 h. The average high temperature trend during spring sowing was 25–32 °C, and the average low temperature was 16–25 °C. During summer sowing, the average high temperature trend was 32–25 °C, and the average low temperature was 25–16 °C. The soil of the experimental field in Hainan was sandy, and the sunshine duration was approximately 12 h. The average high temperature trend was 30–28 °C, and the average low temperature was 21–17 °C. The RILs underwent planting in a randomised complete-block design, where the experiment was replicated twice for enhanced reliability.

Genotype recombinants were planted in experimental fields in the winter of 2020 (Hainan), the winter of 2021 (Hainan), the spring of 2022 (Yangzhou University) and the summer of 2022 (Zhenjiang City, N: 32.22°, E: 119.26°), respectively, for fine mapping. Ten maize plants were sown in the field where the row length was 2 m, the plant spacing was 0.2 m and the row spacing was 0.6 m. Field irrigation, fertilisation, weeding, and pest protection and management are the same as in the general field.

2.2. Investigation of Maize LA Phenotype

LA between stem and leaf above the ear was measured using a protractor at 15 day after pollination [2]. Five randomly selected plants from each line of the RIL population were measured for QTL mapping. The LA above the ear of all plants in the fine mapping plot was recorded, and the average LA was used for analysis.

2.3. Data Analysis

The descriptive statistical analysis, ANOVA and correlation analysis of the phenotypic data of the LA of the RIL population were performed using Excel 2016 and SPSS 21.0 software. The computation of broad-sense heritability (H^2) is formulated as $H^2 = Vg/(Vg + Vge/l + V\varepsilon/rl)$, encompassing the variance components of genotype (Vg), genotype–environment interaction (Vge), and random error ($V\varepsilon$), with r standing for the number of replicates and l for the number of environments.

2.4. QTL Mapping

A high-density linkage map of 2624 bin markers with an average genetic distance of 0.9 cM between the markers was constructed in our previous study [23,24]. QTL mapping of average LA was performed using WinQTL Cartographer 2.5 software. The recombination rate was converted to genetic distance using the Kosambi function. With a walking step of 1.00 cM, the composite interval mapping method was used to obtain the logarithm of odds (LOD) of QTL by 1000 permutations ($p < 0.05$). LOD thresholds greater than 2.5 were considered significant QTL, and the prediction confidence interval was based on a drop interval of 1.5 LOD. The naming rules for QTLs were as follows: 'q' stands for QTL, followed by the LA abbreviation (LA), followed by the sequence number of the chromosomes, and finally, sorted according to their physical location on the chromosome.

2.5. Genome Re-Sequencing

Genomic DNA was extracted from two parental inbred lines, LDC-1 and YS501, and randomly interrupted by a Covaris crusher with a growth rate of 350 bp. The DNA library was constructed using the TruSeq Library Construction Kit and sequenced on the Illumina platform in strict accordance with the instruction manual (Novogene Co., Ltd., Beijing, China). Clean data obtained by removing adapters, poly-N sequences and low-quality reads were then mapped to the reference genome B73 RefGen_v4 (ftp://ftp.ncbi.nlm.nih.gov/genomes/all/GCF/000/005/005/GCF_000005005.2_B73_RefGen_v4, accessed on 29 August 2024) using BWA software (v0.7.8-r455) with default parameters. SAMtools (v1.3.1, parameters as mpileup -m 2 -F 0.002 -d 1000) was used to filter SNP/InDel detection criteria as follows: (1) the depth of the variate position is not less than 4; (2) the mapping quality is not less than 20 [25].

2.6. Fine Mapping of qLA2-3

InDel markers in or near the *qLA2-3* region were designed using a primer design tool (https://www.ncbi.nlm.nih.gov/tools/primer-blast/, accessed on 29 August 2024) to screen for genotype recombinant plants in the residual heterozygous lines (Table S1).

The recombinants in the *qLA2-3* interval were planted into plots after self-crossing for fine mapping in each generation. Each plot has over 100 plants used for measuring leaf angle data. The *qLA2-3* locus was mapped by analysing the genotype of each plant and its corresponding average LA in the plot. The significant difference in LA between the two parental homozygous genotypes indicates that the candidate gene is located in the heterozygous region; otherwise, it is located in the homozygous region.

2.7. RNA-Seq and qPCR

RNA samples were collected from the newly formed ligular region of the third leaf above the ear. The ligular regions of 15 plants were mixed into one sample, and three biological replicates were prepared for $qLA2\text{-}3\text{-}NIL^{YS501}$ and $qLA2\text{-}3\text{-}NIL^{LDC\text{-}1}$. The total RNA in each sample was extracted with a Plant Total RNA Isolation Kit (Vazyme Biotech, Nanjing, China), as referenced in the user manual. The sequencing libraries were constructed using the NEBNext Ultra II RNA Library Prep Kit for Illumina (New England Biolabs, Ipswich, MA, USA) with reference to the standard operating manual. Clean reads were obtained from raw reads after filtering, such as removing $3'$ end adapters (the removed part had at least a 10 bp overlap with the known adapter, allowing for 20% base mismatches) and low-quality data (reads with an average quality score lower than Q20) using Cutadapt (v1.11). Reads were mapped to the maize B73_RefGen_v4 reference genome (https://www.maizegdb.org/, accessed on 29 August 2024) using the HISAT2 (v2.1.0) software (http://ccb.jhu.edu/software/hisat2/index.shtml, accessed on 29 August 2024) under the condition of default parameters. The Read Count values mapped to each gene were counted and regarded as the original expression levels of the genes using HTSeq (v0.9.1). In order to make the gene expression levels among different genes/samples comparable, FPKM (Fragments Per Kilobases per Million fragments) was used to normalise the expression levels after excluding rRNA and tRNA (Personalbio Co., Ltd., Shanghai, China). Genes with expression fold changes exceeding 1.5 ($p < 0.05$) were considered significant differentially expressed genes by DESeq (v1.38.3).

For the qPCR analysis, 1 μg of total RNA was used to synthesise cDNA with an ABScript III RT Master Mix for qPCR (ABclonal, Wuhan, China) in accordance with the manufacturer's instructions. Gene fragments were amplified with Universal SYBR Green Fast qPCR Mix (Abclonal, Wuhan, China) on an ABI StepOnePlus Real-Time PCR system. The ΔCt (threshold cycle) method was used to determine the gene expression levels, and the expression of Maize *ZmGAPDH* (GRMZM2G046804) was taken as the internal control.

3. Results

3.1. Phenotypic Variation in LA Traits

Maize LDC-1 and YS501 inbred lines have similar plant heights but contain a large difference in upright plant architecture. Compared to LDC-1, YS501 showed 4.3°, 5.8°, 17.6°, 16.0°, 18.4°, 18.1° and 17.2° reductions in the first leaf below ear, leaf at ear position, first leaf above ear, second leaf above ear, third leaf above ear, fourth leaf above ear and fifth leaf above ear, respectively (Figure 1A,B). Significant differences in LA between the two parents were observed in all three environments, with parent YS501 containing a larger LA than parent LDC-1 (Table 1). Based on 186 RILs constructed with these two parents, the LA of the genetic mapping population was measured and analysed.

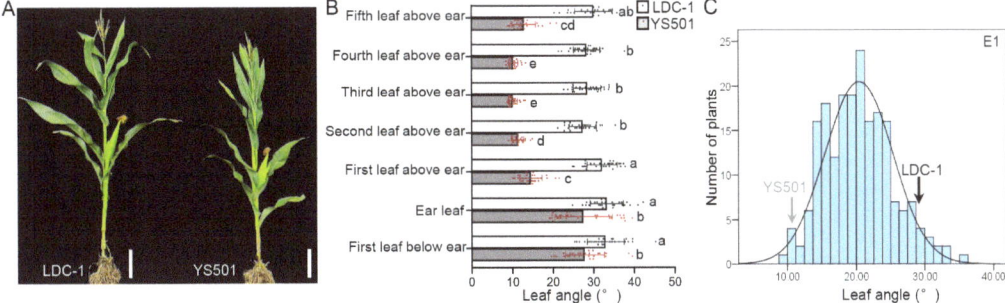

Figure 1. Phenotypic distributions of leaf angles in LDC-1, YS501 and RIL populations: (**A**) Comparison of leaf angles between LDC-1 and YS501. The scale bar was 20 cm. (**B**) Statistical data on leaf angle size between LDC-1 and YS501. $n = 30$; mean compassion letters represent significant differences at $p < 0.05$ level through Tukey's Honest Significant Difference Test. (**C**) Frequency distributions of mean leaf angle above the ear in the RIL population. The data in the E1 environment were used as an example. The X-axis represents the range of leaf angle distribution on the ear within the RIL population, the Y-axis represents the frequency, and the black line represents the normal distribution fitting curve.

Table 1. Phenotypic statistics of the leaf angle in parents and RIL population.

Environment	Parents			RIL Population						
	YS501	LDC-1	*t*-Test *p* Value	Mean ± SD	Range	Kurtosis	Skewness	CV (%)	*F* Value	H^2
E1	11.83	28.87	3.56×10^{-30}	20.33 ± 4.84	9.22–35.02	−0.043	0.331	23.81		
E2	10.25	26.97	1.01×10^{-8}	19.96 ± 6.41	9.07–37.54	−0.389	0.429	32.11	8.216 **	89.15%
E3	11.34	23.89	3.48×10^{-7}	22.50 ± 5.09	9.62–38.69	0.15	0.109	22.62		

SD, standard deviation; CV, coefficient of variation; H^2, broad-sense heritability. ** indicates extremely significant correlation at the $p < 0.01$ level.

The correlation analysis of the LA of the RIL population in three environments showed a considerably significant positive correlation ($p < 0.01$). In E1, E2 and E3 environments, the LA correlation coefficients of each leaf were 0.456–0.749, 0.515–0.812 and 0.450–0.729, respectively (Figure 2). The results of correlation analysis indicated that the genes regulating the LA of five leaves on the ear might have similar genetic regulatory mechanisms, so the average LA was used for further analysis.

Descriptive statistics show that the absolute values of skewness and kurtosis of LA in all environments were less than 1, suggesting that the data of LA present the characteristics of a normal distribution (Table 1). The coefficients of variation in LA in the three environments were 23.81%, 32.11% and 22.62%, respectively, with a large range of variation. The average LA of the RIL population was between that of parents, but there were also a number of lines whose LA was lower or higher than that of parents, showing transgressions (Figures 1C and S1). The genotype *F* value indicated that the genetic difference of LA reached a considerably significant level, and the broad sense heritability of LA was high, reaching 89.15%. These results indicate that the LA has a quantitative character and can be used for QTL mapping.

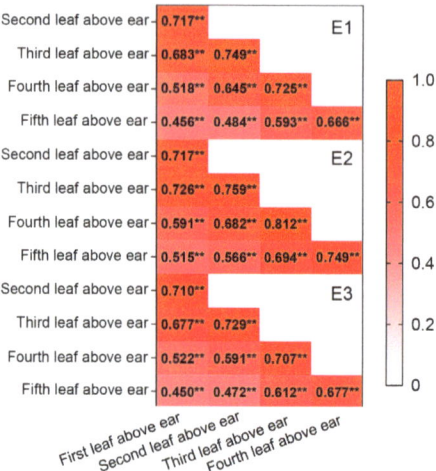

Figure 2. Correlation analysis of leaf angles on ears in the RIL population under three environments. E1: Ledong Experimental Base in Hainan in the winter of 2019; E2: Maize Genetic Breeding Experimental Field in the Agricultural College of Yangzhou University in the spring of 2020; E3: Yangzhou University Experimental Farm in the summer of 2020. ** indicates an extremely significant correlation at the $p < 0.01$ level.

3.2. Identification of QTLs for LA

The average LA of each line was used as phenotypic data for the initial mapping of QTL regulating LA based on a high-density linkage map [24]. A total of 13 QTLs were identified in the three environments, which were distributed on chromosomes 1, 2, 3, 4, 5, 6, 7 and 10 (Table 2). Each QTL explained 3.93%–12.64% of phenotypic variation and had LOD values between 2.7 and 7.21.

Table 2. Quantitative trait loci (QTLs) for leaf angle detected using 186 recombinant inbred lines in three environments.

Trait	QTL	Chr	Position (cM)	Physical Location (bp)	E1 LOD	E1 Add	E1 R^2	E2 LOD	E2 Add	E2 R^2	E3 LOD	E3 Add	E3 R^2
LA	qLA1-1	1	115.1–135.1	39490381–53752442				7.21	2.33	12.64			
	qLA1-2	1	193.0–209.4	105544639–166628038	4.50	1.32	7.22						
	qLA1-3	1	349.7–371.2	281040428–292905589							3.45	1.20	5.27
	qLA2-1	2	55.8–67.6	12495796–16391990	3.86	1.30	6.98						
	qLA2-2	2	144.5–152.5	187673664–191462633							4.83	1.46	7.09
	qLA2-3	2	180.1–196.6	214012471–225807350	6.88	1.65	10.90	7.03	2.37	12.31	4.30	1.39	6.25
	qLA3-1	3	13–18.9	2050621–3087150	3.08	1.12	4.87						
	qLA3-2	3	147.2–153.8	181135621–184419019							3.12	1.09	4.57
	qLA4-1	4	180.4–195.1	208732552–236286534				3.69	1.67	6.25			
	qLA5-1	5	113.3–116.6	61422142–67028398							3.27	−1.07	4.34
	qLA6-1	6	82.6–95.0	116217531–136571957							5.31	1.47	7.94
	qLA7-1	7	231.3–238.7	177668009–179104806							2.70	−1.02	3.93
	qLA10-1	10	53.8–69.6	19031133–90669531	5.97	−1.54	9.45						

LA, leaf angle; Chr, chromosome; LOD, logarithm of odds; Add, additive effect value; R^2, phenotypic contribution rate.

Among QTLs, *qLA2-3* on chromosome 2 was detected in all three environments (Figure 3A). The genetic effect of *qLA2-3* was the largest, accounting for 10.90%, 12.31% and 6.25% of the phenotypic variation, and the synergistic allele originated from LDC-1 (Table 2). The physical location of *qLA2-3* was 214.0–225.8 Mb (genetic locus 180.1–196.6 cM), and there was no reported gene related to LA in this region.

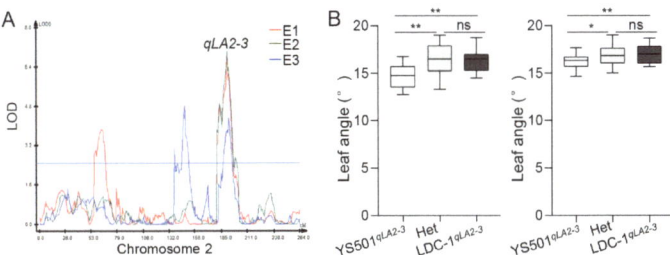

Figure 3. The main QTL locus, *qLA2-3*, for leaf angle detected on chromosome 2: (**A**) Logarithm of odds (LOD) figure for *qLA2-3*. The X-axis is the genetic distance on each chromosome, the Y-axis is the LOD value of the QTL, and the middle horizontal line represents the threshold (LOD = 2.5). (**B**) Allele effects of *qLA2-3* for leaf angle in two plots. * indicates significant difference at the $p < 0.05$ level; ** indicates significant difference at the $p < 0.01$ level; ns indicates no significant difference.

3.3. Validation and Fine Mapping of qLA2-3

To obtain progeny plants with heterozygous genotypes in the major QTL *qLA2-3* region, recombinants with the LDC-1 allele were backcrossed to YS501 and planted into plots after self-crossing. The genetic effect of *qLA2-3* was validated by identifying the genotypes and average LAs of each plant (Figure 3B). Compared to the LDC-1 allele, the YS501 allele at *qLA2-3* reduced the LA.

The major QTL, *qLA2-3*, was located between the initial location markers AX-90526710 and AX-86258763. Within or near the two markers, 12 InDel markers were developed to determine the recombination sites of the genotype (Figure 4). Based on the genotype and phenotype data of six recombinant plant progeny plots, the mapping interval was narrowed to between markers YS54 and YS11. According to reference genome information for B73 RefGen_v4, the physical distance was approximately 3.87 Mb (Figure 4B). Subsequently, the offspring of seven identified recombinants were planted in isolated plots. Based on new InDel markers such as YS4, YS5, A0, LD21, B10 and A9 developed in the interval, the location interval is further narrowed to between markers YS4 and YS5, with a physical distance of approximately 338.46 kb (Figure 4C).

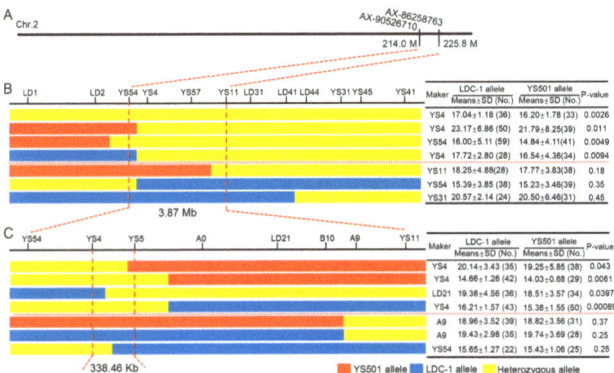

Figure 4. Fine mapping of *qLA2-3* based on recombinant progeny: (**A**) Schematic diagram of the pre-mapping region of *qLA2-3*. (**B**) The genotypes and LA phenotypes of recombinant in the spring of 2022 in Yangzhou. (**C**) The genotypes and LA phenotypes of recombinant in the summer of 2022 in Zhen-jiang. Red, blue and yellow represent homozygous YS501, homozygous LDC-1 and heterozygous YS501/LDC-1 alleles, respectively. Marker, genotype information for statistical analysis; No, the number of individuals in the planting plot. The red dashed line is used to mark the positioning interval. The red dotted line is used to distinguish between significant and insignificant data.

3.4. Evaluation of the Influence of qLA2-3 on Upright Plant Architecture

To verify the allelic effects of *qLA2-3* in the localised region, a pair of near-isogenic lines (*qLA2-3*-NIL^{YS501} and *qLA2-3*-NIL^{LDC-1}) was developed from a RIL that was heterozygous in the 338.46 kb region. Compared to *qLA2-3*-NIL^{LDC-1}, *qLA2-3*-NIL^{YS501} decreased by 0.97°, 2.46°, 3.64° and 4.68° from the first to the fourth leaf above the ear, respectively, and the average LA decreased by 2.51° (Figure 5A,B). There were significant differences in every LA between the two NILs, and alleles from YS501 had a smaller LA.

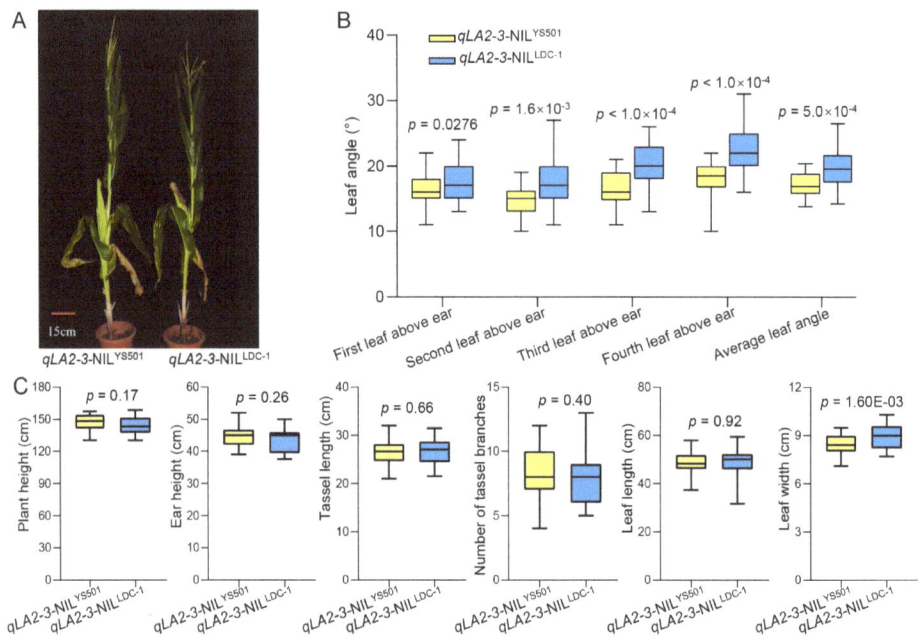

Figure 5. Comparisons of the leaf angle and agronomic traits between *qLA2-3*-NIL^{YS501} and *qLA2-3*-NIL^{LDC-1}. (**A**) The appearance of *qLA2-3*-NIL^{YS501} and *qLA2-3*-NIL^{LDC-1} at the maize filling period. Bar = 15 cm. (**B**) Statistical data of leaf angle size between *qLA2-3*-NIL^{YS501} and *qLA2-3*-NIL^{LDC-1}. (**C**) Effect analysis of *qLA2-3* on architecture traits between *qLA2-3*-NIL^{YS501} and *qLA2-3*-NIL^{LDC-1}. $p < 0.05$ indicates a significant difference; $p < 0.01$ indicates an extremely significant difference.

In addition to LA, plant height, ear height, leaf length and leaf width are relevant traits of upright plant architecture. Six traits, including plant height, ear height, tassel length, number of tassel branches, leaf length and leaf width, were also investigated in the *qLA2-3*-NIL^{YS501} and the *qLA2-3*-NIL^{LDC-1}. Compared to *qLA2-3*-NIL^{LDC-1}, the leaf width of *qLA2-3*-NIL^{YS501} was significantly reduced by 0.51 cm, but no significant difference was detected in the other five plant architecture traits (Figure 5C).

3.5. Predicted Candidate Genes by Genome Re-Sequencing Analysis and Transcriptome Analysis

To narrow down the candidate genes in *qLA2-3*, the genomic variation was compared using re-sequencing data from the two parental lines. The comparison rate of samples ranged from 98.82% to 98.92%; the average coverage depth of reference genomes was 30.38× (LDC-1) to 31.92× (YS501), and the 1× coverage (covering at least one base) was above 91.34%, which could be used for mutation detection. After rigorous filtering, 906 polymorphic SNPs and InDel were detected between the two parents, YS501 and LDC-1, in this candidate region (Figure 6A and Data Set S1). There are 6.53% (59), 5.20% (47), 3.77% (34), 5.87% (53) and 0.11% (1) variants distributed in the exon, upstream, 5′-UTR, 3′-UTR and splicing regions, respectively, except 75.16% in the inter-genic regions (55.04%, 497), introns

(16.28%, 147) and downstream sequences of coding regions (4.10%, 37), which are difficult to cause functional variation.

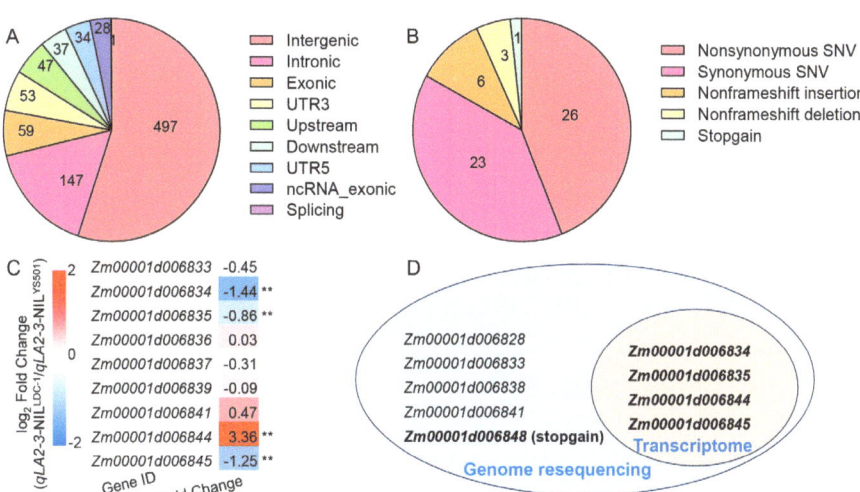

Figure 6. Genomic variation and gene expression changes in the *qLA2-3* candidate regions: The genomic distribution of variations in the leaf angle main QTL *qLA2-3* candidate regions (**A**) and mutation type in exonic regions (**B**). (**C**) Heatmap showing the gene expression changes in *qLA2-3*-NIL^{YS501} and *qLA2-3*-NIL^{LDC-1}. ** indicates significant difference at the $p < 0.01$ level. (**D**) Venn diagrams showing the predicted candidate genes in genome re-sequencing and transcriptome analysis. Nine genes had exons with variable sequences caused by SNPs or Indel in genome re-sequencing, and four of them were differentially expressed genes in transcriptional sequencing.

Among the sequence variations in the coding region, 23, 6 and 3 SNPs or InDel caused synonymous mutations, non-frameshift insertion or non-frameshift deletion, respectively. In addition, 27 SNPs led to non-synonymous mutations, and 1 SNP led to premature termination of transcription due to the presence of a stop codon (stop gain) (Figure 6B and Data Set S2). Within the candidate region, 16 genes were predicted based on the B73 V4 version reference genome of the MaizeGDB website (https://www.maizegdb.org/, accessed on 29 August 2024), including 9 genes with exons of variable sequence caused by SNPs or InDel (Figure 6D and Data Set S2).

To further identify the candidate genes of *qLA2-3*, transcriptome sequencing was performed to compare gene expression changes in the leaf auricle region of *qLA2-3*-NIL^{YS501} and *qLA2-3*-NIL^{LDC-1}. The thresholds of fold change >1.5 and $p < 0.05$ were used to identify significant differentially expressed genes. The results demonstrated that the expression of seven genes was not detected, the expression levels of five genes were not significantly changed, and only four genes (*Zm00001d006834*, *Zm00001d006835*, *Zm00001d006844*, and *Zm00001d006845*) had significant differences in expression levels between the two near-isogenic lines (Figures 6C and S2 and Table S2). The function of *Zm00001d006834* is unknown, but there is a significant difference in its expression level between the two near-isogenic lines, and it may play a role in the regulation of maize leaf angle. *Zm00001d006835* encodes a NUCLEAR FACTOR Y (NF-Y) A-type nuclear transcription factor, which is involved in numerous biological processes, including plant development and photoperiodic regulation [26–28]. *Zm00001d006844* and *Zm00001d006845* encode a serine/threonine–protein kinase and a nuclear export receptor protein (Exportin-t), respectively. The homologous genes in other species play a role in regulating plant development, including leaf development [29–31].

In addition, an SNP emerged in the exon region of *Zm00001d006848* in the LDC-1 inbred line, causing the premature occurrence of a stop codon (Data Set S2). Although it was not detected due to its low expression level (Table S2), it encoded an MYB transcription factor and was listed as a candidate gene because it may regulate the expression of other genes. Considering DNA sequence variation and gene expression differences, *Zm00001d006848*, *Zm00001d006834*, *Zm00001d006835*, *Zm00001d006844* and *Zm00001d006845* were considered as possible candidate genes (Figure 6D).

4. Discussion

The effect of LA on upright plant architecture is closely related to maize planting density [2,3]. To determine the main effect of QTLs controlling LA, RIL populations constructed by maize-inbred lines YS501 and LDC-1, which have a large LA variation, were used to identify QTLs controlling LA. The difference in LA between the two inbred lines was similar (Figure 1A,B), and the correlation of LA between different leaves was high (Figure 2), so the average LA was used for QTL mapping. The genetic architecture of average LA showed that the heritability of LA (89.15%) was high across three environments (Table 1), suggesting that LA was not susceptible to environmental influences. A total of 13 QTLs controlling the size of LA were detected, among which *qLA2-3* was stably expressed in three environments, and the remaining 12 QTLs were only detected in a single environment (Table 2).

Compared to QTL mapping results from other populations, *qLA1-1*, *qLA1-2*, *qLA1-3*, *qLA2-1*, *qLA2-2*, *qLA3-1*, *qLA3-2* and *qLA6-1* identified in this study overlapped with previously reported QTL loci [16,32–34]. *qLA1-1* overlapped with the *qLA-E1-1* interval situated between the markers bnlg439 and bnlg1803 [32]. *qLA1-2* coincided with the *qLA1-2* interval located between the markers umc2112 and umc1703 [16]. *qLA1-3* overlapped with *LA1b* located between the markers bnlg1331 and phi308707 [33]. *qLA2-1* was consistent with the positioning outcomes of B73 × Mo17 population [14]. *qLA2-2* overlapped with the *qFirLA2-1* interval located between the markers umc1065 and umc1637 [34]. *qLA3-1* overlapped with the *qLA-E3-1* interval positioned between the markers umc1394 and umc2257 [32]. *qLA3-2* was in line with the results of *LA3b* [33] and *qFirLA3-2* [34]. *qLA6-1* overlapped with the *qFirLA6-2* interval located between the markers bnlg1732 and umc2162 [34]. These QTLs detected in only one environment in this study were also located in the same location in different populations, suggesting that the QTL between these physical locations is relatively stable. *qLA2-3*, *qLA4-1*, *qLA5-1*, *qLA7-1* and *qLA10-1* were not reported in previous studies, which was due to differences in population genetic background or environment. The contribution rate of *qLA2-3* phenotypic variation reached 12.31%, and it was detected in all three environments, suggesting that this QTL is a stable and major QTL controlling maize LA.

Although many genes regulating maize LA have been cloned by using mutants, these mutants have extremely severe phenotypic variation and are difficult to use for maize genetic improvement [4,5,7–12]. Only a few genes have been cloned through fine cloning strategies, and these alleles have mild LA changes and can be applied to maize production [2,18–21]. The superior allele of the *UPA2* gene has been used to improve maize varieties used in production to achieve higher yields at higher planting densities [2]. However, the genes available for maize LA improvement are still limited, and it is still necessary to explore new gene alleles.

To narrow the interval of *qLA2-3*, InDel markers were developed based on the results of parental whole genome re-sequencing, and the genetic effects of *qLA2-3* were confirmed by genotype and phenotype analysis, which verified the authenticity and reliability of the QTL (Figure 3B). Molecular markers were continuously designed, and new recombinants were screened in the candidate interval. Finally, the target interval was narrowed to between the markers YS4 and YS5 with an interval size of 338.46 kb (B73 RefGen_v4 version). Between the two NILs of *qLA2-3*, the YS501 alleles reduced average LA by 2.51° (Figure 5), suggesting that this QTL may serve as a potential site for improving maize architecture.

Considering that maize is grown as a hybrid, this QTL effect needs to be further verified at the hybridisation stage.

The fine mapping interval contains a total of 16 genes based on the B73 reference genome (Table S2). Four differentially expressed genes were detected in the transcriptomic analysis of the near-isogenic lines $qLA2$-3-NIL^{YS501} and $qLA2$-3-NIL^{LDC-1} (Figure 6 and Table S2). *Zm00001d006835* encodes NF-Y A-type nuclear transcription factors and can co-form heterotrimeric complexes with NF-YB- and NF-YC-type transcription factors [26]. NF-Y transcription factors are involved in a variety of biological processes, such as abiotic stress response, seed germination, root development and photoperiodic regulation [27,28,35,36]. The NF-YC transcription factor ZmNF-YC13 is highly expressed at the base of maize leaves and is involved in the regulation of LA [37]. In addition, overexpression of certain NF-Y (OsNF-YB7, OsHAP3E and ZmNF-YC13) genes will affect the establishment of plant types, resulting in plant dwarf and leaf upright phenotypes [37–39]. In the present study, the expression of *Zm00001d006835* in $qLA2$-3-NIL^{YS501} with small LA was also significantly higher than that of $qLA2$-3-NIL^{LDC-1} with large LA. *Zm00001d006844* encodes a serine/threonine–protein kinase whose homolog in Arabidopsis is involved in chloroplast differentiation of leaves [29]. *Zm00001d006845* encodes a nuclear export receptor protein (Exportin-t), a homologous protein that primarily mediates the tRNA export pathway from the nucleus to the cytoplasm and regulates the development of leaf and inflorescence [30,31]. In addition, the function of *Zm00001d006834* is unknown and cannot be excluded, which may be involved in the regulation of plant LA. Although *Zm00001d006848* is not detected due to low expression (Table S2), the re-sequencing results show that the variation in the exon of *Zm00001d006848* in the LDC-1 inbred line leads to premature termination (Figure 6D and Data Set S2), and it encodes a MYB transcription factor that may affect other genes expression and regulate the angle size of maize leaves. These candidate genes need to be further confirmed through fine mapping and transgenic verification.

5. Conclusions

In the present study, 13 QTLs controlling maize LA were identified using a RIL population, in which *qLA2-3* was a stable and major QTL, explaining up to 12.31% phenotypic variation. InDel markers were developed based on parental re-sequencing data and were used to screen recombinants for fine mapping. The candidate region was finely mapped to the range of 338.46 kb (B73 RefGen_v4 version), which contains 16 predicted genes. The re-sequencing data and transcriptomics suggested that five candidate genes may be involved in the regulation of LA. To further analyse the role of the *qLA2-3* locus in maize genetic improvement, it is necessary to clone the gene through map-based cloning, study its biological functions, and undertake the evaluation of breeding utilisation in future research.

Supplementary Materials: The following supporting information can be downloaded at: https://www.mdpi.com/article/10.3390/agronomy14091978/s1, Figure S1: Frequency distributions of mean leaf angle above ear in the RIL population in E2 and E3 environment. The X-axis represents the range of leaf angle distribution on the ear within the RIL population, the Y-axis represents the frequency and the black line represents the normal distribution fitting curve; Figure S2: The verified results of qRT-PCR in $qLA2$-3-NIL^{YS501} and $qLA2$-3-NIL^{LDC-1}. Values are means ± SE; n = 3 (*, $p < 0.05$; **, $p < 0.01$, Student's t test); Table S1: List of primers; Table S2: Gene expression levels in $qLA2$-3 interval in $qLA2$-3-NIL^{YS501} and $qLA2$-3-NIL^{LDC-1}; Data Set S1: All SNPs and InDel between the LDC-1 and YS501 in $qLA2$-3 candidate region; Data Set S2: SNPs and InDel in the coding region between the LDC-1 and YS501 in $qLA2$-3 candidate region.

Author Contributions: Formal analysis, H.L.; Investigation, Y.H. (Yonghui He), C.W., X.H., Y.H. (Youle Han) and F.L.; Resources, X.Z.; Supervision, Z.Y.; Validation, H.L.; Visualisation, F.L. and X.Z.; Writing—original draft, Y.H. (Yonghui He); Writing—review and editing, Z.Y. All authors have read and agreed to the published version of the manuscript.

Funding: This work was supported by the National Key Research and Development Program (2022YFD1201801-4), the National Natural Science Foundation of China (NSFC; 32101727, 32172054), the JBGS [2021]002 project from the Jiangsu Government, the Natural Science Foundation of Jiangsu Province (Grants No BK20210794), the China Postdoctoral Science Foundation (2022M722701), and A Project Funded by the Priority Academic Program Development of Jiangsu Higher Education Institutions (PAPD).

Data Availability Statement: The National Center for Biotechnology Information Sequence Read Archive (https://www.ncbi.nlm.nih.gov/sra, accessed on 29 August 2024) provides RNA-Seq data and re-sequencing data under accession PRJNA1137054 and PRJNA1137052.

Conflicts of Interest: The authors declare no conflicts of interest.

References

1. Duvick, D.N. The contribution of breeding to yield advances in maize (*Zea mays* L.). In *Advances in Agronomy*; Sparks, D.L., Ed.; Elsevier Academic Press Inc.: San Diego, CA, USA, 2005; Volume 86, pp. 83–145.
2. Tian, J.; Wang, C.; Xia, J.; Wu, L.; Xu, G.; Wu, W.; Li, D.; Qin, W.; Han, X.; Chen, Q.; et al. Teosinte ligule allele narrows plant architecture and enhances high-density maize yields. *Science* **2019**, *365*, 658–664. [CrossRef]
3. Cao, Y.; Zhong, Z.; Wang, H.; Shen, R. Leaf angle: A target of genetic improvement in cereal crops tailored for high-density planting. *Plant Biotechnol. J.* **2022**, *20*, 426–436. [CrossRef] [PubMed]
4. Mantilla-Perez, M.B.; Salas Fernandez, M.G. Differential manipulation of leaf angle throughout the canopy: Current status and prospects. *J. Exp. Bot.* **2017**, *68*, 5699–5717. [CrossRef] [PubMed]
5. Wang, B.; Smith, S.M.; Li, J. Genetic Regulation of Shoot Architecture. *Annu. Rev. Plant Biol.* **2018**, *69*, 437–468. [CrossRef] [PubMed]
6. Tian, J.; Wang, C.; Chen, F.; Qin, W.; Yang, H.; Zhao, S.; Xia, J.; Du, X.; Zhu, Y.; Wu, L.; et al. Maize smart-canopy architecture enhances yield at high densities. *Nature* **2024**, *632*, 576–584. [CrossRef] [PubMed]
7. Bauer, P.; Lubkowitz, M.; Tyers, R.; Nemoto, K.; Meeley, R.B.; Goff, S.A.; Freeling, M. Regulation and a conserved intron sequence of liguleless3/4 knox class-I homeobox genes in grasses. *Planta* **2004**, *219*, 359–368. [CrossRef]
8. Walsh, J.; Waters, C.A.; Freeling, M. The maize gene liguleless2 encodes a basic leucine zipper protein involved in the establishment of the leaf blade–sheath boundary. *Genes Dev.* **1998**, *12*, 208–218. [CrossRef]
9. Moreno, M.A.; Harper, L.C.; Krueger, R.W.; Dellaporta, S.L.; Freeling, M. liguleless1 encodes a nuclear-localized protein required for induction of ligules and auricles during maize leaf organogenesis. *Genes Dev.* **1997**, *11*, 616–628. [CrossRef]
10. Strable, J.; Wallace, J.G.; Unger-Wallace, E.; Briggs, S.; Bradbury, P.J.; Buckler, E.S.; Vollbrecht, E. Maize *YABBY* Genes drooping leaf1 and drooping leaf2 Regulate Plant Architecture. *Plant Cell* **2017**, *29*, 1622–1641. [CrossRef]
11. Lewis, M.W.; Bolduc, N.; Hake, K.; Htike, Y.; Hay, A.; Candela, H.; Hake, S. Gene regulatory interactions at lateral organ boundaries in maize. *Development* **2014**, *141*, 4590–4597. [CrossRef]
12. Makarevitch, I.; Thompson, A.; Muehlbauer, G.J.; Springer, N.M. *Brd1* gene in maize encodes a brassinosteroid C-6 oxidase. *PLoS ONE* **2012**, *7*, e30798. [CrossRef] [PubMed]
13. Best, N.B.; Hartwig, T.; Budka, J.; Fujioka, S.; Johal, G.; Schulz, B.; Dilkes, B.P. *nana plant2* Encodes a Maize Ortholog of the Arabidopsis Brassinosteroid Biosynthesis Gene DWARF1, Identifying Developmental Interactions between Brassinosteroids and Gibberellins. *Plant Physiol.* **2016**, *171*, 2633–2647. [CrossRef] [PubMed]
14. Mickelson, S.M.; Stuber, C.S.; Senior, L.; Kaeppler, S.M. Quantitative Trait Loci Controlling Leaf and Tassel Traits in a B73 × Mo17 Population of Maize. *Crop Sci.* **2002**, *42*, 1902–1909. [CrossRef]
15. Li, C.; Li, Y.; Shi, Y.; Song, Y.; Zhang, D.; Buckler, E.S.; Zhang, Z.; Wang, T.; Li, Y. Genetic control of the leaf angle and leaf orientation value as revealed by ultra-high density maps in three connected maize populations. *PLoS ONE* **2015**, *10*, e0121624. [CrossRef]
16. Ding, J.; Zhang, L.; Chen, J.; Li, X.; Li, Y.; Cheng, H.; Huang, R.; Zhou, B.; Li, Z.; Wang, J.; et al. Genomic Dissection of Leaf Angle in Maize (*Zea mays* L.) Using a Four-Way Cross Mapping Population. *PLoS ONE* **2015**, *10*, e0141619. [CrossRef]
17. Tian, F.; Bradbury, P.J.; Brown, P.J.; Hung, H.; Sun, Q.; Flint-Garcia, S.; Rocheford, T.R.; McMullen, M.D.; Holland, J.B.; Buckler, E.S. Genome-wide association study of leaf architecture in the maize nested association mapping population. *Nat. Genet.* **2011**, *43*, 159–162. [CrossRef]
18. Ren, Z.; Wu, L.; Ku, L.; Wang, H.; Zeng, H.; Su, H.; Wei, L.; Dou, D.; Liu, H.; Cao, Y.; et al. ZmILI1 regulates leaf angle by directly affecting liguleless1 expression in maize. *Plant Biotechnol. J.* **2020**, *18*, 881–883. [CrossRef]
19. Cao, Y.; Zeng, H.; Ku, L.; Ren, Z.; Han, Y.; Su, H.; Dou, D.; Liu, H.; Dong, Y.; Zhu, F.; et al. ZmIBH1-1 regulates plant architecture in maize. *J. Exp. Bot.* **2020**, *71*, 2943–2955. [CrossRef]
20. Zhang, J.; Ku, L.X.; Han, Z.P.; Guo, S.L.; Liu, H.J.; Zhang, Z.Z.; Cao, L.R.; Cui, X.J.; Chen, Y.H. The *ZmCLA4* gene in the qLA4-1 QTL controls leaf angle in maize (*Zea mays* L.). *J. Exp. Bot.* **2014**, *65*, 5063–5076. [CrossRef]
21. Ku, L.X.; Zhang, J.; Guo, S.L.; Liu, H.Y.; Zhao, R.F.; Chen, Y.H. Integrated multiple population analysis of leaf architecture traits in maize (*Zea mays* L.). *J. Exp. Bot.* **2012**, *63*, 261–274. [CrossRef]
22. Ku, L.; Wei, X.; Zhang, S.; Zhang, J.; Guo, S.; Chen, Y. Cloning and characterization of a putative TAC1 ortholog associated with leaf angle in maize (*Zea mays* L.). *PLoS ONE* **2011**, *6*, e20621. [CrossRef] [PubMed]

23. Chen, J.Y.; Lu, F.; Chen, W.Y.; He, Y.H.; Liu, H.H.; Yin, Z.T. Mapping of growth period-related traits in maize. *Jiangsu Agric. Sci* **2022**, *50*, 63–68. (In Chinese) [CrossRef]
24. Liu, H.; Wang, H.; Shao, C.; Han, Y.; He, Y.; Yin, Z. Genetic Architecture of Maize Stalk Diameter and Rind Penetrometer Resistance in a Recombinant Inbred Line Population. *Genes* **2022**, *13*, 579. [CrossRef]
25. Li, H.; Handsaker, B.; Wysoker, A.; Fennell, T.; Ruan, J.; Homer, N.; Marth, G.; Abecasis, G.; Durbin, R.; 1000 Genome Project Data Processing Subgroup. The Sequence Alignment/Map format and SAMtools. *Bioinformatics* **2009**, *25*, 2078–2079. [CrossRef] [PubMed]
26. Petroni, K.; Kumimoto, R.W.; Gnesutta, N.; Calvenzani, V.; Fornari, M.; Tonelli, C.; Holt, B.F.; Mantovani, R. The Promiscuous Life of Plant NUCLEAR FACTOR Y Transcription Factors. *Plant Cell* **2012**, *24*, 4777–4792. [CrossRef]
27. Yang, Y.; Wang, B.; Wang, J.; He, C.; Zhang, D.; Li, P.; Zhang, J.; Li, Z. Transcription factors ZmNF-YA1 and ZmNF-YB16 regulate plant growth and drought tolerance in maize. *Plant Physiol.* **2022**, *190*, 1506–1525. [CrossRef]
28. Su, H.; Chen, Z.; Dong, Y.; Ku, L.; Abou-Elwafa, S.F.; Ren, Z.; Cao, Y.; Dou, D.; Liu, Z.; Liu, H.; et al. Identification of *ZmNF-YC2* and its regulatory network for maize flowering time. *J. Exp. Bot.* **2021**, *72*, 7792–7807. [CrossRef]
29. Lamberti, G.; Gügel, I.L.; Meurer, J.; Soll, J.; Schwenkert, S. The Cytosolic Kinases STY8, STY17, and STY46 Are Involved in Chloroplast Differentiation in Arabidopsis. *Plant Physiol.* **2011**, *157*, 70–85. [CrossRef]
30. Park, M.Y.; Wu, G.; Gonzalez-Sulser, A.; Vaucheret, H.; Poethig, R.S. Nuclear processing and export of microRNAs in *Arabidopsis*. *Proc. Natl. Acad. Sci. USA* **2005**, *102*, 3691–3696. [CrossRef]
31. Hunter, C.A.; Aukerman, M.J.; Sun, H.; Fokina, M.; Poethig, R.S. *PAUSED* Encodes the Arabidopsis Exportin-t Ortholog. *Plant Physiol.* **2003**, *132*, 2135–2143. [CrossRef]
32. Zhang, Z.; Liu, P.; Jiang, F.; Chen, Q.; Zhang, Y.; Wang, X.; Wang, H. QTL mapping for leaf angle and leaf orientation in maize using a four-way cross population. *J. China Agric. Univ.* **2014**, *19*, 7–16. (In Chinese)
33. Lu, M.; Zhou, F.; Xie, C.; Li, M.; Xu, Y.; Warburton, M.; Zhang, S. Construction of a SSR linkage map and mapping of quantitative trait loci (QTL) for leaf angle and leaf orientation with an elite maize hybrid. *Hereditas* **2007**, *29*, 1131–1138. (In Chinese) [CrossRef]
34. Chang, L.; He, K.; Liu, J.; Xue, J. Mapping of QTLs for leaf angle in maize under different environments. *J. Maize Sci.* **2016**, *24*, 49–55. (In Chinese) [CrossRef]
35. Zhang, M.; Zheng, H.; Jin, L.; Xing, L.; Zou, J.; Zhang, L.; Liu, C.; Chu, J.; Xu, M.; Wang, L. miR169o and ZmNF-YA13 act in concert to coordinate the expression of ZmYUC1 that determines seed size and weight in maize kernels. *New Phytologist* **2022**, *235*, 2270–2284. [CrossRef] [PubMed]
36. Su, H.; Cao, Y.; Ku, L.; Yao, W.; Cao, Y.; Ren, Z.; Dou, D.; Wang, H.; Ren, Z.; Liu, H.; et al. Dual functions of ZmNF-YA3 in photoperiod-dependent flowering and abiotic stress responses in maize. *J. Exp. Bot.* **2018**, *69*, 5177–5189. [CrossRef]
37. Mei, X.; Nan, J.; Zhao, Z.; Yao, S.; Wang, W.; Yang, Y.; Bai, Y.; Dong, E.; Liu, C.; Cai, Y. Maize transcription factor ZmNF-YC13 regulates plant architecture. *J. Exp. Bot.* **2021**, *72*, 4757–4772. [CrossRef] [PubMed]
38. Das, S.; Parida, S.K.; Agarwal, P.; Tyagi, A.K. Transcription factor OsNF-YB9 regulates reproductive growth and development in rice. *Planta* **2019**, *250*, 1849–1865. [CrossRef]
39. Ito, Y.; Thirumurugan, T.; Serizawa, A.; Hiratsu, K.; Ohme-Takagi, M.; Kurata, N. Aberrant vegetative and reproductive development by overexpression and lethality by silencing of OsHAP3E in rice. *Plant Sci.* **2011**, *181*, 105–110. [CrossRef]

Disclaimer/Publisher's Note: The statements, opinions and data contained in all publications are solely those of the individual author(s) and contributor(s) and not of MDPI and/or the editor(s). MDPI and/or the editor(s) disclaim responsibility for any injury to people or property resulting from any ideas, methods, instructions or products referred to in the content.

Article

Yield Performance of RD6 Glutinous Rice near Isogenic Lines Evaluated under Field Disease Infection at Northeastern Thailand

Jirapong Yangklang [1], Jirawat Sanitchon [1,*], Jonaliza L. Siangliw [2], Tidarat Monkham [1], Sompong Chankaew [1], Meechai Siangliw [2], Kanyanath Sirithunya [3] and Theerayut Toojinda [2,4]

[1] Department of Agronomy, Faculty of Agriculture, Khon Kaen University, Khon Kaen 40002, Thailand; y_jirapong@kkumail.com (J.Y.); tidamo@kku.ac.th (T.M.); somchan@kku.ac.th (S.C.)
[2] National Center for Genetic Engineering and Biotechnology (BIOTEC), National Science and Technology Development Agency (NSTDA), Thailand Science Park, Phahonyothin, Khlong Nueng, Khlong Luang, Pathum Thani 12120, Thailand; jonaliza.sia@biotec.or.th (J.L.S.); meechai@biotec.or.th (M.S.); theerayut@biotec.or.th (T.T.)
[3] Lampang Campus, Rajamangala University of Technology Lanna, Amphoer Mueang, Lampang 52000, Thailand; psirithunya999@gmail.com
[4] Rice Science Center, Kamphaengsaen Campus, Kasetsart University, Nakhon Pathom 73140, Thailand
* Correspondence: jirawat@kku.ac.th; Tel.: +6681-567-4364

Abstract: RD6, the most popular glutinous rice in Thailand, is high in quality but susceptible to blast and bacterial blight disease. It was thus improved for disease resistance through marker-assisted backcross selection (MAS). The objective of this study was to evaluate the performance of improved near isogenic lines. Eight RD6 rice near isogenic lines (NILs) derived from MAS were selected for evaluation with RD6, a standard susceptible check variety, as well as recurrent parent for a total of nine genotypes. The experiment was conducted during the wet season under six environments at three locations, Khon Kaen, Nong Khai, and Roi Et, which was repeated at two years from 2019 to 2020. Nine genotypes, including eight RD6 rice near isogenic lines (NILs) selected from two in-tandem breeding programs and the standard check variety RD6, were evaluated to select the high-performance new improved lines. The first group, including four NILs G1–G4, was gene pyramiding of blast and BB resistance genes, and the second group, including another four NILs G5–G8, was gene pyramiding of blast resistance and salt tolerance genes. Field disease screening was observed for all environments. Two disease occurrences, blast (leaf blast) and bacterial blight, were found during the rainy season of all environments. The NILs containing blast resistance genes were excellent in gene expression. On the other hand, the improved lines containing the $xa5$ gene were not highly resistant under the severe stress of bacterial blight (Nong Khai 2020). Notwithstanding, G2 was greater among the NILs for yield maintenance than the other genotypes. The agronomic traits of most NILs were the same as RD6. Interestingly, the traits of G2 were different in plant type from RD6, specifically photosensitivity and plant height. Promising rice RD6 NILs with high yield stability, good agronomic traits, and disease resistance were identified in the genotypes G1, G2, and G7. The high yield stability G1 and G7 are recommended for widespread use in rain-fed areas. The G2 is specifically recommended for use in the bacterial blight (BB) disease prone areas.

Keywords: yield stability; specific adaptation; gene pyramiding; marker-assisted backcrossing; rice disease resistance

Citation: Yangklang, J.; Sanitchon, J.; Siangliw, J.L.; Monkham, T.; Chankaew, S.; Siangliw, M.; Sirithunya, K.; Toojinda, T. Yield Performance of RD6 Glutinous Rice near Isogenic Lines Evaluated under Field Disease Infection at Northeastern Thailand. *Agronomy* **2024**, *14*, 1871. https://doi.org/10.3390/agronomy14081871

Academic Editors: Jiezheng Ying, Zhiyong Li and Chaolei Liu

Received: 10 July 2024
Revised: 15 August 2024
Accepted: 18 August 2024
Published: 22 August 2024

Copyright: © 2024 by the authors. Licensee MDPI, Basel, Switzerland. This article is an open access article distributed under the terms and conditions of the Creative Commons Attribution (CC BY) license (https://creativecommons.org/licenses/by/4.0/).

1. Introduction

Glutinous rice (sticky rice) is a type of rice found mainly in Southeast Asia countries (ASEAN), especially along the Me Kong subregion as Thailand, Lao PDR, Myanmar, Vietnam, and Cambodia, including the Southwestern area of China [1]. The unique characteristics of glutinous rice are its opaque appearance (white color grain), low amylose content, and soft texture because of the high amylopectin content, which make it a common staple for food, desserts, and drinks [2]. Amylose content is related to starch, and the loose network structures found in glutinous rice result in a low amylose content [3]. Even though glutinous rice is of low economic value, it is a strategic crop that more than 20 million people consume on a daily basis [1], and it is especially believed to provide basic nutritional security to rice farmers and their families [4]. In Thailand, glutinous rice is omnipresent and mostly consumed in the Northeastern regions where approximately 17% [5] of the land is affected by salt caused by underground salt rock [6]. The glutinous rice variety most famous in Thailand is the RD6 cultivar due to greater cooking and eating quality, like softness, stickiness, and fragrant aroma [7]. The RD6 variety is only one glutinous rice from three cultivars, including Khaw Dwak Mali 105 and RD15, which occupies lowland rain-fed areas the comprise over 70 percent of Northeastern Thailand. The RD6 rice variety has been approved since 1977, which, given its long time in use, is developed from KDML105 by a mutation method using gamma ray treatment. The mutant RD6 variety was changed to be glutinous rice with a low amylose content, but it still presents most of the agronomic traits of the parent, such as plant height, grain shape, and photo sensitivity, while disease susceptibility, including bacterial blight (BB) and blast disease, and saline soil influence plant growth and yield.

RD6 was susceptible to most Thailand local blast isolates collected from a diverse outbreak area [8], which genotype is confirmed by microsatellite franking marker. In addition, blast susceptibility of the RD6 variety was demonstrated under field conditions by natural infection with mixed isolates during 2015–2018 at Khon Kaen province, Thailand. The result indicated that RD6 was susceptible to blast disease [9]. Meanwhile, RD6 is also susceptible to BB disease infection in both greenhouse and field conditions. The greenhouse condition was confirmed during 2015–2016, which evaluated through disease scoring based on lesion length on symptomatic rice leaves at day 14 post-inoculation; the result clearly showed that RD6 varieties were more susceptible to BB disease infection of various isolates compared with resistance varieties [10]. RD6 presented high susceptibility to salinity conditions, which affected plant growth at the seedling stage, confirmed under greenhouse conditions by validating twenty-one day seedlings on a salt solution and artificial soil salinity; the result showed that RD6 was highly susceptible for both conditions with a saline concentration of 12 dS m^{-1} [7]. After a breeding program, similar to each other, salinity evaluation under paddy field conditions at Northeastern Thailand, the yield of a salinity-susceptible rice variety decreased with increasing salinity levels above 8 dS m^{-1} [11].

To solve these constraints, Rice Project, Khon Kaen University located at Northeastern of Thailand has developed the glutinous rice RD6 for disease resistance specially focused on BB, blast resistance, and salt tolerance. The RD6 cultivar was bred for resistance to BB and blast disease with tolerance for saline soil by gene pyramiding into the RD6 parent. RD6 blast resistance was improved by pyramiding various resistance genes on chromosomes 1, 2, 11, and 12 into the RD6 cultivar through marker-assisted selection (MAS); the product has strength RD6 NILs [8]. Continuously, RD6 development for various disease resistance was the destination for BB and blast resistance with gene pyramiding using MAS. RD6 NILs were produced from this research and possessed a total of five genes with four blast and one BB resistance genes; the NILs showed a greater resistance level for blast and BB diseases than its parent at the seedling stage [12]. Furthermore, RD6 gene pyramiding was also developed for blast resistance plus saline soil tolerance, using the developed method as mentioned before. MAS was used for genotypic and phenotypic selection through the breeding steps to deliver promising lines. The experiments were conducted under greenhouse conditions by artificial inoculation for blast via the upland-short row and salt tolerance validated in a salt solution and artificial soil salinity. RD6 NILs were satisfied for both blast resistance and salt tolerance validation, and the NILs were greater than RD6, including other susceptible check varieties [7]. After a breeding program, the plant breeder will investigate yield and its contributed performance of new improved genotypes across several environments (multi-environment trials; MET) to select and recommend a superior genotype for a target environment. This is because genotypes respond differently to changing environments, which causes yield instability [13]. This yield instability is referred to as genotype by environment (GE) interaction. To define the presence of GE interaction in a MET, a data visualization tool, namely genotype by environment interaction (GGE biplot), is applied. GGE biplot is an effective tool to evaluate genotype evaluation and mega-environment identification of each genotype [14]. GGE biplot was successfully used for GE interaction and yield stability in several crops [15–17], including rice [18]. Therefore, in this study, the NILs were selected for field evaluation in Northeastern Thailand. Field evaluation was conducted at multiple locations in the vicinity of the area. The experiment was specially focused on the performance of NILs, including yield, agronomic traits, and field disease resistance by natural disease infection. The objective of this study was to evaluate RD6 NILs for yield performance under pressure of disease infections.

2. Materials and Methods

2.1. Plant Materials

Rice RD6 NILs obtained from MAS were selected from two breeding programs, with entry numbers 1–4 selected from pyramiding genes of blast and bacterial blight resistance. The line numbers 1–3 contained blast resistance genes on 4 chromosomes [*qBLch1, 2, 11, 12*] and *xa5* recessive gene for BB resistance, whereas line number 4 carried only 4 blast resistance genes [*qBLch1* and 2 are major, while *qBLch11* and 12 are minor QTLs]. Other groups with entry numbers 5–8 were selected from pyramiding genes of blast resistance plus salt tolerance; the NILs were common in containing salt tolerance gene *Salt1* but differed in the number of blast resistance genes, as number 5 contained four genes, *qBLch1, 2, 11*, and 12; number 6 contained three genes, *qBLch1, 11*, and 12; number 7 contained two genes, *qBLch11*, and 12; and number 8 contained two genes, *qBLch1*, and 11 [7,12]. The eight NILs were evaluated together with the standard check variety RD6. A total of 9 rice genotypes were evaluated in this experiment (Table 1).

Table 1. Rice genotypes with trait of improvement and genes/QTLs contained.

Entry Number	Line/Variety Name	Trait of Improvement	Genes/QTLs
1	Blast + BLB-36	Blast and bacterial blight resistance	*qBLch1, 2, 11, 12* and *xa5*
2	BC$_2$F$_5$ 2-8-2-26-1	Blast and bacterial blight resistance	*qBLch1, 2, 11, 12* and *xa5*
3	BC$_2$F$_5$ 2-8-2-52	Blast and bacterial blight resistance	*qBLch1, 2, 11, 12* and *xa5*
4	BC$_2$F$_5$ 9-1/15-1-27	Blast and bacterial blight resistance	*qBLch1, 2, 11, 12*
5	BC$_4$F$_5$ 132-12-61	Blast resistance and saline soil tolerance	*Salt1 qBLch1, 2, 11, 12*
6	BC$_4$F$_5$ 132-25-18	Blast resistance and saline soil tolerance	*Salt1 qBLch1, 11, 12*
7	BC$_4$F$_5$ 132-98-87	Blast resistance and saline soil tolerance	*Salt1 qBLch11, 12*
8	BC$_4$F$_5$ 132-174-54	Blast resistance and saline soil tolerance	*Salt1 qBLch1, 11*
9	RD6 (Standard check)	-	No QTLs

2.2. Experimental Sites, Experimental Design, and Field Management

The experiment was conducted in Northeastern Thailand under paddy field conditions in the wet season of the years 2019 and 2020. The experiment site was spread over three diverse locations in each year. The first location was Khon Kaen province (16°45′10″ N 102°37′58″), almost the heartland of the region, which reported natural blast disease infection [9]. The second site was Nong Khai province (17°51′0″ N 102°35′6″ E), the uppermost area of the region close to the Mekong River. This area has high rainfall amounts. The last location was Roi Et province (15°36′33″ N 103°48′1″ E), the largest area for high quality jasmine rice production with an arid and saline territory [19].

The experimental design was laid out by a randomized complete block (RCB) design with 3 replications of each individual site. Seedling preparation of individual sites followed the layout of each site; one thousand seeds per plot were prepared in a labeled zip-lock bag. Seeds were soaked for 24 h, drained, and incubated for another 24 h to induce germination. The nursery soil preparation was performed fifteen days before sowing; the soil was prepared by sieving and mixed well, weighed equally, and put into 30 cm diameter submerged pots and soaked until the sowing date. The 30-day seedlings were transplanted into the field plot with three seedlings per hill with a spacing of 25 × 25 cm between plant and row inside a plot size of 2 × 4 m (8 m^2). Field management, including weeding, pest, and pathogen, was controlled manually. Fertilizer was applied following the recommendation of the Department of Agriculture of Thailand. The fertilizer formula used was 16-20-0 for clay soil at a rate of 50 kg ha^{-1}:62.5 kg ha^{-1}:0 kg ha^{-1} (N:P$_2$O$_5$:K$_2$O) and 16-8-8 for sandy loam soil at a rate of 50 kg ha^{-1}:25 kg ha^{-1}:25 kg ha^{-1} (N:P$_2$O$_5$:K$_2$O).

2.3. Soil Evaluation and Rainfall Measurements

Soil sampling was randomized following soil sampling guidelines [20]. A soil core was used for sampling 30 cm of soil depth. The randomized area was nine points spread around the experimental area of each site. Soil samples were analyzed by the Soil Laboratory of Khon Kaen University; the soil analysis focused on soil texture, EC, and soil nutrients.

The rainfall data used were manually recorded by a rain gauge at each individual experimental site. The rainfall data were an accumulation of twenty-four hours of a daily record for individual rainy days; the rainfall record duration of individual sites started from the transplanting date and finished on the harvest date.

2.4. Morphological Evaluation of Rice Traits

Morphological recording was focused on flowering date, plant height, and tiller number. Flowering date was recorded when the plant of each plot would start flowering; half the amount of flowered plants of each plot were recorded at days to 50% flowering. Plant height of each plot was measured from 4 plants randomly from the plot center at ripening stage, measuring from the ground surface along to the plant tip. Tiller number was recorded at the maximum tiller stage, which averaged 4 hills randomized from the plot center.

2.5. Yield and Yield Components

Yield evaluation data were collected from the data from the harvesting area of the plot inside 4 m^2 or 64 hills without border effects. The harvest date of the individual plot was based on 30 days after the 50% flowering date of each plot. Rice grains were collected together with a label inside a net bag, and seed moisture was decreased by sundry under a translucent air flow greenhouse; fourteen percent seed moistures of individual plots were threshed and processed, and perfectly filled seed of each plot were measured by a digital weighing scale, which converted into weight per hectare.

Yield components were recorded for panicle number, harvest index, seed per panicle, sample seed weight, and seed size. Panicle number was recorded by the average from 4 hills randomized on the center of each plot without border effects; record time was conducted on grain filling stage by manual numerate observation. The harvest index was the harvest of 4 whole plants randomized from the harvest area of each plot collected in a net bag together with a label inside; biological yield and economic yield after sundry were measured by a digital weighing scale (HI = Biological yield/Economic yield; Dry grain yield/Total dry weight). Seed per panicle was collected from 4 perfect panicles randomized from each plot after harvest. The yield, harvest index, and number of seeds per panicle were averaged from filled seed of four panicles collected. Sample seed weight was conducted after yield weighing by preparing 1000 perfectly filled seeds per plot from seed yield, which was measured by a digital weighing scale. Seed size was averaged from the measurement of 10 paddy seeds randomized from the seed yield and measured by a vernier caliper.

2.6. Disease Screening/Index

Prior to conducting NILs evaluation, all improved lines were tested for disease resistance via artificial inoculation against some isolate pathogens. This research was in stage of multilocational trial to confirm their resistance to different combinations of natural pathogens in individual sites. The improved lines were preliminarily evaluated by artificial inoculation and then evaluated under natural infection in this study. Disease evaluation followed the severity scale of the Standard Evaluation System for rice (SES) [21,22]. Disease severity scores were recorded from slight symptom to higher severity from ten hills of each plot. The lowest severity scores were based to analyze and categorize resistance levels (Tables 2 and 3) [23].

Table 2. Scale for blast disease observation.

Scale	Disease Observation
0	No lesions observed
1	Small brown specks of pin-point size or larger brown specks without sporulation center, small roundish to slightly elongated, necrotic gray spots, approximately 1–2 mm in diameter, with a distinct brown margin
3	Lesion type is the same as in scale 2, but a significant number of lesions are on the upper leaves, typical susceptible blast lesions 3 mm or longer, infecting less than 4% of the leaf area
5	Typical blast lesions infecting 4–25% of the leaf area
7	Typical blast lesions infecting 26–50% of the leaf area
	Typical blast lesions infecting 51–75% of the leaf area, and many leaves are dead
9	More than 75% leaf area affected

Table 3. Disease severity score for bacterial blight infection on rice under field conditions.

Disease Score	Section Rate	Lesion Area (%)	Disease Reaction
0	0.00–0.50	0	Highly Resistant (HR)
1	0.51–2.50	1–10	Resistant (R)
3	2.51–4.50	11–30	Moderately Resistant (MR)
5	4.51–6.50	31–50	Moderately Susceptible (MS)
7	6.51–8.50	51–75	Susceptible (S)
9	8.50–9.00	76–100	Highly Susceptible (HS)

2.7. Statistical Analysis

Individual data were analyzed via analysis of variance (ANOVA) following a randomized complete block (RCB) design. Prior to combined analysis, the normality and homogeneity of the data from all environments were verified. A combined analysis was used to analyze the interaction of genotype and environment (GxE interaction). A least significance difference (LSD) at $p < 0.05$ [24] was used for mean comparison for quantitative data. Yield stability and yield response were analyzed consistent with the AMMI model (cluster dendrogram and GGE-biplot); yield stability and yield response analyses were carried out using R software Version 4.0.2 [25,26].

3. Results

3.1. Environment Characterizations

Environmental data were collected for climate data and disease occurrence. The climate data collected were rainfall with distribution, soil properties, and altitude; the data were used together from other experiments, such as the field evaluation of KDML105 introgression lines [27] (Figure 1). Disease occurrence of blast and bacterial blight (BB) was found and recorded.

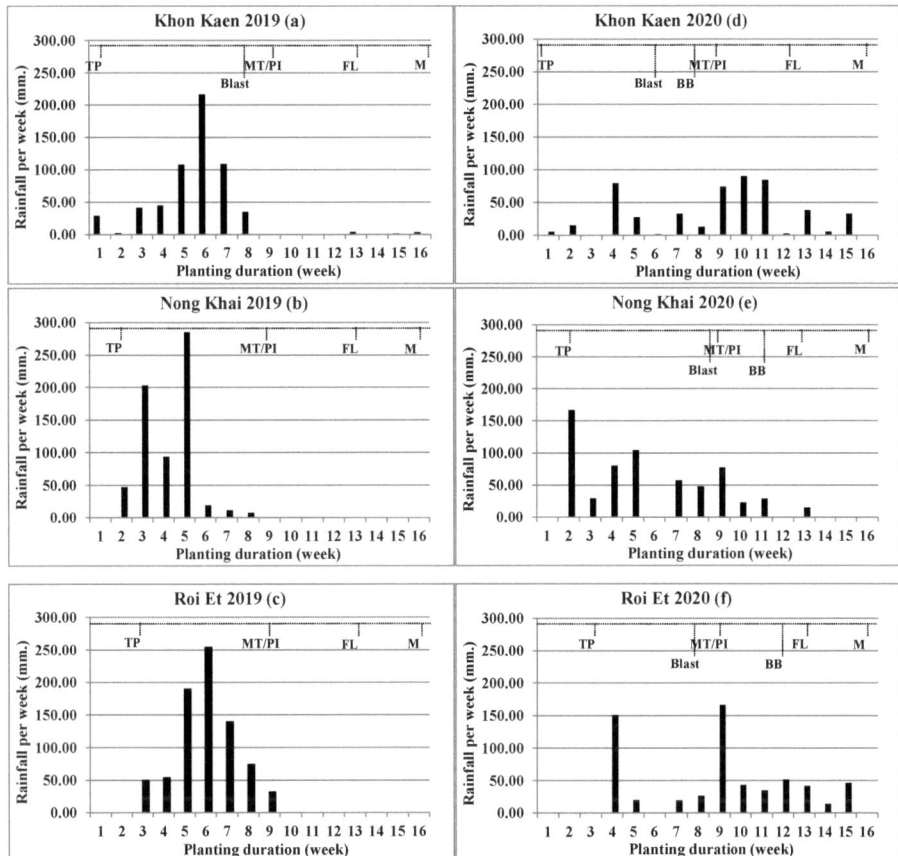

Figure 1. Rainfall (mm) per week during experimental planting of individual six environments: Khon Kaen 2019 (**a**), Nong Khai 2019 (**b**), Roi Et 2019 (**c**), Khon Kaen 2020 (**d**), Nong Khai 2020 (**e**), and Roi Et 2020 (**f**). TP = Transplant, MT/PI = Maximum tiller number/Panicle primodia initiation, FL = Flowering, M = Maturity, Blast = Blast disease screening, BB = Bacterial blight disease screening.

The altitudes of the experimental area ranged from 120 to 220 m above sea level (MASL); the Khon Kaen site was higher than Nong Khai and Roi Et, respectively. Soil textures were different among the locations; Khon Kaen and Roi Et had sandy loam soil, and Nong Khai had clay soil. Rainfall data among the environments were different for the number of rainy days, number of rainy weeks, and total rainfall value (Table 4).

Table 4. Environment data: altitude, soil texture, and rainfall during experiments.

Environment	KK2019	NK2019	RE2019	KK2020	NK2020	RE2020
Altitude (meter)	220.0	200.0	120.0	220.0	200.0	120.0
Number of rainy days	28.0	29.0	33.0	41.0	34.0	33.0
Number of rainy weeks	9.0	7.0	7.0	14.0	11.0	12.0
Total rainfall accumulative	594.2	667.5	740.5	501.1	629.7	613.7
Soil texture	Sandy loam	Clay	Sandy loam	Sandy loam	Clay	Sandy loam

Rainfall of all environments had an equivocal difference among the sites. The rain accumulates ranged from 501.1 to 740.5 mm; the higher rainfall accumulate environment is

Roi Et 2019, followed by Nong Khai 2019, Nong Khai 2020, Roi Et 2020, Khon Kaen 2019, and Khon Kaen 2020, respectively. Overall, rain accumulates for all three locations in 2019 were higher than those in 2020. However, rainfall distributions were not related to rainfall accumulations. In 2020, there was a wider rain distribution than in 2019. The number of rainy days for Khon Kaen 2020 was higher than 2019 with 41 and 28 days, respectively; the number of rainy days for Nong Khai location in 2020 was higher than 2019 with 34 and 29 days, respectively; but for the Roi Et location, there was no statistically significant difference, with 33 rainy days in both years. In addition, the number of rainy weeks was evidently concerned, which, in 2020, all locations had a wider rain distribution than 2019. The result indicated, in year 2019, there was a short rainfall distribution with heavy rain; the rain accumulation per week showed the highest peak at above 200 mm/week, with less rainfall after the maximum tiller (MT) stage. In year 2020, rainfall was substantially widely distributed, with rainfall accumulation per week showing a wide distribution from the start of planting (TP) until the flowering-grain filling stage.

In this study, two disease occurrences, blast and BB, were encountered. Blast disease infections were found at four environments: Khon Kaen 2019, Khon Kaen 2020, Nong Khai 2020, and Roi Et 2020. BB disease infections were found at three environments conducted in 2020: Khon Kaen 2020, Nong Khai 2020, and Roi Et 2020. The comparison among experiment years showed that the year 2020 was found to be greater than the year 2019 in disease occurrence, which was related to longer rainfall distributions, an important factor for both disease spread and occurrence.

3.2. Blast and Bacterial Blight Disease Virulence

Blast disease score of virulence was found at four environments: Khon Kaen 2019, Khon Kaen 2020, Nong Khai 2020, and Roi Et 2020. The grand mean of severity score was found at Khon Kaen 2020, with an average score of 2.69, followed by Khon Kaen 2019, with an average score of 1.89, while Nong Khai 2020 and Roi Et 2020 had average scores of 1.03 and 0.81, respectively. The effect of experimental site on severity score was highly significant ($p \leq 0.01$). The RD6 variety was found to be blast susceptible, as a higher disease score than that of the NILs was observed. All of the blast resistance NILs were great in gene expression for blast resistance, with scores lower than RD6. At the Khon Kaen 2019 site, the NILs were strong in resistance to blast, with a slight score of 1.12–1.50, which is still significant among the NILs due to its difference in genetic introgressed. Meanwhile, the RD6 variety appeared as highly susceptible, with a blast score of 6.42. For the Khon Kaen 2020 site, the rice NILs had scores that ranged from 1.97 to 3.01, which was not different among the lines, while the RD6 variety showed a higher blast score than the NILs at 4.29. At the Nong Khai 2020 site, slight blast disease symptom was shown on the NILs, with scores ranging from 0.57 to 1.17, which slightly differed among the lines, and the RD6 variety showed a higher blast score of 3.13. The last site of blast observation was Roi Et 2020, also having slight blast symptom; the rice NILs had scores ranging from 0.50 to 0.80, which is not different among the lines, and the RD6 variety showed a higher blast score of 2.90 (Table 5).

BB virulence was found at only three sites evaluated in 2020. The BB scoring comparison of each site showed a significant difference ($p \leq 0.05$) for Khon Kaen 2020 and highly significant difference ($p \leq 0.01$) for Nong Khai 2020 and Roi Et 2020. The highest average BB score was found at Nong Khai, with an average score of 4.44, whereas at the inferior sites, Khon Kaen and Roi Et had average scores of 4.17 and 3.30, respectively. Interestingly, the NIL BC_2F_5 2-8-2-26-1 was the lowest in severity score over the other lines at all three sites. In contrast, RD6 was the most susceptible over all entries at all three sites. Moreover, NIL BC_2F_5 2-8-2-26-1 was greater than the other NILs carrying the *xa5* gene for BB resistance. It demonstrated the expression of horizontal BB resistance (Table 5).

Table 5. Disease score for blast and bacterial blight observed from 4 experimental sites in the year 2020.

Entry	Line/Variety Name	Blast Score					BB Score			
		KK2019	KK2020	NK2020	RE2020	Mean	KK2020	NK2020	RE2020	Mean
1	Blast + BLB-36	1.50 b	2.69 bc	0.63 c	0.47 b	1.38	4.47 ab	4.07 d	3.93 ab	3.70
2	BC_2F_5 2-8-2-26-1	1.23 b	2.11 bc	0.90 bc	0.40 b	1.18	3.13 d	3.53 d	2.00 e	2.44
3	BC_2F_5 2-8-2-52	1.47 b	2.44 bc	0.63 c	0.43 b	1.30	3.80 bcd	4.07 d	2.47 de	2.86
4	BC_2F_5 9-1/15-1-27	1.26 b	2.82 bc	0.57 c	0.53 b	1.31	4.20 bcd	4.20 cd	3.40 abc	3.15
5	BC_4F_5 132-12-61	1.43 b	2.43 bc	1.17 b	0.63 b	1.44	3.27 cd	3.53 d	3.27 bcd	2.99
6	BC_4F_5 132-25-18	1.12 b	3.01 b	0.88 bc	0.80 b	1.53	4.47 ab	4.20 cd	3.80 ab	3.90
7	BC_4F_5 132-98-87	1.16 b	1.97 c	0.72 bc	0.60 b	1.14	4.47 ab	5.00 bc	3.67 abc	3.40
8	BC_4F_5 132-174-54	1.45 b	2.42 bc	0.60 c	0.50 b	1.24	4.33 abc	5.13 b	2.93 cd	3.05
9	RD6	6.42 a	4.29 a	3.13 a	2.90 a	4.02	5.40 a	6.20 a	4.20 a	4.07
Average		1.89	2.69	1.03	0.81		4.17	4.44	3.30	
CV (%)		30.76	21.00	27.49	31.19		15.07	10.61	14.28	
F-test		**	**	**	**		*	**	**	

KK 2019 = Khon Kaen 2019, NK 2019 = Nong Khai 2019, RE 2019 = Roi Et 2019, KK 2020 = Khon Kaen 2020, NK 2020 = Nong Khai 2020, RE 2020 = Roi Et 2020. * = significant at $p \leq 0.05$, ** = significant at $p \leq 0.01$. Letter a–e is a range of mean difference between a–e.

3.3. Yield Evaluation on Multi-Environments

Yield evaluation over six environments showed a high significance among sites ($p \leq 0.01$) but not significant between genotypes. Moreover, an interaction between environment and genotype (GxE effect) was detected at $p \leq 0.05$ (Table 6).

Table 6. Combined analysis for grain yield of rice RD6 NILs.

Source	DF	MS	F
Environment (E)	5	9,241,919	18.53 **
Reps with in E	12	498,755	
Genotype (G)	8	350,750	1.79 ns
G*E interaction	40	322,209	1.65 *
Pooled error	96	195,857	
Total	161		

DF = degree of freedom, MS = mean square error, ns = not significant different, * = significant at 95% level, ** = significant at 99% level.

Yield comparison among the environments showed that Khon Kaen 2019 had a higher yield, followed by Nong Khai 2020, Khon Kaen 2020, Roi Et 2020, Roi Et 2019, and Nong Khai 2019, with grain yields of 3535, 3484, 3195, and 3087 kg/ha^{-1}, respectively, which are not different among the five lines group (Table 7). Khon Kaen 2019 is, thus, the greatest in grain yield compared with the other environments.

Table 7. Rice grain yield comparison between environments.

Environment	Grain Yield (kg ha^{-1})
Khon Kaen 2019	4651 a *
Nong Khai 2019	3087 c
Roi Et 2019	3149 bc
Khon Kaen 2020	3484 bc
Nong Khai 2020	3535 b
Roi Et 2020	3195 bc
Mean	3517
CV (%)	20.08
F-test	**

* In the column, alphabet a–z arranges the bigger to smaller data, and means with the same alphabet are not different significant. ** means the F-test are high significant in grain yield (at $p \leq 0.01$)

Environment (E) and genotype (G) interaction for yield of the nine rice genotypes under six evaluation sites was significant ($p \leq 0.05$). Yield comparison among genotype under a particular environment was different. At Khon Kaen 2019, the rice yield among genotype was not significantly different, with an average of 4651 kg ha^{-1} and a yield ranging from 4174 to 4913 kg ha^{-1}. Whereas, Nong Khai 2019, Roi Et 2019, Khon Kaen 2020, Roi Et 2020, and Nong Khai 2020 had location mean yield and yield ranges of 3087, 2490–3419 kg ha^{-1}; 3149 kg ha^{-1}, 2599–3737 kg ha^{-1}; 3484 kg ha^{-1}, 3206–3745 kg ha^{-1}; 3195 kg ha^{-1} 2669–3467 kg ha^{-1}, and 3535 kg ha^{-1}, 2842–4771 kg ha^{-1}, respectively. Yield among genotype in these environments was not significantly different. Except for the Nong Khai 2020 environment, rice yield among genotype was significantly different at $p \leq 0.05$ level. The NIL BC$_2$F$_5$ 2-8-2-26-1 gave a yield of 4771 kg ha^{-1} that was significantly different from the other genotypes. Those genotypes were not significantly different among themselves. In addition, the environment Nong Khai 2020 was found to have a considerable yield range value by yield range of 1929 kg ha^{-1}, and its range was higher than the other five environments (Table 8).

Table 8. Grain yield of rice 8 RD6 NILs comparison with original RD6 at individual environments.

Entry	Lines/Variety	Grain Yield (kg ha^{-1})						
		KK2019	NK2019	RE2019	KK2020	NK2020	RE2020	Combined
1	Blast + BLB-36	4484	2490	3228	3336	3343 b	3241	3353
2	BC$_2$F$_5$ 2-8-2-26-1	4174	3419	2900	3637	4771 a	3138	3673
3	BC$_2$F$_5$ 2-8-2-52	4718	3083	3212	3373	3432 b	3395	3536
4	BC$_2$F$_5$ 9-1/15-1-27	4745	2956	3317	3745	3519 b	2918	3533
5	BC$_4$F$_5$ 132-12-61	4910	3359	3453	3308	3623 b	3268	3654
6	BC$_4$F$_5$ 132-25-18	4827	3150	2599	3560	3449 b	3247	3472
7	BC$_4$F$_5$ 132-98-87	4699	3323	3187	3555	3204 b	2669	3440
8	BC$_4$F$_5$ 132-174-54	4913	2813	3737	3640	3630 b	3415	3691
9	RD6 (standard check)	4389	3186	2709	3206	2842 b	3467	3300
	Mean	4651	3087	3149	3484	3535	3195	3517
	CV (%)	8.88	13.61	14.3	9.14	15.12	15.28	12.58
	F-test	ns	ns	ns	ns	*	ns	ns
	Range value	739	929	1138	539	1929	798	391

KK2019 = Khon Kaen 2019, NK2019 = Nong Khai 2019, RE2019 = Roi Et 2019, KK2020 = Khon Kaen 2020, NK2020 = Nong Khai 2020, RE2020 = Roi Et 2020, ns = not significant different, * = significant at 95% level. Letter a and b means the two means are significant different.

3.4. Yield Stability and Adoption

Yield stability analyzed by GGE-biplot showed stability of the genotype G1-G8 compared with RD6 under environments E1-E6 (arrow lines). GGE-biplot explained for PC1 with 46.4% and PC2 with 25.9%. The stability map shows that the yields of NILs G1

(Blast + BLB-36), G3 (BC$_2$F$_5$ 2-8-2-52), G4 (BC$_2$F$_5$ 9-1/15-1-27), G5 (BC$_4$F$_5$ 132-12-61), G6 (BC$_4$F$_5$ 132-25-18), and G7 (BC$_4$F$_5$ 132-98-87) were of greater stability than RD6, and NILs G2 (BC$_2$F$_5$ 2-8-2-26-1) and G8 (BC$_4$F$_5$ 132-174-54) were of low stability. However, specific adaptation of crop yield was showing for NILs adaptation to various environments: G2 (BC$_2$F$_5$ 2-8-2-26-1) was high yield for environment E2 (Nong Khai 2019), E4 (Khon Kaen 2020), and E5 (Nong Khai 2020); G8 (BC$_4$F$_5$ 132-174-54) had greater yield for environment E1 (Khon Kaen 2019) and E3 (Roi Et 2019); and RD6 had greater yield for environment E6 (Roi Et 2020) (Figure 2).

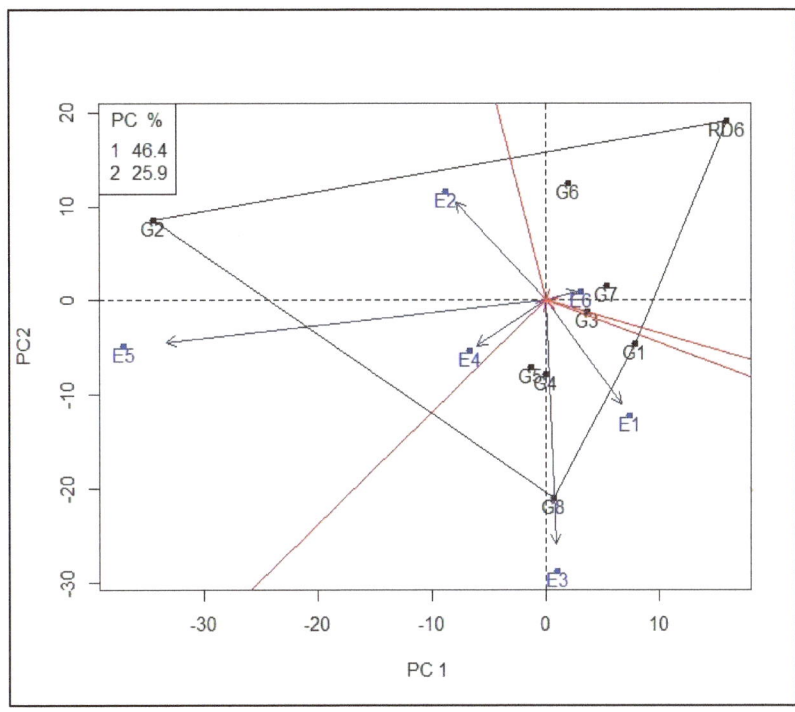

Environment
E1 = Khon Kaen 2019
E2 = Nong Khai 2019
E3 = Roi Et 2019
E4 = Khon Kaen 2020
E5 = Nong Khai 2020
E6 = Roi Et 2020

Genotype
G1 = Blast + BLB-36
G2 = BC$_2$F$_5$ 2-8-2-26-1
G3 = BC$_2$F$_5$ 2-8-2-52
G4 = BC$_2$F$_5$ 9-1/15-1-27
G5 = BC$_4$F$_5$ 132-12-61
G6 = BC$_4$F$_5$ 132-25-18
G7 = BC$_4$F$_5$ 132-98-87
G8 = BC$_4$F$_5$ 132-174-54
RD6 = RD6 variety

Figure 2. Yield stability analysis by GGE-biplot.

3.5. Yield Components and Agronomic Traits Evaluation

Yield components and agronomic traits means of individual genotypes for comparison were averaged over six environments. Plant height (PH) average among genotype was highly significant ($p \leq 0.01$). RD6 was the highest with 161 cm of plant height, and G2 (BC$_2$F$_5$ 2-8-2-26-1) was the lowest plant height with 142 cm; the plant heights of the other NILs were similar among genotypes. Days to flowering (DTF) was highly significant

($p \leq 0.01$); almost all genotypes were not significant, with the range of 106–110 days, but only one of its G2 (BC$_2$F$_5$ 2-8-2-26-1) was shorter in day to flowering with an average 99 days. The tiller number (TN) and panicle number (PN) were both not significant among genotypes, with means of 10.4 and 10.2 for tiller number and panicle number, respectively. Seed per panicle (SP) was not significantly different, with a mean of 164 seeds per panicle. Thousand seed weight (1000 SW) was significant at 95%; G6 (BC$_4$F$_5$ 132-25-18) was higher in seed weight, but the other NILs did not have a different seed weight than that of RD6. Harvest index (HI) was highly significant at $p \leq 0.01$; all of the NILs had a higher harvest index than that of RD6. The NILs' harvest index ranged between 0.37 and 0.40, whereas RD6 was 0.33 in harvest index (Table 9).

Table 9. Agronomic traits of rice genotypes.

Entry	Variety	PH (cm.)	DTF (day)	TN	PN	SP	1000 SW	HI
1	Blast + BLB-36	151 abc	110 a	9.8	9.6	176	26.3 bc	0.38 a
2	BC$_2$F$_5$ 2-8-2-26-1	142 c	99 b	10.7	10.5	155	25.5 c	0.40 a
3	BC$_2$F$_5$ 2-8-2-52	152 abc	109 a	10.6	10.4	168	25.6 c	0.37 a
4	BC$_2$F$_5$ 9-1/15-1-27	148 bc	109 a	10.4	10.3	175	26.5 abc	0.38 a
5	BC$_4$F$_5$ 132-12-61	154 ab	107 a	10.8	10.6	165	26.4 abc	0.37 a
6	BC$_4$F$_5$ 132-25-18	158 ab	108 a	10.4	10.2	153	27.5 a	0.37 a
7	BC$_4$F$_5$ 132-98-87	153 ab	107 a	10.0	9.8	172	26.2 bc	0.37 a
8	BC$_4$F$_5$ 132-174-54	158 ab	106 a	10.5	10.1	160	27.2 ab	0.38 a
9	RD6 (standard check)	161 a	110 a	10.7	10.5	155	26.2 bc	0.33 b
	Mean	153	107	10.4	10.2	164	26.36	0.37
	F-test	**	**	ns	ns	ns	*	**
	CV%	11.06	6.54	22.45	21.79	20.47	6.63	14.88

PH = Plant height, DTF = Day to flowering, TN = Tiller number, PN = Panicle number, SP = Seed per panicle, 1000 SW = 1000 seed weight, HI = Harvest index. * = F-test is significant at $p \leq 0.05$, ** = F-test is high significant at $p \leq 0.01$, ns = F-test is not significant.

4. Discussion

4.1. Disease Occurrence under Various Environments

Differentiation among environments played an important role in these experiments and is inclusive of disease infection and rain distribution. Infections of the two diseases, blast and BB, were found (Table 5). Disease occurrence was more frequent in the year 2020 than 2019. In the year 2019, only Khon Kaen had blast occurrence. Meanwhile, on some experiment sites, disease infection frequency in year 2020 was more than 2019; in 2019, blast disease infection was found at only one site, Khon Kaen (Table 4). Meanwhile, the year 2020 had both disease infections at all three experimental sites. Alternatively, blast is an airborne disease, while BB is waterborne; this makes blast easier to spread than BB disease during the period of short rain distribution. Blast was found across four sites, while BB was found at three sites (Figure 1). Blast grows rapidly under cloudy days with intermittent rainfall development, and the disease spreads at temperatures 25–30 °C and relative humidity above 90% [28,29]. In some environments, such as Nong Khai (2020), the most resistant line did not show a strong resistance to BB (Table 5). This might be caused by some isolated pathogens that are still virulent. Identification of a new resistant source for further gene pyramiding is the prospect to achieve more durable resistance.

The experimental site for disease screening evaluation was specifically related to disease frequency occurrence and severity. The only experimental site where blast disease was found for both experimental years was the Khon Kaen location, which was appropriate for natural blast disease screening, and serious BB disease infection severity was significant for Nong Khai because of a higher average disease score (Table 5). Moreover, natural disease infection screenings were needed to be conducted under various locations and repeated for more than two years, and climate changes were affected for disease occurrence. Normally, field disease evaluation was based on natural infection under an uncontrollable environment similar to field evaluation of RD6 introgression lines for yield performance, blast, and BB resistance across five environments and found blast at four environments, whereas neck blast and BB occurred at three environments and one environment, respectively [9].

4.2. Adaptability of Rice Genotypes

Environment affected the rice genotype significantly. In terms of yield, the rice genotypes were not significantly different when compared among the NILs and RD6 (Khon Kaen 2019, Nong Khai 2019, Roi Et 2019, Khon Kaen 2020, and Roi Et 2020). The result indicated that the breeding lines were close to the recurrent parent (RD6). Furthermore, comparing among the NILs, no significant difference was found (Table 5) The result was a resemblance with the study of Aung Nan [9], which evaluated RD6 near isogenic blast and BB resistance lines under various environments in 2015 to 2018 and found that the grain yield of rice genotypes had no significant difference for each of five environments when compared with the RD6 standard check variety, in which the environments were conducted under pressure of disease affected as leaf blast, neck blast, and BB; disease screenings were clearly classified for blast resistance and susceptible genotypes but not clearly among BB resistance and susceptible genotype, and the relevant BB resistant *xa5* recessive gene was excellent for horizontal resistance, which was influenced by the environments [30].

In general, proper utilization of rice genotypes needs to focus on production area. An environment with low disease infection needs to be recommended for the high yield stability rice genotype. On the other hand, a specific adaptation genotype would need to be recommended for a severe disease infection area. Yield stability of most genotypes, including G1, G3, G4, G5, G6, and G7, was higher than that of RD6. In addition, under a severe disease environment such as Nong Khai 2020, severe BB infection (Table 5) was shown after entering a reproductive stage (Figure 1e); the yield of most genotypes decreased, while G2, the BB resistant line, had a higher yield. The result resembled that of Ansari [31] who studied yield loss assessment due to BB at different resistance levels. The susceptible variety had a lower yield, which was inoculated during the reproductive stage. BB affected plant leaves, photosynthesis, and panicle fertilization [32,33].

4.3. Disease Resistance Levels of Rice RD6 NILs

This study focused on the experimentation of two disease infections, which included leaf blast and bacterial blight. Disease resistance levels of rice genotypes were varied depending on the resistance gene and gene frequency. The blast disease resistance level among resistance and susceptible genotypes was compared; all of the evaluated environments showed significant differences, and all of the NILs that contained the blast resistance gene had greater blast resistance than the RD6 standard check variety (Table 5). Meanwhile, BB disease resistance levels among the genotypes were significant but not strongly different; the NILs that contained the single BB resistance gene *xa5* were not highly resistant because the recessive gene *xa5* was excellent for horizontal resistance [30]. Moreover, to increase BB resistance levels into rice NILs, other resistant genes need to be added. The dominance genes *Xa13* and *Xa21* were found to increase the resistance level of the rice genotype, as pyramiding lines harboring multiple BB-resistant genes *Xa21* + *xa13* + *xa5* had higher resistance than those that carried the gene combination of *Xa21* + *xa13*, *Xa21* + *xa5*, *xa13* + *xa5*, and susceptible variety control [34–37].

4.4. Plant Phenotype of Rice RD6 NILs

The RD6 rice NILs mostly presented a phenotype approximately with the RD6 parental variety, including photosensitivity, plant height, tiller number, and panicle number, which was a result of the backcrossing method. This study has two different phenotype groups of BC_2 and BC_4. The group BC_4 showed a phenotype, including yield, yield stability, and agronomic traits, that was closer to RD6 than BC_2. In general, for conveying qualitative traits such as disease resistance, the backcross approach has been widely used for influence over the genetic variance in the segregating population [38]; as the backcrossing approach was used to improve the RD6 variety for blast resistance and salt tolerance through maker-assisted selection, this study had improvements into four cycles (BC$_4$), which the plant phenotype of progeny were not significant with recurrence parent and disease resistance levels were not significant with donor parents [7]. On the other hand, BC$_2$ group was a few significant as G2 showed non-photosensitivity and short plant height. These segregation traits were affected from the donor parent and possible to express on low backcrossing round NILs. Notably, G2 is mostly the same as RD6, except for blast and BB resistance. Furthermore, it is shorter in height and not sensitive to photoperiod. Shorter plant height makes it less in lodging during the harvest period and results in a decreased yield. Moreover, non-photosensitivity makes it able to avoid drought and flooding constraints by fixing suitable planting date as well as harvest date. These are benefits from marker-assisted backcrossing together with modification of conventional selection to deliver a new promising improved line for further utilizing among natural disasters.

5. Conclusions

Rice RD6 NILs evaluation was performed at three locations replicated by two years from 2019 to 2020. NILs promising lines selection for utilization on Northeastern of Thailand was the focus for agronomic traits and yield adaptability under pressure of natural disease infections. The high yield stability genotypes G1, G3, G4, G5, G6, and G7 had greater yield stability than RD6 and are recommended for wider utilization. On the other hand, two genotypes, G2 and G8, had low yield stability; G2 had a greater yield maintained under severe BB disease infection that was appropriate for a severe BB spread area. In contrast, G8 had higher yield under low disease pressure. In addition, G2 expressed segregation traits in non-photosensitivity and lower plant height that could be recommended for an irrigation area or to replace RD6 for early maturity when subjected to drought and flooding avoidance. Furthermore, all of the rice genotypes need more pyramiding BB resistance genes to increase the durable resistance.

Author Contributions: Conceptualization, S.C. and J.S.; methodology, J.Y., S.C. and J.S.; validation, J.Y., T.M., J.L.S., M.S., K.S. and T.T.; writing—original draft preparation, J.Y. and S.C.; writing—review and editing, T.M. and S.C.; supervision, J.S.; project administration, J.S.; funding acquisition, J.Y. and J.S. All authors have read and agreed to the published version of the manuscript.

Funding: Thailand Graduate Institute of Science and Technology (TGIST) from the National Science and Technology Development Agency (NSTDA) for studies scholarship (Contract no. SCA-CO-2562-9717-TH) as well as the research fund by NSTDA grant number P1950205.

Data Availability Statement: The original contributions presented in the study are included in the article, further inquiries can be directed to the corresponding author.

Acknowledgments: This research was supported by the Plant Breeding Research Centre for Sustainable Agriculture, Khon Kaen University, Khon Kaen, Thailand. We wish to express our thanks to Thailand Graduate Institute of Science and Technology (TGIST) for studies scholarship (Contract no. SCA-CO-2562-9717-TH).

Conflicts of Interest: The authors declare no conflicts of interest.

References

1. Sattaka, P. Geographical distribution of glutinous rice in the greater Mekong sub-region. *J. Mekong Soc.* **2016**, *12*, 27–48. [CrossRef]
2. Lian, X.; Wang, C.; Zhang, K.; Li, L. The retrogradation properties of glutinous rice and buckwheat starches as observed with FT-IR, 13C NMR and DSC. *Int. J. Biol. Macromol.* **2014**, *64*, 288–293. [CrossRef]
3. Tian, J.; Qin, L.; Zeng, X.; Ge, P.; Fan, J.; Zhu, Y. The role of amylose in gel forming of rice flour. *Foods* **2023**, *12*, 1210. [CrossRef]
4. Naivikul, O. Enhancing food (glutinous rice) and nutritional security through products development and diversification. In Proceedings of the Consultative Meeting on GMS Cooperative Research and Networking on Food Security, Food Safety and Nutritional Security (focusing on Glutinous Rice, Good Agriculture Practice and Food Quality Assurance), Kasetsart University Chalermphrakiet, Sakon Nakhon Province Campus, Sakon Nakhon, Thailand, 4–5 October 2013; pp. 4–5.
5. Sahunalu, P. Rehabilitation of salt affected lands in Northeast Thailand. *Tropics* **2003**, *13*, 39–51. [CrossRef]

6. Gardner, L.S. *Salt Resources of Thailand. Report of Investigation-Thailand, Department of Mineral Resources*; Thailand Department of Mineral Resources: Bangkok, Thailand, 1967; Volume 11, pp. 1–100. Available online: https://pubs.usgs.gov/publication/70205789 (accessed on 17 August 2024).
7. Thanasilungura, K.; Kranto, S.; Monkham, T.; Chankaew, S.; Sanitchon, J. Improvement of a RD6 variety for blast resistance and salt tolerance through maker-assisted backcrossing. *Agronomy* **2020**, *10*, 1118. [CrossRef]
8. Suwannual, T.; Chankaew, S.; Monkham, T.; Saksirirat, W.; Sanitchon, J. Pyramiding of four blast resistance QTLs into Thai rice cultivar RD6 through marker-assisted selection. *Czech J. Genet. Plant Breed.* **2017**, *53*, 1–8. [CrossRef]
9. Aung Nan, M.S.; Janto, J.; Sribunrueang, A.; Monkham, T.; Sanitchon, J.; Chankaew, S. Field evaluation of RD6 introgression lines for yield performance, blast, bacterial blight resistance, and cooking and eating qualities. *Agronomy* **2019**, *9*, 825. [CrossRef]
10. Chumpol, A.; Monkham, T.; Saepaisan, S.; Sanitchon, J.; Falab, S.; Chankaew, S. Phenotypic broad spectrum of bacterial blight disease resistance from Thai indigenous upland rice germplasm implies novel genetic resource for breeding program. *Agron. J.* **2022**, *12*, 1930. [CrossRef]
11. Yang, Y.; Ye, R.; Srisutham, M.; Nontasri, T.; Sritumboon, S.; Maki, M.; Yoshida, K.; Oki, K.; Homma, K. Rice production in farmer fields in soil salinity classified areas in Khon Kaen, Northeast Thailand. *Sustainability* **2022**, *14*, 9873. [CrossRef]
12. Pinta, W.; Toojinda, T.; Sanitchon, J. Pyramiding of blast and bacterial leaf blight resistance gene into rice cultivar RD6 using marker assisted selection. *Afr. J. Biotechnol.* **2013**, *12*, 4432–4438. [CrossRef]
13. Oladosu, Y.; Rafii, M.Y.; Abdullaha, N.; Hussin, G.; Ramlie, A.; Rahimf, H.A.; Miah, G.; Usman, M. Principle and application of plant mutagenesis in crop improvement: A review. *Biotechnol. Biotechnol. Equip.* **2016**, *30*, 1–16. [CrossRef]
14. Yan, W.; Hunt, L.A.; Sheng, Q.; Szlavnics, Z. Cultivar evaluation and mega-environment investigation based on the GGE biplot. *Crop Sci.* **2000**, *40*, 597–605. [CrossRef]
15. Alwala, S.; Kwolek, T.; McPherson, M.; Pellow, J.; Meyer, D. A comprehensive comparison between Eberhart and Russell joint regression and GGE biplot analyses to identify stable and high yielding maize hybrids. *Field Crop. Res.* **2010**, *119*, 225–230. [CrossRef]
16. Jalata, Z. GGE-Biplot Analysis of Multi-Environment Yield Trials of Barley (*Hordeium vulgare* L.) Genotypes in Southeastern Ethiopia Highlands. *Int. J. Plant Breed. Genet.* **2010**, *5*, 59–75. [CrossRef]
17. Zurweller, B.A.; Xavier, A.; Tillman, B.L.; Mahan, J.R.; Payton, P.R.; Puppala, N.; Rowland, D.L. Pod yield performance and stability of peanut genotypes under differing soil water and regional conditions. *J. Crop Improv.* **2018**, *32*, 532–551. [CrossRef]
18. Abdelrahman, M.; Alharbi, K.; El-Denary, M.E.; Abd El-Megeed, T.; Naeem, E.-S.; Monir, S.; Al-Shaye, N.A.; Ammar, M.H.; Attia, K.; Dora, S.A.; et al. Detection of superior rice genotypes and yield stability under different nitrogen levels using AMMI model and stability statistics. *Plants* **2022**, *11*, 2775. [CrossRef]
19. Kong-ngern, K.; Buaphan, T.; Tulaphitak, D.; Phuvongpha, N.; Wongpakonkul, S.; Threerakulpisut, P. Yield, yield components, soil minerals and aroma of KDML 105 rice in Tungkularonghai, Roi-Et, Thailand. *Int. Sch. Sci. Res. Innov.* **2011**, *5*, 204–209. Available online: https://zenodo.org/record/1082841/files/14365.pdf (accessed on 17 August 2024).
20. Jason, P.A. *Soil Sampling Guidelines*; Purdue Extension: West Lafayette, IN, USA, 2018; Volume 368, pp. 1–6. Available online: https://extension.purdue.edu/county/vanderburgh/_media/collecting-soil-samples-for-testing-ho-71-w.pdf (accessed on 17 August 2024).
21. International Rice Research Institute (IRRI). *Standard Evaluation System for Rice (SES)*, 5th ed.; International Rice Research Institute: Manila, Philippines, 2013; pp. 12–17. Available online: https://www.scribd.com/document/333585255/SES-5th-Edition (accessed on 17 August 2024).
22. Banerjee, A.; Roya, S.; Bagb, M.K.; Bhagata, S.; Karb, M.K.; Mandala, N.P.; Mukherjeeb, A.K.; Maiti, D. A survey of bacterial blight (*Xanthomonas oryzae* pv. *oryzae*) resistance in rice germplasm from eastern and northeastern India using molecular markers. *Crop Prot.* **2018**, *122*, 168–176. [CrossRef]
23. Chaudhary, R.C. Internationalization of elite germplasm for farmers: Collaborative mechanisms to enhance evaluation of rice genetic resources. *New Approaches Improv. Use Plant Genet. Resour.* **1996**, *26*, 1–26.
24. Gomez, K.A.; Gomez, A.A. *Statistical Procedures for Agricultural Research*, 2nd ed.; John Wiley and Sons: New York, NY, USA, 1984; 680p. Available online: https://books.google.co.th/books?hl=en&lr=&id=PVN7_XRhpdUC&oi=fnd&pg=PA1&ots=Ht559inuq5&sig=agimlLxykrXUsP4EA8gmM23AKXU&redir_esc=y#v=onepage&q&f=false (accessed on 17 August 2024).
25. Onofi, A.; Ciricfolo, E. Using R to perform the AMMI analysis on agriculture variety trials. *R News* **2007**, *7*, 14–19.
26. R Development Core Team. *R: A Language and Environment for Statistical Computing*; R Foundation for Statistical Computing: Vienna, Austria, 2020. Available online: www.R-project.org (accessed on 24 April 2020).
27. Yangklang, J.; Sanitchon, J.; Siangliw, J.L.; Monkham, T.; Chankaew, S.; Siangliw, M.; Sirithunya, K.; Toojinda, T. Yield performance evaluation of KDML105 rice introgression lines, developed through maker-assisted selection, under BB and blast disease infection in Northeashern Thailand. *HSOA J. Agron. Agric. Sci.* **2023**, *6*, 1–9. [CrossRef]
28. Nagaraja, V.; Matucci-Cerinic, M.; Furst, D.E.; Kuwana, M.; Allanore, Y.; Denton, C.P.; Raghu, G.; Mclaughlin, V.; Rao, P.S.; Seibold, J.R.; et al. Current and future outlook on disease modification and defining low disease activity in systemic sclerosis. *Arthritis Rheumatol.* **2020**, *72*, 1049–1058. [CrossRef]
29. Magarey, R.D.; Sutton, T.B.; Thayer, C.L. A simple generic infection model for foliar fungal plant pathogens. *Phytopathology* **2005**, *95*, 92–100. [CrossRef]

30. Han, J.; Xia, Z.; Liu, P.; Li, C.; Wang, Y.; Guo, L.; Jiang, G.; Zhai, W. TALEN-based editing of TFIIAy5 changes rice response to *Xanthomonas oryzae* pv. *Oryzae*. *Sci. Rep.* **2020**, *10*, 2036. [CrossRef] [PubMed]
31. Ansari, T.H.; Ahmed, M.; Ara, A.; Khan, M.A.I.K.; Mian, M.S.; Zahan, Q.S.A.; Tomita, M. Yield loss assessment of rice due bacterial blight at different resistance level. *Bangladesh J. Plant Pathol.* **2018**, *34*, 71–76.
32. Reddy, A.P.K.; Mackenzie, D.R.; Rouse, D.I.; Roa, A.V. Relationship of bacterial leaf blight severity to grain yield of rice. *Phytopathology* **1979**, *69*, 967–969. Available online: https://www.apsnet.org/publications/phytopathology/backissues/Documents/1979Articles/Phyto69n09_967.pdf (accessed on 17 August 2024). [CrossRef]
33. Elings, A.; Rossing, W.A.H.; Van Der Werf, W. Virtual lesion extension: A measure to quantify the effects of bacterial blight on rice leaf CO_2 exchange. *Phytopathology* **1999**, *89*, 789–795. [CrossRef]
34. Huang, N.; Angeles, E.R.; Domingo, J.; Magpantay, G.; Singh, S.; Zhang, G.; Kumaravadivel, N.; Bennett, J.; Khush, G.S. Pyramiding of bacterial blight resistance genes in rice: Marker assisted selection using RFLP and PCR. *Theor. Appl. Genet.* **1997**, *95*, 313–320. [CrossRef]
35. Petpisit, V.; Khush, G.S.; Kauffman, H.E. Inheritance of resistance to bacterial blight in rice. *Crop Sci.* **1977**, *17*, 551–554. [CrossRef]
36. Huang, S.; Antony, G.; Li, T.; Liu, B.; Obasa, K.; Yang, B.; White, F.F. The broadly effective recessive resistance gene *xa5* of rice is a virulence effector-dependent quantitative trait for bacterial blight. *Plant J.* **2016**, *86*, 186–194. [CrossRef]
37. Pradhan, K.C.; Mohapatra, S.R.B.; Nayak, D.K.; Pandit, E.; Jena, B.K.; Sangeeta, S.; Pradhan, A.; Samal, A.; Meher, J.; Behera, L.; et al. Incorporation of two bacterial blight resistance genes into the popular rice variety, Ranidhan through marker-assisted breeding. *Agriculture* **2022**, *12*, 1287. [CrossRef]
38. Léon, J. An overview of backcross breeding in plants. *Int. Sch. J.* **2021**, *8*, 1.

Disclaimer/Publisher's Note: The statements, opinions and data contained in all publications are solely those of the individual author(s) and contributor(s) and not of MDPI and/or the editor(s). MDPI and/or the editor(s) disclaim responsibility for any injury to people or property resulting from any ideas, methods, instructions or products referred to in the content.

Article

Transcriptome-Wide Genetic Variations in the Legume Genus *Leucaena* for Fingerprinting and Breeding

Yong Han [1,2,*], Alexander Abair [3], Julian van der Zanden [1], Madhugiri Nageswara-Rao [4], Saipriyaa Purushotham Vasan [2], Roopali Bhoite [1], Marieclaire Castello [1], Donovan Bailey [3], Clinton Revell [1], Chengdao Li [1,2] and Daniel Real [1]

[1] Department of Primary Industries and Regional Development, South Perth, WA 6151, Australia; clinton.revell@dpird.wa.gov.au (C.R.); c.li@murdoch.edu.au (C.L.); daniel.real@dpird.wa.gov.au (D.R.)
[2] Western Crop Genetics Alliance, College of Environmental and Life Sciences, Murdoch University, Perth, WA 6150, Australia
[3] Department of Biology, New Mexico State University, Las Cruces, NM 88003, USA; dbailey@nmsu.edu (D.B.)
[4] USDA Agriculture Research Service, Subtropical Horticulture Research Station, Miami, FL 33158, USA
* Correspondence: yong.han@dpird.wa.gov.au; Tel.: +61-893607590

Abstract: *Leucaena* is a versatile legume shrub/tree used as tropical livestock forage and in timber industries, but it is considered a high environmental weed risk due to its prolific seed production and broad environmental adaptation. Interspecific crossings between *Leucaena* species have been used to create non-flowering or sterile triploids that can display reduced weediness and other desirable traits for broad use in forest and agricultural settings. However, assessing the success of the hybridisation process before evaluating the sterility of putative hybrids in the target environment is advisable. Here, RNA sequencing was used to develop breeding markers for hybrid parental identification in *Leucaena*. RNA-seq was carried out on 20 diploid and one tetraploid *Leucaena* taxa, and transcriptome-wide unique genetic variants were identified relative to a *L. trichandra* draft genome. Over 16 million single-nucleotide polymorphisms (SNPs) and 0.8 million insertions and deletions (indels) were mapped. These sequence variations can differentiate all species of *Leucaena* from one another, and a core set of about 75,000 variants can be genetically mapped and transformed into genotyping arrays/chips for the conduction of population genetics, diversity assessment, and genome-wide association studies in *Leucaena*. For genetic fingerprinting, more than 1500 variants with even allele frequencies (0.4–0.6) among all species were filtered out for marker development and testing *in planta*. Notably, SNPs were preferable for future testing as they were more accurate and displayed higher transferability within the genus than indels. Hybridity testing of ca. 3300 putative progenies using SNP markers was also more reliable and highly consistent with the field observations. The developed markers pave the way for rapid, accurate, and cost-effective diversity assessments, variety identification and breeding selection in *Leucaena*.

Keywords: crossing; genetic variants; hybridity; molecular marker; next-generation sequencing (NGS); phylogeny

1. Introduction

The mimosoid legume genus *Leucaena* includes a diverse set of New World shrub and tree species with a long history of human use [1]. A subset of these versatile species has been widely spread in tropical regions across the globe. In particular, the value of *Leucaena leucocephala* (Lam.) de Wit. as a vigorous nitrogen-fixing species has been broadly recognised, resulting in its widespread usage in erosion management, multipurpose cropping systems [2], phytoremediation, as a source of paper [3] and even biofuel [4]. *L. leucocephala* foliage is highly nutritious, with a 15–18% higher protein content than tropical grasses and cereal straws [5]. Therefore, the crop has been extensively planted for livestock feeding in some regions, such as Queensland in Australia [6]. Nevertheless,

L. leucocephala is regulated as an invasive weed in over 50 countries [7], and in northern Western Australia, its use is constrained by its weediness status [8].

Developing non-flowering or sterile *Leucaena* by the interspecific crossing of species has long been of interest [8,9]. One of the more promising approaches involves crosses between parents of different ploidy levels, creating odd-numbered polyploid offspring that can be seed sterile. Recently, over 1400 interploidal crosses were made to generate seedless triploids and thus promote the forage crop in Australia and elsewhere [10]. Massive efforts are going into this work (and related efforts elsewhere), including the process of crossing, seed collection, seeding, transplantation to the field, nursing plants for at least 2.5 years prior to the onset of flowering [10], and the subsequent phenotyping of flowering traits across individuals. Hence, an accurate, reliable, rapid, and high-throughput diagnostic approach for hybrid detection is required to identify *Leucaena* triploids in the early seedling stage to avoid wasting downstream efforts on non-hybrid individuals.

Molecular markers and marker-assisted selection (MAS) have been successfully implemented for plant breeding over the decades to rapidly identify favourable species/genotypes and pinpoint the genomic regions in plants that express traits of interest, thus driving from phenotype to genotype-based selection [11]. Markers developed against chloroplast, nuclear ribosomal DNA internal transcribed spacers [12], sequence-characterised amplified regions within nuclear-encoded loci [13], and simple sequence repeats (SSRs; [14]) have been applied for phylogenetic and evolutionary studies in *Leucaena*. However, such sets of markers do not represent the genome broadly and are therefore of limited utility in modern high-throughput plant genotyping and breeding practices. Moreover, most species of *Leucaena* are likely to have many heterozygous loci because of self-incompatibility and therefore an out-crossing habit (i.e., *L. diversifolia*, [15]), hindering the selection of homozygous and polymorphic markers between parental lines for interspecific hybrid detection.

The advent of next-generation sequencing (NGS) technology allows the detection of millions of sequence variations within the genome and transcriptome of a plant genotype, which can be utilised to assess and characterise genetic diversity within populations, species, or germplasm collections. Molecular markers based on NGS have been developed in many tree species worldwide and used as valuable genetic tools for non-model organisms [16,17]. Indeed, SNP (single nucleotide polymorphism) markers are the most abundant, with genome-wide coverage, high stability and Mendelian inheritance [18], and they are compatible with high-throughput genotyping technologies such as microarray and KASP (kompetitive allele-specific PCR). However, such a platform has not been developed for *Leucaena* to aid in genetic studies and breeding practices.

In the present study, mRNA-based transcriptome-wide genetic variants among 21 *Leucaena* taxa have been identified for the first time. The large-scale sequence variations revealed the molecular identity of *Leucaena* species. As a proof of concept, a core set of SNPs with an even allele frequency among *Leucaena* species was selected for KASP marker development and genotyping. Over 3000 *Leucaena* individuals from interspecific crossings were tested for hybridity using polymorphic markers between parental lines. The developed high-throughput and cost-effective platform can rapidly identify triploids at the seedling stage, saving time and cost for breeding programs. Moreover, the variant database can be adopted for future genetic diversity, mapping and fingerprinting studies, MAS and breeding in *Leucaena*.

2. Materials and Methods

2.1. Plant Materials

For sequencing and variant calling, 20 diploids (*L. collinsii*, *L. cruziana*, *L. cuspidata*, *L. esculenta*, *L. greggii*, *L. lanceolata*, *L. lempirana*, *L. macrophylla* ssp. *macrophylla* and ssp. *istmensis*, *L. magnifica*, *L. multicapitula*, *L. matudae*, *Leucaena pueblana*, *L. pulverulenta*, *L. retusa*, *L. salvodorensis*, *L. shannonii*, *L. trichandra*, *L. trichodes*, and *L. zacapana*) and one tetraploid species *L. diversifolia* were used (Table 1). Plant samples were collected from the Leucaena living collection at New Mexico State University, USA.

Table 1. SNP and indel counts in 20 diploid and one tetraploid *Leucaena* taxa mapping against an *L. trichandra* draft genome.

Taxon	nRefHom	nNonRefHom	nHets	nTransitions	nTransversions	nIndels	NCBI Accession Number
Leucaena collinsii Britton & Rose (2×)	8428728	5,228,368	2,806,318	4,440,773	3,593,913	450,406	SRX2719653
Leucaena cruziana Britton & Rose (2×)	8,284,936	5,330,027	2,839,134	4,515,216	3,653,945	459,723	SRX2719651
Leucaena cuspidata Standl. (2×)	7,554,107	5,839,938	3,033,407	4,891,943	3,981,402	486,368	SRX2719650
Leucaena diversifolia (Schltdl.) Benth. (4×)	7,686,304	4,977,243	3,791,795	4,830,917	3,938,121	458,478	SRX2719649
Leucaena esculenta (DC.) Benth. (2×)	7,024,119	6,160,890	3,218,074	5,159,900	4,219,064	510,737	SRX2719648
Leucaena greggii S. Watson (2×)	7,282,246	6,090,881	3,048,606	5,037,017	4,102,470	492,087	SRX2719647
Leucaena lanceolata S. Watson (2×)	8,312,577	5,360,102	2,782,799	4,505,372	3,637,529	458,342	SRX2719645
Leucaena lempirana C.E. Hughes (2×)	8,429,399	5,192,417	2,842,191	4,444,316	3,590,292	449,813	SRX2719644
Leucaena macrophylla subsp. *istmensis* C.E. Hughes (2×)	8,244,333	5,462,404	2,743,421	4,536,361	3,669,464	463,662	SRX2719641
Leucaena macrophylla subsp. *macrophylla* Benth. (2×)	8,208,257	5,382,254	2,857,991	4,554,355	3,685,890	465,318	SRX2719640
Leucaena magnifica (C.E. Hughes) C.E. Hughes (2×)	8,465,544	5,150,898	2,848,108	4,415,835	3,583,171	449,270	SRX2719639
Leucaena matudae (Zarate) C.E. Hughes (2×)	6,982,398	6,187,703	3,231,807	5,181,555	4,237,955	511,912	SRX2719638
Leucaena multicapitula Schery (2×)	8,372,304	5,299,403	2,788,487	4,474,205	3,613,685	453,626	SRX2719637
Leucaena pueblana Britton & Rose (2×)	5,793,382	8,948,739	1,578,880	5,790,542	4,737,077	592,819	SRX2719610
Leucaena pulverulenta (Schltdl.) Benth. (2×)	7,251,662	6,145,143	3,024,462	5,046,850	4,122,755	492,553	SRX2719635
Leucaena retusa Benth. (2×)	7,248,430	6,151,778	3,021,196	5,057,547	4,115,427	492,416	SRX2719634
Leucaena salvadorensis Standl. ex Britton & Rose (2×)	8,350,557	5,157,831	2,950,991	4,481,168	3,627,654	454,441	SRX2719633
Leucaena shannonii Donn.Sm. (2×)	8,391,792	5,170,550	2,899,176	4,459,365	3,610,361	452,302	SRX2719632
Leucaena trichandra (Zucc.) Urb. (2×)	8,773,909	4,785,556	2,925,521	4,257,895	3,453,182	428,834	SRX2719631
Leucaena trichodes (Jacq.) Benth. (2×)	8,285,059	5,390,552	2,777,911	4,516,372	3,652,091	460,298	SRX2719630
Leucaena zacapana (C.E. Hughes) R. Govind. & C.E. Hughes (2×)	8,419,838	5,192,514	2,851,655	4,448,285	3,595,884	449,813	SRX2719629
Total number of unique variants		SNPs 16,396,328		9,024,214	7,372,114	816,282	

Abbreviations: nRefHom, number of homozygous alleles that are same as the reference; nNonRefHom, number of homozygous alleles that are different from the reference; nHets, number of heterozygous alleles; nTransitions, number of nucleotide transitions; nTransversions, number of nucleotide transversions; nIndels, number of insertion/deletion alleles. Transcriptome sequencing data and voucher information have been deposited at NCBI (https://www.ncbi.nlm.nih.gov/sra) and are available by searching corresponding accession numbers.

At the Western Australia Department of Primary Industries and Regional Development (DPIRD), 224 accessions of *Leucaena* from nine diploids (*L. pulverulenta*, *L. collinsii*, *L. zacapana*, *L. shannonii*, *L. macrophylla*, *L. retusa*, *L. greggii*, *L. trichandra*, and *L. trichodes*) and three tetraploids (*L. leucocephala* ssp. *leucocephala* and ssp. *glabrata*, and *L. diversifolia*) were selected for interspecific crossing in a glasshouse in 2018 (Generation 1) and 2019 (Generation 2), respectively [10]. About 3000 seedlings of the cross progeny were raised in a naturally lit glasshouse at DPIRD, Perth, and subsequently sampled for genotyping with molecular markers.

2.2. RNA-seq and Variant Calling

Given the evolutionary history of hybridisation/polyploidisation in *Leucaena*, the next-generation transcript sequencing-based approach was employed [19], designed to account for the problem of paralogy by selecting for SNPs only from orthologous genes. For all species except *L. pueblana*, the total RNA was extracted from whole seedlings. Whole seedlings with three true leaves were sampled to obtain a wide range of expressed genes. RNA quality was assessed using Nanodrop (Thermo Fisher Scientific Inc., Waltham, MA, USA), Qubit (Thermo Fisher Scientific Inc., USA), and agarose gel electrophoresis before library preparation. Illumina RNA-TruSeq libraries (Illumina Inc., USA) were generated for each sample using 4 μg total RNA following the manufacturer's instructions. Libraries were quantified using the Qubit and checked using Bioanalyzer 2100 (Agilent Technologies, Santa Clara, CA, USA). Each library was sequenced to a depth of 40 M read pairs (100 bp each) using 101 bp reads (Axeq Technologies, Seoul, the Republic of Korea). No living material was available for *L. pueblana*, so Illumina-based gDNA genome skimming was employed using the same approach but with DNA extracted with a Qiagen DNeasy Kit (Qiagen, Germantown, MD, USA) from silica-dried leaves, Illumina TruSeq library, and sequenced via Illumina HiSeq 2000. For all samples, paired-end 100 bp paired reads were trimmed of their Illumina adaptors and filtered for a minimum read length of 65 bases using Trimmomatic v0.34 [20].

This study utilised a chromosomal-scale genome assembly for *Leucaena trichandra* (Bailey et al., unpublished) as the reference genome. Trimmed sequence data were aligned to the reference genome using STAR (v2.4.0.1, [21]). Using the SAMtools (v1.9; [22]) *view* with the "-q10" option, uniquely mapped reads were extracted from the total set of accepted hits. This step filtered out paralogous genes. The SAMtools varFilter.pl utility was used for variant calling. The first step in this process was the generation of an index from the genome using the SAMtools *faidx* command. All accepted hit files were then set up for variant calling using the SAMtools *mpileup* command. VCFtools [23] generated a BCF file that was converted to a human-readable VCF file containing all the variants detected.

Preliminary filtering on the VCF file was conducted using VCFtools. Reads with mapping quality Phred scores below 30 were removed from the VCF file. A fraction of the results were manually verified in Geneious (Version R11). A visual inspection of the first 15 positions from a randomly selected contig (tig00000853) from *L. collinsii* in the original STAR mapping showed positive results. All positions compared between the visual output from the Geneious alignment viewer and the filtered VCF file were concordant.

Variant statistics were visualised in charts by SigmaPlot 14.0.

2.3. Variant Filtration, Marker Development, and Phylogenetic Analysis

To test the differentiation power in *Leucaena* species, sets of variants were filtered out by PLINK 1.9 (https://www.cog-genomics.org/plink/, accessed on 14 February 2019) analysis according to allele frequency and genome location, including a pool of 4,492,396 variants after trimming monomorphic variants with an allele frequency below 0.1 and missingness over 0.25 in the whole variant database; a pool of 75,001 variants evenly distributed in each 10 kb region within the *L. trichandra* reference genome; a pool of 48 indels with a minimum 6 bp difference from a random 100 Mb sequence interval (tig4612_575177–tig5770_674183; Table S1); and a pool of 1533 homozygous SNP variants with an allele frequency of 0.4–0.6 (Table S2). Principal component analysis (PCA) was performed using PLINK 1.9, and plots were drawn with the R package "ggplot2" [24]. Molecular markers were developed based on the indel/SNP variations and then validated across species of *Leucaena*. Indel markers that resulted in different amplicon sizes were visualised by 2% agarose or 6% polyacrylamide gel electrophoresis. For SNP screening, the KASP genotyping assay, a novel competitive allele-specific PCR for SNP scoring based on dual FRET (fluorescent resonance energy transfer), was employed. The highly conserved variants in Western Australian crosses were further selected (Table S3) and KASP markers were designed for about 100 bp length amplicons with sequences referring to *L. trichandra* using Geneious Prime (Version 2019.1).

An SNP-based phylogenetic tree was developed to illustrate the ability of the marker system to reflect the phylogenetic history of the genus. The all-species VCF file was first filtered for biallelic SNPs with in-house scripts and then converted into phylip format using vcf2phylip.py [25]. The interspecific hybrid tetraploid taxon *L. diversifolia* was removed from the matrix to avoid violating the assumption of a bifurcating phylogenetic tree. A maximum likelihood tree was created using raxml-ng (v0.9.0, [26]) with GTR + G to build both the best tree and to conduct convergent bootstrap analysis.

2.4. Genomic DNA Isolation

Leucaena leaves can be rich in polyphenols and polysaccharides that inhibit polymerase chain reaction (PCR), so different DNA extraction methods were used according for experimental purposes (see "Results and Discussion"). For polymorphic marker screening between parental accessions, high-quality genomic DNA was extracted from about 400 mg of soft root tissue using the cetyltrimethylammonium bromide (CTAB) method [27] after sampling, cleaning and crushing with beads using TissueLyser II (Qiagen, USA).

Quick DNA isolation from the young leaves was also conducted for all F_1 seedlings using the AquaGenomic kit (MoBiTec, Göttingen, Germany), following the manufacturer's instructions, or a modified sodium dodecyl sulphate (SDS) method described below.

For the SDS method, about 100 mg of fresh leaf tissues were collected into 1.2 mL strip tubes (SSIbio, Lodi, CA, USA) assembled in 96-well racks and lysed by a tissue lyser with beads in 300 µL of AP1 solution (100 mM Tris-HCl, 50 mM EDTA, 0.5 M NaCl, and 1.5% SDS), incubated at 65 °C for 30 min, and then mixed well with 100 µL of AP2 solution (5 M potassium acetate, 2% polyvinyl pyrrolidone mol. wt. 10,000, pH 6.2). After 5 min on ice, the 96-well extraction plates were centrifuged at 4000 rcf for 10 min at 4 °C (Allegra X-15R, Beckman Coulter, Brea, CA, USA). DNA was precipitated by mixing 100 µL of supernatant with an equal volume of isopropanol and incubated at −20 °C for 15 min. After centrifugation (4000 rcf), the DNA pellet was washed with 70% ethanol twice and finally dissolved in 50 µL of Milli-Q water. DNA concentration and quality were checked using NanoDrop One (Thermo Scientific, USA) at the Western Australia State Agricultural Biotechnology Centre.

2.5. Polymerase Chain Reaction (PCR) and Scoring

Indel and KASP markers were screened for polymorphisms and picked to differentiate the expected genotype for each cross (Table S4). Markers were then applied to both a set of positive control DNA (consisting of the two parental accessions, an artificial DNA mix (1:1 by weight), and the corresponding F_1 individuals (test cases).

For indel markers, the PCR reaction mixture was composed of 1 µL of 10× Bioline buffer, 1 µL of GC buffer, 0.8 µL of Bioline 50 mM $MgCl_2$, 0.2 µL of 10 mM dNTPs, 0.5 µL of forward and reverse primer at 10 µM, 0.04 µL of BioTaq DNA polymerase, 3 µL of Cresol red dye, 2.5 µL of MilliQ water, and 50 ng (1 µL) of template DNA in a 10 µL reaction volume. PCR was carried out in the Applied Biosystems Veriti Dx 96-Well Fast Thermal Cycler, starting with a denaturation step at 94 °C for 3 min, followed by 35 cycles at 94 °C for 30 s, 56 °C for 30 s, 72 °C for 30 s, and finishing with an extension step at 72 °C for 5 min. PCR products were analysed on 6% polyacrylamide gel (PAGE).

For SNP markers, a KASP master mix (LGC Biosearch Technologies, Hoddesdon, UK) consisting of Taq polymerase, nucleotides, $MgCl_2$, universal FRET (fluorescence resonant energy transfer) cassettes, and ROX™ passive reference dye were used. The forward primers are allele-specific, each accommodating a unique tail sequence corresponding to the FRET cassette. The PCR reaction and procedure were performed as described by Real et al. [10]. Results were scored using QuantStudio Real-Time PCR software (Version 1.3).

2.6. Field Trials and F_1 Phenotyping

Young seedlings germinated from two crossing generations were transplanted to the field plots at Kununurra (latitude—15.65, longitude 128.72) and Carnarvon (latitude—24.86, longitude 113.73) in 2020 (Generation 1) and 2021 (Generation 2), respectively. The trial arrangement, management, and irrigation were described by Real et al. [10]. The flowering and fruiting behaviour of plants were observed one or two years after transplanting at both sites.

3. Results and Discussion
3.1. Transcriptome-Wide Genetic Variants in Leucaena Detected by RNA Sequencing

Overall, 16396328 unique SNPs and 816282 indels were identified from 21 taxa, including 20 diploids *L. collinsii*, *L. cruziana*, *L. cuspidata*, *L. esculenta*, *L. greggii*, *L. lanceolata*, *L. lempirana*, two subspecies of *L. macrophylla*, *L. magnifica*, *L. multicapitula*, *L. matudae*, *L. pulverulenta*, *L. retusa*, *L. salvodorensis*, *L. shannonii*, *L. trichandra*, *L. trichodes*, *L. zacapana*, *L. pueblana* (genomic DNA), and one tetraploid, *L. diversifolia* (Table 1). Variants were relatively evenly distributed across species, with an average of 5.6 million SNPs and 0.47 million indels per accession, respectively. The alleles were displayed at diverse frequencies, and the results showed that variants at the 0.68–0.77 window only occupied about 0.2% of the total for both SNP and indels (Table S5). Interestingly, total RNA-seq mapping rates ranged from 87 to 95% with uniquely mapped read rates ranging from 69 to 77% among all species, indicating no distinct difference between samples. The mapping parameters of RNA-seq reads to the genome vary in plants, depending on the genome complexity, reference availability and quality, and the mapping quality threshold. The high mapping rate in *Leucaena* was comparable to the values in the model plants Arabidopsis [28] and wheat [29], suggesting the high quality of RNA-seq and thus the reliability of variant calling.

Among the point mutations, transitions (Ts) were more abundant than transversions (Tv), with a Ts/Tv ratio of 1.22, which was similar to the findings in *Rhododendron* species [30] and sunflower [31] based on RNA-seq. In particular, interchanges between purines (A/G) and pyrimidines (C/T) accounted for 27.5% each among the SNPs, while the transversions of C to G or vice versa were the rarest and had a percentage of only 3.7% (Figure 1a). For indels, changes of 1–3 bp were dominant and made up about 74% of the total variations (Figure 1b). Relative to the *L. trichandra* reference genome, the number of indel sites generally decreased with increasing indel lengths up to ±20 base pairs. Notably, larger deletions (over 15 bp) were more common than insertions with the same length, which were six times more in the *Leucaena* population. Indels are more prevalent than other structural variants such as SSRs, and the genotyping of indel markers is technologically less demanding compared with SNP detection, which usually requires expensive reagents and equipment [32]. Consequently, indels are preferred for cost-effective gel electrophoresis-based testing where small sequence polymorphisms are less likely to score accurately. However, both sequencing techniques and bioinformatics tools used for NGS analysis affect the sensitivity and specificity of indel detection in silico [33]. Hence, the reliability of diagnostic markers needs to be further confirmed through experimentation.

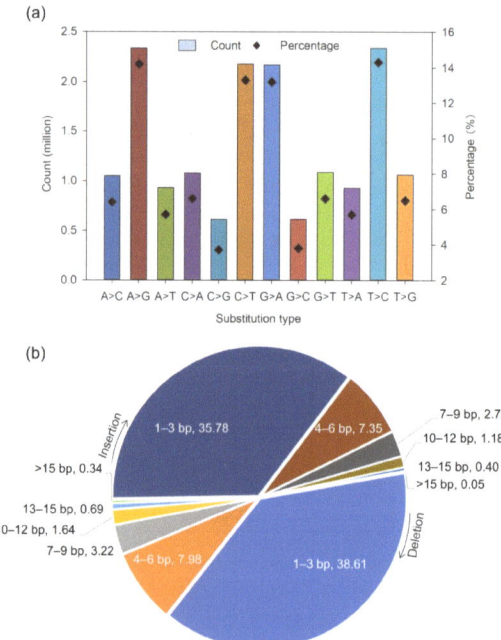

Figure 1. Representation of different types of SNPs and indels detected across 21 *Leucaena* taxa. (**a**) Counts and percentages of each substitution type; (**b**) percentages of various indel sizes among *Leucaena* species. Variant calling was completed after RNA sequencing and mapping against an *L. trichandra* reference genome.

In crop science and breeding, genomic resources can significantly promote the identification of genetic controls of key agronomic traits and the development of breeding tools to accelerate the selection process [34]. For *Leucaena*, *L. trichandra* is the common putative progenitor of four tetraploid species, *L. confertiflora*, *L. pallida*, *L. involucrata* and *L. diversifolia* [35], and its draft genome has been completed by the Bailey Lab at New Mexico State University based on PacBio and Illumina sequencing (Bailey et al., unpublished). Together with other transcriptomic data from a few species and a mitochondrial genome of *L. trichandra* (summarised by Abair et al. [36]), such resources pave the way for future applied and basic research on *Leucaena*. Notably, this study presented near genus-wide RNA-seq data, which provides comprehensive molecular information on the species identities and, as proof of concept, can be used for phylogenetic analysis and breeding selection in *Leucaena*.

3.2. The Genus-Wide Sequence Variations Reflect the Phylogenetic History of Leucaena and Serve as a Genotyping Reference for Research and Breeding

Species diversity and diversification in *Leucaena* have been investigated through a variety of phylogenetic and population genetic analyses that include *Leucaena* and relevant outgroup analyses. Most recently, these include markers for SCAR-based nuclear loci, nrDNA ITS, cpDNA regions, AFLPs, and SSRs [13,14,35]. Previous results for *Leucaena* phylogenetics recovered three diploid clades (Clades 1, 2, and 3) that correlate well with morphology and geographic distribution (reviewed by Abair et al. [36]). Here, an SNP-derived variant phylogeny was developed to confirm its consistency with *Leucaena*'s current phylogeny, population biology, geography, and morphology and the suitability of generating breeding markers based on genus-wide sequence variation. The phylogeny, rooted internally with *L. cuspidata*, is interpretable as consistent with prior results, with each of the three clades potentially diagnosable along the tree (Figure 2a). Each of those putative groups has the same species and nearly all of the same potential species' relationships therein. The reconstruction of relationships among species using the SNPs developed here supports the idea that these markers are appropriate for fingerprinting species in plant breeding studies.

The ability to correctly discriminate every species from one another was investigated through PCA analysis of detected variations (Figure 2b). Principal components 1 and 2 were sufficient to separate the three clades with allopatric geographic distributions (*sensu* Govindarajulu et al. [13]). Removing the monomorphic variants with frequencies less than 0.1 and missing values over 0.25 helped distinguish the most widespread Clade 1 species (Figure 2c). To investigate whether there were conserved or highly polymorphic genomic regions across the genus, we filtered out 75,001 variants spanning 10 kb each and regrouped all species (Figure S1). Even though less than 0.5% of total variants were captured, the species grouping by PC1 and PC2 was identical to that using the whole dataset, indicating that the core set of variants is comprehensive for further genetic analysis and fingerprinting. This set of variants can be further developed into molecular markers that differentiate every sampled species with the correct capture of identities and speciation for breeding practices.

The results revealed the similarities and relationships among the clades and species. Notably, taking advantage of the reference genome, a core set of about 75,000 variants can be genetically mapped and transformed into genotyping arrays/chips for the conduction of population genetics, diversity assessments, and genome-wide association studies (GWAS) to assess the prospect of it occurring in other crops like biennial caraway (*Carum carvi* [37]). Despite the limited population size in this study and the heterozygous background in *Leucaena*, the detected variants help predict the linkage disequilibrium loci across genus/species, which is vital in implementing genome-wide association studies for prediction accuracy [38].

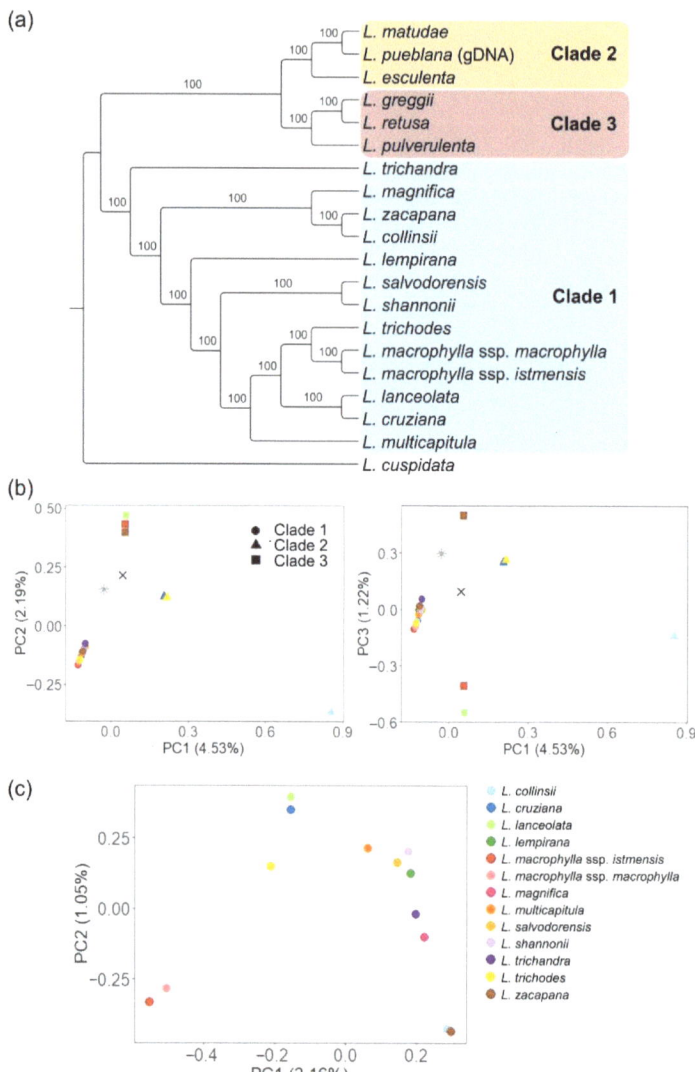

Figure 2. Phylogenetic and principal component analysis using transcriptome-wide genetic variants in *Leucaena* following the clade nomenclature of Govindarajulu et al. [13]. (**a**) Maximum likelihood (ML) tree for species topology in *Leucaena* using all SNPs; (**b**) PCA plots for species clustering using all mapped variants. The hybrid tetraploid *L.diversifolia* was indicated by "*" and *L. cuspidata* by "×", respectively. The proportion of variance explained by the principal components was shown on the axis labels. (**c**) PCA plots of 13 Clade 1 diploids based on 4,492,396 variants after trimming monomorphic variants with an allele frequency of 0.1 and missingness of 0.25.

3.3. SNP Markers Outperformed Indels for Leucaena Genotyping

It has long been known that extracting high-quality DNA from plants can be seriously complicated by the presence of various secondary metabolites and phenolics [39]. In this study, we tested different methods to purify genomic DNA from *Leucaena* tissues (Figure S2). DNA extracted from leaves with the AquaGenomic kit or CTAB was not of high quality and had polysaccharide/polyphenol contamination (Figure S2a). The presence of those mucilaginous exudates led to inaccurate NanoDrop and flow cytometric readings and inhibited PCR amplification efficiency. Therefore, the leaf tissues were replaced by roots to obtain pure and clean DNA after CTAB extraction (Figure S2b). However, digging up roots is tedious, and sometimes the woody tissues cannot be completely lysed. A quick and high-throughput SDS method was developed to isolate DNA with acceptable quality for PCR from young *Leucaena* leaves (Figure S2c), by which 800 samples were processed by a single person daily.

Indels can be easily visualised and scored through agarose gel electrophoresis. An indel pool with 48 variants over 5 bp sequence differences in a random 100 Mb interval was filtered, which can distinguish all 21 sequenced taxa in silico (Figure S3; Table S1). The PAGE analysis showed that selected indel markers can distinguish species such as *L. diversifolia* and *L. pulverulenta* (Figure 3a indel 30). However, they were not necessarily selective for accessions in the same species, such as in *L. pulverulenta* using indels 9 and 30 (Figure 3a, Lane 10–15). Although it was expected that the hybrid tetraploid species *L. diversifolia* displayed bands consistent with higher heterozygosity (Figure 3a indel 30), some *L. pulverulenta* accessions also showed multiple bands on the gel (Figure 3a indels 6 and 9), in line with the out-crossing character of diploid *Leucaena*, thus posing challenges in hybridity scoring. Given that the indels have lower levels of variation among closely related species (*L. pulverulenta* and *L. diversifolia*), this finding is consistent with previous reports that suggest larger indels tend to be conserved at higher phylogenetic levels [40,41].

For hybridity testing, the heterozygous nature of the genus makes it increasingly complex to distinguish the origin of triploid samples from the parents as both parents need to be homozygous. A very limited number of indels with polymorphisms between individuals (i.e., indel 6) were further selected from the 48 indel pool for hybridity screening to confirm the heterozygous characteristics of both male and female parent bands (Figure 3b). Generally, indels are not highly polymorphic between the species, and the sizes are not consistent with in silico predictions. Possible reasons for this include the short-read nature of NGS, which is false positive when aligned with the reference genome, and the bioinformatic tools used for variation mining such as SAMtools [22] and GATK [42] are not tailored for indel discovery, where the algorithms are restricted to identify indels less than 10 bp [32]. Notably, long indels (>15 bp) and structural variations can be more effectively detected through high-coverage whole-genome resequencing or pan-genome sequencing to facilitate the fine mapping of loci for traits of interest, as evident in another perennial tree plant tea (*Camellia sinensis* (L.) Kuntze, [43]). However, it still requires high costs and calls for international research collaboration to achieve this in *Leucaena*.

RNA-seq offers a high coverage of SNP discovery compared to whole-genome or whole-exome sequencing [44]. Therefore, genotyping by SNPs was pursued as the more accurate and reliable approach, while short-read Illumina sequencing was used for variant detection and calling. In the present study, a total of 1533 homozygous SNP variants with an allele frequency of 0.4–0.6 were selected for genotyping (Table S2). The core set can effectively distinguish most *Leucaena* taxa, especially *L. diversifolia*, which was intensively used as a tetraploid parent in Western Australia's crosses (Figure 3c). Overall, the 57 most popular SNPs were further selected for the crossings between nine diploid (*L. pulverulenta*, *L. collinsii*, *L. zacapana*, *L. shannonii*, *L. macrophylla*, *L. retusa*, *L. greggii*, *L. trichandra*, and *L. trichodes*) and two tetraploid species (*L. leucocephala* ssp. *leucocephala* and ssp. *glabrata*, and *L. diversifolia*) in Western Australia (Table S3). Among those, 39 KASP markers were successfully developed based on the *L. trichandra* reference sequence (Table S4).

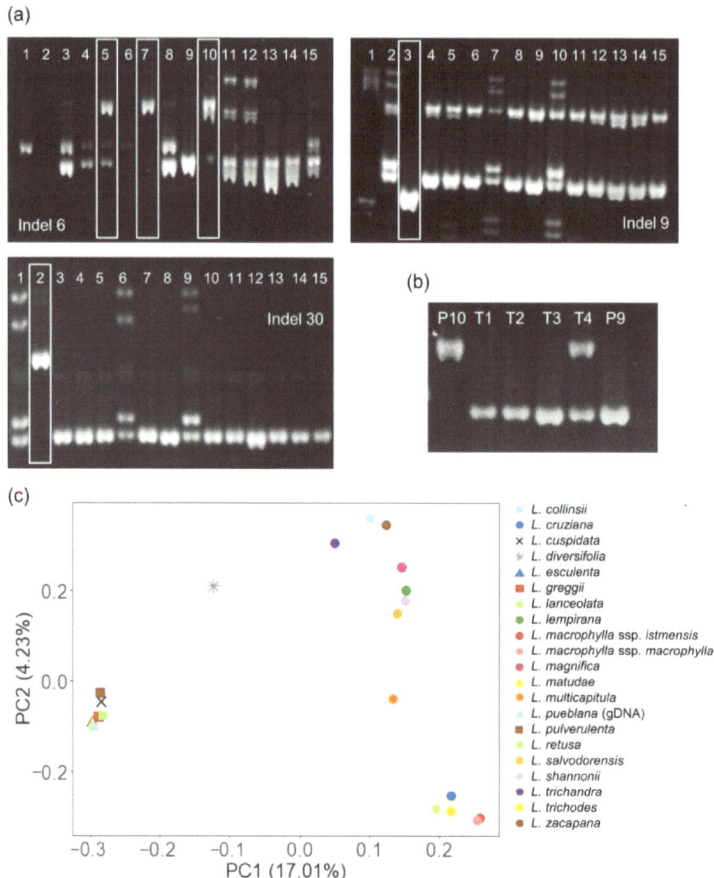

Figure 3. Development of indel markers and a core SNP set for fingerprinting. (**a**) Testing indels with different *L. diversifolia* and *L. pulverulenta* accessions for reciprocal crosses. Lane 1, *L. diversifolia* 71.4; Lane 2–5, *L. pulverulenta* 63.5, 84.3, 87.2 and 84.4; Lane 6, *L. diversifolia* 72.3; Lane 7–8, *L. pulverulenta* 75.2 and 83.2; Lane 9, *L. diversifolia* 80.1; Lane 10–15, *L. pulverulenta* 83.3, 86.2, 88.3, 77.2, 87.3 and 87.4. White boxes indicated distinct genotypes. (**b**) PAGE gel of indel marker 6 for a selection of parental lines and triploids generated. P10, *L.pulverulenta* 83.3; P9, *L.diversifolia* 80.1; T1 to T4, F1 individuals. (**c**) Grouping of 21 sequenced *Leucaena* taxa with 1533 homozygous SNPs (Table S2).

3.4. Reliability of Marker Fingerprinting Validated by Field Observations

KASP markers were classified as effective when they distinguished the parental lines used in crossings and grouped the triploids either with heterozygous genotypes from bi-parents or close to the pollen donor allele (Figure 4a and Figure S4). For example, KASP marker 9 (variant tig2946_pilon_26087) successfully distinguished female parent *L.pulverulenta* accession 84.3, allele 2-specific, and male parent *L.diversifolia* accession 80.1, specific to allele 1 (Figure S4). Triploids were dominant to allele 1 due to their polyploid nature, and more gene copies were received from the pollen donor allele, which was a tetraploid. An artificial hybrid that mixed an equal amount of DNA from bi-parents was used as a genotype scoring reference. The DNA mix dominated allele 1, like other triploids generated through the crossing. KASP genotyping also identified F1 individuals carrying female alleles only (*L. leucocephala* ssp. *glabrata*) that could be selfings (Figure 4a). Every offspring generated from a crossing and the corresponding parental lines were treated as a unique entity.

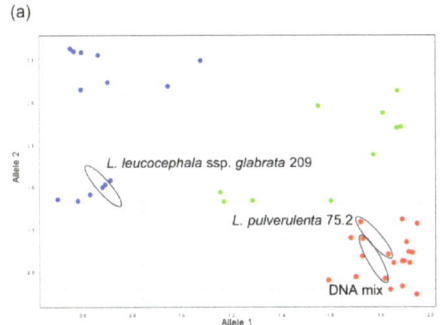

Figure 4. Development of KASP markers based on SNPs and genotyping *Leucaena* populations. (**a**) Genotyping a cross from *L. leucocephala* ssp. *glabrata* (pollen receiver) and *L. pulverulenta* (pollen donor). KASP marker 24 was used, and the DNA mix contained an equal amount of both parental accessions. Each sample had two technical repeats, as shown in the ovals. (**b**) A summary of genotyping results compared with phenotype observation two years after transplanting at both Kununurra and Carnarvon, WA. Sterile or partially sterile refers to plants that never flowered, flowered without pods, flowered with few pods or flowered with aborted seeds. The phenotype results were recalculated from Real et al. [10].

From the 285 (Generation 1) and 3144 (Generation 2) plants derived from two generations of crosses, 235 and 1444 F1 individuals were confirmed as triploids, respectively (Figure 4b). They accounted for 82.5% and 45.9% of the total tested plants and the values were very close to the field observations of their flowering characteristics. The average percentages of sterile or partially sterile F1s planted at Kununurra and Carnarvon, Western Australia, were 81.7% and 48.5% for two crossing generations, respectively (Figure 4b). The results indicated that the selected SNPs and developed KASP markers are robust and reliable for *Leucaena* identification and fingerprinting.

The transcriptomic approach (RNA-Seq) has been used to generate large-scale and high-density molecular markers in non-model plants, such as sunflower (*Helianthus annuus* L., [31]), *Rhododendron* species [30], turnip canola (*Brassica napus* L., [45]) and grapes (*Vitis vinifera* L., [46]), for genotyping arrays, genetic mapping and breeding studies. These commonly used SNPs have a relatively high level of cross-species transferability compared with indels, suggesting that they are ideal for constructing high-resolution genetic maps and analysing genetic diversity and population structure in *Leucaena*. However, the conversion of SNPs into robust KASP markers has been primarily restricted to major crops, such as maise [47], rice [48], and wheat [49], identifying that the KASP markers developed here are an important step forward in this non-model system for multipurpose agriculture across the tropics. It is likely that more reference genomes for each *Leucaena* species will be captured to promote robust marker development and PCR amplification efficiency.

4. Conclusions

Leucaena represents a complex non-model natural system with deep interest for use in a variety of tropical agricultural practices [1,9]. High levels of interspecific crossability provide extensive possibilities for trait improvements using classical plant breeding methods [50]. However, the invasive nature of some *Leucaena* species creates an urgent need to combine classical breeding with modern molecular genetic breeding practices to efficiently identify offspring of interest for both their potential hybrid sterile nature (e.g., triploid) and useful parental characteristics. Here, relatively easily generated RNA-seq data are used to develop both SNP and indel markers to aid genetic studies and molecular breeding for

Leucaena. The 39 KASP markers generated and tested here represent a major step forward for *Leucaena* breeding work.

These new tools have considerable potential to provide rapid advances in *Leucaena* genotyping, but those adopting these makers will want to pay attention to two potentially confounding factors that can complicate the usage of such tools in many plant species, including 1) the quality of the DNA extracted [39] and 2) the heterozygous nature of parental taxa used in the specific crosses being investigated [10,15]. Clean DNA extractions, free of difficult *Leucaena* mucilage, are critical for reliable PCR results. The root-derived DNA and/or rapid extraction method noted herein are further advances along these lines. In relation to heterozygosity, it is critical that KASP markers are tested on the specific parent accession each time new crosses are conducted so that heterozygosity for a parental marker is not confused with a successful crossing between species.

Supplementary Materials: The following supporting information can be downloaded at: https://www.mdpi.com/article/10.3390/agronomy14071519/s1, Figure S1: Principal component analysis of 21 *Leucaena* taxa based on 75,001 variants evenly distributed in each 10 kb region within the *L. trichandra* reference; Figure S2: *Leucaena* species DNA extracted with different methods; Figure S3: Principal component analysis of 21 *Leucaena* taxa based on 48 indel variants filtered from a random 100 Mb interval in the *L. trichandra* reference with length over five base pairs; Figure S4: Genotyping a cross from *L. pulverulenta* (pollen receiver) and *L.diversifolia* (pollen doner); Table S1: 48 Indels and the variations in Western Australia's representative crossing species; Table S2: 1533 SNP homozygous variants filtered out for fingerprinting; Table S3: SNPs selected for Western Australia's interspecific crossings; Table S4: KASP and indel marker primers used for genotyping; Table S5: Counts of variants at different allele frequencies among 21 *Leucaena* taxa.

Author Contributions: Conceptualisation, D.R. and C.L.; methodology, Y.H., D.B. and C.L.; formal analysis, Y.H., A.A., M.N.-R., S.P.V. and D.B.; investigation, A.A., M.N.-R., S.P.V., R.B. and C.R.; resources, D.R., C.R., M.C. and D.B.; data curation, Y.H., A.A. and D.B.; writing—original draft preparation, Y.H.; writing—review and editing, J.v.d.Z., D.B., D.R. and C.L.; supervision, D.R.; funding acquisition, D.R., C.R. and C.L. All authors have read and agreed to the published version of the manuscript.

Funding: This research was funded by Meat & Livestock Australia (MLA) and the WA Department of Primary Industries and Regional Development (DPIRD) for the 'Sterile *Leucaena* project'. The *Leucaena trichandra* genome and broader transcriptome data were supported by the US National Science Foundation grant 1238731 (to D.B.).

Data Availability Statement: Transcriptome sequencing data and *Leucaena* voucher information have been deposited at NCBI (https://www.ncbi.nlm.nih.gov/sra) and are available by searching corresponding accession numbers listed in Table 1.

Acknowledgments: We thank Gaofeng Zhou and Hoang Viet Dang (Western Crop Genetics Alliance, Murdoch University) for instructions on KASP genotyping and PCA analysis, respectively, and Mengistu Yadete (DPIRD) for helping with the sample collection.

Conflicts of Interest: The authors declare no conflicts of interest.

References

1. Hughes, C.E. *Leucaena: A Genetic Resources Handbook*; Tropical Forestry Paper No 37; Oxford Forestry Institute: Oxford, UK, 1998.
2. Sithole, N.; Tsvuura, Z.; Kirkman, K.; Magadlela, A. Nitrogen source preference and growth carbon costs of *Leucaena leucocephala* (Lam.) de Wit saplings in South African grassland soils. *Plants* **2021**, *10*, 2242. [CrossRef]
3. Khanna, N.K.; Shukla, O.P.; Gogate, M.G.; Narkhede, S.L. *Leucaena* for paper industry in Gujarat, India: Case study. *Trop. Grassl.-Forrajes. Trop.* **2019**, *7*, 200–209. [CrossRef]
4. Alemán-Ramirez, J.; Okoye, P.U.; Torres-Arellano, S.; Mejía-Lopez, M.; Sebastian, P. A review on bioenergetic applications of *Leucaena leucocephala*. *Ind. Crop. Prod.* **2022**, *182*, e114847. [CrossRef]
5. Jube, S.; Borthakur, D. Development of an Agrobacterium-mediated transformation protocol for the tree-legume *Leucaena leucocephala* using immature zygotic embryos. *Plant Cell Tiss. Org.* **2009**, *96*, 325–333. [CrossRef]
6. Buck, S.; Rolfe, J.; Lemin, C.; English, B. Adoption, profitability and future of *Leucaena* feeding systems in Australia. *Trop. Grassl.-Forrajes. Trop.* **2019**, *7*, 303–314. [CrossRef]

7. Campbell, S.; Vogler, W.; Brazier, D.; Vitelli, J.; Brooks, S. Weed *Leucaena* and its significance, implications and control. *Trop. Grassl.-Forrajes. Trop.* **2019**, *7*, 280–289. [CrossRef]
8. Real, D.; Han, Y.; Bailey, C.D.; Vasan, S.; Li, C.; Castello, M.; Broughton, S.; Abair, A.; Crouch, S.; Revell, C. Strategies to breed sterile *Leucaena* for Western Australia. *Trop. Grassl.-Forrajes. Trop.* **2019**, *7*, 80–86. [CrossRef]
9. Brewbaker, J.L. Breeding *Leucaena*: Tropical multipurpose leguminous tree. In *Plant Breeding Reviews*; Janick, J., Ed.; Wiley-Blackwell Inc.: Hoboken, NJ, USA, 2016; Volume 40, pp. 43–123. [CrossRef]
10. Real, D.; Revell, C.; Han, Y.; Li, C.; Castello, M.; Bailey, C.D. Successful creation of seedless (sterile) *Leucaena* germplasm developed from interspecific hybridisation for use as forage. *Crop Pasture Sci.* **2022**, *74*, 783–796. [CrossRef]
11. Lema, M. Marker-assisted selection in comparison to conventional plant breeding. *Agric. Res. Technol.* **2018**, *14*, e555914. [CrossRef]
12. Hughes, C.E.; Bailey, C.D.; Harris, S.A. Divergent and reticulate species relationships in *Leucaena* (Fabaceae) inferred from multiple data sources: Insights into polyploid origins and nrDNA polymorphism. *Am. J. Bot.* **2002**, *89*, 1057–1073. [CrossRef]
13. Govindarajulu, R.; Hughes, C.E.; Bailey, C.D. Phylogenetic and population genetic analyses of diploid *Leucaena* (Leguminosae; Mimosoideae) reveal cryptic species diversity and patterns of divergent allopatric speciation. *Am. J. Bot.* **2011**, *98*, 2049–2063. [CrossRef]
14. Rajarajan, K.; Uthappa, A.R.; Handa, A.K.; Chavan, S.B.; Vishnu, R.; Shrivastava, A.; Handa, A.; Rana, M.; Sahu, S.; Humar, N.; et al. Genetic diversity and population structure of *Leucaena leucocephala* (Lam.) de Wit genotypes using molecular and morphological attributes. *Genet. Resour. Crop. Evol.* **2022**, *69*, 71–83. [CrossRef]
15. Walton, C.S. *Leucaena (Leucaena leucocephala) in Queensland*; Queensland Department of Natural Resources and Mines: Brisbane, Australia, 2003. Available online: https://www.daf.qld.gov.au/__data/assets/pdf_file/0009/57294/IPA-Leucaena-PSA.pdf (accessed on 13 November 2023).
16. Russell, J.R.; Hedley, P.E.; Cardle, L.; Dancey, S.; Morris, J.; Booth, A.; Odee, D.; Mwaura, L.; Omondi, W.; Angaine, P.; et al. tropiTree: An NGS-based EST-SSR resource for 24 tropical tree species. *PLoS ONE* **2014**, *9*, e102502. [CrossRef]
17. Tan, J.; Guo, J.J.; Yin, M.Y.; Wang, H.; Dong, W.P.; Zeng, J.; Zhou, S.L. Next generation sequencing-based molecular marker development: A case study in *Betula alnoides*. *Molecules* **2018**, *23*, 2963. [CrossRef]
18. Mammadov, J.; Aggarwal, R.; Buyyarapu, R.; Kumpatla, S. SNP markers and their impact on plant breeding. *Int. J. Plant. Genomics* **2012**, *2012*, e728398. [CrossRef]
19. Nagy, I.; Barth, S.; Mehenni-Ciz, J.; Abberton, M.T.; Milbourne, D. A hybrid next generation transcript sequencing-based approach to identify allelic and homeolog-specific single nucleotide polymorphisms in allotetraploid white clover. *BMC Genomics* **2013**, *14*, e100. [CrossRef]
20. Bolger, A.M.; Lohse, M.; Usadel, B. Trimmomatic: A flexible trimmer for Illumina sequence data. *Bioinformatics* **2014**, *30*, 2114–2120. [CrossRef]
21. Dobin, A.; Davis, C.A.; Schlesinger, F.; Drenkow, J.; Zaleski, C.; Jha, S.; Batut, P.; Chaisson, M.; Gingeras, T.R. STAR: Ultrafast universal RNA-seq aligner. *Bioinformatics* **2013**, *29*, 15–21. [CrossRef]
22. Li, H.; Handsaker, B.; Wysoker, A.; Fennell, T.; Ruan, J.; Homer, N.; Marth, G.; Abecasis, G.; Durbin, R. The Sequence Alignment/Map format and SAMtools. *Bioinformatics* **2009**, *25*, 2078–2079. [CrossRef]
23. Danecek, P.; Auton, A.; Abecasis, G.; Albers, C.A.; Banks, E.; DePristo, M.A.; Handsaker, R.E.; Lunter, G.; Marth, G.T.; Sherry, S.T.; et al. The variant call format and VCFtools. *Bioinformatics* **2011**, *27*, 2156–2158. [CrossRef]
24. Wickham, H. ggplot2: Elegant Graphics for Data Analysis. 2016. Available online: https://ggplot2.tidyverse.org/ (accessed on 11 March 2024).
25. Ortiz, E.M. vcf2phylip v2.0: Convert a VCF Matrix into Several Matrix Formats for Phylogenetic Analysis. Available online: https://zenodo.org/records/2540861 (accessed on 9 September 2019).
26. Kozlov, A.M.; Darriba, D.; Flouri, T.; Morel, B.; Stamatakis, A. RAxML-NG: A fast, scalable and user-friendly tool for maximum likelihood phylogenetic inference. *Bioinformatics* **2019**, *35*, 4453–4455. [CrossRef]
27. Murray, M.G.; Thompson, W.F. Rapid isolation of high molecular weight plant DNA. *Nucleic Acids Res.* **1980**, *8*, 4321–4325. [CrossRef]
28. Conesa, A.; Madrigal, P.; Tarazona, S.; Gomez-Cabrero, D.; Cervera, A.; McPherson, A.; Szcześniak, M.W.; Gaffney, D.J.; Elo, L.L.; Zhang, X.; et al. A survey of best practices for RNA-seq data analysis. *Genome Biol.* **2016**, *17*, 13. [CrossRef]
29. Pearce, S.; Vazquez-Gross, H.; Herin, S.Y.; Hane, D.; Wang, Y.; Gu, Y.Q.; Dubcovsky, J. WheatExp: An RNA-seq expression database for polyploid wheat. *BMC Plant Biol.* **2015**, *15*, 299. [CrossRef]
30. Wang, S.; Li, Z.; Guo, X.; Fang, Y.; Xiang, J.; Jin, W. Comparative analysis of microsatellite, SNP, and InDel markers in four *Rhododendron* species based on RNA-seq. *Breeding Sci.* **2018**, *68*, 536–544. [CrossRef]
31. Bachlava, E.; Taylor, C.A.; Tang, S.; Bowers, J.E.; Mandel, J.R.; Burke, J.M.; Knapp, S.J. SNP discovery and development of a high-density genotyping array for sunflower. *PLoS ONE* **2012**, *7*, e29814. [CrossRef]
32. Lv, Y.; Liu, Y.; Zhao, H. mInDel: A high-throughput and efficient pipeline for genome-wide InDel marker development. *BMC Genomics* **2016**, *17*, e290. [CrossRef]
33. Sehn, J.K. Insertions and deletions (Indels). In *Clinical Genomics*; Kulkarni, S., Pfiefer, J., Eds.; Elsevier Inc.: Amsterdam, The Netherlands, 2015; pp. 129–150. [CrossRef]

34. Karunarathne, S.; Walker, E.; Sharma, D.; Li, C.; Han, Y. Genetic resources and precise gene editing for targeted improvement of barley abiotic stress tolerance. *J. Zhejiang Univ. Sci. B* **2023**, *24*, 1069–1092. [CrossRef]
35. Govindarajulu, R.; Hughes, C.E.; Alexander, P.J.; Bailey, C.D. The complex evolutionary dynamics of ancient and recent polyploidy in *Leucaena* (Leguminosae; Mimosoideae). *Am. J. Bot.* **2011**, *98*, 2064–2076. [CrossRef]
36. Abair, A.; Hughes, C.E.; Bailey, C.D. The evolutionary history of *Leucaena*: Recent research, new genomic resources and future directions. *Trop. Grassl.-Forrajes. Trop.* **2019**, *7*, 65–73. [CrossRef]
37. von Maydell, D.; Beleites, C.; Stache, A.; Riewe, D.; Krähmer, A.; Marthe, F. Genetic variation of annual and biennial caraway (*Carum carvi*) germplasm offers diverse opportunities for breeding. *Ind. Crop. Prod.* **2024**, *208*, e117798. [CrossRef]
38. Uffelmann, E.; Huang, Q.Q.; Munung, N.S.; De Vries, J.; Okada, Y.; Martin, A.R.; Martin, H.C.; Lappalainen, T.; Posthuma, D. Genome-wide association studies. *Nat. Rev. Methods Primers* **2021**, *1*, 59. [CrossRef]
39. Porebski, S.; Bailey, L.G.; Baum, B.R. Modification of a CTAB DNA extraction protocol for plants containing high polysaccharide and polyphenol components. *Plant Mol. Biol. Rep.* **1997**, *15*, 8–15. [CrossRef]
40. Nagy, L.G.; Kocsubé, S.; Csanádi, Z.; Kovács, G.M.; Petkovits, T.; Vágvölgyi, C.; Papp, T. Re-mind the gap! Insertion—Deletion data reveal neglected phylogenetic potential of the nuclear ribosomal internal transcribed spacer (ITS) of fungi. *PLoS ONE* **2012**, *7*, e49794. [CrossRef]
41. Ashkenazy, H.; Cohen, O.; Pupko, T.; Huchon, D. Indel reliability in indel-based phylogenetic inference. *Genome Biol. Evol.* **2014**, *6*, 3199–3209. [CrossRef]
42. McKenna, A.; Hanna, M.; Banks, E.; Sivachenko, A.; Cibulskis, K.; Kernytsky, A.; Garimella, K.; Altshuler, D.; Gabriel, S.; Daly, M.; et al. The Genome Analysis Toolkit: A MapReduce framework for analysing next-generation DNA sequencing data. *Genome Res.* **2010**, *20*, 1297–1303. [CrossRef]
43. Liu, S.; An, Y.; Tong, W.; Qin, X.; Samarina, L.; Guo, R.; Xia, X.; Wei, C. Characterization of genome-wide genetic variations between two varieties of tea plant (*Camellia sinensis*) and development of InDel markers for genetic research. *BMC Genomics* **2019**, *20*, 935. [CrossRef]
44. Piskol, R.; Ramaswami, G.; Li, J.B. Reliable identification of genomic variants from RNA-seq data. *Am. J. Hum. Genet.* **2013**, *93*, 641–651. [CrossRef]
45. Huang, Z.; Peng, G.; Gossen, B.D.; Yu, F. Fine mapping of a clubroot resistance gene from turnip using SNP markers identified from bulked segregant RNA-Seq. *Mol. Breed.* **2019**, *39*, 131. [CrossRef]
46. Muñoz-Espinoza, C.; Di Genova, A.; Sánchez, A.; Correa, J.; Espinoza, A.; Meneses, C.; Maass, A.; Orellana, A.; Hinrichsen, P. Identification of SNPs and InDels associated with berry size in table grapes integrating genetic and transcriptomic approaches. *BMC Plant Biol.* **2020**, *20*, e365. [CrossRef]
47. Chen, Z.; Tang, D.; Ni, J.; Li, P.; Wang, L.; Zhou, W.; Li, C.; Lan, H.; Li, L.; Liu, J. Development of genic KASP SNP markers from RNA-Seq data for map-based cloning and marker-assisted selection in maize. *BMC Plant. Biol.* **2021**, *21*, 157. [CrossRef] [PubMed]
48. Yang, G.; Chen, S.; Chen, L.; Sun, K.; Huang, C.; Zhou, D.; Huang, Y.; Wang, J.; Liu, Y.; Wang, H.; et al. Development of a core SNP arrays based on the KASP method for molecular breeding of rice. *Rice* **2019**, *12*, 21. [CrossRef] [PubMed]
49. Kaur, B.; Mavi, G.S.; Gill, M.S.; Saini, D.K. Utilization of KASP technology for wheat improvement. *Cereal Res. Commun.* **2020**, *48*, 409–421. [CrossRef]
50. Tao, D.; Kalendar, R.; Paterson, A.H. Editorial: Interspecific hybridisation in plant biology. *Front. Plant Sci.* **2022**, *13*, e1026492. [CrossRef] [PubMed]

Disclaimer/Publisher's Note: The statements, opinions and data contained in all publications are solely those of the individual author(s) and contributor(s) and not of MDPI and/or the editor(s). MDPI and/or the editor(s) disclaim responsibility for any injury to people or property resulting from any ideas, methods, instructions or products referred to in the content.

Article

Authenticity Identification of F₁ Hybrid Offspring and Analysis of Genetic Diversity in Pineapple

Panpan Jia [1,2,3,†], Shenghui Liu [1,3,†], Wenqiu Lin [1,3,4,*], Honglin Yu [1,3], Xiumei Zhang [1,3], Xiou Xiao [1,3], Weisheng Sun [1,3,4], Xinhua Lu [1,3] and Qingsong Wu [1,3,4,*]

1. South Subtropical Crop Research Institute, Chinese Academy of Tropical Agricultural Sciences, Zhanjiang 524091, China
2. School of Horticulture and Forestry, Huazhong Agricultural University, Wuhan 430070, China
3. Laboratory of Tropical Fruit Biology, Ministry of Agriculture, Zhanjiang 524091, China
4. Hainan Provincial Engineering Research Center for Pineapple Germplasm Innovation and Utilization, Zhanjiang 524091, China
* Correspondence: linwenqiu1989@163.com (W.L.); hnwuqs@163.com (Q.W.)
† These authors contributed equally to this work.

Citation: Jia, P.; Liu, S.; Lin, W.; Yu, H.; Zhang, X.; Xiao, X.; Sun, W.; Lu, X.; Wu, Q. Authenticity Identification of F₁ Hybrid Offspring and Analysis of Genetic Diversity in Pineapple. *Agronomy* 2024, *14*, 1490. https://doi.org/10.3390/agronomy14071490

Academic Editor: Matthew Hegarty

Received: 21 May 2024
Revised: 28 June 2024
Accepted: 5 July 2024
Published: 9 July 2024

Copyright: © 2024 by the authors. Licensee MDPI, Basel, Switzerland. This article is an open access article distributed under the terms and conditions of the Creative Commons Attribution (CC BY) license (https:// creativecommons.org/licenses/by/ 4.0/).

Abstract: Breeding is an effective method for the varietal development of pineapple. However, due to open pollination, it is necessary to conduct authentic identification of the hybrid offspring. In this study, we identified the authenticity of offspring and analyzed the genetic diversity within the offspring F₁ hybrids resulting from crosses between 'Josapine' and 'MD2' by single nucleotide polymorphism (SNP) markers. From the resequencing data, 26 homozygous loci that differentiate between the parents have been identified. Then, genotyping was performed on both the parents and 36 offspring to select SNP markers that are suitable for authentic identification. The genotyping results revealed that 2 sets of SNP primers, namely SNP4010 and SNP22550, successfully identified 395 authentic hybrids out of 451 hybrid offspring. We randomly selected two true hybrids and four pseudohybrids for sequencing validation, and the results have shown that two true hybrids had double peaks with A/G, while pseudohybrids had single peaks with base A or G. Further study showed that the identification based on SNP molecular markers remained consistent with the morphological identification results in the field, with a true hybridization rate of 87.58%. K-means clustering and UPGMA tree analysis revealed that the hybrid offspring could be categorized into two groups. Among them, 68.5% of offspring aggregated with MD2, while 31.95% were grouped with Josapine. The successful application of SNP marker to identify pineapple F₁ hybrid populations provides a theoretical foundation and practical reference for the future development of rapid SNP marker-based methods for pineapple hybrid authenticity and purity testing.

Keywords: pineapple; hybrid offspring; authenticity identification; SNP markers; genetic diversity

1. Introduction

Pineapple (*Ananas comosus* (L.) Merr) is indigenous to southern Brazil and Paraguay, and is one of the most economically significant tropical fruit in the world; it is also the only edible fruit of the Bromeliaceae family. The fruit is the sold fresh, dried, or in fruit juices, and is generally used as a source of flavors and fragrances. It has been cultivated in more than 80 countries or regions all around the world [1]. In recent years, the total world production for pineapples has increased, reaching about 29,361,138 metric tons in 2022 (FAOSTAT). China is one of the ten top pineapple-producing countries all over the world [2]. It is predominantly cultivated in provinces, such as Guangdong, Hainan, Yunnan, Guangxi and Fujian, which plays a pivotal role in the tropics and significantly contributes to rural revitalization in China [3]. 'Comte de Paris' is the major cultivar in China, with a history spanning nearly a century, which resulting in concentrated harvesting and diminished quality, thereby impeding the development of the pineapple industry. The selection and promotion of new varieties represent effective measures for adjusting varietal structures and enhancing industrial quality. Crossbreeding can effectively improve the objective trait, such as high yield, high quality, strong disease resistance, and good adaptability, and this is the key way to cultivate new varieties of pineapple. 'MD-2' (PRI hybrid 73-114), a hybrid bred by the Hawaiian Pineapple Research Institute in the United States with 'Smooth Cayenne' as the parent, is among the most important fresh pineapple varieties in the world and has a comparable yield and a good sugar profile to balance acid during the winter months [4]. Cabral et al. have bred new variety 'Imperia' from a crossing between 'Perolera' and 'Smooth Cayenne' [5]. In China, crossbreeding has made remarkable progress. New varieties of pineapple bred through hybrid selection in mainland China include 'Yue Cui', 'Yue Tong', 'Yue tian', and 'Renong56' [6,7].

Given the self-incompatibility of pineapples, pollination between of the same variety does not result in seed production [8]. Therefore, the crossbreeding of pineapple is mainly carried out through artificial pollination directly on the mother plants without the removal of stamens. Moreover, due to prolonged developmental pollination patterns, bees and other insects are involved in cross-pollination, leading to the outcome that the offspring might not come from the intended parents [9]. Pineapple, a perennial monocotyledonous herbaceous fruit tree, typically requires a minimum of 2.5 years from seed sowing to fruiting. Usually, the authenticity identification of hybrid offspring is investigated in field, but this is time-consuming and labor-intensive. Identifying early hybrid seedlings proves effective in preventing the wastage of resources during the subsequent breeding of individuals that do not align with the breeding objectives. Furthermore, seedling identification enables the early detection and elimination of potentially undesirable gene combinations, fostering variety improvement and supporting genetic studies [10].

Morphological characteristics observation, fluorescence in situ hybridization, chromosome counting, and molecular labeling are the common identification methods for the identification of offspring. Currently, hybrid authenticity identification primarily relies on observing the morphological characteristics of hybrid offspring in pineapple. However, the concentrated use of parental material results in diminishing morphological differences among hybrid offspring. Furthermore, morphological identification is vulnerable to the impact of external environmental factors and subjective breeder judgment. It also faces challenges, such as being time-consuming and suffering from difficulties in quantification [11,12]. Fluorescence in situ hybridization (FISH) and chromosome counting are also used to authenticate the plants. However, due to the fact that these techniques are complex, technically demanding, time-consuming, and labor-intensive, requiring strict sample preparation and processing and relying on professional equipment, they are not suitable for large-scale applications [13]. Additionally, the biochemical labeling identification method is constrained by expensive experimental techniques, time-consuming processes, and the need for professional knowledge and high-quality samples. These shortcomings restrict its efficiency in large-scale applications [13]. With the development of sequencing, molecular markers are widely use in hybrid authenticity and purity identification; this method has the advantages of high accuracy, high efficiency, not being affected by environmental conditions, and scientific data quantification, which can fully reveal the genetic information between offspring and parents from the genome level [14].

Molecular markers have been successfully applied to the identification of hybrid authenticity and genetic diversity analysis in maize, gourd, mango, rape, and citrus [15–20], including random amplified polymorphic DNA (RAPD), amplified fragment length polymorphism (AFLP), restricted fragment length polymorphism (RFLP), simple repeat sequence (SSR), and inter simple sequence repeat (ISSR) [15–20]. However, these molecular markers still have shortcomings in terms of stability, polymorphism and automation level of operation and hard to data sharing [21–25]. The single nucleotide polymorphism marker (SNP) is the latest generation of molecular markers, which refers to the DNA sequence polymorphism caused by a single nucleotide variation at the genome level [26]. This technique relies on allele-specific oligonucleotide elongation and fluorescence resonance energy transfer for signal detection [27]. SNP markers have the advantages of high polymorphism, easy detection, good stability, and high throughput, and they are suitable for genome-wide association studies [28–30]. They have also been recognized by the International Union for the Protection of New Varieties of Plants (UPOV) BMT Molecular Detection Guidelines as a standard method for determining variety or seed purity.

The advancement of whole-genome sequencing technology has significantly facilitated the application of SNP markers in verifying hybrid authenticity in several crops, such as maize, rice, melon, and cowpea [31–34]. Josia et al. used 92 SNP markers to assess the genetic purity of 26 inbred lines, 4 doubled haploid lines, and 158 single-cross maize hybrids, revealing that 67% of the inbred lines were pure, while 33% showed heterozygosity levels exceeding 5% [31]. In rice, 41 SNP markers were used for both Inpari Blas (i.e., line number 16, 21 and 22) and Inpari HDB (i.e., line number 10, 15 and 18) identification by inter-varietal genotyping [32]. Kishor et al. employed 96 genome-wide SNP markers and high-throughput Fluidigm genotyping technology to successfully distinguish 85 melon F_1 hybrids, their parental lines, and 6 PT melon breeding lines [33]. In cowpea F_1 plants, 79% of the putative were true hybrids, 14% were selfed plants, and 7% were undetermined by 17 SNP markers [34]. However, there is currently no research on the application of SNP markers to authenticate the hybrid F_1 generation population of pineapples.

In this study, both homozygous and discrepant loci of parental were screened according to the genome resequencing data of 'Josapine' and 'MD2' to develop SNP markers. SNP markers were used to identify the hybrid offspring of 'Josapine' and 'MD2', and we verified the results of SNP typing by using DNA sequencing. Subsequently, K-means and UPGMA methods were used to analyze the genetic diversity of the offspring. Fast and accurate identification of target offspring at the seedling stage saves time and money, improves selection efficiency, increases genetic diversity, and ensures excellent traits in the planted varieties. This development not only promotes the breeding process of new varieties, but also makes genetic analysis of key traits possible.

2. Materials and Methods

2.1. Plant Materials and DNA Extraction

The hybrid population was constructed with 'MD2' as the male parent and 'Josapine' as the female parent (Figure 1A,B). A total of 451 hybrid offspring (named as JM1–JM451) resulting from the crossbreeding through artificial pollination was obtained in March–April 2020. These parents and offspring were cultivated in the South Subtropical Crop Research Institute Zhanjiang, Guangdong, China (21_1002″ N; 110_16034″ E).

Figure 1. The fruits of hybrid parents. (**A**) Josapine, (**B**) MD2, and (**C**) Leaf base.

The white part of the base of fresh young pineapple leaves were extracted using the improved CTAB method [35] (Figure 1C), and the DNA of two parents and hybrid F_1 generation leaves were extracted. The DNA sample concentration was uniformly diluted with ddH$_2$O to 10–100 ng/μL. The uniform concentration DNA was stored in a 96-well plate and placed in a −20 °C refrigerator for subsequent PCR amplification.

2.2. Establishment of System for Authenticity Verification

In this study, SNP markers were obtained from pineapple germplasm resources by resequencing. Homozygous SNP loci with different parental genotypes ('MD2' and 'Josapine') were screened to authenticate the hybrid offspring. Seventeen SNPs exhibiting strong polymorphisms were selected for genetic diversity analysis, considering four parameters: minimum allele frequency (MAF) > 0.43, deletion rate < 3%, PIC > 0.37, and heterozygosity < 0.4. Primers were designed based on the SNP loci information using the primer design tool (http://www.snpway.com (accessed on 6 May 2023)) on the webpage of Wuhan Jingpeptide Biotechnology Co., Wuhan, China). For each SNP loci, two allele-specific primers and one universal primer were designed. The primers were synthesized by Wuhan Sangon Biological Co., Ltd. (Wuhan, China) with FAM- or HEX-tails (FAM tail: 5-GAAGGTGACCAAGTTCATGCT-3'; HEX tail: 5'-GAAGGTCGGAGTCAACGGATT-3'). The 2×PARMS mix reagent was procured from Wuhan Jingpeptide Biotechnology Co. (Wuhan, China). The PARMS assay was performed in a 6 μL PCR system/condition that consisted of 3 μL of PARMS master mix (Wuhan Jingpeptide Biotechnology Co., Wuhan, China), approximately equal to 0.45 μL of primer, 1.55 μL of H_2O, and 1 μL of DNA at a concentration of 10–100 ng/μL. The sample DNA of 'Josapine', 'MD2', and 451 offspring, along with the specified system, were dispensed into 384-well plates using an INTEGR pipette. The PCR program was as follows: 15 min at 94 °C, 10 touchdown cycles of 94 °C for 20 s and 65–55 °C for 60 s (decreasing by 0.7 °C percycle), and 26 cycles of 94 °C for 20 s and 57 °C for 60 s. After the PCR, fluorescence data were read and analyzed using the ABI QuantStudio6 QS6 instrument (USA, AppliedBiosystems, Waltham, MA, USA).

2.3. Authenticity Verification of Hybrids

PARMS utilizes FAM and HEX as the reporter fluorescence for the 2 alleles, with ROX fluorescence serving as the internal reference fluorescence. Genotyping is fluorescence-based, and the corresponding signals appear when the allele undergoes amplification. SNP markers are employed for genotyping parents and hybrid progenies through competitive allele-specific PCR, detecting the fluorescence signal value of the measured locus [36,37]. The genomic DNA of 'Josapine', 'MD2', and 451 hybrid offspring were genotyped using the selected primers. The genotypes of each sample were tallied. If the sample's genotype was pure, it was categorized as a pseudohybrid, and if the sample had a heterozygous genotype, it was classified as a true hybrid.

Ensembl Plant (https://plants.ensembl.org/index.html (accessed on 30 May 2023)) was utilized to fetch before and after 250 bp sequences of the 5578047 SNP site from contig18, which was from the SNP22550 marker. Specific primers were designed using NCBI Primer-BLAST (https://www.ncbi.nlm.nih.gov/tools/primer-blast/ (accessed on 16 June 2023)) to demonstrate genotyping reliability. The primers' information was followed: (F22550: 5'-ATCATTCTCGCTTGCCTCCG-3'; R22550: 5'-TCC ATGTAACTCCAGCATTTCAGA-3'). The DNA of two offspring of each genotypes used as a template were randomly selected for PCR amplification, followed by electrophoresis detection. Subsequently, these PCR products were sent to Sangon Biotech (Shanghai, China) for sequencing. If the sequencing peak graph of the SNP locus exhibited a single peak, this indicated a pure genotype. Conversely, if these sequencing peak graphs showed overlapping peaks, this signified a heterozygous genotype.

2.4. Genetic Diversity Analysis

'MD2', 'Josapine', and their hybrid offspring underwent genotyping using 18 sets of SNP markers. Hybrid offspring with a high deletion rate were excluded. Subsequently, the genotypic data of the parents and 313 hybrid offspring were analyzed. The SNP genotype data were converted into binary coded data using Excel 2016 software, with the wild type indicated as (1, 1), the mutant as (2, 2), heterozygous genotypes as (1, 2), and deletion sites recorded as (0, 0). Genetic diversity parameters, such as minimum allele frequency (MAF), gene diversity (GD), heterozygosity (He), and polymorphism information content

(PIC) were calculated using the Powermarker V3.25 software [38]. We used R language (# install. Packages ("factoextra") and # install. Packages ("cluster")) calculated the value of the K-means to carry out the cluster analysis. The two-by-two genetic distance matrix of the genotyping data was calculated, and a neighbor-joining tree (NJ) was constructed by Nei (1973) [39] using the standard genetic distance; it was visualized on MEGA 11(version 11.0.13) [40].

3. Results

3.1. Screening of Pure Co-Dominant SNP Markers in Hybrid Parents

Based on the information of pineapple whole-genome resequencing SNP data, we screened for SNP markers that exhibited different pure genotypes between 'Josapine' and 'MD2'. According to this criterion, a total of 26 sets of primers were obtained by screening between the parents of 'Josapine' and 'MD2'. These primers were distributed across LG1, LG2, LG3, LG5, LG6, LG7, LG8, LG11, LG13, LG14, LG15, LG18, LG19, LG21, and LG24, covering a total of 15 chromosomes (Supplementary Materials, Table S1). Primer screenings for the 26 sets of SNP markers involved 'Josapine', 'MD2', and 36 F_1 progenies, and only 6 representative typing results were cited in this paper (Figure 2).

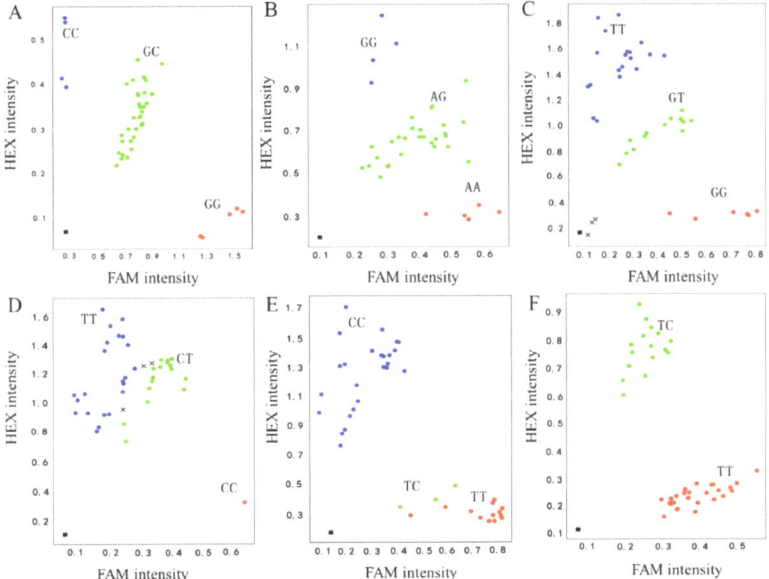

Figure 2. Typing results of some primers. (**A**) The genotyping of SNP4010, (**B**) the genotyping of SNP22550, (**C**) the genotyping of SNP10432, (**D**) the genotyping of SNP16328, (**E**) the genotyping of SNP20777, and (**F**) the genotyping of SNP25886. Note: • represents the genotype of 'Josapine', • represents the genotype of 'MD2', and • represents the genotype of the F_1 generation of the cross.

Out of 26 polymorphic SNP markers, only 18 were able to distinguish between 2 parents and 36 hybrids. Notably, SNP4010 and SNP22550 had the highest success rate in distinguishing between parents and hybrids, both at 87.5%. SNP1247, SNP12371, SNP2640, SNP7124, SNP22550, and SNP4010 surpassed a 50% success rate in hybrid identification. Furthermore, SNP12223 and SNP4855 exhibited success rates of less than 10% in identifying hybrids (Figure 3).

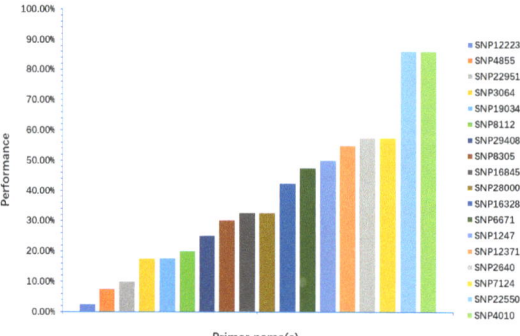

Figure 3. Efficiency of SNP markers in identification of hybrid authenticity.

3.2. Identification of Hybrid Authenticity

In order to authenticate the authenticity of the hybrid offspring, 451 hybrid offspring, 'Josapine', and 'MD2' underwent PCR amplification by SNP4010 and SNP22550. The typing results showed that 28 and 17 individual plants exhibiting genotypic consistency with 'Josapine' were identified using SNP4010 and SNP22550, with genotypes GG and AA, respectively. Additionally, 19 and 27 individual plants exhibiting genotypic consistency 'MD2' were identified using SNP4010 and SNP22550, respectively. Furthermore, a total of 404 and 407 individual plants exhibiting hybrid genotypes GC and AG were identified by SNP4010 and SNP22550, respectively (Figure 4). Combining the two markers, only those that showed both true hybrids were recognized as true hybrids, and a total of 395 true hybrids and 56 false hybrids were identified, with a true hybridization rate of 87.58% (Table 1).

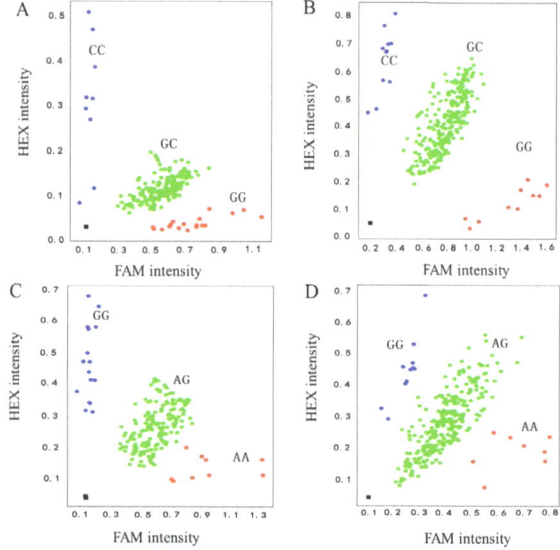

Figure 4. Genotyping results of 451 single plants. (**A**,**B**) Genotyping of SNP4010; (**C**,**D**) genotyping of SNP22550. Note: • represents the genotype of 'Josapine', • represents the genotype of 'MD2', and • represents the genotype of the F_1 generation of the cross.

Table 1. Statistical results of 'MD2' × 'Josapine' hybrid authenticity identification.

Hybrid Combination (♀ × ♂)	Total Number of F₁ (Plants)	SNP4010 Typing Statistics (Plants)		SNP22550 Typing Statistics (Plants)		True Hybrids (Plants)	Pseudohybrids (Plants)	True Hybrid Rate (%)
'MD2' × 'Josapine'	451	GG	28	AA	17	395	56	87.58
		GC	404	AG	407			
		CC	19	GG	27			

3.3. Sequencing Verification of True Hybrids and Pseudohybrids

In order to verify the accuracy of genotyping, specific primers of the SNP22550 (5578047) position, which is located in chromosome 18, were designed, namely F22550 and R22550. These primers were used to amplify the PCR of the pseudohybrids MJ83, MJ11, MJ227, MJ270, and the true hybrids MJ129, MJ173, respectively, which were identified from genotyping. The PCR products were sent for sequencing by Sangon Biotech (Shanghai, China). These results showed that MJ11 and MJ270 had single peaks at position 5578047 with base A; MJ227 and MJ83 had single peaks with base G; and MJ129 and MJ173 had peak sets with base AG (Figure 5). These results indicated that MJ11, MJ270, MJ83, and MJ227 were pseudohybrids, and that MJ129 and MJ173 were true hybrids. The DNA sequencing results were consistent with the SNP typing results, indicating the stability and credibility of the SNP typing results.

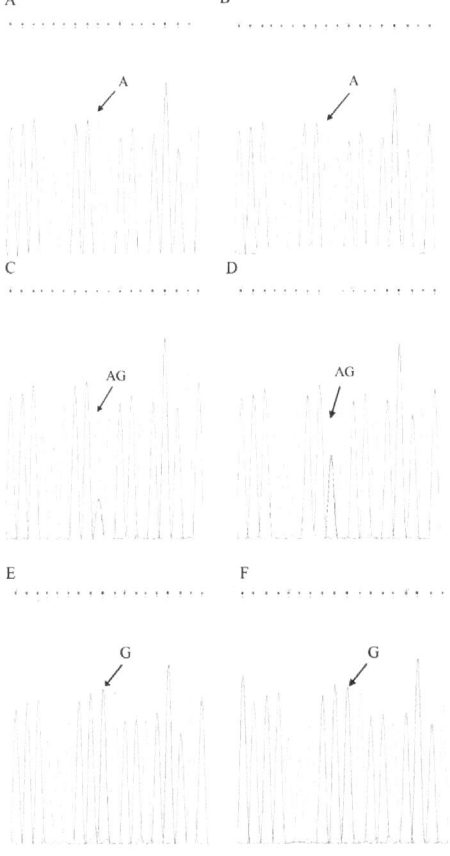

Figure 5. Leaf phenotypes and sequencing results of some true- and pseudohybrids. (**A**) MJ11, (**B**) MJ270, (**C**) MJ129, (**D**) MJ173, (**E**) MJ227, and (**F**) MJ83.

3.4. Genetic Diversity Analysis

A total of 18 SNP markers were used to analyze the genetic diversity of 313 offspring, which remained after we removed the offspring with a high deletion rate (Table S2). The statistical information on genetic diversity revealed that the mean minor allele frequency (MAF) was 0.334, with a minimum of 0.164 (SNP28156), and a maximum of 0.500 (SNP4010) (Table 2). The prevalence of a MAF between 0.20 and 0.30 was 47.06%, followed by from 0.40 to 0.50 and from 0.30 to 0.40 at 35.29% and 11.70%, respectively (Table 2, Figure 6A), indicating a relatively low overall frequency of minor alleles in the markers used. The mean gene diversity index was 0.422, ranging from 0.274 (SNP28156) to 0.500 (SNP4010), with the majority falling between 0.40 and 0.50 (58.82%), followed by 0.30 and 0.40 (35.29%) (Table 2, Figure 6B). This result suggests a relatively high genetic diversity in the markers. The mean observed heterozygosity was 0.429, the maximum value was 0.994 (SNP4010), and the minimum value was 0.191 (SNP30909) (Table 2). Due to the use of SNP markers with two alleles, heterozygosity was primarily distributed between 0.40 and 0.50 (35.29%), followed by 0.20 to 0.30, 0.30 to 0.40, and 0.50 to 0.60, all relatively evenly distributed, and with each accounting for 17.65% (Figure 6C). Polymorphic information content (PIC) was mainly below 0.40, averaging 0.331, and ranging from 0.236 (SNP28156) to 0.750 (SNP5172, SNP20855, and SNP4010) (Table 2). It was evenly distributed in the ranges of 0.30 to 0.35 and 0.35 to 0.40, followed by 0.25 to 0.30 at 17.65% (Figure 6D).

Table 2. Genetic diversity statistics for 18 SNP markers in 313 hybrid offspring.

Statistical Information	Maximum Value	Corresponds to Marker	Minimum Value	Corresponds to Marker	Average Value
Frequency of secondary effector loci	0.500	SNP4010	0.164	SNP28156	0.334
Gene diversity	0.500	SNP4010	0.274	SNP28156	0.422
Heterozygosity	0.994	SNP4010	0.191	SNP30909	0.429
Polymorphic information content	0.375	SNP5172, P20855 SNP4010	0.236	SNP28156	0.331

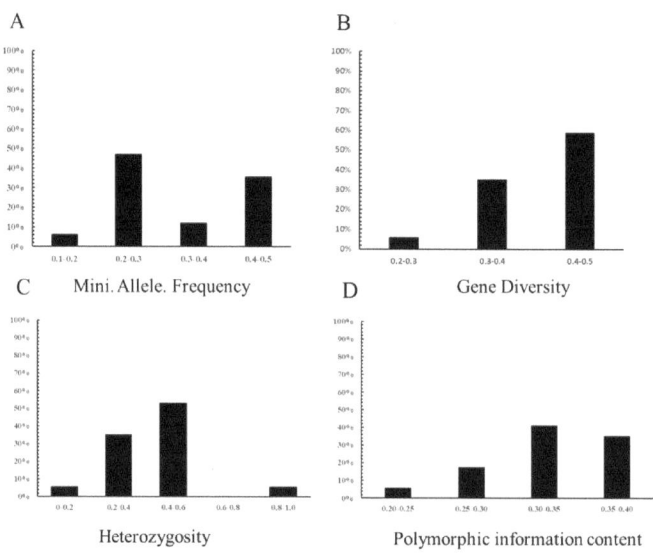

Figure 6. 18 SNP markers for 'Josapine' and 'MD2' and the distribution of 313 hybrid offsprings' genetic diversity data. The ordinate represents the proportion of the total number of 18 marks; the abscissa represents the distribution of genetic information content.

3.5. Clustering Patterns and Class Composition among Genotypes of Parents and Offspring

Here, 18 sets of SNP markers were chosen for a clustering analysis involving the parents and 313 hybrid offspring. The value of the K-means was calculated by using R language (#install.packages ("factoextra") and #install.packages ("cluster")). The results have shown that when the number of clusters K was two, the value of the gap statistic K was at its maximum. According to the K-means value, 313 hybrid offspring were clustered into two groups. Cluster I was associated with the parent 'MD2', and cluster II was associated with the parent 'Josapine' (Figure 7).

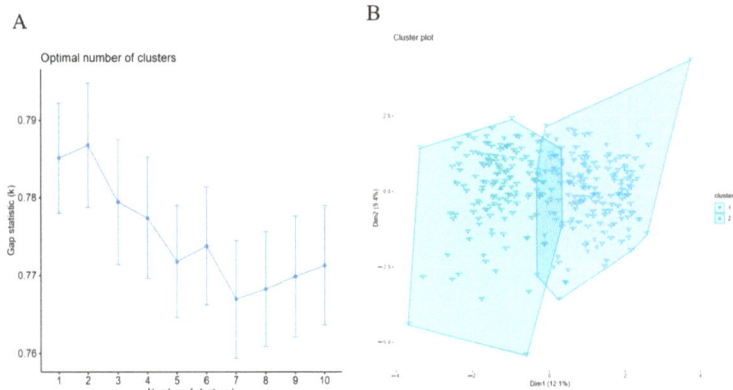

Figure 7. K-means clustering of parental and hybrid offspring genotypes (K = 2). The green color represent Cluster I and the blue color represent cluster II. (**A**) Optimal number of clusters, (**B**) K-means cluster plot.

The UPGMA clustering revealed that the 313 hybrid offspring were divided into 2 classes, consistent with the results of K-means clustering analysis. Among these, 213 hybrid offspring were grouped with the parent 'MD2', representing 68.05% of the total, while 100 hybrid offspring were associated with the parent 'Josapine', representing 31.95% of the total (Figure 8). The genetic composition of the hybrid offspring consists of two-thirds maternal genetic information and one-third paternal genetic information.

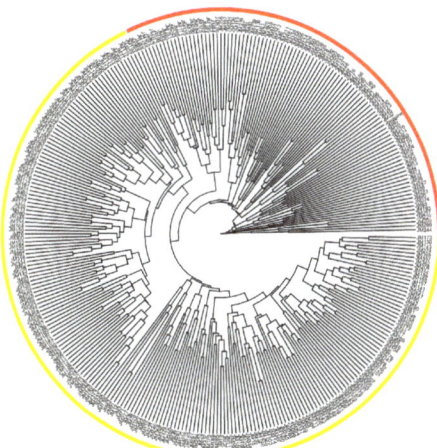

Figure 8. UPGMA tree of parental and hybrid offspring. the red color represent 213 hybrid offspring were grouped with the parent 'MD2', the yellow color represent 100 hybrid offspring were grouped with the parent 'Josapine'.

4. Discussion

The authentication of hybrid offspring plays an important role in maintaining the precision of plant genetic background, advancing gene function research, and facilitating the breeding of new varieties. In pineapple, the artificial cross-pollination process may lead to the generation of pseudohybrids, attributed to different factors, such as an unclear pollen source, a naturally heterogametic pollination, and irregular operations [32]. The presence of pseudohybrids disrupts the accuracy of genetic analysis and gene mapping and reduces breeding efficiency within the population. Usually, the authenticity of the hybrid offspring of pineapples is determined by multi-year and multi-site morphological analysis, which can time-consuming, laborious, and potentially influenced by environmental factors due to certain characteristics, leading to biased results.

Molecular markers are widely used for the authentication of hybrid offspring. In sweet tea, only one pair is required to identify all common-type hybrid offspring by RAPD, regardless of whether co-dominant or pure dominant markers are used [41]. However, there are at least five parental markers to identify true hybrids when heterozygous dominant markers are employed; the hybridization rate is 97% [42]. Eight pairs of highly polymorphic SSR markers were employed to identify sixty-five mango F_1 hybrids as true. A total of 62 true hybrids were identified, resulting in a hybridization rate of 95.38%. Although these markers can be used for authenticating offspring, they have fewer detection sites, lower throughput, and are not easily conducive to data sharing [43].

SNP molecular markers, representing the latest generation of molecular markers, can be typed using high-throughput and low-cost technology, are suitable for large-scale high-throughput testing platforms, and significantly shorten the identification time. Previous studies have shown that only one pair of pure co-dominant SNP markers is required to screen pure SNP loci with parental differences for hybrid authenticity [44]. Utilizing competitive allele-specific PCR based on single nucleotide polymorphisms to identify hybrid authenticity in cowpea, KASP-SNP markers detected true hybrids in 72% of the population with a 100% success rate [34]. In this study, we screened for the pure SNP markers SNP4010 and SNP22550, with both exhibiting differences between parents. Subsequently, we utilized these two markers to identify 395 true hybrid plants, resulting in a true hybrid rate of 87.5%.

The absence or nullity of alleles in the maternal locus has been reported to be a cause of pseudohybrids. Our study revealed the detection of 19 and 23 individual plants in pseudohybrids with the same genotype as the maternal plant for SNP4010 and SNP22550, respectively. The probable reason for this absence of the bi-parental locus may stem from abnormal chromosome exchanges, recombination, or mutations due to DNA modification. Similar results were found in mangoes, i.e., the absence of maternal loci was exhibited in the hybrid offspring by 14 AFLP markers [43]. Additionally, it is suggested that there may be interference in the binding of certain sites and individual primers, leading to site deletions [42]. The phenomenon of null alleles also exists in the identification of early poplar hybrid offspring using co-dominant SNP markers, and such hybrid offspring are not included in the final count [45]. Whether the deletion of parental loci can be used as a basis for preliminary determination of hybrids requires more in-depth research. In the present study, combined with field phenotyping, some plants with the same genotype as the parents were indeed not hybrid offspring of 'Josapine' and 'MD2'. Therefore, it is considered that only hybrid offspring with heterozygous genotypes are true hybrids.

5. Conclusions

In this study, 26 homozygous loci were used to develop SNP markers to screen the authentic hybrids offspring of 'Josapine' and 'MD2'. SNP4010 and SNP22550 were successfully used to distinguish between true- and pseudohybrid offspring. After verification of the genotyping and DNA sequencing, the true hybridization rate was 87.58%. Furthermore, these offspring were clustered into two groups, 68.5% of which were related to 'MD2', and 31.95% of which were related to 'Josapine'. These findings serve as a valuable reference for the early discernment of genuine and spurious hybrids in future crossbreeding endeavors involving pineapple and other fruit trees.

Supplementary Materials: The following supporting information can be downloaded at: https://www.mdpi.com/article/10.3390/agronomy14071490/s1, Table S1 The information of 26 primers; Table S2 The information of 18 primers.

Author Contributions: Conceptualization, P.J., S.L. and W.L.; Methodology, W.L. and Q.W.; Software, P.J. and H.Y.; Validation, P.J. and X.X.; Formal analysis, X.Z., W.S. and X.L.; Investigation, S.L. and X.L.; Resources, W.S. and X.L.; Data curation, P.J. and H.Y.; Writing—original draft, P.J. and S.L.; Writing—review and editing, W.L. and Q.W.; Visualization, X.Z. and X.L.; Supervision, W.L., H.Y. and X.X.; Project administration, X.Z. and Q.W.; Funding acquisition, Q.W. All authors have read and agreed to the published version of the manuscript.

Funding: The work was supported by the Central Public-interest Scientific Institution Basal Re-search Fund (No. 1630062024010); the National Tropical Plants Germplasm Resource Center (NTPGRC2023N TPGRC2024-027); and the National Key Research and Development Special Program of the People's Republic of China (No. 2019YFD1000505).

Data Availability Statement: The original contributions presented in the study are included in the article and Supplementary Material, further inquiries can be directed to the corresponding authors.

Conflicts of Interest: The authors declare that they have no competing interests.

References

1. Fassinou, H.V.N.; Lommen, W.J.; Agbossou, E.K.; Struik, P.C. Trade-offs of flowering and maturity synchronisation for pineapple quality. *PLoS ONE* **2015**, *10*, e0143290. [CrossRef]
2. Li, D.; Jing, M.; Dai, X.; Chen, Z.; Ma, C.; Chen, J. Current status of pineapple breeding, industrial development, and genetics in China. *Euphytica* **2022**, *218*, 85. [CrossRef]
3. Deng, C.; Yuping, L.I.; Liang, W.; Lu, Y.E. Present situation and countermeasures of pineapple industry in China. *J. Agric. Sci* **2018**, *46*, 1031–1034.
4. Chan, Y.K. Breeding of seed and vegetatively propagated tropical fruits using papaya and pineapple as examples. *Acta Hortic.* **2008**, *787*, 69–76. [CrossRef]
5. Cabral, J.R.S.; de Matos, A.P.; Junghans, D.T.; Souza, F.V.D. Pineapple genetic improvement in Brazil. *VI Int. Pineapple Symp.* **2007**, *822*, 39–46. [CrossRef]
6. Liu, Y.; Zhong, Y.; Meng, X.C. Investigation on the variability of tissue culture seedling of "Yue crisp" pineapple. *South China Fruits* **2006**, *35*, 38.
7. Liu, C.H.; Liu, Y. Phenotype Analysis of Pineapple Hybrid Line Obtained by Mixed-pollen Cross and Its Paternal Origin Analysis. *J. Agric.* **2018**, *8*, 39–45.
8. Brewbaker, J.L.; Gorrez, D.D. Genetics of Self-Incompatibility in the Monocot Genera, Ananas (Pineapple) and Gasteria. *Am. J. Bot.* **1967**, *54*, 611–616.
9. Cascante-Marín, A.; Núñez-Hidalgo, S. A Review of Breeding Systems in the Pineapple Family (Bromeliaceae, Poales). *Bot. Rev.* **2023**, *89*, 308–329. [CrossRef]
10. Jin, S.B.; Yun, S.H.; Park, J.H.; Park, S.M.; Koh, S.W.; Lee, D.H. Early identification of citrus zygotic seedlings using pollen-specific molecular markers. *Hortic. Sci. Technol.* **2015**, *33*, 598–604. [CrossRef]
11. Su, M.; Zhang, C.; Feng, S. Identification and genetic diversity analysis of hybrid offspring of azalea based on EST-SSR markers. *Sci. Rep.* **2022**, *12*, 15239. [CrossRef]
12. Hong, H.; Lee, J.; Chae, W. An economic method to identify cultivars and elite lines in radish (Raphanus sativus L.) for small seed companies and independent breeders. *Horticulture* **2023**, *9*, 140. [CrossRef]
13. Nadeem, M.A.; Nawaz, M.A.; Shahid, M.Q.; Doğan, Y.; Comertpay, G.; Yıldız, M.; Baloch, F.S. DNA molecular markers in plant breeding: Current status and recent advancements in genomic selection and genome editing. *Biotechnol. Biotechnol. Equip.* **2018**, *32*, 261–285. [CrossRef]
14. Zhou, W.; Tian, Q.Q.; Li, T.; Huang, B.; Wen, Q. Phenotypic traits and SSR molecular identification of hybrid progenies of Camellia chekiangoleosa × C. semiserrata. *Guihaia* **2014**, 1–11.
15. Cholastova, T.; Soldanova, M.; Pokorny, R. Random amplified polymorphic DNA (RAPD) and simple sequence repeat (SSR) marker efficacy for maize hybrid identification. *Afr. J. Biotechnol.* **2011**, *10*, 4794–4801.
16. Ali, A.; Jin, D.W.; Yong, B.P.; Zu, H.D.; Zhi, W.C.; Ru, K.C.; San, J.G. Molecular identification and genetic diversity analysis of Chinese sugarcane (Saccharum spp. hybrids) varieties using SSR markers. *Trop. Plant Biol.* **2017**, *10*, 194–203. [CrossRef]

17. Wang, L.P.; Dai, D.L.; Wu, X.H.; Wang, B.G.; Li, G.J. Application of AFLP markers in fast determination of seed purity in gourd, Lagenaria siceraria cv. Zhepu No. 2. *Acta Agriculturae Zhejiangensis* **2008**, *20*, 84–87.
18. Zhang, Y.; An, R.; Song, M.; Xie, C.; Wei, S.; Wang, D.; Mu, J. A set of molecular markers to accelerate breeding and determine seed purity of CMS three-line hybrids in Brassica napus. *Plants* **2023**, *12*, 1514. [CrossRef] [PubMed]
19. Golein, B.; Fifaei, R.; Ghasemi, M. Identification of zygotic and nucellar seedlings in citrus interspecific crosses by inter simple sequence repeats (ISSR) markers. *Afr. J. Biotechnol.* **2011**, *10*, 18965–18970.
20. Krishna, T.A.; Maharajan, T.; Roch, G.V.; Ramakrishnan, M.; Ceasar, S.A.; Ignacimuthu, S. Hybridization and hybrid detection through molecular markers in finger millet [Eleusine coracana (L.) Gaertn.]. *J. Crop Improv.* **2020**, *34*, 335–355. [CrossRef]
21. Agarwal, M.; Shrivastava, N.; Padh, H. Advances in molecular marker techniques and their applications in plant sciences. *Plant Cell Rep.* **2008**, *27*, 617–631. [CrossRef] [PubMed]
22. Bardakci, F. Random amplified polymorphic DNA (RAPD) markers. *Turk. J. Biol.* **2001**, *25*, 185–196.
23. Althoff, D.M.; Gitzendanner, M.A.; Segraves, K.A. The utility of amplified fragment length polymorphisms in phylogenetics: A comparison of homology within and between genomes. *Syst. Biol.* **2007**, *56*, 477–484. [CrossRef] [PubMed]
24. Sarwat, M. ISSR: A reliable and cost-effective technique for detection of DNA polymorphism. *Plant DNA Fingerprint. Barcod. Methods Protoc.* **2012**, *862*, 103–121.
25. Santhy, V.; Sandra, N.; Ravishankar, K.V.; Chidambara, B. Molecular Techniques for Testing Genetic Purity and Seed Health. In *Seed Science and Technology*; Springer: Berlin/Heidelberg, Germany, 2023; pp. 365–389. [CrossRef]
26. Al-Samarai, F.R.; Al-Kazaz, A.A. Molecular markers: An introduction and applications. *Eur. J. Mol. Biotechnol.* **2015**, *9*, 118–130. [CrossRef]
27. Ott, A.; Liu, S.; Schnable, J.C.; Yeh, C.T.E.; Wang, K.S.; Schnable, P.S. tGBS® genotyping-by-sequencing enables reliable genotyping of heterozygous loci. *Nucleic Acids Res.* **2017**, *45*, e178. [CrossRef]
28. Rafalski, A. Applications of single nucleotide polymorphisms in crop genetics. *Curr. Opin. Plant Biol.* **2002**, *5*, 94–100. [CrossRef] [PubMed]
29. Paux, E.; Sourdille, P.; Mackay, I.; Feuillet, C. Sequence-based marker development in wheat: Advances and applications to breeding. *Biotechnol. Adv.* **2012**, *30*, 1071–1088. [CrossRef]
30. Song, L.; Wang, R.; Yang, X.; Zhang, A.; Liu, D. Molecular markers and their applications in marker-assisted selection (MAS) in bread wheat (*Triticum aestivum* L.). *Agriculture* **2023**, *13*, 642. [CrossRef]
31. Josia, C.; Mashingaidze, K.; Amelework, A.B.; Kondwakwenda, A.; Musvosvi, C.; Sibiya, J. SNP-based assessment of genetic purity and diversity in maize hybrid breeding. *PLoS ONE* **2021**, *16*, e0249505. [CrossRef]
32. Utami, D.W.; Rosdianti, I.; Dewi, I.S.; Ambarwati, D.; Sisharmini, A.; Apriana, A.; Somantri, I.H. Utilization of 384 SNP genotyping technology for seed purity testing of new Indonesian rice varieties Inpari Blas and Inpari HDB. *SABRAO J. Breed. Genet.* **2016**, *48*, 416–424.
33. Kishor, D.S.; Noh, Y.; Song, W.H.; Lee, G.P.; Jung, J.K.; Shim, E.J.; Chung, S.M. Identification and purity test of melon cultivars and F1 hybrids using fluidigm-based snp markers. *Hortic. Sci. Technol.* **2020**, *38*, 686–694. [CrossRef]
34. Ongom, P.O.; Fatokun, C.; Togola, A.; Salvo, S.; Oyebode, O.G.; Ahmad, M.S.; Boukar, O. Molecular fingerprinting and hybridity authentication in cowpea using single nucleotide polymorphism based kompetitive allele-specific PCR assay. *Front. Plant Sci.* **2021**, *12*, 734117. [CrossRef]
35. Aboul-Maaty, N.A.F.; Oraby, H.A.S. Extraction of high-quality genomic DNA from different plant orders applying a modified CTAB-based method. *Bull. Natl. Res. Cent.* **2019**, *43*, 25. [CrossRef]
36. Ayalew, H.; Tsang, P.W.; Chu, C.; Wang, J.; Liu, S.; Chen, C.; Ma, X.F. Comparison of TaqMan, KASP and rhAmp SNP genotyping platforms in hexaploid wheat. *PLoS ONE* **2019**, *14*, e0217222. [CrossRef] [PubMed]
37. Dipta, B.; Sood, S.; Mangal, V.; Bhardwaj, V.; Thakur, A.K.; Kumar, V.; Singh, B. KASP: A high-throughput genotyping system and its applications in major crop plants for biotic and abiotic stress tolerance. *Mol. Biol. Rep.* **2024**, *51*, 508. [CrossRef] [PubMed]
38. Liu, K.; Muse, S.V. PowerMarker: An integrated analysis environment for genetic marker analysis. *Bioinformatics* **2005**, *21*, 2128–2129. [CrossRef]
39. Liu, C.; Zhao, N.; Jiang, Z.C.; Zhang, H.; Zhai, H.; He, S.Z.; Gao, S.P.; Liu, Q.C. Analysis of genetic diversity and population structure in sweetpotato using SSR markers. *J. Integr. Agric.* **2023**, *22*, 3408–3415. [CrossRef]
40. Kumar, S.; Stecher, G.; Li, M.; Knya, C.; Tamura, K. MEGA X: Molecular evolutionary genetics analysis across computing platforms. *Mol. Biol. Evol.* **2018**, *35*, 1547. [CrossRef]
41. Kai, C.Z.; Rong, Q.L.; Xiao, Y.B.; Shu, P.Y.; Lu, P.W.; Shi, X.J. Sexual hybrid identification in apomictic PingYiTianCha seedlings using RAPD markers. *J. Agric. Biotechnol.* **1997**, *5*, 392–396.
42. Han, G.; Xiang, S.; Wang, W.; Wei, X.; He, B.; Li, X.; Liang, G. Identification and genetic diversity of hybrid progenies from Shatian pummelo by SSR. *Sci. Agric. Sin.* **2010**, *43*, 4678–4686.
43. Li, X.; Zheng, B.; Xu, W.; Ma, X.; Wang, S.; Qian, M.; Wu, H. Identification of F1 hybrid progenies in mango based on Fluorescent SSR markers. *Horticulture* **2022**, *8*, 1122. [CrossRef]

44. Liu, W.; Xiao, Z.X.; Jiang, N.H.; Yang Xiao, Y.; Yuan P, Y.; Qiu, Y.P.; Fan, C.F.; Xiang, X. Identification of Litchi (*Litchi chinensis* Sonn) Hybrids by SNP Markers. *Mol. Plant Breed.* **2016**, *14*, 647–654.
45. Isabel, N.; Lamothe, M.; Thompson, S.L. A second-generation diagnostic single nucleotide polymorphism (SNP)-based assay, optimized to distinguish among eight poplar (*Populus* L.) species and their early hybrids. *Tree Genet. Genomes* **2013**, *9*, 621–626. [CrossRef]

Disclaimer/Publisher's Note: The statements, opinions and data contained in all publications are solely those of the individual author(s) and contributor(s) and not of MDPI and/or the editor(s). MDPI and/or the editor(s) disclaim responsibility for any injury to people or property resulting from any ideas, methods, instructions or products referred to in the content.

Article

Combining Ability, Heritability, and Heterosis for Seed Weight and Oil Content Traits of Castor Bean (*Ricinus communis* L.)

Mu Peng [1,†], Zhiyan Wang [2,†], Zhibiao He [3], Guorui Li [2], Jianjun Di [2], Rui Luo [2], Cheng Wang [2] and Fenglan Huang [2,4,5,6,7,*]

[1] Hubei Key Laboratory of Biological Resources Protection and Utilization, Hubei Minzu University, Enshi 445000, China; pengmu1025@hotmail.com
[2] College of Life Science and Food Engineering, Inner Mongolia University for the Nationalities, Tongliao 028000, China; 18247637531@163.com (Z.W.); liguorui@imun.edu.cn (G.L.); dijianjun@imun.edu.cn (J.D.); luorui@imun.edu.cn (R.L.); wangcheng@imun.edu.cn (C.W.)
[3] Tongliao Institute of Agriculture and Animal Husbandry, Tongliao 028000, China; hezhibiao@139.com
[4] Key Laboratory of Castor Breeding of the State Ethnic Affairs Commission, Tongliao 028000, China
[5] Inner Mongolia Industrial Engineering Research Center of Universities for Castor, Tongliao 028000, China
[6] Inner Mongolia Key Laboratory of Castor Breeding and Comprehensive Utilization, Tongliao 028000, China
[7] Inner Mongolia Engineering Research Center of Industrial Technology Innovation of Castor, Tongliao 028000, China
* Correspondence: huangfenglan@imun.edu.cn
† These authors contributed equally to this work and shared co-first authorship.

Citation: Peng, M.; Wang, Z.; He, Z.; Li, G.; Di, J.; Luo, R.; Wang, C.; Huang, F. Combining Ability, Heritability, and Heterosis for Seed Weight and Oil Content Traits of Castor Bean (*Ricinus communis* L.). *Agronomy* 2024, 14, 1115. https://doi.org/10.3390/agronomy14061115

Academic Editor: Ryan Whitford

Received: 8 May 2024
Revised: 20 May 2024
Accepted: 22 May 2024
Published: 23 May 2024

Copyright: © 2024 by the authors. Licensee MDPI, Basel, Switzerland. This article is an open access article distributed under the terms and conditions of the Creative Commons Attribution (CC BY) license (https:// creativecommons.org/licenses/by/ 4.0/).

Abstract: Hybridization is an important evolutionary force, and heterosis describes the phenomenon where hybrids exhibit superior traits compared to their parents. This study aimed to evaluate the one-hundred-seed weight and fatty acid content in F_1 generations, investigating the effects of different parental crosses using a 9×3 incomplete diallel design (NCII). One of the challenges faced in this study was the complexity of accurately determining the influence of both genetic and environmental factors on trait inheritance. A total of 36 F_1 crosses were analyzed for general combining ability (GCA), specific combining ability (SCA), and heritability. The results showed that the level of each index in F_1 is closely related to its parents. Significant differences in GCA and SCA were observed among parental traits in most crosses. The ratio of GCA to SCA ranged from 0 to 3, indicating the pivotal role of SCA over GCA in castor breeding efforts. High narrow-sense heritability was recorded in palmitic acid (30.98%), oleic acid (28.68%), and arachidonic acid (21.34%), suggesting that these traits are predominantly under the control of additive gene action, and hence these characters can be improved by selection. Additionally, heterosis exhibited diverse patterns across traits. Based on the evaluated combining ability, heritability, and heterosis, the inbred lines CSR181 and 20111149 were recommended for castor crossbreeding due to their potential to yield progeny with optimal oil-related traits. This research contributes valuable knowledge to the field of castor breeding, providing a foundation for developing superior castor cultivars.

Keywords: castor; hybrid combination; offspring; fatty acid

1. Introduction

One-hundred-seed weight is an important trait which reflects seed fullness and size and has important evolutionary ecological value. Traditionally, seed weight has been considered a stable trait in species; however, several research studies have shown that there are significant differences in seed weight among species or individuals. Such differences might affect seed germination or seedling traits. Generally, larger seeds have higher germination and emergence rates than smaller seeds, and could produce more vigorous seedlings to improve survival. Vandamme et al. [1] indicated that seed size greatly affected shoot and root growth beyond the seedling stage. At the same time, larger seedlings were not sensitive to density pressure. On the other hand, smaller seeds harbored relatively faster germination that larger seeds, so they could gain a better competitive advantage [2]. Recently, Zhang et al. [3] thoroughly investigated seed weight in soybean through a genome-wide association study, genomic prediction, and marker-assisted selection methods. Their results suggested that soybean seed size is controlled and regulated by many genes. These findings would help us to better understand the genetic basis of soybean seed size and promote identify of gene regulation. Richardson et al. [4] used one-hundred-seed weight as a subspecies diagnosis, seed purity, and identification, and pointed out that seed size was mainly influenced by genetic factors and limited environmental factors. However, only a few studies have reported whether the role of parental combinations in offspring has considerable implications on the seed weight and oil content of oil crops [5,6].

The type of parents in hybridization and backcrossing leads to the recombination of alleles. The interaction between environment and genetic structure could cultivate offspring that are different from their parents. Hybridization is regarded as an important evolutionary force; gene transfer among species could result in more genetic material than mutants [7]. To date, at least 30–80% of species might originate from hybridization [8]. In general, the hybrid is unfavorable to its ancestors, as offspring may exist in reproductive isolation, leading to hybrid weakness, lethality, or sterility. The first generation of hybrids (F_1), however, are the exception, harboring higher biological performance than their parents. The nature and magnitude of gene action, general combining ability (GCA), and specific combining ability (SCA) are important factors in selecting the desirable parent and crosses for the exploration of heterosis [9,10]. Combining ability analysis is one of the powerful tools to assess combining ability effects and also helps in identifying the nature of gene action involved in various quantitative characters [11]. This information is helpful to plant breeders for developing hybrid castor varieties with good quality. Therefore, the present investigation was undertaken for isolation of better combining parents for suitable hybrids.

Castor (*Ricinus communis* L.) is a special industrial oil crop. Ricinus oleic acid, one of the major contents in castor bean, is the main chemical raw material in the fatty acid of castor seed [12,13]. At present, seed weight and fatty acid content in castor seed are two main indexes that have been reported to evaluate the quality of varieties [12–14], whereas few reports have used the ratio of ricinoleic acid in crude fat. Thus far, researchers have determined the seed weight, fatty acid composition, and content in different castor germplasms [15,16]. However, little work has been conducted thus far on the heterosis and combining ability of fatty acid compositions in castor bean. Therefore, in this study, we estimated the combining ability, heritability, and heterosis of parents based on fatty acid composition. These findings will provide an in-depth understanding of the inheritance pattern of castor.

2. Materials and Methods

2.1. Plant Materials

A total of twelve ripe and healthy seeds of castor cultivars were collected and cultivated at the Academy of Agricultural Science in Tongliao, China, based on the their performance in terms of yield index, quality index, and stress resistance index [17]. During flowering, the twelve female lines (aLmAB1, aLmAB2, aLmAB3, aLmAB4, aLmAB5, aLmAB6, aLmAB7, aLmAB8, aLmAB9, aLmAB10, aLmAB11, aLmAB12), and three male lines (20102189, CSR181, 20111149) were crossed in a 9 × 3 incomplete diallel design (NCII). Thus, 36 cross-combinations were obtained. The parental combinations are listed in Table 1. Each block was 10 m × 10 m, and the plants were transplanted with 6 × 8 per block. The design of random block method was repeated three times. After emasculation, bagging, and artificial pollination, the fruits were harvested at 60 days. About 100 seeds were randomly selected from parents and F_1 for character investigation and measurement. Husk from each seed was carefully removed. Seeds were dried in an oven at 60 °C ± 2 °C for 3 h. A total of 14 indexes were determined, including 100-seed weight, 100-seed weight (dehusked seeds), 100-seed weight (dehusked oven-dried seeds), crude fatty content (dehusked oven-dried seed), crude fatty content (100-seed), behenic acid, palmitic acid, stearic acid, oleic acid, linoleic acid, linolenic acid, arachidic acid, and arachidonic acid.

Table 1. Incomplete diallel cross-table for castor.

Female \ Male	CSR181	20111149	20102189
aLmAB1	F1-4-1	F1-5-1	F1-6-1
aLmAB2	F1-4-2	F1-5-2	F1-6-2
aLmAB3	F1-4-3	F1-5-3	F1-6-3
aLmAB4	F1-4-4	F1-5-4	F1-6-4
aLmAB5	F1-4-5	F1-5-5	F1-6-5
aLmAB6	F1-4-6	F1-5-6	F1-6-6
aLmAB7	F1-4-7	F1-5-7	F1-6-7
aLmAB8	F1-4-8	F1-5-8	F1-6-8
aLmAB9	F1-4-9	F1-5-9	F1-6-9
aLmAB10	F1-4-10	F1-5-10	F1-6-10
aLmAB11	F1-4-11	F1-5-11	F1-6-11
aLmAB12	F1-4-12	F1-5-12	F1-6-12

2.1.1. Measurement of Fatty Acid Composition

The fatty acid composition ratio was calculated based on the corresponding chromatographic peaks [13]. By applying computer automatic and manual retrieval with NIST98 and the Wiley Registry of Mass Spectral Data, the fatty acid composition of castor oil was measured and analyzed [14]. Fatty acid composition and absolute content were measured according to our previous report: for more detailed information, see [12,14].

2.1.2. Statistical Data Analysis

All the measurements were taken in triplicate. Analysis of variance (ANOVA) for the recorded traits, combining ability and diallel analysis, were performed using DPS v 7.05 [18]. The GCA and SCA effect and genetic parameter measurements were conducted based on Model I and Method II of Griffing's method [19]. The relative importance of the additive or non-additive genes (σ^2GCA/σ^2SCA ratio) were conducted by Baker [20].

Mid-parent heterosis (MPH) and high-parent heterosis (HPH) were calculated according to Su et al. [21].

3. Results

3.1. Analysis of Variance between Parents and Hybrids

Firstly, we conducted variance analysis on the 14 characters of the hybrid combinations of castor (Table 2). The ANOVA results revealed significant differences ($p < 0.05$) among the 26 cross-combinations, indicating the existence of inherent variation among the 14 traits in the crosses. Therefore, the next step was to conduct variance analysis to determine combining ability. The mean performance of 9 parents and 36 F_1 crosses is listed in Tables S1 and 3. Behenic acid had the highest coefficient of variation (CV/%) (131.15), while crude fatty content (dehusked oven-dried seed) harbored the lowest, with an average of 32.85 across all crosses. aLmAB3 had the highest value of 100-seed weight (dehusked seeds) (22.55 g), 100-seed weight (dehusked oven-dried seeds) (22.11 g), crude fatty content (dehusked oven-dried seed) (67.4%), crude fatty content (100-seed) (14.99 g), palmitic acid (0.17%), linoleic acid (0.74%), linolenic acid (0.08%), and ricinoleic acid (13.18%), whereas aLmAB7 and aLmAB10 showed the lowest performances for most of the oil-related traits. For most oil-related traits, 20102189 and aLmAB3 had the higher scores. Progenies from F1-4-12 had the highest 100-seed weight, 100-seed weight (dehusked seeds), 100-seed weight (dehusked oven-dried seeds), crude fatty content (100-seed), and ricinoleic acid. F1-4-4, F1-4-6, and F1-4-7 expressed the lowest mean values for most of the traits with the exception of stearic acid. F1-5-8 yielded the highest values of palmitic acid, oleic acid, linoleic acid, and arachidic acid. Overall, the progenies from crosses involving CSR181 and 20111149 performed better for the studied traits. Moreover, the CV% in fatty acid composition was higher than that in seed oil content, indicating that hybridization is a major cause of fatty acid composition. The overall mean performance of the crosses with CSR181 and 20111149 as the male line exceeded that of the parents for the 14 traits.

Table 2. Analysis of variance (ANOVA) for the 14 characters in castor.

	Source	Blocks	Combination	Female	Male	Female × Male	Error
	Df	1	26	8	2	16	26
	100-seed weight (g)	0.0054 **	20.1208 **	7.9662	25.1849	25.5651 **	-1.7×10^{-12}
	100-seed weight (dehusked seeds) (g)	0.0054 **	13.7443 **	5.9175	21.8771	16.6411 **	-8.4×10^{-13}
	100-seed weight (dehusked oven-dried seeds) (g)	0.0054 **	13.0894 **	5.8971	22.6935	15.4851 **	-2.8×10^{-13}
	Crude fatty content (dehusked oven-dried seed) (%)	0.0054 **	9.7961 **	7.1622	12.0189	10.8353 **	5.6×10^{-12}
Mean squares	Crude fatty content (100-seed) (g)	0.0054 **	7.7347 **	3.4761	10.6526	9.4992 **	7×10^{-14}
	Behenic acid (%)	0.0014 **	0.0001 **	0.0001	0.0002 *	0	0
	Palmitic acid (%)	0.0054 **	0.001 **	0.0013	0.0004	0.0009 **	0
	Stearic acid (%)	0.0054 **	0.0019 **	0.0009	0.0033	0.0022 **	0
	Oleic acid (%)	0.0054 **	0.025 **	0.0228	0.035	0.0248 **	0
	Linoleic acid (%)	0.0054 **	0.018 **	0.0135	0.0179	0.0203 **	1×10^{-15}
	Linolenic acid (%)	0.005 **	0.0005 **	0.0005	0.0003	0.0005 **	0
	Arachidic acid (%)	0.0016 **	0 **	0	0	0	0
	Arachidonic acid (%)	0 **	0.0002 **	0.0004	0.0002	0.0002 **	0
	Ricinoleic acid (%)	0.0122 **	6.204 **	2.7411	8.2146	7.6842 **	1.75×10^{-13}

* Significant at the 5% level of significance and ** significant at the 1% level of significance.

Table 3. Mean performance of the parents and F_1 in castor.

	Mean$_{parents}$	Mean$_{F1}$	Mean$_{Total}$	CV/%
100-seed weight (g)	26.01	28.15	27.08	12.99
100-seed weight (dehusked seeds) (g)	19.09	20.86	19.975	16.2
100-seed weight (dehusked oven-dried seeds) (g)	18.64	20.38	19.51	16.19
Crude fatty content (dehusked oven-dried seed) (%)	64.13	64.07	64.1	3.63
Crude fatty content (100-seed) (g)	12.02	13.11	12.565	18.55
Behenic acid (%)	0	0.01	0.005	131.15
Palmitic acid (%)	0.14	0.15	0.145	23.04
Stearic acid (%)	0.12	0.17	0.145	67.49
Oleic acid (%)	0.45	0.49	0.47	35.34
Linoleic acid (%)	0.61	0.69	0.65	18.4
Linolenic acid (%)	0.07	0.08	0.075	16.47
Arachidic acid (%)	0.01	0.01	0.01	39.67
Arachidonic acid (%)	0.04	0.05	0.045	42.25
Ricinoleic acid (%)	10.55	11.45	11	18.53

3.2. Combining Ability Performance

GCA refers to the average performance of hybrid offspring in a certain trait after crossing one parent with multiple parents. It is primarily determined by the additive effects of genes and represents the stable heritable portion [15]. Significant differences in GCA effects were observed between different parents and traits, indicating variations in the magnitude of additive genetic effects for different parents in relation to the trait.

Among the 12 female lines, aLmAB8 had the largest GCA value for the 100-seed weight (g), 100-seed weight (dehusked seeds), 100-seed weight (dehusked oven-dried seeds), crude fatty content (dehusked oven-dried seed), palmitic acid, oleic acid, linoleic acid, arachidic acid, and ricinoleic acid, and had positive GCA effects for all the traits (Table 4). aLmAB6 and aLmAB7 had the lowest negative GCA effects for the majority of traits with the exception of crude fatty content (100-seed) and arachidonic acid, respectively. aLmAB1, aLmAB6, and aLmAB11 exhibited the fewest positive effects for 13 traits. The highest GCAs for stearic acid, behenic acid, and linolenic acid, and arachidonic acid were observed in aLmAB4, aLmAB10, aLmAB9, and aLmAB7, respectively.

The vitality of hybrids is a direct manifestation of heterozygosity, which is due to the fact that hybrids contributed by both parents may have superior gene content [22]. For 100-seed weight (g), 100-seed weight (dehusked seeds), 100-seed weight (dehusked oven-dried seeds), linoleic acid, and ricinoleic acid, the largest SCA effect was found in F-1-4-12 (Table 4). Positive SCA effects of crude fatty content (dehusked oven-dried seed) were detected in 19 of the 36 crosses, of which F1-4-2, F1-4-12, and F1-5-5 ranked in the top three. The highest and lowest SCA effects for crude fatty content (100-seed) were found in F1-4-10 (31.96) and F1-4-1 (−23.84). F1-5-8 showed the highest positive SCA effects for palmitic acid, oleic acid, and arachidic acid. The SCA effects of behenic acid ranged from 340.98 (F1-5-10) to −220.46 (F1-4-10), and stearic acid ranged from 271.96 (F1-4-4) to −123.39 (F1-5-4). F1-5-7 exhibited the highest positive SCA effects for linolenic acid and arachidonic acid. For all oil-related traits, positive SCA effects existed in ≥50% of the overall crosses, except for arachidic acid. F1-4-12, F1-6-2, and F1-6-3 were shown to be the best cross-combinations as 13 of the 14 traits had positive SCA effects. Conversely, all traits in F1-4-7 had negative SCA effects.

From the above conclusions, it can be observed that the general combining ability (GCA) of any trait varies with variation of the parents. Different traits of any parent show significant differences in their GCA. Therefore, it is challenging to find a universally good combination for all traits. Hence, information about parental genetic effects is crucial.

3.3. Estimation of Genetic Parameters

All GCA/SCA ratios were greater than 1.0, except for 100-seed weight, behenic acid, stearic acid, linolenic acid, and arachidic acid, as indicated in Table 5. The variance across all environments for the 14 traits was 0, suggesting that genetic factors might be so dominant in determining the trait that any minor environmental differences are not significant enough to cause detectable variability. Palmitic acid exhibited the highest broad-sense heritability (100%) and the highest narrow-sense heritability (30.98%), indicating that both additive and non-additive genes influenced this trait. Most traits displayed high broad-sense heritability (100%), except for arachidic acid (45.09%), indicating that these traits are predominantly controlled by additive genes. Nonetheless, broad-sense heritability exceeded narrow-sense heritability in all traits, with narrow-sense heritability for linolenic acid being zero, suggesting a predominant or exclusive role of non-additive gene action in their inheritance. The high narrow-sense heritability was recorded in palmitic acid (30.98%), oleic acid (28.68%), and arachidonic acid (21.34%), suggesting that these traits are predominantly under the control of additive gene action, and hence these characters can be improved by selection.

Table 4. Estimates of general combining ability (GCA) and specific combining ability (SCA) effects of parents and crosses for different traits in castor.

	Line	100-Seed Weight (g)	100-Seed Weight (Dehusked Seeds) (g)	100-Seed Weight (Dehusked Oven-Dried Seeds) (g)	Crude Fatty Content (Dehusked Oven-Dried Seed) (%)	Crude Fatty Content (100-Seed) (g)	Behenic Acid (%)	Palmitic Acid (%)	Stearic Acid (%)	Oleic Acid (%)	Linoleic Acid (%)	Linolenic Acid (%)	Arachidic Acid (%)	Arachidonic Acid (%)	Ricinoleic Acid (%)
Parents	20102189	−2.21	−5.97	−5.91	−2.01	2.24	−20.02	−13.78	13.04	−14.56	−8.12	−3.7	9.76	−19.31	−7.08
	20111149	2.55	3.31	3.77	1.62	−3.76	60.13	13.98	−3.04	22.68	7.91	3.7	9.76	32.41	4.07
	CSR181	−0.34	2.66	2.14	0.39	6.52	−40.11	−0.2	−10	−8.12	0.2	0	−19.51	−13.1	3.01
	aLmAB1	−2.35	−0.94	−3.32	−0.26	7.08	−20.15	−0.79	−16.43	−11.6	−2.61	−1.23	−12.2	−4.83	−3.52
	aLmAB2	6.53	4.27	5.09	1.63	−8.06	−20.15	−12.6	−10	1.62	−1.14	−1.23	−12.2	3.45	7.61
GCA effects for parents	aLmAB3	6.24	7.05	5.87	0.83	0.03	−46.63	6.3	−10	−3.25	8.16	3.7	−12.2	−13.1	7.79
	aLmAB4	−8.06	−6.91	−6.58	−2.22	−19.13	60.23	6.3	142.14	23.2	1.31	3.7	−12.2	20	−10.37
	aLmAB5	−1.64	−2.01	−1.31	1.74	−11.3	−20.15	3.94	−10	−6.73	−2.12	−1.23	−12.2	−13.1	0.47
	aLmAB6	−13.1	−15.72	−15.99	−3.43	15.45	−46.77	−22.05	48.57	−38.05	−16.8	−11.11	−12.2	−29.66	−18.3
	aLmAB7	−7.04	−10.2	−10.13	−1.26	−2.21	33.35	−10.24	−25	−3.94	−10.93	−1.23	−12.2	61.38	−12.73
	aLmAB8	11.99	13.26	13.66	1.93	9.67	6.73	37.01	24.29	55.22	19.41	3.7	104.88	3.45	13.18
	aLmAB9	2.42	−0.01	0.29	−2.2	−1.27	−20.15	−7.87	−16.43	−15.08	−3.59	8.64	−12.2	−4.83	−1.37
	aLmAB10	6.01	7.5	8.51	1.4	7	140.48	8.66	−5.71	7.19	10.11	−6.17	−12.2	−4.83	9.86
	aLmAB11	−4.2	−2.62	−2.41	1.5	−5.52	−73.38	−10.24	−12.14	−6.03	−6.53	−1.23	−12.2	−29.66	−0.55
	aLmAB12	3.18	6.33	6.3	0.33	15.35	6.6	1.57	−12.14	−2.55	4.73	3.7	17.07	11.72	7.93
	F1-4-1	−3.56	−5.11	−3.83	−1.71	−23.84	20.02	−0.39	−38.75	−4.93	−5.59	8.64	−9.76	11.03	−5.84
	F1-4-2	5.47	9.78	10.79	4.81	−20.91	20.02	18.5	−0.18	27.78	14.97	8.64	−9.76	2.76	14.69
	F1-4-3	−9.52	−20.64	−19.8	−4.02	5.88	−33.35	21.65	−51.61	−23.72	−17.82	−11.11	−9.76	−5.52	−24.84
	F1-4-4	−13.91	−17.71	−17.7	−4.38	0.84	180.38	−14.57	271.96	6.21	−19.78	3.7	−9.76	35.86	−22.86
	F1-4-5	6.71	7.91	7.02	−0.75	−15.09	20.02	9.06	−19.46	8.99	7.14	23.46	−9.76	−5.52	5.48
	F1-4-6	−1.65	1.7	0.74	−0.04	7.64	46.63	6.69	−25.89	11.08	2.73	3.7	−9.76	11.03	0.2
	F1-4-7	−10.88	−12.93	−13.56	−2.43	−12.64	−33.48	−12.2	−49.46	−31.38	−11.95	−20.99	−9.76	−80	−13.59
	F1-4-8	3.2	6.69	6.88	1.34	12.35	−86.71	−24.02	−28.04	−29.99	−2.65	−11.11	−39.02	2.76	10.61
	F1-4-9	−10.07	−13.28	−13.68	0.84	3.97	20.02	−14.57	−38.75	−18.16	−14.89	−16.05	−9.76	−13.79	−12.45
	F1-4-10	7.37	9.76	10.12	2.52	31.96	−220.46	11.42	−10.89	13.86	12.52	−1.23	78.05	11.03	12.71
	F1-4-11	7.17	6.21	5.13	−0.73	−4.56	−6.6	9.06	−17.32	8.29	5.67	8.64	−9.76	11.03	3.52
SCA effects for crosses	F1-4-12	19.66	27.62	27.89	4.56	−18.2	73.52	32.68	8.39	31.96	29.65	3.7	48.78	19.31	32.38
	F1-5-1	−4.38	−6.52	−5.91	−2.01	12.84	−60.13	13.98	9.46	−21.29	−5.46	−13.58	−9.76	−15.86	−3.43
	F1-5-2	−11.95	−11.88	−4.43	0.03	23.71	−60.13	−23.43	−16.25	−34.51	−18.68	−13.58	−9.76	−24.14	−17.19
	F1-5-3	6.12	7.5	−13.08	−5.15	−8.89	46.6	7.28	28.75	10.03	7.26	−3.7	−9.76	−7.59	13.86
	F1-5-4	17.6	20.55	8.37	4.11	−7.06	−60.27	14.37	−123.39	27.44	18.52	11.11	−9.76	−15.86	24.67
	F1-5-5	−5.34	−6.7	19.79	4.35	4.72	−60.13	−11.61	3.04	−32.42	−7.42	1.23	−9.76	−7.59	−7.77
	F1-5-6	−2.93	−4.65	−6.49	−2.51	2.64	−113.36	−21.06	3.04	−21.98	−14.76	−18.52	−9.76	−15.86	−5.53
	F1-5-7	2.02	2.98	−5.06	−2.12	8.31	46.87	2.56	24.46	29.52	5.79	30.86	−9.76	166.21	1.58
	F1-5-8	0.57	0.1	3.02	2.17	−3.19	73.48	54.53	33.04	83.12	15.09	−3.7	136.59	0.69	−2.38
	F1-5-9	6.55	9.2	−0.25	2.18	6.45	−60.13	7.28	22.32	3.07	11.66	20.99	−9.76	−15.86	8.49
	F1-5-10	−4.11	−3.6	8.99	−0.47	−16.78	340.98	4.92	11.61	1.68	−2.04	−8.64	−9.76	−15.86	−3.95

Table 4. Cont.

Parents	Line	100-Seed Weight (g)	100-Seed Weight (Dehusked Seeds) (g)	100-Seed Weight (Dehusked Oven-Dried Seeds) (g)	Crude Fatty Content (Dehusked Oven-Dried Seed) (%)	Crude Fatty Content (100-Seed) (g)	Behenic Acid (%)	Palmitic Acid (%)	Stearic Acid (%)	Oleic Acid (%)	Linoleic Acid (%)	Linolenic Acid (%)	Arachidic Acid (%)	Arachidonic Acid (%)	Ricinoleic Acid (%)
	F1-5-11	3.51	4.9	−3.69	0.29	10.08	−6.9	−4.53	18.04	−10.15	7.26	1.23	−9.76	−15.86	7.5
	F1-5-12	−7.67	−11.88	5.06	1.1	2.85	−86.88	−16.34	−14.11	−34.51	−17.21	−3.7	−39.02	−32.41	−15.85
	F1-6-1	7.94	11.63	−12.23	−3.98	11	40.11	14.37	29.29	26.22	11.05	4.94	19.51	4.83	9.27
	F1-6-2	6.48	2.1	3.77	1.62	−2.79	40.11	4.92	16.43	6.73	3.71	4.94	19.51	21.38	2.5
	F1-6-3	3.4	13.13	8.25	1.68	3.01	−13.25	14.37	22.86	13.69	10.56	14.81	19.51	13.1	10.99
	F1-6-4	−3.69	−2.84	2.3	0.34	6.21	−120.12	0.2	−148.57	−33.64	1.26	−14.81	19.51	−20	−1.81
	F1-6-5	−1.37	−1.21	11.43	−0.09	10.36	40.11	2.56	16.43	23.43	0.29	−24.69	19.51	13.1	2.3
	F1-6-6	4.58	2.95	−2.09	0.03	−10.28	66.73	14.37	22.86	10.9	12.03	14.81	19.51	4.83	5.33
	F1-6-7	8.86	9.95	−0.54	3.27	4.33	−13.39	9.65	25	1.86	6.16	−9.88	19.51	−86.21	12.01
	F1-6-8	−3.78	−6.79	4.32	2.16	−9.16	13.23	−30.51	−5	−53.13	−12.44	14.81	−97.56	−3.45	−8.23
	F1-6-9	3.52	4.08	10.54	0.26	−10.43	40.11	7.28	16.43	15.08	3.22	−4.94	19.51	29.66	3.96
	F1-6-10	−3.27	−6.17	−6.62	−3.52	−15.19	−120.52	−16.34	−0.71	−15.55	−10.48	9.88	−68.29	4.83	−8.75
	F1-6-11	−10.68	−11.11	4.68	−0.36	2.24	13.5	−4.53	−0.71	1.86	−12.93	9.88	19.51	4.83	−11.02
	F1-6-12	−11.98	−15.74	−6.43	−2.8	−3.76	13.36	−16.34	5.71	2.55	−12.44	0	−9.76	13.1	−16.53

Table 5. Estimates of genetic variances and heritability for the 14 traits in castor.

Genetic Parameter	100-Seed Weight (g)	100-Seed Weight (Dehusked Seeds) (g)	100-Seed Weight (Dehusked Oven-Dried Seeds) (g)	Crude Fatty Content (Dehusked Oven-Dried Seed) (%)	Crude Fatty Content (100-Seed) (g)	Behenic Acid (%)	Palmitic Acid (%)	Stearic Acid (%)	Oleic Acid (%)	Linoleic Acid (%)	Linolenic Acid (%)	Arachidic Acid (%)	Arachidonic Acid (%)	Ricinoleic Acid (%)
σ^2 GCA/σ^2 SCA	0.0000:1.4794	0.4894:0.2573	0.4858:0.4221	1.0036:0.0000	0.4067:0.2068	0.0000:0.0000	0.0003:0.0001	0.0000:0.0008	0.0070:0.0032	0.0021:0.0007	0.0000:0.0000	0.0000:0.0000	0.0001:0.0000	0.2155:0.1559
σ^2 GCA/σ^2 SCA	0.0000	1.9021	1.1509	1.0036	1.9666	0.0000	3	0.0000	2.1875	3	0.0000	0.0000	1	1.3823
σ^2E	0.0000	0.0000	0.0000	0.0000	0.0000	0.0000	0.0000	0.0000	0.0000	0.0000	0.0000	0.0000	0.0000	0.0000
h^2B/%	100	100	100	100	100	100	100	100	100	100	100	45.09	100	100
h^2N/%	15.13	8.42	10.73	18.93	12.46	10.66	30.98	5.03	28.68	19.60	0.00	13.28	21.34	9.97

σ^2 GCA, σ^2 SCA, and σ^2E are estimates of GCA, SCA, and environment variance, respectively; h^2B is broad-sense heritability; h^2N is narrow-sense heritability.

3.4. Heterosis in Cross-Combination

The magnitude of heterosis (MPH and HPH) in the F_1 hybrids exhibited a wide range of variation (−100% to 1900%) (Table 6). In general, the mean HPH values were lower compared to the MPH values across all traits. Approximately 73.26% of the crosses demonstrated positive MPH, while around 21.36% showed positive HPH. F1-4-12 showed the highest positive MPH and HPH for the 100-seed weight (g), 100-seed weight (dehusked seeds), 100-seed weight (dehusked oven-dried seeds), crude fatty content (100-seed), and ricinoleic acid, whereas the lowest negative MPH and HPH for these traits were observed in F1-4-4. For the crude fatty content (dehusked oven-dried seed), 50% of MPH values were positive, while only one HPH value was positive, which was detected in F1-5-3. The largest positive MPH and HPH of behenic acid were found in F1-5-10. Palmitic acid, oleic acid, linoleic acid, and arachidic acid showed similar MPH and HPH trends, of which F1-5-8 and F1-4-7 harbored the highest and lowest values, respectively. F1-4-4 had the highest positive MPH and HPH values for stearic acid. F1-5-7 had the highest MPH and HPH of linolenic acid and arachidonic acid.

Table 6. Heterosis of 14 traits in cross-combination in castor.

Cross-Combination	100-Seed Weight (g)	100-Seed Weight (Dehusked Seeds) (g)	100-Seed Weight (Dehusked Oven-Dried Seeds) (g)	Crude Fatty Content (Dehusked Oven-Dried Seed) (%)	Crude Fatty Content (100-Seed) (g)	Behenic Acid (%)	Palmitic Acid (%)	Stearic Acid (%)	Oleic Acid (%)	Linoleic Acid (%)	Linolenic Acid (%)	Arachidic Acid (%)	Arachidonic Acid (%)	Ricinoleic Acid (%)
MPH														
F1-4-1	−0.54	−3.88	−4.94	−4.07	−9.23	150	92.86	−16.67	−24.44	−4.92	14.29	0	0	−9.29
F1-4-2	18.84	18.07	20.23	4.33	24.71	150	100	41.67	24.44	19.67	14.29	0	0	25.02
F1-4-3	2.31	−12.1	−12.34	−5.29	−17.39	−100	78.57	−25	−35.56	−6.56	0	0	−25	−17.63
F1-4-4	−17.92	−24.15	−23.66	−8.69	−30.7	900	85.71	591.67	24.44	−16.39	14.29	0	50	−35.17
F1-4-5	11.34	9.17	9.12	−1.11	7.32	150	107.14	16.67	−4.44	9.84	28.57	0	−25	7.3
F1-4-6	−10.11	−12.57	−13.79	−5.57	−19.05	150	78.57	−41.67	−35.56	−11.48	0	0	−25	−18.77
F1-4-7	−13.53	−22.52	−23.02	−5.79	−27.87	150	71.43	−41.67	−44.44	−21.31	−14.29	0	−25	−27.68
F1-4-8	22.3	24.52	25.32	1.17	26.04	−100	107.14	50	20	22.95	0	100	0	26.64
F1-4-9	−2.42	−11.79	−11.75	−3.46	−15.31	150	71.43	−16.67	−42.22	−16.39	0	0	−25	−14.12
F1-4-10	20.34	21.58	23.23	1.81	24.88	−100	114.29	33.33	15.56	29.51	0	100	0	25.31
F1-4-11	9.07	6.65	5.85	−1.33	3.83	−100	92.86	16.67	−4.44	3.28	14.29	0	−25	4.08
F1-4-12	30.57	39.81	40.24	2.79	43.34	400	128.57	50	24.44	42.62	14.29	100	25	44.55
F1-5-1	3.73	4.71	4.99	1.29	5.74	150	107.14	25	−2.22	13.11	0	0	25	5.4
F1-5-2	5.15	4.56	4.72	−1.98	2.08	150	85.71	0	−2.22	0	0	0	25	2.56
F1-5-3	24.38	28.76	29.02	6.47	36.52	400	135.71	58.33	40	39.34	14.29	0	25	36.4
F1-5-4	21.34	27.76	27.9	3.66	31.86	400	142.86	58.33	86.67	44.26	28.57	0	50	28.44
F1-5-5	3.46	3.35	4.94	0.76	5.16	150	114.29	25	−8.89	11.48	14.29	0	25	5.02
F1-5-6	−6.34	−9.38	−9.55	−4.01	−13.73	−100	78.57	−25	−31.11	−13.11	−14.29	0	0	−12.89
F1-5-7	5.57	4.98	5.69	2.45	7.65	650	114.29	33.33	60	16.39	42.86	0	275	0.85
F1-5-8	24.61	27.45	28.11	5.64	34.53	650	214.29	108.33	180	60.66	14.29	0	50	24.64
F1-5-9	20.72	22.89	23.61	−1.14	21.46	150	121.43	41.67	20	31.15	42.86	200	25	20.66
F1-5-10	13.07	17.13	18.72	3.21	21.88	1900	135.71	41.67	42.22	31.15	0	0	25	19.34
F1-5-11	10.27	15.35	16.36	4.13	20.47	150	107.14	41.67	15.56	22.95	14.29	0	25	20.47
F1-5-12	6.15	6.81	6.97	−2.11	4.16	150	107.14	0	−6.67	8.2	14.29	0	25	4.36
F1-6-1	13.92	23.83	17.06	1.72	18.39	150	121.43	41.67	15.56	22.95	14.29	0	25	18.01
F1-6-2	21.95	19.12	19.74	2.26	21.71	150	100	33.33	8.89	16.39	14.29	0	25	22.75
F1-6-3	18.3	34.21	30.58	1.04	31.2	−100	128.57	41.67	11.11	34.43	28.57	0	0	32.13
F1-6-4	−4.84	1.52	2.2	−1.89	−0.33	−100	114.29	16.67	−11.11	16.39	0	0	0	−1.42
F1-6-5	4.61	8.64	9.66	5.3	14.81	150	100	33.33	17.78	11.48	−14.29	0	0	14.79
F1-6-6	−1.35	−1.78	−1.07	−0.97	−2.58	150	100	−8.33	−28.89	8.2	14.29	0	−25	−2.27
F1-6-7	9.84	11.89	12.12	−0.7	10.48	150	107.14	25	−2.22	8.2	0	0	−25	11
F1-6-8	16.76	19.22	19.37	−1.29	17.14	150	114.29	50	2.22	21.31	28.57	0	0	17.16
F1-6-9	14.3	16.61	17.11	−2.26	13.81	150	107.14	25	0	13.11	14.29	25	25	14.6
F1-6-10	10.84	13.62	13.95	−1.11	12.06	150	100	16.67	−8.89	13.11	14.29	−100	0	12.99
F1-6-11	−8.23	−2.83	−2.09	1.42	−1.25	−100	92.86	8.33	−4.44	−8.2	0	0	−25	−0.76
F1-6-12	−1.65	1.89	1.45	0.05	2.58	150	92.86	16.67	0	4.92	14.29	0	25	2.46

Table 6. Cont.

Cross-Combination	100-Seed Weight (g)	100-Seed Weight (Dehusked Seeds) (g)	100-Seed Weight (Dehusked Oven-Dried Seeds) (g)	Crude Fatty Content (Dehusked Oven-Dried Seed) (%)	Crude Fatty Content (100-Seed) (g)	Behenic Acid (%)	Palmitic Acid (%)	Stearic Acid (%)	Oleic Acid (%)	Linoleic Acid (%)	Linolenic Acid (%)	Arachidic Acid (%)	Arachidonic Acid (%)	Ricinoleic Acid (%)
HPH														
F1-4-1	−13.48	−18.63	−19.86	−9.32	−27.22	0	−23.53	−44.44	−47.69	−21.62	0	0	−33.33	−27.39
F1-4-2	3.38	−0.04	1.36	−1.37	0	0	−17.65	−5.56	−13.85	−1.35	0	0	−33.33	0.08
F1-4-3	−11	−25.59	−26.1	−10.47	−33.76	−100	−35.29	−50	−55.38	−22.97	−12.5	0	−50	−34.07
F1-4-4	−28.6	−35.79	−35.64	−13.68	−44.43	300	−29.41	361.11	−13.85	−31.08	0	0	0	−48.1
F1-4-5	−3.14	−7.58	−8.01	−6.52	−13.94	0	−11.76	−22.22	−13.85	−9.46	12.5	0	−50	−14.11
F1-4-6	−21.81	−25.99	−27.32	−10.73	−35.09	0	−35.29	−61.11	−33.85	−27.03	−12.5	0	−50	−34.98
F1-4-7	−24.78	−34.41	−35.1	−10.94	−42.16	0	−41.18	−61.11	−55.38	−35.14	−25	0	−50	−42.11
F1-4-8	6.39	5.41	5.65	−4.36	1.07	0	−11.76	0	−61.54	1.35	−12.5	0	−33.33	1.37
F1-4-9	−15.12	−25.32	−25.6	−8.74	−32.09	−100	−41.18	−44.44	−16.92	−31.08	−12.5	0	−50	−31.26
F1-4-10	4.68	2.93	3.89	−3.76	0.13	0	−5.88	−11.11	−20	6.76	−12.5	100	−33.33	0.3
F1-4-11	−5.12	−9.71	−10.76	−6.72	−16.74	−100	−23.53	−22.22	−33.85	−14.86	0	0	−50	−16.69
F1-4-12	13.58	18.36	18.23	−2.83	14.94	100	5.88	0	−13.85	17.57	0	100	−16.67	15.71
F1-5-1	−9.77	−11.35	−11.49	−4.25	−15.21	0	−11.76	−16.67	−32.31	−6.76	−12.5	0	−16.67	−15.63
F1-5-2	−8.53	−11.49	−11.71	−7.34	−18.15	0	−29.41	−33.33	−32.31	−17.57	−12.5	0	−16.67	−17.91
F1-5-3	8.19	9	8.77	0.65	9.47	100	11.76	5.56	−3.08	14.86	0	0	−16.67	9.18
F1-5-4	5.55	8.16	7.82	−2	5.74	100	17.65	5.56	29.23	18.92	12.5	0	0	2.81
F1-5-5	−10	−12.51	−11.53	−4.75	−15.68	0	−5.88	−16.67	−36.92	−8.11	0	0	−16.67	−15.93
F1-5-6	−18.53	−23.28	−23.74	−9.26	−30.82	−100	−35.29	−50	−52.31	−28.38	−25	0	−33.33	−30.27
F1-5-7	−8.16	−11.13	−10.9	−3.15	−13.68	200	−5.88	−11.11	10.77	−4.05	25	0	150	−19.27
F1-5-8	8.39	7.89	8.01	−0.13	7.87	200	76.47	38.89	93.85	32.43	0	200	0	−0.23
F1-5-9	5.02	4.04	4.21	−6.54	−2.6	0	0	−5.56	−16.92	8.11	25	0	−16.67	−3.41
F1-5-10	−1.64	−0.84	0.09	−2.43	−2.27	700	11.76	−5.56	−1.54	8.11	−12.5	0	−16.67	−4.48
F1-5-11	−4.08	−2.35	−1.9	−1.56	−3.4	0	−5.88	−5.56	−20	1.35	0	0	−33.33	−3.57
F1-5-12	−7.66	−9.58	−9.81	−7.46	−16.48	0	−11.76	−33.33	−35.38	−10.81	0	0	−16.67	−16.46
F1-6-1	−0.9	4.83	−1.31	−3.85	−5.07	0	−11.76	−20	−20	−10.81	0	0	−33.33	−5.54
F1-6-2	6.09	0.84	0.95	−3.33	−2.4	0	−17.65	−5.56	−24.62	−4.05	0	0	−16.67	−1.75
F1-6-3	2.91	13.61	10.09	−4.48	5.2	−100	5.88	−5.56	−23.08	10.81	12.5	0	−33.33	5.77
F1-6-4	−17.22	−14.06	−13.84	−7.25	−20.08	−100	−5.88	−22.22	−38.46	−4.05	−12.5	0	−33.33	−21.09
F1-6-5	−9	−8.03	−7.55	−0.46	−7.94	0	−5.88	−11.11	−18.46	−8.11	−25	0	−33.33	−8.12
F1-6-6	−14.18	−16.85	−16.6	−6.38	−21.88	0	−17.65	−38.89	−50.77	−10.81	0	0	−50	−21.78
F1-6-7	−4.45	−5.28	−5.47	−6.13	−11.41	0	−11.76	−16.67	−32.31	−10.81	−12.5	0	−50	−11.15
F1-6-8	1.57	0.93	0.63	−6.69	−6.07	0	−5.88	0	−29.23	0	12.5	0	−33.33	−6.22
F1-6-9	−0.57	−1.29	−1.27	−7.61	−8.74	0	−11.76	−16.67	−30.77	−6.76	0	0	−16.67	−8.27
F1-6-10	−3.58	−3.81	−3.93	−6.52	−10.14	0	−17.65	−22.22	−36.92	−6.76	0	−100	−33.33	−9.56
F1-6-11	−20.17	−17.74	−17.46	−4.13	−20.81	−100	−23.53	−27.78	−33.85	−24.32	−12.5	0	−50	−20.56
F1-6-12	−14.45	−13.75	−14.47	−5.42	−17.75	0	−23.53	−22.22	−30.77	−13.51	0	0	−16.67	−17.98

4. Discussion

Castor oil currently has more than 700 uses, and its market is limitless due to its diverse applications, ranging from industrial to pharmacological uses [23]. The world demand for castor oil and its derivative, castor biodiesel, is continuously growing at the rate of about 3 to 5% per annum, especially as they are essential for preventing the freezing of fuels and lubricants used in aircraft and space rockets at extremely low temperatures [24]. With the development of biodiesel, castor oil's primary market is expanding into the energy sector [25].

Heterosis is a general phenomenon of living nature in which heterozygote F_1-hybrid plants are superior to their parents in one or more traits. Tang [26] analyzed heterosis using different hybridized castor combinations, and their results showed that superiority was mainly manifested in the aspects of single plant yield, soluble sugar content, and 100-seed weight. However, no information is available about heterosis in castor oil content. In this study, a comprehensive approach was used to measure castor seed weight and oil content traits, including 100-seed weight, crude fatty content, and fatty acid composition. This related method has been successfully applied in our previous studies to screen the correlation between different castor seed sizes and oil content [13]. Our results indicated significant variations in seed weight and oil content traits among 15 parents and 36 F_1 hybrids (Tables 2 and S1), suggesting the possibility of selecting superior parents and hybrid varieties with high oil content. Specific hybrids, such as F1-4-4, F1-4-6, and F1-4-7, exhibited the lowest mean values for most of the traits with the exception of stearic acid, while F1-4-12 had the highest values for several traits, including 100-seed weight, crude fatty content, and ricinoleic acid. Both of the hybrids involving the parent CSR181 exhibited a variation in seed weight. Additionally, all detected traits in F1-4-12 were even higher than the average values of CSR181, indicating the presence of heterosis in F1-4-12. In contrast, the averages of the other 11 hybrids typically varied with the parents, indicating the complexity of oil content inheritance and the potential influence of factors such as non-additive genetic effects, gene interactions, and environmental influences.

The results from the analysis of different hybrids reveal a complex pattern of heterosis in castor plants. Approximately 73.26% of the hybrids exhibited positive MPH, signifying that these hybrids were superior to their parents in various traits. This phenomenon indicates the potential for increased productivity and desirable characteristics in these hybrids [27]. Interestingly, around 21.36% of the hybrids showed positive HPH. Despite the negative values of HPH for crude fatty content, some specific crosses had higher values than their parents. This suggests that while overall heterosis might not favor higher crude fatty content, specific combinations have the potential to produce improved varieties in terms of crude fatty content. These crosses could be crucial for developing castor plants with enhanced oil quality. Furthermore, most of the crosses showed positive MPH for essential fatty acids such as palmitic acid, behenic acid, stearic acid, and oleic acid, indicating that these hybrids outperformed their parents in these indicators, showcasing their potential for commercial applications, particularly in the production of high-quality castor oil. However, arachidonic acid and crude fatty content exhibited mainly negative or low positive MPH and HPH, suggesting that these traits might be primarily controlled by recessive effects in the parents [21].

In castor, traditional cross-breeding is one of the most effective methods to select improved varieties by exploiting heterosis [28]. This study found variations in GCA and SCA effects between the parents and crosses, with both positive and negative values for all seed weight and oil content traits. However, no evidence showed a correlation between GCA and SCA. Similar findings were reported in other studies [29]. At the same time, parents with higher GCA effect were more likely to produce crosses with excellent traits [30]. The hybrids F1-4-12, F1-6-2, and F1-6-3 exhibited positive SCA effects for 13 out of 14 seed weight and oil content traits, with F1-4-12 especially showing the highest SCA effect for most features. In these crosses, at least one parent has a high GCA effect for traits, for example, 20102189 and aLmAB12. Therefore, in breeding practices, while emphasizing the selection of parents with high GCA, it is also necessary to strengthen screening for SCA [31]. Only by combining GCA and SCA can the heterosis of castor be effectively utilized. Additionally, F1-4-12 showed positive MPH and HPH effects for most traits, indicating its high value in castor breeding. These findings provide new insights into castor breeding and contribute to the development of new castor varieties with ideal high oil content.

Many studies have reported that the significance of GCA effects indicates that parents can pass more favorable alleles, thereby transmitting these traits onto their offspring [32]. Significance of SCA effects suggests deviations in the behavior of hybrids compared to expectations based on parental GCA [33]. GCA is attributed to genes with additive effects, while SCA is associated with non-additive genetic effects [34]. Therefore, the presence of significant GCA and SCA effects implies the importance of both additive and non-additive genetic components in controlling the studied traits. In this study, GCA mean squares for 100-seed weight (dehusked seeds), crude fatty content, palmitic acid, oleic acid, linoleic acid, arachidonic acid, and behenic acid content were higher than the SCA mean squares, indicating the predominance of additive effects in controlling these traits. Non-additive gene action played a significant role in the 100-seed weight content, where the intensity of SCA effects significantly surpassed GCA effects. The GCA/SCA ratios for 100-seed weight (dehusked seeds) (1.90), 100-seed weight (dehusked oven-dried seeds) (1.15), crude fatty content (dehusked oven-dried seed) (1.00), crude fatty content (1.97), palmitic acid (3), oleic acid (2.19), linoleic acid (3), arachidonic acid (1), and behenic acid (1.38) were all greater than 1, indicating that these traits are primarily controlled by additive genetic effects. On the other hand, the GCA/SCA for 100-seed weight was less than 1 (0.56), indicating that non-additive gene effects predominantly control this trait. Similar results were reported by others [32,35], where substantial differences between GCA and SCA allow the improvement of specific traits by selecting parents that pass on the desired traits to their offspring with additive genetic effects. Genetic analysis of fatty acid content in seed oil and studies on yield-forming traits have demonstrated significant differences between GCA and SCA in the F_1 generation, indicating the dominance of additive variability over non-additive variability [36].

In our study, a limitation is that we did not collect environmental impact data across multiple years and locations. Temporal variability can significantly affect the consistency of our data, as environmental conditions may fluctuate seasonally and annually [37]. Spatial heterogeneity further complicates our analysis, as different locations have distinct environmental characteristics and practices. Additionally, ensuring data consistency and standardization across various regions and years is challenging, potentially leading to inconsistencies in our impact estimates. We also encountered gaps in the data due to accessibility issues in certain areas. Furthermore, while our findings are valuable at a local level, they may not fully capture global environmental trends. Lastly, the influence of local context-specific factors, such as economic conditions [38], may not be fully accounted for in our analysis. These limitations should be considered when interpreting our results and drawing broader conclusions.

5. Conclusions

This study represents the first comprehensive report on parental general combining ability (GCA), specific combining ability (SCA), and heritability in castor for the selection of optimal oil-related traits. Both additive (GCA) and non-additive (SCA) genetic effects significantly contributed to the variation in fatty acid composition in castor. Moreover, heterosis exhibited diverse patterns across different traits, underscoring the complexity of trait inheritance in castor hybrids. Based on the evaluation of combining ability, heritability, and heterosis, the lines CSR181 and 20111149 have been identified as promising candidates for castor crossbreeding programs. These lines show potential to produce progeny with superior oil-related traits, thereby contributing to the advancement of castor breeding efforts. This research provides valuable insights and a solid foundation for future breeding strategies aimed at improving oil yield and quality in castor plants.

Supplementary Materials: The following supporting information can be downloaded at: https://www.mdpi.com/article/10.3390/agronomy14061115/s1, Table S1: Mean performance of the parents and F1 progenies for the oil contact traits.

Author Contributions: Conceptualization, M.P. and F.H.; methodology, M.P. and F.H.; validation, M.P. and F.H.; formal analysis, M.P., Z.W., Z.H., G.L., J.D., R.L. and C.W.; investigation, M.P. and Z.W.; resources, Z.H. and F.H.; data curation, M.P. and Z.W.; writing—original draft preparation, M.P.; writing—review and editing, M.P., Z.W., Z.H., G.L., J.D., R.L., C.W. and F.H. All authors have read and agreed to the published version of the manuscript.

Funding: This work was supported by the following agencies: National Natural Science Foundation of China (2021MS03008); Grassland Talent Innovation Team of Inner Mongolia Autonomous Region—Castor Molecular Breeding Research Innovative Talent Team (2022); In 2023 and 2024, the Department of Science and Technology of Inner Mongolia Autonomous Region approved the construction project of Inner Mongolia Autonomous Region Key Laboratory of Castor Breeding and Comprehensive Utilization; Inner Mongolia University for Nationalities 2022 Basic Research Business Funds for Universities Directly Under the Autonomous Region (237); Inner Mongolia Autonomous Region Castor Industry Collaborative Innovation Center Open Fund Project (MDK2021010; MDK2022010; MDK2023001; MDK2023002).

Data Availability Statement: All data included in this study are available upon request by contact with the corresponding author.

Conflicts of Interest: The authors declare no conflicts of interest.

References

1. Vandamme, E.; Pypers, P.; Smolders, E.; Merckx, R. Seed weight affects shoot and root growth among and within soybean genotypes beyond the seedling stage: Implications for low P tolerance screening. *Plant Soil* **2016**, *401*, 65–78. [CrossRef]
2. Hendrix, S.D. Variation in Seed Weight and Its Effects on Germination in *Pastinaca sativa* L. (Umbelliferae). *Am. J. Bot.* **1984**, *71*, 795–802. [CrossRef]
3. Zhang, J.; Song, Q.; Cregan, P.B.; Jiang, G.L. Genome-wide association study, genomic prediction and marker-assisted selection for seed weight in soybean (Glycinemax). *Theor. Appl. Genet.* **2016**, *129*, 117–130. [CrossRef] [PubMed]
4. Richardson, B.A.; Ortiz, H.G.; Carlson, S.L.; Jaeger, D.M.; Shaw, N.L. Genetic and environmental effects on seed weight in subspecies of big sagebrush: Applications for restoration. *Ecosphere* **2016**, *6*, 1–13. [CrossRef]
5. Siddiqi, M.H.; Ali, S.; Bakht, J.; Khan, A.; Khan, S.A.; Khan, N. Evaluation of sunflower lines and their crossing combinations for morphological characters, yield and oil contents. *Pak. J. Bot* **2012**, *44*, 687–690.
6. Thompson, T.; Fick, G.; Cedeno, J. Maternal Control of Seed Oil Percentage in Sunflower 1. *Crop Sci.* **1979**, *19*, 617–619. [CrossRef]
7. Anderson, E. Introgressive hybridization. *Biol. Rev.* **1953**, *28*, 280–307. [CrossRef]
8. Wendel, J.F.; Rettig, J.H. Molecular Evidence for Homoploid Reticulate Evolution among Australian Species of Gossypium. *Evolution* **1991**, *45*, 694–711. [CrossRef] [PubMed]
9. Rahaman, M.A. Study of nature and magnitude of gene action in hybrid rice (*Oryza sativa* L.) through experiment of line x tester mating design. *Int. J. Appl. Res.* **2016**, *2*, 405–410.
10. Alam, A.; Ahmed, S.; Begum, M.; Sultan, M.K. Heterosis and combining ability for grain yield and its contributing characters in maize. *Bangladesh J. Agric. Res.* **2008**, *33*, 375–379. [CrossRef]
11. Mandal, A.B.; Majumder, N.D.; Bandyopadhyay, A.K. Nature and magnitude of genetic parameters of few important agronomic traits of rice under saline vis-à-vis normal soil. *J. Andaman Sci. Assoc.* **1990**, *6*, 109–114.
12. Chen, X.; Peng, M.; Huang, F.; Luo, R.; Zhao, Y.; Bao, C.; Lei, X.; Li, Y. A Quantitative Assay for Fatty Acid Composition of Castor Seed in Different Developmental Stages. *Mol. Plant Breed.* **2016**, *7*, 1–8.
13. Huang, F.; Bao, C.; Peng, M.; Zhu, G.; He, Z.; Chen, X.; Luo, R.; Zhao, Y. Chromatographic analysis of fatty acid composition in differently sized seeds of castor accessions. *Biotechnol. Biotechnol. Equip.* **2015**, *29*, 892–900. [CrossRef]
14. Huang, F.L.; Zhu, G.L.; Chen, Y.S.; Meng, F.J.; Peng, M.; Chen, X.F.; He, Z.B.; Zhang, Z.Y.; Chen, Y.J. Seed characteristics and fatty acid composition of castor (*Ricinus communis* L.) varieties in Northeast China. *Phyton* **2015**, *84*, 26–33.
15. Alves, A.A.C.; Manthey, L.; Isbell, T.; Ellis, D.; Jenderek, M.M. Diversity in oil content and fatty acid profile in seeds of wild cassava germplasm. *Ind. Crops Prod.* **2014**, *60*, 310–315. [CrossRef]
16. Román-Figueroa, C.; Cea, M.; Paneque, M.; González, M.E. Oil content and fatty acid composition in castor bean naturalized accessions under Mediterranean conditions in Chile. *Agronomy* **2020**, *10*, 1145. [CrossRef]

17. Kim, H.; Lei, P.; Wang, A.; Liu, S.; Zhao, Y.; Huang, F.; Yu, Z.; Zhu, G.; He, Z.; Tan, D. Genetic diversity of castor bean (*Ricinus communis* L.) revealed by ISSR and RAPD markers. *Agronomy* **2021**, *11*, 457. [CrossRef]
18. Tang, Q.Y.; Feng, M.G. *DPS Data Processing System for Practical Statistics*; Science: Beijing, China, 2002.
19. Griffing, B. Concept of General and Specific Combining Ability in Relation to Diallel Crossing Systems. *Aust. J. Biol. Sci.* **1955**, *9*, 463–493. [CrossRef]
20. Baker, R. Issues in diallel analysis. *Crop Sci.* **1978**, *18*, 533–536. [CrossRef]
21. Su, J.; Zhang, F.; Yang, X.; Feng, Y.; Yang, X.; Wu, Y.; Guan, Z.; Fang, W.; Chen, F. Combining ability, heterosis, genetic distance and their intercorrelations for waterlogging tolerance traits in chrysanthemum. *Euphytica* **2017**, *213*, 42. [CrossRef]
22. Nisha, S.; Veeraragavathatham, D. Heterosis and combining ability for fruit yield and its component traits in pumpkin (*Cucurbita moschata* Duch. ex Poir.). *Adv. Appl. Res.* **2014**, *6*, 158–162. [CrossRef]
23. Chakrabarty, S.; Islam, A.; Yaakob, Z.; Islam, A. Castor (*Ricinus communis*): An underutilized oil crop in the South East Asia. In *Agroecosystems—Very Complex Environmental Systems*; IntechOpen: London, UK, 2021; Volume 61.
24. Saadaoui, E.; Martín, J.J.; Tlili, N.; Cervantes, E. Castor bean (*Ricinus communis* L.) Diversity, seed oil and uses. In *Oilseed Crops: Yield and Adaptations under Environmental Stress*; Wiley: Hoboken, NJ, USA, 2017; pp. 19–33.
25. Shrirame, H.Y.; Panwar, N.; Bamniya, B. Bio diesel from castor oil-a green energy option. *Low Carbon Econ.* **2011**, *2*, 1. [CrossRef]
26. Tang, Y. Analysis on Hereterosis of Castor Agronomic Characters. *J. Inn. Mong. Univ. Natl.* **2015**, *30*, 119–125.
27. Kaleri, M.; Jatoi, W.; Baloch, M.; Mari, S.; Memon, S.; Khanzada, S.; Rajput, L.; Lal, K. Heterotic effects in sunflower hybrids for earliness and yield traits under well-watered and stressed conditions. *SABRAO J. Breed. Genet.* **2023**, *55*, 609–622. [CrossRef]
28. Senthilvel, S.; Manjunatha, T.; Lavanya, C. Castor breeding. In *Fundamentals of Field Crop Breeding*; Springer: Berlin/Heidelberg, Germany, 2022; pp. 945–970.
29. Anusha, G.; Rao, D.S.; Jaldhani, V.; Beulah, P.; Neeraja, C.; Gireesh, C.; Anantha, M.; Suneetha, K.; Santhosha, R.; Prasad, A.H. Grain Fe and Zn content, heterosis, combining ability and its association with grain yield in irrigated and aerobic rice. *Sci. Rep.* **2021**, *11*, 10579. [CrossRef] [PubMed]
30. Teodoro, L.P.R.; Bhering, L.L.; Gomes, B.E.L.; Campos, C.N.S.; Baio, F.H.R.; Gava, R.; da Silva Júnior, C.A.; Teodoro, P.E. Understanding the combining ability for physiological traits in soybean. *PLoS ONE* **2019**, *14*, e0226523. [CrossRef] [PubMed]
31. Aryana, I.G.P.M.; Wangiyana, W. Combining Ability Analyses of Diallel Crosses among Black, Red, and White Rice of Javanica and Japonica Types for Developing New Promising Lines of Rice. *Nongye Jixie Xuebao/Trans. Chin. Soc. Agric. Mach.* **2023**, *54*, 210–220.
32. Begna, T. Combining ability and heterosis in plant improvement. *Open J. Political Sci.* **2021**, *6*, 108–117.
33. Pavan, R.; Prakash, G.; Mallikarjuna, N. General and specific combining ability studies in single cross hybrids of maize (*Zea mays* L.). *Curr. Biot.* **2011**, *5*, 196–208.
34. Sughroue, J.R. Proper Analysis of the Diallel Mating Design. Ph.D. Thesis, Iowa State University, Ames, IA, USA, 1995.
35. Al-Mamun, M.; Rafii, M.Y.; Misran, A.B.; Berahim, Z.; Ahmad, Z.; Khan, M.M.H.; Oladosu, Y. Combining ability and gene action for yield improvement in kenaf (*Hibiscus cannabinus* L.) under tropical conditions through diallel mating design. *Sci. Rep.* **2022**, *12*, 9646. [CrossRef]
36. Góral, H.; Jasienski, M.; Zajac, T. Zdolność kombinacyjna odmian lnu oleistego pod względem cech plonotwórczych. *Biul. Inst. Hod. I Aklim. Roślin* **2006**, *240*, 237–242.
37. Tonkin, J.D.; Bogan, M.T.; Bonada, N.; Rios-Touma, B.; Lytle, D.A. Seasonality and predictability shape temporal species diversity. *Ecology* **2017**, *98*, 1201–1216. [CrossRef] [PubMed]
38. Hahn, R.W. The impact of economics on environmental policy. *J. Environ. Econ. Manag.* **2000**, *39*, 375–399. [CrossRef]

Disclaimer/Publisher's Note: The statements, opinions and data contained in all publications are solely those of the individual author(s) and contributor(s) and not of MDPI and/or the editor(s). MDPI and/or the editor(s) disclaim responsibility for any injury to people or property resulting from any ideas, methods, instructions or products referred to in the content.

Communication

Creation of Bacterial Blight Resistant Rice by Targeting Homologous Sequences of *Xa13* and *Xa25* Genes

Yiwang Zhu [1,2,†], Xiaohuai Yang [3,†], Peirun Luo [3,†], Jingwan Yan [1], Xinglan Cao [2,4], Hongge Qian [2], Xiying Zhu [2,4], Yujin Fan [2,4], Fating Mei [1], Meiying Fan [1], Lianguang Shang [2], Feng Wang [1,*] and Yu Zhang [5,*]

1. Institute of Biotechnology, Fujian Academy of Agricultural Sciences/Fujian Provincial Key Laboratory of Genetic Engineering for Agriculture, Fuzhou 350003, China; zhuyiwang@caas.cn (Y.Z.); yjw@fjage.org (J.Y.); 13516377413@163.com (F.M.); fanmy1127@163.com (M.F.)
2. Shenzhen Branch, Guangdong Laboratory of Lingnan Modern Agriculture, Genome Analysis Laboratory of the Ministry of Agriculture and Rural Affairs, Agricultural Genomics Institute at Shenzhen, Chinese Academy of Agricultural Sciences, Shenzhen 518124, China; xinglancao2000@163.com (X.C.); qianhg0706@stu.hunau.edu.cn (H.Q.); 18338242630@163.com (X.Z.); 17319781516@163.com (Y.F.); shanglianguang@caas.cn (L.S.)
3. Shenzhen Agricultural Science and Technology Promotion Center, Shenzhen 518000, China; yangxh1021yy@163.com (X.Y.); 15799031119@163.com (P.L.)
4. State Key Laboratory of Crop Stress Adaptation and Improvement, School of Life Sciences, Henan University, Kaifeng 475004, China
5. Food Crops Research Institute, Yunnan Academy of Agricultural Sciences (YAAS), Yunnan Seed Laboratory, Yunnan Key Laboratory for Rice Genetic Improvement, Kunming 650200, China

* Correspondence: wf@fjage.org (F.W.); zhangyu_rice@163.com (Y.Z.)
† These authors contributed equally to this work.

Abstract: Bacterial blight is a destructive disease in rice caused by *Xanthomonas oryzae* pv. *oryzae* (Xoo). Single resistance genes often have limitations in providing broad-spectrum resistance, as pathogens continuously evolve and vary. Breeding rice varieties with multiple disease resistance genes has proven to be an effective strategy for controlling bacterial blight. In this study, a single Cas9/gRNA construct was used to target the homologous sequences of *Xa13* and *Xa25* genes through destroying the target gene function, creating bacterial blight resistance in five rice varieties. These materials provide promising germplasm resources for the development of rice varieties with durable resistance to bacterial blight.

Keywords: bacterial blight resistant; CRISPR/Cas9; *Xa13*; *Xa25*; rice

Citation: Zhu, Y.; Yang, X.; Luo, P.; Yan, J.; Cao, X.; Qian, H.; Zhu, X.; Fan, Y.; Mei, F.; Fan, M.; et al. Creation of Bacterial Blight Resistant Rice by Targeting Homologous Sequences of *Xa13* and *Xa25* Genes. *Agronomy* **2024**, *14*, 800. https://doi.org/10.3390/agronomy14040800

Academic Editor: Yong-Bao Pan

Received: 15 February 2024
Revised: 4 April 2024
Accepted: 8 April 2024
Published: 12 April 2024

Copyright: © 2024 by the authors. Licensee MDPI, Basel, Switzerland. This article is an open access article distributed under the terms and conditions of the Creative Commons Attribution (CC BY) license (https://creativecommons.org/licenses/by/4.0/).

1. Introduction

Bacterial blight (BB) of rice, caused by the Gram-negative bacterium *Xanthomonas oryzae* pv. *oryzae* (*Xoo*), is one of the most devastating bacterial diseases in rice production [1]. This pathogen infects rice plants and disrupts their normal physiological functions, particularly photosynthesis. By colonizing the vascular tissues, the bacterium obstructs the movement of water and nutrients, impairs the plant's ability to absorb sunlight, and hampers the synthesis of essential carbohydrates. As a result, infected plants exhibit characteristic symptoms such as leaf wilting, chlorosis, and necrosis, ultimately leading to significant yield reductions that can reach up to 50% [2]. Evidence from crop research has shown that plants have co-evolved resistance (*R*) genes that can specifically recognize pathogen effectors to activate effector-triggered immunity [3,4]. Therefore, breeding rice varieties with major disease resistance genes is an effective and economical strategy for controlling bacterial blight disease in rice production [5].

Genome editing, facilitated by engineered nucleases, has revolutionized basic and applied biology. This technology offers significant advantages in both fundamental research and crop improvement by enabling precise modifications at specific target sequences. The emergence of CRISPR/Cas9 has greatly contributed to the widespread use of genome editing in plant breeding. It has been instrumental in exploring gene function and enhancing desirable traits in plants [6–9]. CRISPR/Cas9 is particularly effective in introducing mutations through non-homologous end joining of site-specific double–stranded DNA breaks [10,11]. Deletions were found to be the most common type of mutation, followed by insertions, with the majority of mutations being single base changes. Occasionally, base replacements and combined mutations were also observed at the target sequence. Notably, numerous agronomic trait-related genes have been successfully edited in rice using this technology [12,13].

To date, many genes that confer dominant or recessive host resistance to *Xoo* have been identified and some of them have been molecularly cloned, including *Xa1*, *Xa2*, *Xa3/Xa26*, *Xa5*, *Xa7*, *Xa10*, *Xa13*, *Xa21*, *Xa23*, *Xa25*, *Xa26*, *Xa27*, *Xa31(t)*, *Xa41(t)*, *Xa45(t)*, and *Xa46(t)* [4,14–18]. Of these, most are effective resistance (*R*) genes against bacterial blight which have been integrated into cultivated rice varieties for genetic improvement mainly through continuously hybridized or transgenic means. Nevertheless, crossbreeding is time-consuming and transgenesis introduces exogenous genes, resulting in limited applications. In contrast with the numerous examples of dominant *R* gene-mediated resistance, only a few susceptibility (*S*) genes, such as *Xa5*, *Xa13*, *Xa25*, and *Xa41(t)*, have been identified in effector-triggered susceptibility [4,14,19,20]. *Xa5* encodes a γ subunit of the transcription factor IIA (TFIIAγ), inhibiting the transfer of pathogens to disrupt disease progression [4,21]. Three *S* genes, *Xa13*, *Xa25*, and *Xa41(t)*, belong to the SWEET multiple gene family, leading to sucrose exportation into the xylem vessels and facilitating the pathogen's proliferation in rice [14,20,22]. Thus, *S* genes can serve as targets for genome editing to create new materials resistant to bacterial blight disease. However, single resistance genes generally have limitations in broad-spectrum resistance due to the continuous evolution and variation of pathogens. Therefore, the development of durable and broad-spectrum resistant materials would be of significant importance for research on the control of bacterial blight in rice.

2. Materials and Methods

2.1. Plant Material and Growth Conditions

Two indica varieties, YuXiangYouZhan (YXYZ) and WuXiangSiMian (WXSM), an indica restorer line, Shuhui143 (S143), an indica sterile line, ZhiNongS (ZNS), and a maintainer line, GuFengB (GFB), of rice were used as the wild type (WT) control and transformation host. Most of the WT and transgenic plants were cultivated in a standard greenhouse at Fuzhou Experimental Station (26.08° N, 119.28° E), Fujian Province, China. The growing season in this province begins in May and extends to mid-October.

2.2. Vectors Construction and Rice Transformation

A single-guide RNA (sgRNA) sequence targeting the flanking sequences of the 30 bp homologous sequences in the *Xa13* and *Xa25* alleles was designed. The oligonucleotides corresponding to the designed sgRNA sequences were synthesized (Table S2), and oligonucleotide dimers were cloned into a CRISPR/Cas9 plant expression vector VK005-01 (View-Solid Biotech, Beijing, China) according to the manufacturer's instructions. The resulting constructs contained a *Cas9* gene driven by the maize ubiquitin promoter and a designed sgRNA sequence under control of the rice *U6* promoter.

2.3. Rice Transformation

The Cas9/sgRNA construct was transfected into *Agrobacterium tumefaciens* EHA105 by means of electroporation. Rice calli of YXYZ, WXSM, S143, ZNS and GFB were transformed with *Agrobacterium* strains harboring the Cas9/sgRNA construct. Generation of transgenic rice plants was carried out as previously described [23].

2.4. Genotype Analysis

Rice leaf samples were subjected to genomic DNA extraction using the CTAB method [24]. PCR amplification was performed to determine the genotypes of *Xa13*/*Xa25* alleles in YXYZ, WXSM, S143, ZNS, and GFB. The annealing temperature was 58 °C and PCR amplification was performed after 35 cycles. The primers were purchased from the Sangon Biological Engineering Technology Company (Shanghai, China). Sanger sequencing was employed to analyze the PCR products. Sequencing chromatograms were deciphered following the protocol described [25]. The DNA sequences were aligned using Clustal Omega [26]. To assess the presence or absence of the Cas9/sgRNA T-DNA, PCR amplification was conducted using specific primers targeting the hygromycin phosphotransferase *(Hpt)* gene and the *Cas9* gene.

2.5. Disease Assays

Xoo populations PXO99 were kept in a −80 °C refrigerator at the Institute of Biotechnology, Fujian Academy of Agricultural Sciences. To rejuvenate the bacteria, we incubated the strains on TSA (tryptic soy agar) plates containing appropriate antibiotics and stored them at 28 °C for 2–4 days to allow the bacteria to grow adequately. The bacteria were then collected from the TSA plates and resuspended in sterilized distilled water to form a suspension. The optical density of the suspension was measured at OD_{600} = 0.5. The scissor blades were dipped into the Xoo suspension and cut at about 2 cm from the leaf tip, and then the fully expanded leaves of rice plants (6–8 weeks old) were inoculated, with 5 leaves per plant. Three or more mutants were inoculated at a time. The length of the spots was measured 15 days after inoculation. The spots were measured on each test plant.

3. Results

3.1. Selection of the Targeted Genes and sgRNA Recognition Site

Previous studies demonstrated that knockdown of the *S* genes, *Xa13* or *Xa25*, resulted in enhanced resistance to *Xoo*. We discovered a 30 bp homologous sequence in the third exon of these two genes (Figure 1; Figure S1). Hence, we designed a single Cas9/gRNA within the shared sequence to target *Xa13* and *Xa25* genes simultaneously. To test our hypothesis, we selected an elite rice variety, YuXiangYouZhan (YXYZ), which is known for its high yield and eating quality but is susceptible to bacterial blight. Sequence analysis revealed that the YXYZ contained the same 30 bp homologous fragments as described above.

3.2. Efficient CRISPR/Cas9-Mediated Targeted Mutagenesis in T_0 Transgenic Rice

Next, the Cas9/sgRNA was constructed and transformed into calli via *Agrobacterium*-mediated transformation. A total of 32 independent T_0 transgenic plants were generated from the calli of YXYZ. Subsequent genotyping of T_0 transgenic plants identified 29, 26 and 18 plants harboring *xa13*, *xa25*, and *xa13*/*xa25* double gene mutations, respectively (Table 1; Figure 2). Among the T_0 edited lines, most of the mutants obtained were bi-allelic and homozygous mutations (Table S1). These results indicate that the *xa13*, *xa25*, and *xa13*/*xa25* double mutants were successfully obtained by the CRISPR/Cas9 system relying on a single Cas9/sgRNA.

3.3. The Homozygous Lines Increased the Bacterial Blight Resistance of Rice

To characterize the bacterial blight resistance phenotype of the mutant lines, eight-week-old plants of the homozygous *xa13*, *xa25*, and *xa13*/*xa25* mutants (K4-#14, K4-#16) were inoculated with PXO99, which is a strain of *Xanthomonas oryzae* pv. *oryzae*. The results show that *xa13*/*xa25* variants had short lesions on the inoculated leaves, whereas the leaves of wild-type plants exhibited typical *Xoo* infection with longer water-soaked lesions (Figure 3). Simultaneously, the obtained *xa13* single-gene mutations of YXYZ were also inoculated with pathogens, which displayed shorter lesions on the inoculated leaves than the wild-type plants (Figure 3). These findings demonstrate that the *xa13*, *xa25*, and *xa13/xa25* double mutants possessed enhanced bacterial blight resistance compared with wild-type plants.

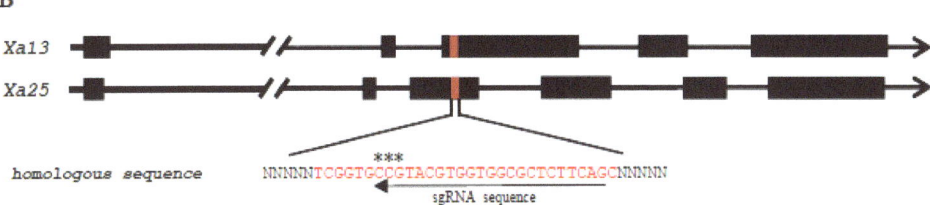

Figure 1. Homology alignment of genes *Xa13* and *Xa25*. (**A**) Sequence alignment revealed a 30 bp homologous sequence containing a gRNA recognition site. (**B**) Gene structures of *Xa13* and *Xa25*. The 30 bp homologous sequence and protospacer adjacent motif (PAM) in *Xa13* and *Xa25* are indicated by red letters and asterisks, respectively.

Table 1. T_0 plants transformed with Cas9/sgRNA constructs targeting the homologous sequences of *Xa13* and *Xa25* genes.

Rice Variety	No. of Transgenic Plants	No. of Plants with Mutations	No. of Plants with Single Gene Mutations		No. of Plants with *xa13/xa25* Double Gene Mutations
			Mutations of *xa13* (%)	Mutations of *xa25*	
YXYZ	55	32	29	26	18
WXSM	63	33	27	23	16
S143	43	26	19	19	12
ZNS	34	20	13	15	7
GFB	61	35	29	24	16
Total	256	146	127	107	69

3.4. Putative Off-Target Analysis

To identify whether the gRNA would edit a non-matching genomic sequence, the bi-allelic and homozygous *xa13/xa25* mutants YXYZ-#14, YXYZ-#16, YXYZ-#16, and YXYZ-#26 were evaluated for potential off-target effects. Considering that the greater number of mismatched bases and those closer to the PAM region are more likely to interfere with gRNA recognition, we selected possible off-target sequences based on the following two criteria: firstly, there is at least one base mismatch near the PAM region, and the total number of mismatched bases is between 1 and 5 bp. Secondly, seven groups of Cas9/sgRNA candidate targets were selected using CRISPR-P (http://skl.scau.edu.cn/targetdesign/, accessed on 14 February 2023) and BLASTN online tools. As a result, a 21 bp target point sequence was obtained (Table 2; Table S2). No obvious off-target events were found in the transgenic T_0 mutants (Table 2).

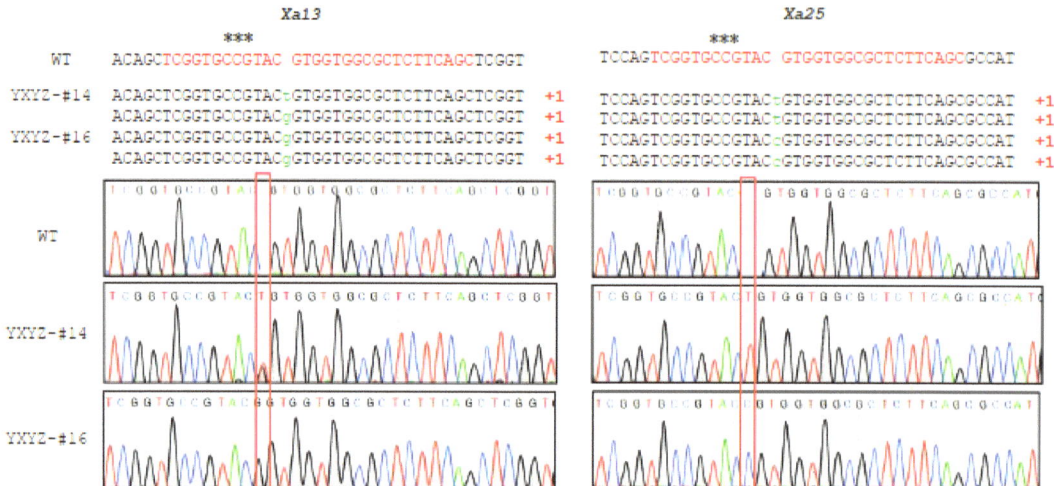

Figure 2. Mutations and corresponding sequencing chromatograms of *xa13/xa25* T$_0$ double mutants in the YXYZ. The insertions are shown in green letters. The sgRNAs are indicated by asterisks. Numbers on the right indicate the insertion length compared with *Xa13* and *Xa25*. YXYZ-#14 and YXYZ-#16 were xa13/xa25 double mutants in the YXYZ.

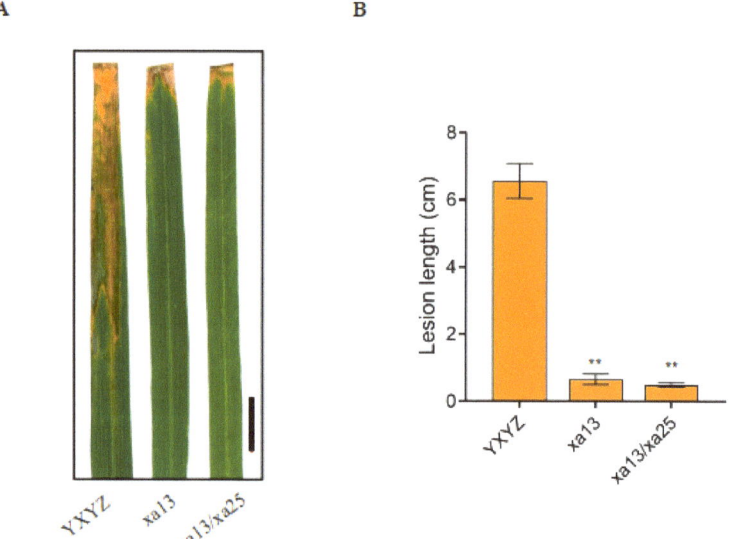

Figure 3. Resistance identification of wild-type and double-gene knockout lines. (**A**) Phenotypic characteristics of double-gene knockout lines *xa13/xa25*, the single-gene mutants *xa13* and *xa25*, and the corresponding wild-type YXYZ after inoculation with PXO99. Scale bar, 1 cm. (**B**) Statistical analysis of associated lesion lengths ($n = 5$ leaves). p values were generated by means of Student's t test. Error bars, SEM (** $p < 0.01$).

3.5. Xa13/xa25 Double Mutants Show Increased Resistance to Bacterial Blight Disease under the Background of Four High-Quality Varieties

In addition, we selected four other rice varieties to confirm whether *xa13*, *xa25*, and *xa13/xa25* double mutants displayed increased bacterial blight resistance, including an indica variety, WuXiangSiMian (WXSM), an indica restorer line, Shuhui143 (S143), an indica sterile line, ZhiNongS (ZNS), and a maintainer line, GuFengB (GFB). The genotyping of T_0 transgenic plants identified 33, 26, 20, and 35 plants with mutations in the target site in WXSM, S143, ZNS, and GFB, respectively (Figure 4; Table S1). Among the T_0 mutant plants, the mutation rate of single genes ranged from 38.2% to 75%. Overall, 25.4% (16/63), 27.9% (12/43), 20.6% (7/34), and 26.2% (16/61) were $xa13/xa25$ double mutants (Table S1). Likewise, the $xa13/xa25$ double mutants in these four cultivars also displayed more resistance to bacterial blight disease than the corresponding wild-type plants after inoculation with PXO99 (Figure 5).

Table 2. Evaluation of potential off-target sites.

Target	Name of Putative Off-Target Sites	Putative Off-Target Locus	Putative Off-Target Sequence *	No. of Mismatch Bases	No. of Plants Examined	No. of Indel Mutation
Cas9/sgRNA	OFF1	ch02: 18455705	GCTGAAGAGCGTCACCACGTACGG	2	4	0
	OFF2	ch05: 15644350	GTCGAGGAGCGCCACCACGTGCGG	4	4	0
	OFF3	ch09: 13253210	GCTGAAGGCCGTCACCACGTCCGG	4	4	0
	OFF4	ch08: 25248560	GCTGAAGCACACCACCATGTACGG	4	4	0
	OFF5	ch09:14373426	GCTGAACAGCTCCCCACGTCCGG	4	4	0
	OFF6	Ch03: 7709411	GCTGGAGAGCTCCACCACGGACGG	4	4	0
	OFF7	Ch03: 13007484	GCTCAGCAGCGCCACCGCGTACGG	5	4	0

* PAM sequence NGG is indicated in blue. Mismatch nucleotides are marked in red.

Figure 4. Mutations and corresponding sequencing chromatograms of *xa13*, *xa25*, and *xa13/xa25* T_0 double mutants in the WXSM, S143, ZNS and GFB plants. The insertions are shown in green letters. The sgRNAs are indicated by asterisks. Numbers on the right indicate the insertion length compared with *Xa13* and *Xa25*.

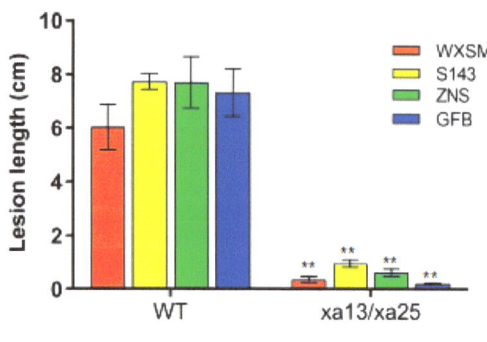

Figure 5. Phenotype of *xa13*, *xa25*, and *xa13/xa25* double mutants in the WXSM, S143, ZNS, and GFB plants infected with PXO99, and statistical analysis of the lesion lengths of these four cultivars after inoculation with PXO99 (n = 5 leaves). Data are presented as mean ± SD. Scale bar, 1 cm. *p* values were generated by means of Student's *t* test. Error bars, SEM (** $p < 0.01$).

4. Discussion

Rice bacterial blight is a serious disease that causes huge economic losses in global rice production. Breeding rice varieties resistant to bacterial blight is an important goal, but it faces challenges. Gene editing approaches in rice breeding for bacterial blight resistance in rice primarily rely on the utilization of single resistance genes or the knockout of a single susceptible gene [27,28]. However, single-gene resistance is susceptible to being overcome by the pathogen [29]. Therefore, combining or stacking multiple resistance genes in a rice variety is performed to enhance resistance stability and durability and reduce the pathogen's ability to adapt to resistance. This approach can generate progeny with multiple resistance genes, thereby improving resistance against bacterial blight. However, due to the complexity of genetic inheritance and gene interactions, as well as the labor-intensive nature of hybridization and stacking efforts, selecting suitable parents and implementing effective hybrid combinations for resistance remain technical challenges.

With the development of genomics and gene editing technology, opportunities have been created for the precision genetic improvement of crops. In particular, gene function research and genome editing technologies progressing rapidly. A large number of negative regulation genes involved in rice quality had been modified by genome engineering technologies. For instance, the knockout of *OsBADH2* using the TALEN technology produced fragrant rice, while a new glutinous rice variety was created by means of the targeted knockout of the *Waxy* gene using the CRISPR/Cas9 system [30]. Low-Cd-accumulating indica rice was generated by means of the CRISPR/Cas9-targeted mutagenesis of *OsNramp5* [31], while a reduction in seed chalkiness was achieved by editing *OsGS3* using CRISPR/Cas9, influencing the grain length–width ratio [32]. It is generally recognized that genome engineering technologies can be used to knock out negative regulatory genes efficiently for crop breeding.

5. Conclusions

In summary, we engineered a single Cas9/gRNA within a shared sequence to target both the *Xa13* and *Xa25* genes. Then, *xa13*, *xa25*, and *xa13/xa25* double mutants were simultaneously obtained in a high-quality elite rice YXYZ strain using *Agrobacterium*-mediated genetic transformation. As expected, the inoculation of leaves with PXO99 indicated that the *xa13*, *xa25*, and *xa13/xa25* double mutants displayed a markedly increased resistance to bacterial blight. In parallel, the mutants in the four other elite rice cultivars also displayed enhanced bacterial blight resistance. Taken together, these results provide an efficient and potential strategy for developing improved rice varieties with bacterial blight resistance.

Supplementary Materials: The following supporting information can be downloaded at: https://www.mdpi.com/article/10.3390/agronomy14040800/s1. Figure S1. The CDS of *Xa 13* and *Xa 25* in varieties YXYZ. The 30-bp homologous sequence is indicated by Red boxes. Table S1. Genotypes of *xa13/xa25* T$_0$ mutant plants in five rice varieties. Table S2. List of primers used in this study.

Author Contributions: Original draft preparation, Y.Z. (Yiwang Zhu), X.Y. and P.L.; data curation, X.C., J.Y. and H.Q.; investigation, X.Z., F.M., M.F. and Y.F.; data curation, visualization, L.S., F.W. and Y.Z. (Yu Zhang). All authors have read and agreed to the published version of the manuscript.

Funding: This work was supported by grants from the '5511' Collaborative Innovation Project of Fujian Academy of Agricultural Sciences (XTCXGC2021002), the Special Project for Public Welfare of the Research Institute of Fujian Province (2021R1027005) and the Shenzhen Science and Technology Program (JCYJ20230807145759008).

Data Availability Statement: Data is contained within the article or Supplementary Materials.

Conflicts of Interest: The authors declare that they have no conflicts of interest.

References

1. Mew, T.W. Current status and future prospects of research on bacterial blight of rice. *Annu. Rev. Phytopathol.* **1987**, *25*, 359–382. [CrossRef]
2. Liu, W.; Liu, J.; Triplett, L.; Leach, J.E.; Wang, G. Novel insights into rice innate immunity against bacterial and fungal pathogens. *Annu. Rev. Phytopathol.* **2014**, *52*, 213–241. [CrossRef] [PubMed]
3. Jones, J.D.G.; Dangl, J.L. The plant immune system. *Nature* **2006**, *444*, 323–329. [CrossRef] [PubMed]
4. Zhang, J.; Coaker, G.; Zhou, J.; Dong, X. Plant immune mechanisms: From reductionistic to holistic points of view. *Mol. Plant* **2020**, *13*, 1358–1378. [CrossRef] [PubMed]
5. Jiang, N.; Yan, J.; Liang, Y.; Shi, Y.; He, Z.; Wu, Y.; Zeng, Q.; Liu, X.; Peng, J. Resistance genes and their interactions with bacterial blight/leaf streak pathogens (*Xanthomonas oryzae*) in rice (*Oryza sativa* L.)—An Updated Review. *Rice* **2020**, *13*, 3. [CrossRef] [PubMed]
6. Jiang, W.; Zhou, H.; Bi, H.; Fromm, M.; Yang, B.; Weeks, D.P. Demonstration of CRISPR/Cas9/sgRNA-mediated targeted gene modification in Arabidopsis, tobacco, sorghum and rice. *Nucleic Acids Res.* **2013**, *41*, e188. [CrossRef] [PubMed]
7. Mali, P.; Yang, L.; Esvelt, K.M.; Aach, J.; Guell, M.; DiCarlo, J.E.; Church, G.M. RNA-guided human genome engineering via Cas9. *Science* **2013**, *339*, 823–826. [CrossRef]
8. Nekrasov, V.; Staskawicz, B.; Weigel, D.; Jones, J.D.; Kamoun, S. Targeted mutagenesis in the model plant Nicotiana benthamiana using Cas9 RNA-guided endonuclease. *Nat. Biotechnol.* **2013**, *31*, 691–693. [CrossRef] [PubMed]
9. Shan, Q.; Wang, Y.; Li, J.; Zhang, Y.; Chen, K.; Liang, Z.; Zhang, K.; Liu, J.; Xi, J.J.; Qiu, J.-L.; et al. Targeted genome modification of crop plants using a CRISPR-Cas9 system. *Nat. Biotechnol.* **2013**, *31*, 686–688. [CrossRef]
10. Cong, L.; Ran, F.A.; Cox, D.; Lin, S.; Barretto, R.; Habib, N.; Hsu, P.D.; Wu, X.; Jiang, W.; Marraffini, L.A.; et al. Multiplex genome engineering using CRISPR/Cas systems. *Science* **2013**, *339*, 819–823. [CrossRef]
11. Feng, Z.; Zhang, B.; Ding, W.; Liu, X.; Yang, D.-L.; Wei, P.; Cao, F.; Zhu, S.; Zhang, F.; Mao, Y.; et al. Efficient genome editing in plants using a CRISPR/Cas system. *Cell Res.* **2013**, *23*, 1229–1232. [CrossRef] [PubMed]
12. Feng, Z.; Mao, Y.; Xu, N.; Zhang, B.; Wei, P.; Yang, D.-L.; Wang, Z.; Zhang, Z.; Zheng, R.; Yang, L.; et al. Multigeneration analysis reveals the inheritance, specificity, and patterns of CRISPR/Cas-induced gene modifications in Arabidopsis. *Proc. Natl. Acad. Sci. USA* **2014**, *111*, 4632–4637. [CrossRef] [PubMed]
13. Liu, X.; Wu, S.; Xu, J.; Sui, C.; Wei, J. Application of CRISPR/Cas9 in plant biology. *Acta Pharm. Sin. B* **2017**, *7*, 292–302. [CrossRef] [PubMed]
14. Yang, B.; Sugio, A.; White, F.F. Os8N3 is a host disease susceptibility gene for bacterial blight of rice. *Proc. Natl. Acad. Sci. USA* **2006**, *103*, 10503–10508. [CrossRef] [PubMed]
15. Liu, Q.; Yuan, M.; Zhou, Y.; Li, X.; Xiao, J.; Wang, S. A paralog of the MtN3/saliva family recessively confers race-specific resistance to *Xanthomonas oryzae* in rice. *Plant Cell Environ.* **2011**, *34*, 1958–1969. [CrossRef] [PubMed]
16. Hutin, M.; Sabot, F.O.; Ghesquiere, A.; Koebnik, R.; Szurek, B. A knowledge-based molecular screen uncovers a broad-spectrum *OsSWEET14* resistance allele to bacterial blight from wild rice. *Plant J.* **2016**, *84*, 694–703. [CrossRef] [PubMed]
17. Ji, C.; Ji, Z.; Liu, B.; Cheng, H.; Liu, H.; Liu, S.; Yang, B.; Chen, G. *Xa1* allelic R genes activate rice blight resistance suppressed by interfering TAL effectors. *Plant Commun.* **2020**, *1*, 100087. [CrossRef]
18. Chen, X.; Liu, P.; Mei, L.; He, X.; Chen, L.; Liu, H.; Shen, S.; Ji, Z.; Zheng, X.; Zhang, Y.; et al. *Xa7*, a new executor *R* gene that confers durable and broad-spectrum resistance to bacterial blight disease in rice. *Plant Commun.* **2021**, *2*, 100143. [CrossRef] [PubMed]

19. Jiang, G.; Xia, Z.; Zhou, Y.; Wan, J.; Li, D.; Chen, R.; Zhai, W.; Zhu, L. Testifying the rice bacterial blight resistance gene *xa5* by genetic complementation and further analyzing *xa5* (*Xa5*) in comparison with its homolog *TFIIAg1*. *Mol. Genet. Genom.* **2006**, *275*, 354–366. [CrossRef]
20. Kim, Y.-A.; Moon, H.; Park, C.-J. CRISPR/Cas9-targeted mutagenesis of *Os8N3* in rice to confer resistance to *Xanthomonas oryzae* pv. *oryzae*. *Rice* **2019**, *12*, 67. [CrossRef]
21. Iyer, A.S.; McCouch, S.R. The rice bacterial blight resistance gene *xa5* encodes a novel form of disease resistance. *Mol. Plant-Microbe Interact.* **2004**, *17*, 1348–1354. [CrossRef]
22. Chu, Z.; Yuan, M.; Yao, J.; Ge, X.; Yuan, B.; Xu, C.; Li, X.; Fu, B.; Li, Z.; Bennetzen, J.L.; et al. Promoter mutations of an essential gene for pollen development result in disease resistance in rice. *Genes Dev.* **2006**, *20*, 1250–1255. [CrossRef]
23. Hiei, Y.; Ohta, S.; Komari, T.; Kumashiro, T. Efficient transformation of rice (*Oryza sativa* L.) mediated by Agrobacterium and sequence analysis of the boundaries of the T-DNA. *Plant J.* **1994**, *6*, 271–282. [CrossRef]
24. Porebski, S.; Bailey, L.G.; Baum, B.R. Modification of a CTAB DNA extraction protocol for plantscontaining high polysaccharide and polyphenol components. *Plant Mol. Biol. Rep.* **1997**, *15*, 8–15. [CrossRef]
25. Liu, W.; Xie, X.; Ma, X.; Li, J.; Chen, J.; Liu, Y.G. DSDecode: A webbased tool for decoding of sequencing chromatograms for genotyping of targeted mutations. *Mol. Plant* **2015**, *8*, 1431–1433. [CrossRef] [PubMed]
26. Sievers, F.; Wilm, A.; Dineen, D.; Gibson, T.J.; Karplus, K.; Li, W.; Lopez, R.; McWilliam, H.; Remmert, M.; Söding, J.; et al. Fast, scalable generation of high-quality protein multiple sequence alignments using Clustal Omega. *Mol. Syst. Biol.* **2011**, *7*, 539. [CrossRef] [PubMed]
27. Wang, M.; Li, S.; Li, H.; Song, C.; Xie, W.; Zuo, S.; Zhou, X.; Zhou, C.; Ji, Z.; Zhou, H. Genome editing of a dominant resistance gene for broad-spectrum resistance to bacterial diseases in rice without growth penalty. *Plant Biotechnol. J.* **2024**, *22*, 529–531. [CrossRef]
28. Zeng, X.; Luo, Y.; Vu, N.T.Q.; Shen, S.; Xia, K.; Zhang, M. CRISPR/Cas9-mediated mutation of *OsSWEET14* in rice cv. Zhonghua11 confers resistance to *Xanthomonas oryzae* pv. *oryzae* without yield penalty. *BMC Plant Biol.* **2020**, *20*, 313. [CrossRef] [PubMed]
29. Xu, Z.; Xu, X.; Wang, Y.; Liu, L.; Li, Y.; Yang, Y.; Liu, L.; Zou, L.; Chen, G. A varied AvrXa23-like TALE enables the bacterial blight pathogen to avoid being trapped by *Xa23* resistance gene in rice. *J. Adv. Res.* **2022**, *42*, 263–272. [CrossRef]
30. Zhang, J.; Zhang, H.; Botella, J.R.; Zhu, J.K. Generation of new glutinous rice by CRISPR/Cas9-targeted mutagenesis of the *Waxy* gene in elite rice varieties. *J. Integr. Plant Biol.* **2018**, *60*, 369–375. [CrossRef]
31. Tang, L.; Mao, B.; Li, Y.; Lv, Q.; Zhang, L.; Chen, C.; He, H.; Wang, W.; Zeng, X.; Shao, Y.; et al. Knockout of *OsNramp5* using the CRISPR/Cas9 system produces low Cd-accumulating indica rice without compromising yield. *Sci. Rep.* **2017**, *7*, 14438. [CrossRef] [PubMed]
32. Li, M.; Li, X.; Zhou, Z.; Wu, P.; Fang, M.; Pan, X.; Lin, Q.; Luo, W.; Wu, G.; Li, H. Reassessment of the four yield-related genes *Gn1a*, *DEP1*, *GS3*, and *IPA1* in rice using a CRISPR/Cas9 system. *Front. Plant Sci.* **2016**, *7*, 377. [CrossRef] [PubMed]

Disclaimer/Publisher's Note: The statements, opinions and data contained in all publications are solely those of the individual author(s) and contributor(s) and not of MDPI and/or the editor(s). MDPI and/or the editor(s) disclaim responsibility for any injury to people or property resulting from any ideas, methods, instructions or products referred to in the content.

Article

Characterization and Transcriptome Analysis Reveal Exogenous GA₃ Inhibited Rosette Branching via Altering Auxin Approach in Flowering Chinese Cabbage

Xinghua Qi [1,†], Ying Zhao [2,†], Ningning Cai [2], Jian Guan [1], Zeji Liu [1], Zhiyong Liu [1], Hui Feng [1] and Yun Zhang [1,*]

1. Department of Horticulture, Shenyang Agricultural University, 120 Dongling Road, Shenhe District, Shenyang 110866, China; qixinghua4858@163.com (X.Q.); philg1212@163.com (J.G.); liuzeji1315@163.com (Z.L.); liuzhiyong99@syau.edu.cn (Z.L.); fenghuiaaa@syau.edu.cn (H.F.)
2. School of Functional Food and Wine, Shenyang Pharmaceutical University, Shenyang 110016, China; zhaoying941013@163.com (Y.Z.); c2827377601@163.com (N.C.)
* Correspondence: zhangyun511@syau.edu.cn; Tel.: +86-024-88487143
† These authors contributed equally to this work.

Citation: Qi, X.; Zhao, Y.; Cai, N.; Guan, J.; Liu, Z.; Liu, Z.; Feng, H.; Zhang, Y. Characterization and Transcriptome Analysis Reveal Exogenous GA₃ Inhibited Rosette Branching via Altering Auxin Approach in Flowering Chinese Cabbage. *Agronomy* **2024**, *14*, 762. https://doi.org/10.3390/agronomy14040762

Academic Editor: Caterina Morcia

Received: 4 March 2024
Revised: 29 March 2024
Accepted: 5 April 2024
Published: 8 April 2024

Copyright: © 2024 by the authors. Licensee MDPI, Basel, Switzerland. This article is an open access article distributed under the terms and conditions of the Creative Commons Attribution (CC BY) license (https://creativecommons.org/licenses/by/4.0/).

Abstract: Branching is an important agronomic trait that is conducive to plant architecture and yield in flowering Chinese cabbage. Plant branching is regulated by a complex network mediated by hormones; gibberellin (GA) is one of the important hormones which is involved in the formation of shoot branching. Research on the regulatory mechanism of GA influencing rosette branch numbers is limited for flowering Chinese cabbage. In this study, the exogenous application of 600 mg/L GA₃ effectively inhibited rosette branching and promoted internode elongation in flowering Chinese cabbage. RNA-Seq analysis further found that these DEGs were significantly enriched in 'the plant hormone signal transduction' pathways, and auxin-related genes were significantly differentially expressed between MB and MB_GA. The upregulation of auxin (AUX) and the upregulation of auxin/indole-3-acetic acid (AUX/IAA), as well as the downregulation of SMALL AUXIN-UPREGULATED RNA (*SAUR*), were found in the negative regulation of the rosette branching. The qRT-PCR results showed that the expression of AUX/IAA and SAUR from IAA gene family members were consistent with the results of transcriptome data. Phytohormone profiling by targeted metabolism revealed that endogenous auxin contents were significantly increased in MB_GA. Transcriptome and metabolome analysis clarified the main plant hormones and genes underlying the rosette branching in flowering Chinese cabbage, confirming that auxin could inhibit rosette branching. In this regard, the results present a novel angle for revealing the mechanism of gibberellin acting on the branching architecture in flowering Chinese cabbage.

Keywords: flowering Chinese cabbage; branching; gibberellin; auxin; transcriptome

1. Introduction

Flowering Chinese cabbage (*Brassica campestris* L. ssp. chinensis [L.] Makino var. utilis Tsen et Lee) is an important and popular vegetable because of its high nutrient content and good flavor [1]. The major edible part of flowering Chinese cabbage is the stalk, the development of which has a direct impact on plant yield. The main species in Guangdong has one remarkable stalk per plant that can be harvested only once in production. A local variety, 'Zengcheng flowering Chinese cabbage', found in the Zengcheng District of Guangdong Province (23°26′ N, 113°8′ E), can be harvested multiple times because of the strong development of axillary branches on the rosette shoot.

The shoot branching of flowering Chinese cabbage is a vital characteristic that is conducive to improving crop yield and adaptability [2]. The branching system includes the main branch developed from the apical meristem, the lateral branch developed from the axillary meristem, and the primary rosette branch that grows from the basal rosette leaves [3]. The bud can immediately grow, hibernate, or remain dormant after the formation of axillary buds. The activity of axillary buds is closely related to different endogenous and environmental stimuli, like plant hormones, nutrients, and light. The hormones involved in the network regulation of shoot branching include auxins, strigolactones (SLs), and cytokinins (CKs). However, further determination is needed on how these hormones interact to regulate the activation and growth of axillary buds in flowering Chinese cabbage.

Auxin is associated with axillary bud outgrowth and shoot branching formation by apical dominance [4]. The proliferation apex generates excess auxin, which is transported downwards to the stem to repress the growth of axillary buds and promote apical elongation. Removing the apex releases its inhibition on axillary buds, triggering the formation of branches. At present, some genes related to auxin transport that affect shoot branching formation have been reported. A mutant of *OsIAA6* in rice showed abnormal tiller outgrowth because of the regulation of the auxin transporter *OsPIN1* and tillering suppressor *OsTB1* [5]. In *Arabidopsis*, the major auxin influx carrier is AUXIN INFLUX CARRIER PROTEIN 1 (AUX1), whereas the main auxin efflux carrier is PIN-FORMED1 (PIN1) because it facilitates efficient auxin export from cells [6–9]. Exploring the genes involved in auxin transport is of great significance for revealing the shoot branching mechanism of flowering Chinese cabbage. Furthermore, other phytohormones also play important roles during the initiation of shoot branching. CKs promote shoot branching by upregulating the expression of *PIN3*, *PIN4*, and *PIN7* on the basal plasma membrane of xylem parenchyma cells [10]. The branching regulation of SL signaling pathways involves the MAX genes, which lead to the polyubiquitination and 26S-proteasome-mediated degradation of D53 by the receptor D14 and the recruitment of the SCF complex in *Arabidopsis* [11]. ABA inhibits the growth of lateral buds by the transcription factor BRANCHED1 (BRC1), which promotes ABA synthesis by activating the expression of the transcription factor family HOMEOBOX PROTEIN (HB), including NCED3 in Arabidopsis [12]. In addition, the SL and BR signaling pathways commonly regulate and control shoot branching via MAX2-induced degradation of *bri1*-EMS-suppressor 1 (BES1) [13].

Bioactive gibberellin (GA) is a diterpenoid plant hormone that undergoes biosynthesis through complex pathways and controls almost all plant development processes throughout the plant life cycle [14]. The bioactive GAs are GA_1, GA_3, GA_4, and GA_7 in higher plants [15]. GAs is also a plant hormone that plays an important role in the regulation of bud outgrowth, the specific role of GAs in branching has been characterized well [16–18]. GAs are often considered to be branching inhibitors because the shoot branching phenotype has been observed in GA biosynthesis mutants of *Arabidopsis* and in genetically modified plants of different species lacking GAs. For example, gibberellin inhibits the formation of axillary buds in *Arabidopsis* by regulating the activity of the DELLA-SPL9 complex [19]. Transgenic rice overexpressing *GA2oxs*, which normally limits bioactive GA levels, exhibited early and increased tillering by inactivating endogenous and exogenous bioactive GAs [20]. In rice, high GA levels can stimulate APC/CTE to facilitate the degradation of *MOC1* in the AM, leading to restricted tillering [21]. In aspen plants, a decrease in the biological activity of GAs leads to significantly higher lateral buds compared to the wild type [22]. GAs negatively regulate the formation of axillary buds by overexpression of the GA metabolic gene GA2ox, leading to an increased number of tillers in turfgrass [23]. In a suppressor of runnerless strawberry mutant (*srl*), FveRGA1, encoding a DELLA protein, negatively regulates stolon formation [24]. Contrary to the above results, some studies have shown that GAs can promote shoot branching. In the perennial woody plant *Jatropha curcas,* GAs and CKs both negatively influence BRC1 and BRC2 expression to synergistically promote lateral bud outgrowth [25]. The growth of axillary buds in a GA biosynthesis mutant is restricted, while the application of bioactive GA rescues the phenotype in strawberries [26]. In sweet cherries, spraying GA_3 can promote the growth of lateral branches [27]. However, GA_3 under light-induced sugar metabolism contributes to bud burst, as been reported in rose [28]. Moreover, when exogenous spraying of GA_3, GA_4, and GR_{24} was performed on aspen trees, GA_3 induced shoot branching from axillary bud abscission whereas GA_4 promoted outgrowth [29].

Plant growth regulators have been widely used to improve crop yield and quality. In recent years, there has been increasing interest in the exogenous application of GA_3 to regulate plant growth and development. In a previous study, we constructed a gene mapping population in flowering Chinese cabbage using the non-branching double haploid line 'CX010' and the multiple branching double haploid line 'CX020' as parents, and the two tandem genes *BraA07g041560.3C* and *BraA07g041570.3C*, which encode gibberellin 2-oxidase that in turn acts on C19 gibberellins, were obtained by a map-based cloning strategy [30]. One of the two genes negatively regulated the formation of shoot branching in flowering Chinese cabbage. Hence, we chose to spray GA_3 to explore the potential molecular mechanism of gibberellin affecting flowering Chinese cabbage branching. In this study, comparative transcriptome analysis was used to characterize the gene expression profiles and high-performance liquid chromatography (HPLC) was used to test the levels of plant hormone content. We analyzed GA_3 to inhibit the branching of rosette shoots in flowering Chinese cabbage by affecting the auxin signaling pathway. Our study provides novel information for clarifying the molecular mechanism of flowering Chinese cabbage varieties.

2. Materials and Methods

2.1. Plant Materials and Exogenous GA_3 Spraying Treatment

CX020 (MB) is a multiple-branch phenotype isolated by microspore culture to obtain a double haploid (DH) line of Zengcheng flowering Chinese cabbage. This is a pure line in the genetic sense. The experimental materials were planted in the experimental field of Shenyang Agricultural University (41°79' N, 123°4' E). In spring 2019, the seeds were directly sowed into a 50-hole tray, and after 20 days, the pots were changed to a red plastic bowl with a diameter of 15 cm to conveniently observe the phenotypic changes. Exogenous GA3 (Solarbio, Beijing, China) with different concentrations (200, 400, 600, 800, and 1000 mg/L) was applied to the whole plant (except for the roots) when the third euphylla had grown and was carried out once every two days for a total of 3 times. Each concentration consisted of 10 plants with 3 replicates. The control material was separately sprayed with distilled water. Finally, 600 mg/L was selected as the optimal spraying concentration. The branching traits appeared after 40 days of spraying GA_3. To reduce differences between plants, 3 plants were selected from each of MB and MB_GA, and basal parts of 3 cm were randomly collected from each group of plants. All of the stems collected from three MB or MB_GA plants were mixed together to obtain a library, named MB_1, MB_2, MB_3, MB_GA_1, MB_GA_2, and MB_GA_3. All samples were frozen in liquid nitrogen and then stored in a $-80\ °C$ freezer to sequence transcriptomes and to measure hormone content.

2.2. RNA Extraction and Transcriptome Sequencing

Total RNA was extracted using a Plant RNAprep Pure Micro Kit (TIANGEN, Beijing, China). The concentration, quality, and integrity were determined using a NanoDrop spectrophotometer (Thermo Fisher Scientific, Waltham, MA, USA). Three micrograms of RNA were used as the input material for the RNA sample preparations. Sequencing libraries were generated using the TruSeq RNA Sample Preparation Kit (Illumina, San Diego, CA, USA), as per the manufacturer's instructions. In order to select cDNA fragments of the preferred 200 bp length, the library fragments were purified using the AMPure XP system (Beckman Coulter, Beverly, CA, USA) [31]. DNA fragments with ligated adaptor molecules on both ends were selectively enriched using the Illumina PCR Primer Cocktail in a 15 cycle PCR reaction. The products were purified (AMPure XP system) and quantified using an Agilent high-sensitivity DNA assay on a Bioanalyzer 2100 system (Agilent, Suzhou, China). The library was sequenced on a Hiseq platform (Illumina, San Diego, CA, USA).

2.3. Transcriptome Analysis

The clean reads were mapped to the brassica reference genome (http://brassicadb.cn/#/ (accessed on 9 March 2020)). The combination of false discovery rate (FDR) ≤ 0.001 and the absolute value of log2Ratio ≥ 1 were used as the threshold for judging significant different gene expression levels [32–34]. In the study, genes with a threshold value of |log2 ratio| ≥ 1.0 and a false discovery rate (FDR) ≤ 0.01 were defined as possessing significant differential gene expression (DEGs). The trend analysis was performed by the genescloud tools (https://www.genescloud.cn, accessed on 9 March 2020).

Gene set enrichment analysis for Gene Ontology (GO) was performed using the topGo 2.54.0 package (Sunovo Hulian, Beijing, China). ClusterProfiler was used to perform KEGG enrichment analysis. The gene list and gene number of each pathway were calculated using the differential genes annotated by the KEGG pathway. The hypergeometric distribution method was used to calculate $p \leq 0.05$, and compared with the background of the entire genome, to determine the differences between the main biological functions of the genes [35].

2.4. Metabolite Profiling of Hormone Content Using UPLC-MS/MS

Next, 3 cm stem parts were sampled with 3 biological replicates for each group of samples to measure hormone content. Fresh plant materials were harvested, weighed, immediately frozen in liquid nitrogen, and stored at $-80\ °C$ until needed. Plant materials (50 mg fresh weight) were frozen in liquid nitrogen, ground into powder, and extracted with 1 mL methanol/water/formic acid (15:4:1, $v/v/v$). The combined extracts were evaporated to dryness under a nitrogen gas stream, reconstituted in 100 µL of 80% methanol (v/v), and filtered through a 0.22 µm pore size, 14 mm diameter filter for further LC-MS analysis. The sample extracts were analyzed using an UPLC-ESI-MS/MS system. The LC column used was a Waters ACQUITY UPLC HSS T3 C18 (100 mm × 2.1 mm i.d. 1.8 µm). The HPLC effluent was connected to an electrospray ionization (ESI)-triple quadrupole-linear ion trap–MS/MS system (Applied Biosystems 4500 Q TRAP) [36]. A specific set of MRM transitions was monitored for each period, as determined by the plant hormones eluted within this period.

2.5. Real-Time Quantitative PCR Analysis (qRT-PCR)

Transcriptome gene expression was analyzed using real-time quantitative PCR (qRT-PCR). cDNA was prepared using an iScript™ cDNA Synthesis Kit according to the manufacturer's protocol. qRT-PCR was performed using a Bio-Rad CFX96 real-time system with SYBR® Green PCR Supermix (Hercules, CA, USA). Each system contained 10.4 µL of SYBR mixture, forward and reverse primers, 8.8 µL of H_2O, and 0.8 µL of cDNA in a total final volume of 20 µL. This technique was repeated three times. The following qRT-PCR program was used: 98 °C for 30 s, followed by 39 cycles of a two-step reaction (98 °C for 15 s and 60 °C for 30 s). The 2-$\Delta\Delta$CT method was used to calculate the relative expression levels [37]. *Actin* gene was used as a control. To verify the expression levels detected by RNA-seq, the RNA-seq data were compared to the data obtained by qRT-PCR. DEGs were identified by two-fold change (log2 ratio ≥ 2). The primers are listed in Supplementary Materials. All reactions were performed with three technical and biological replicates, and three independent biological replicates were conducted for all the qRT-PCR reactions.

2.6. Statistical Analysis

The data were determined using the *t*-test with IBM SPSS Statistics 23 software (IBM, New York, NY, USA) at the 0.05 level to compare the significant differences. Maps were generated in Origin 2018 (OriginLab, Guangzhou, China).

3. Results

3.1. Exogenous GA$_3$ Obviously Inhibited Rosette Branching

'CX020 (MB)' plants formed a multiple-branch phenotype, the cauline branching numbers increased from when sprayed with GA$_3$ (Figure 1A,B) and the rosette branching numbers decreased from when sprayed with GA$_3$ (Figure 1C,D) when the cauline and the rosette branching numbers were counted during harvest. The total number of cauline branches, 15.20, was significantly greater in the MB_GA group than in the MB group. The number of rosette branches were 3.40 and 9.04 in the MB_GA and MB groups, respectively, which was significantly lower in the MB_GA group than in the MB group (Figure 2A). This result revealed that the effect of GA on the branching phenotype of MB was noticeably different. In addition, the length of the internode in the MB_GA group was significantly different from that in the MB group (Figure 2B). These results showed that GAs influence rosette branching from the first to the fourth node and may limit the growth of lateral branches in flowering Chinese cabbage.

Figure 1. Phenotypes of MB and MB with 600 mg/L of GA$_3$ treatment. The entire plant appearance of MB (**A**) and MB after 600 mg/L of GA$_3$ treatment (**B**). The phenotype of rosette branching in MB (**C**) and MB after 600 mg/L of GA$_3$ treatment (**D**). The rosette branching is indicated with red arrows.

3.2. Influences of Exogenous GA$_3$ on Gene Expressions

To determine the DEGs involved in the regulatory mechanisms of branching, we performed an integrated transcriptome analysis of MB and MB_GA and sequenced six cDNA libraries (two samples with three replicates). A total of 379.9 million high-quality reads were generated, constituting 57.4 GB of cDNA sequences. The Q30 value (sequence error rate was 0.1%) of each sample was no less than 93.7%. To further define the quality of sequencing, 90.36% of read coverage was analyzed, representing the percentage of a gene covered by the reads (Tables S1 and S2). The GC content of the six libraries was 47.26%, 47.17%, 47.18%, 47.22%, 47.21%, and 47.21%, respectively. The comparison efficiency between the reads and the reference genome of each sample was between 94.4% and 95.26% (Tables S3 and S4). We first counted the scatter plot to confirm gene expression levels in the samples and generated a heatmap plot to show changes in the six libraries (Figure 3A). The RNA-seq data from the biological replicates were found to co-cluster when assessed using principal component analysis (PCA). The results of the PCA plot were consistent across the biological replicates (Figure 3B). The volcano plot shows that there were more genes with multiple significant differences among the upregulated genes, while the difference multiple of downregulated genes was smaller (Figure 3C).

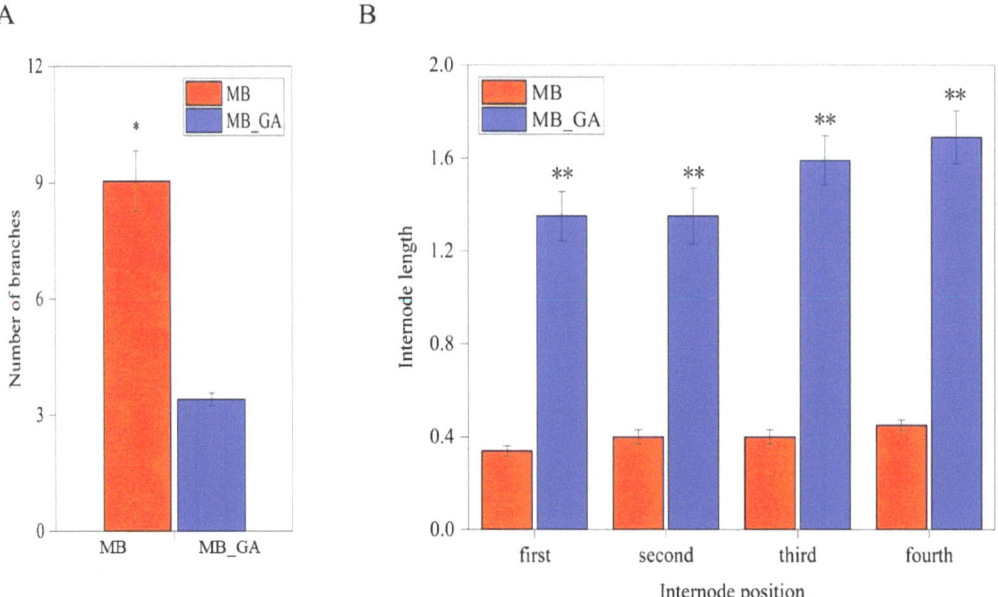

Figure 2. The primary rosette branch numbers of MB and MB_GA (**A**). Relationship between internode position and internode length (**B**). * and ** indicate significant differences in expression levels at $p < 0.05$ and $p < 0.01$ between the two types as determined according to the *t*-test.

To identify the genes involved in the formation and development of branching, significant DEGs (with the filter criteria set as fold change ≥ 2.0 and FDR ≤ 0.001) were obtained for the two types of plants (MB and MB_GA) (Tables S5 and S6). As a result, 2183 DEGs were identified between MB and MB_GA, of which 1318 were upregulated and 865 were downregulated in MB versus MB_GA (Figure 3D). The number of downregulated DEGs was higher than that of the upregulated DEGs.

All 2183 DEGs were classified into eight trend clusters with an algorithm developed from gene expression trends (RPKM ≥ 2, FDR ≤ 0.001 and $|\log2(ratio)| \geq 1$). The results showed that in the clusters 1, 2, 3, and 4, the gene expression decreased obviously from MB to MB_GA, containing 941 genes, and in the clusters 5, 6, 7, 8, and 9, the gene expression increased obviously from MB to MB_GA, containing 1242 genes (Figure 4). These results indicate that the transcriptome data are reliable.

3.3. Enrichment Analysis for the DEGs

To further confirm the biological functions and assign genes related to branching, DEGs were mapped using GO terms to classify their functions ($p \leq 0.05$), and the top 20 enriched pathways were used to make the GO enrichment map (Figure 5A). We identified 144 enriched GO terms that were assigned, including 73 under biological processes, 55 under molecular functions, and 16 under cellular components (Table S7). Under biological processes, the subcategories were related to biological regulation (GO: 0008152; 1043DEG) and regulation of metabolic processes (GO: 0019222; 236DEG), and the subcategories related to hormones were response to hormones (GO: 0009725; 37DEG) and response to auxin (GO: 0009733; 32DEG). Under molecular function, the subcategories were related to catalytic activity (GO: 0003824; 986DEG) and DNA binding (GO: 0003677; 302DEG).

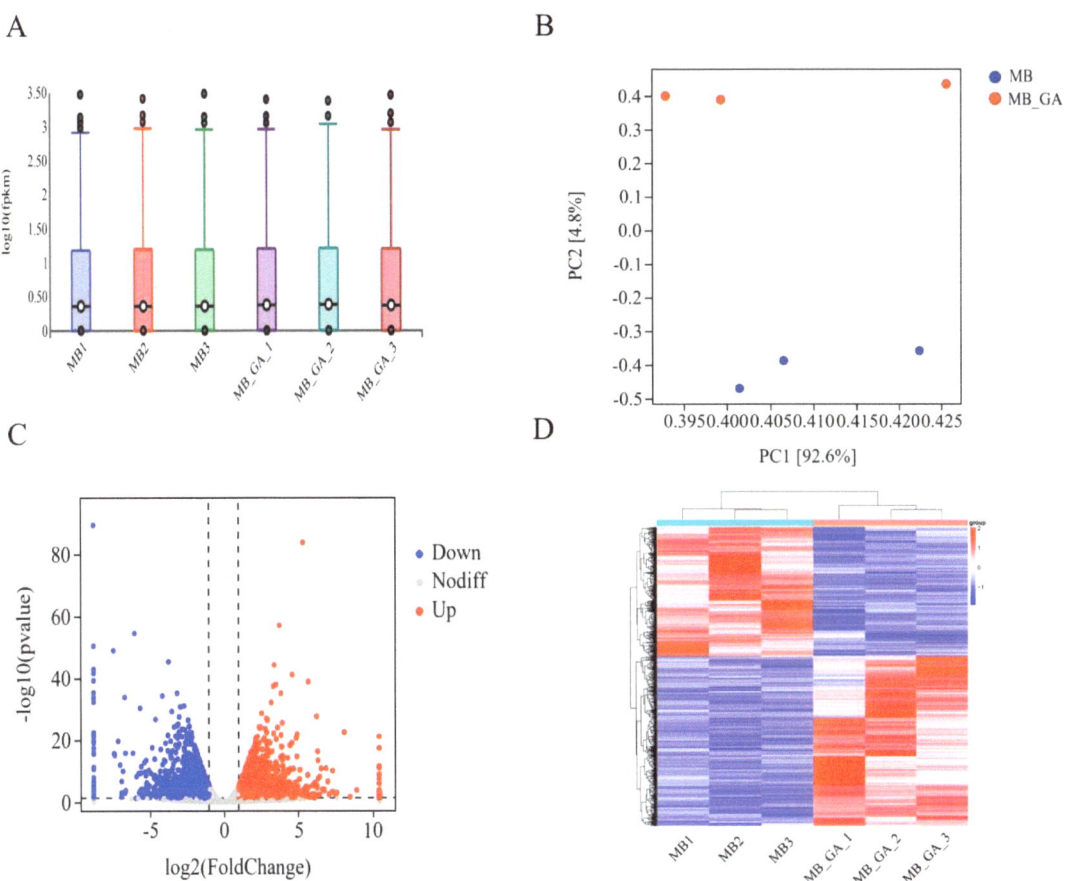

Figure 3. Comparative analysis of expression patterns of differentially expressed genes (DEGs) in six segments of MB. Hierarchical cluster diagram of gene expression level in samples. MB1, MB2, and MB3 are controls; MB_GA_1, MB_GA_2, and MB_GA_3 are the experimental groups treated with 600 mg/L GA$_3$ (**A**). Principal component analysis (PCA) of the differentially expressed genes between MB and MB_GA (**B**). Volcano plot of the differentially expressed genes between MB and MB_GA. Blue indicates downregulation of the gene, red indicates upregulation of the gene, and gray indicates non−regulation of the gene (**C**). The number of up− and down−regulated expressed genes (**D**).

To further identify genes associated with metabolic pathways, a total of 771 DEGs were mapped to 41 KEGG pathways (Table S8). The top 20 KEGG pathways were significantly enriched (Figure 5B, Table 1). The plant hormone signal transduction pathway (brp04075; 74) was the largest category, which was significantly enriched compared to other pathways, followed by phenylpropanoid biosynthesis (brp00940; 51) and starch and sucrose metabolism (brp00500; 39). These results indicate that hormone signal transduction was highly enriched in branching development, suggesting the complexity of the mechanisms underlying the development of branching in plants.

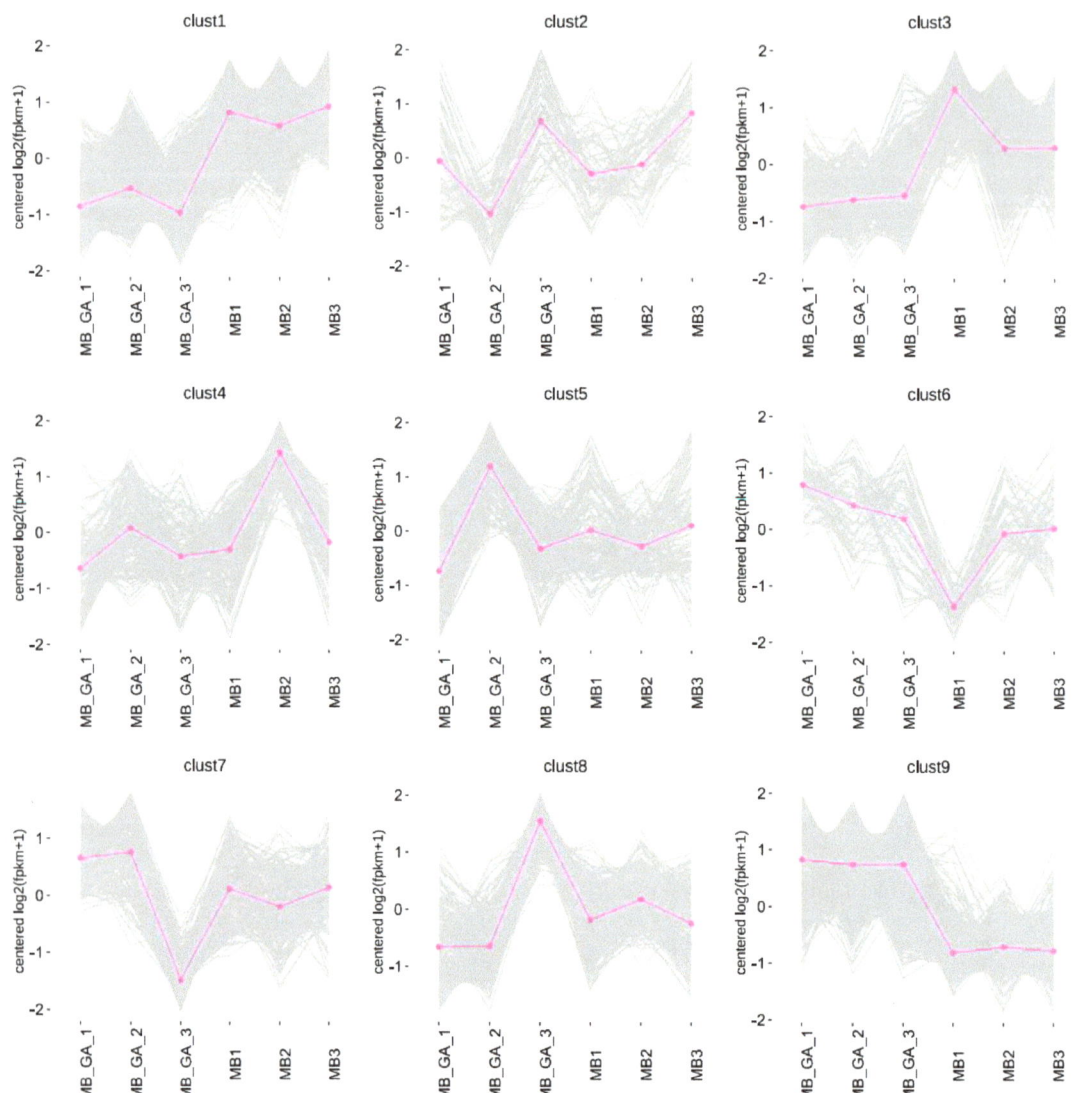

Figure 4. The trend analysis of the expression profile of all DEGs in all six pairwise comparisons. The *y*-axis represents the clustering groups of the gene expression level and the *x*-axis represents the different samples.

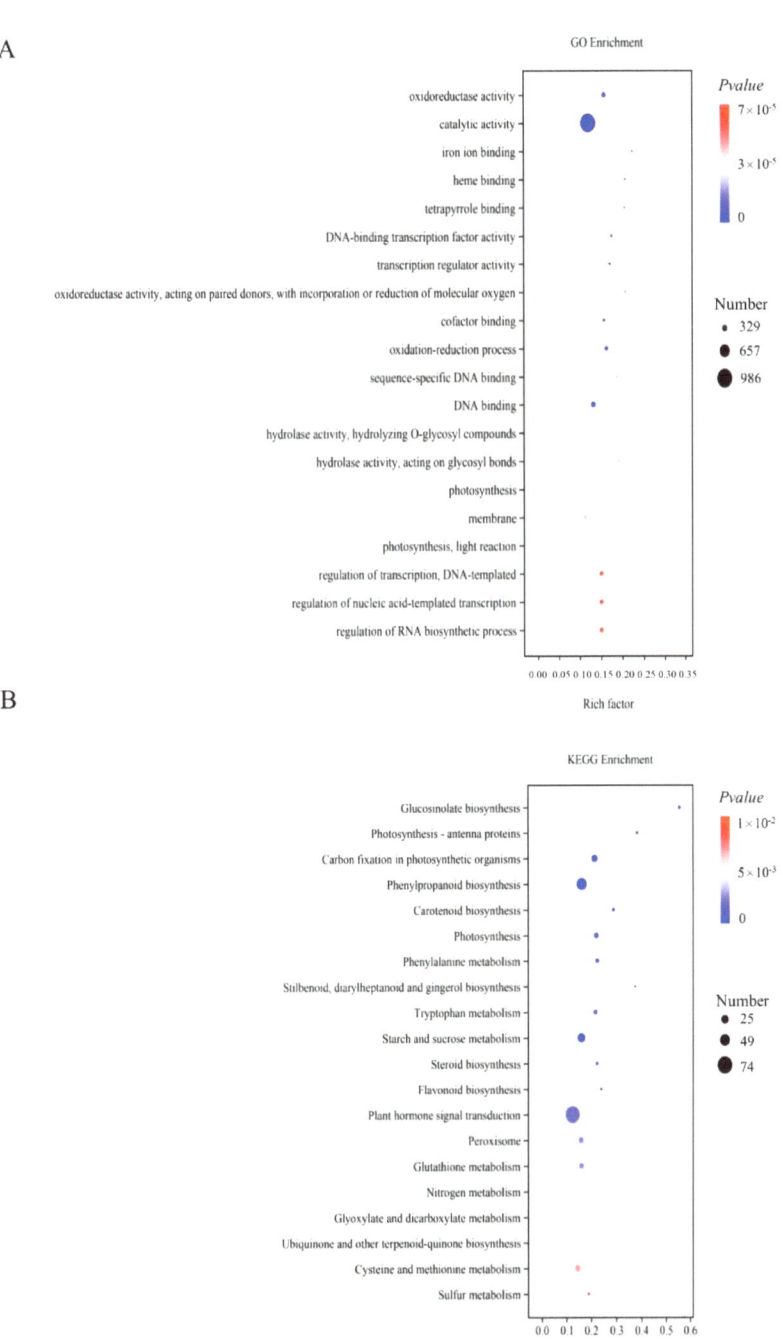

Figure 5. Twenty most significantly enriched GO and KEGG metabolic pathways. (**A**) GO enrichment analysis for MB vs. MB_GA; (**B**) KEGG enrichment analysis for MB vs. MB_GA.

Table 1. Significantly enriched KEGG pathways of DEGs in MB vs. MB_GA.

Pathway ID	Pathway	DEG (%)	Total DEGs	Adjust p Value
brp04075	Plant hormone signal transduction	74 (9.60%)	771	1.73×10^{-2}
brp00940	Phenylpropanoid biosynthesis	51 (6.61%)	771	6.20×10^{-4}
brp00500	Starch and sucrose metabolism	39 (5.06%)	771	2.91×10^{-3}
brp04626	Plant–pathogen interaction	38 (4.93%)	771	3.54×10^{-1}
brp04016	MAPK signaling pathway—plant	32 (4.15%)	771	6.81×10^{-2}
brp00710	Carbon fixation in photosynthetic organisms	30 (3.90%)	771	2.24×10^{-4}
brp00270	Cysteine and methionine metabolism	29 (3.76%)	771	4.43×10^{-2}
brp04146	Peroxisome	27 (3.50%)	771	1.95×10^{-2}
brp00480	Glutathione metabolism	26 (3.37%)	771	1.95×10^{-2}
brp00630	Glyoxylate and dicarboxylate metabolism	24 (3.11%)	771	3.32×10^{-2}
brp00230	Purine metabolism	24 (3.11%)	771	3.19×10^{-1}
brp00195	Photosynthesis	23 (2.98%)	771	8.07×10^{-4}
brp00260	Glycine, serine, and threonine metabolism	21 (2.72%)	771	5.77×10^{-2}
brp00040	Pentose and glucuronate interconversions	21 (2.72%)	771	4.72×10^{-1}
brp00360	Phenylalanine metabolism	20 (2.60%)	771	1.71×10^{-3}
brp00380	Tryptophan metabolism	20 (2.60%)	771	2.16×10^{-3}
brp00130	Ubiquinone and other terpenoid–quinone biosynthesis	16 (2.08%)	771	3.71×10^{-2}
brp00561	Glycerolipid metabolism	16 (2.08%)	771	2.49×10^{-1}
brp00966	Glucosinolate biosynthesis	15 (1.95%)	771	1.41×10^{-7}
brp00906	Carotenoid biosynthesis	15 (1.95%)	771	8.07×10^{-4}

Total DEGs: total number of DEGs with pathway annotations. Percentage (%) = 100% × (number of DEGs)/total number of DEGs.

3.4. DEGs in Plant Hormones Related to Branching

In KEGG enrichment analysis, we observed that the plant hormone signaling pathway involved the most differentially expressed genes (Table S9). Based on this, we further studied plant hormone-related genes. Of these, 40 DEGs were enriched in the IAA signal transduction pathway, indicating a potential connection between auxin and GA. The DEGs of the primary auxin-response factors included auxin1 (*AUX1*), auxin/indole-3-acetic acid (*AUX/IAA*), SMALL AUXIN-UPREGULATED RNA (*SAUR*), and *GH3* involved in IAA biosynthesis [38]. Only one *AUX1* gene (*BraA04g027030.3C*) was annotated which showed log2 ratios of 2.268 in MB_GA/MB. *AUX/IAA* genes (*BraA01g031100.3C, BraA05g024120.3C, BraA03g054630.3C, BraA01g035920.3C, BraA01g008880.3C, BraA03g040660.3C, BraA03g057220.3C, BraA03g017680.3C, BraA03g040670.3C, BraA06g001750.3C*) were annotated of which 10 DEGs may be associated with plant branching. Twenty-four *SAUR* genes were annotated, of which eight genes (*BraA09g050440.3C, BraA05g001150.3C, BraA04g032240.3C, BraA05g009190.3C, BraA10g011930.3C, BraA01g013230.3C, BraA04g025770.3C, BraA03g023650.3C*) were upregulated in MB_GA and sixteen genes (*BraA07g016220.3C, BraA01g006580.3C, BraA10g022010.3C, BraA02g007500.3C, BraA07g002710.3C, BraA07g039120.3C, BraA01g003540.3C, BraA10g022020.3C, BraA01g003530.3C, BraA01g003550.3C, BraA06g044670.3C, BraA01g000130.3C, BraA02g022940.3C, BraA02g007480.3C, BraA01g003510.3C, BraA08g015690.3C*) were downregulated in MB_GA. Five *GH3* genes (*BraA03g053600.3C, BraA10g025440.3C, BraA02g013860.3C, BraA09g054360.3C, BraA06g004330.3C*) were upregulated in MB_GA. The *GH3* upregulated genes showed log2 ratios of 2.255, 1.260, 1.232, 1.873, and 1.823, respectively (Figure 6A).

Cytokinins are often used as the second messenger of auxins to regulate branching development [39]. Two *ARABIDOPSIS RESPONSE REGULATOR* (*B-ARR*) genes (*BraA07g012270.3C, BraA03g046670.3C*) were annotated. Ten genes had relevance to the ABA signal, including *BraA03g029760.3C, BraA03g002100.3C, BraA05g005770.3C, BraA10g000590.3C, BraA03g021430.3C, BraA07g024750.3C, BraA03g020010.3C, BraA06g027110.3C, BraA05g018520.3C*, and *BraA10g016990.3C* (Figure 6A).

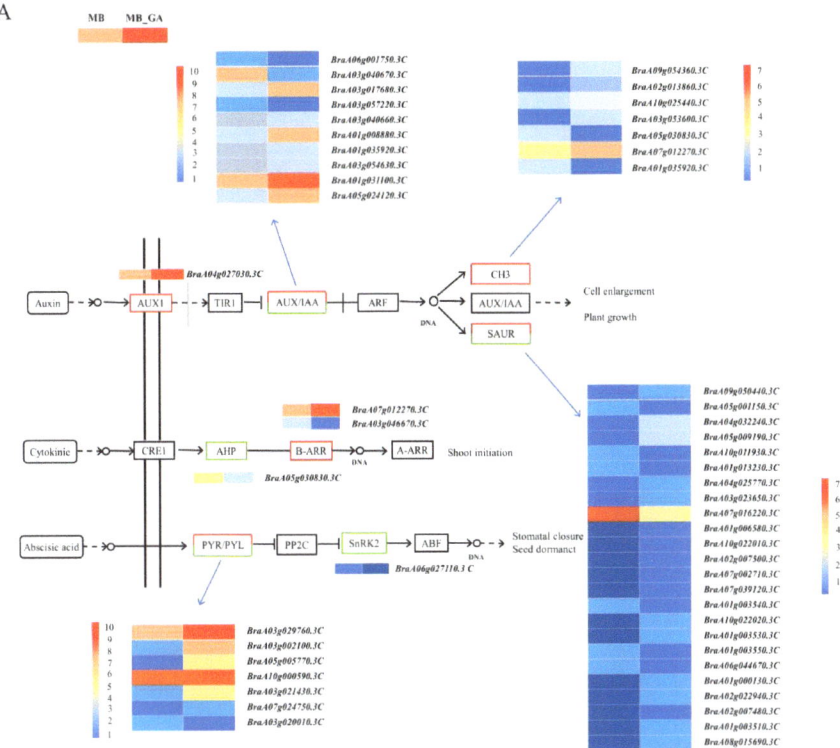

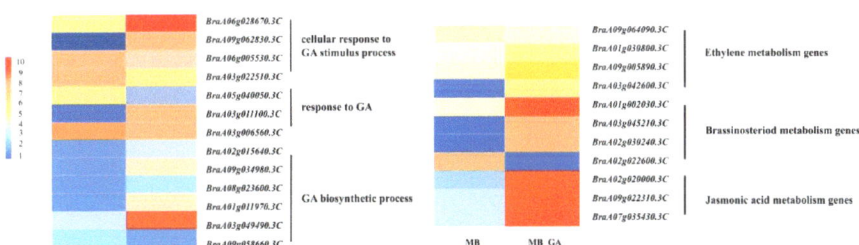

Figure 6. Expression of differentially expressed genes involved in the plant hormone genes (auxin, cytokinin, abscisic acid). Note: The bar on the right represents the relative expression values. Upregulated genes are marked with red borders and downregulated genes with green borders. Unchanged genes are marked with black borders (**A**). Expression of differentially expressed genes involved in the plant hormone genes (gibberellin, ethylene, brassinosteroid, jasmonic acid) (**B**).

Gibberellin has an important effect not only on internode elongation but also on branching development. In rice, auxin and gibberellin can regulate the negative gravity response of the stem by antagonizing the expression of *XYLOGLUCAN ENDOTRANSGLYCOSYLASE* (*XET*) [40,41]. We detected two *XET* genes (*BraA10g015660.3C*, *BraA02g012200.3C*) that were upregulated in the MB. In the GA signal transduction pathway, we detected one differentially expressed *GA INSENSITIVE DWARF 1* (*GID1*) gene (*BraA05g040050.3C*) [42], which was upregulated in MB and its log2 ratio (MB_GA/MB) was −1.142. In the GA biosynthetic process, *GA20ox2* (*BraA02g015640.3C*), *GA2ox2* (*BraA09g034980.3C*, *BraA08g023600.3C*), *GA2ox8* (*BraA01g011970.3C* and *BraA03g049490.3C*), and *GA3ox1* (*BraA09g058660.3C*) were upregulated in the MB_GA. In the cellular response to GA stimulus process, *BraA06g028670.3C*, *BraA09g062830.3C*, and *BraA06g005530.3C* (bZIP transcription factor family protein) were upregulated and *BraA03g042690.3C* (gibberellin-regulated family protein) was downregulated. In response to GA$_3$ treatment, the *GASA10* gene (*BraA03g011100.3C*) was upregulated and the *GASA14* gene (*BraA03g006560.3C*) was downregulated (Figure 6B).

In the BR signal transduction pathway, two were upregulated (*BraA01g002030.3C* and *BraA02g030240.3C*), and two were downregulated (*BraA03g045210.3C* and *BraA02g022600.3C*) in MB_GA. Four genes associated with ethylene metabolism were upregulated, including *BraA09g064090.3C*, *BraA01g030800.3C*, *BraA09g005890.3C*, and *BraA03g042600.3C* in MB_GA compared to MB. Three JA-related genes, including *BraA02g020000.3C*, *BraA09g022310.3C*, and *BraA07g035430.3C*, were upregulated in MB_GA compared to MB (Figure 6B).

3.5. qRT-PCR Analysis on DEGs Related to Auxin

To further confirm the accuracy of the RNA-seq data, 11 DEG-related auxin biosynthesis pathway genes that showed log2 ratio ≥ 2 in expression between MB and MB_GA were selected for qRT-PCR. These genes were *AUX1* (*BraA04g027030.3C*), *IAA29* (*BraA03g057220.3C*), *IAA1* (*BraA06g001750.3C*), *SAUR10* (*BraA09g053610.3C*), *SAUR21* (*BraA02g007500.3C*), *SAUR1* (*BraA01g003530.3C*), *SAUR59* (*BraA09g050440.3C*), *SAUR45* (*BraA05g009190.3C*), *SAUR16* (*BraA06g044670.3C*), *SAUR37* (*BraA04g000440.3C*), and *CH3.5* (*BraA03g053600.3C*) (Table S10). Correlation analysis showed that the high expression patterns of these genes that were selected by RNA-seq were consistent with the qRT-PCR data (Figure 7), thus indicating the dependability of the transcriptome results obtained in this study.

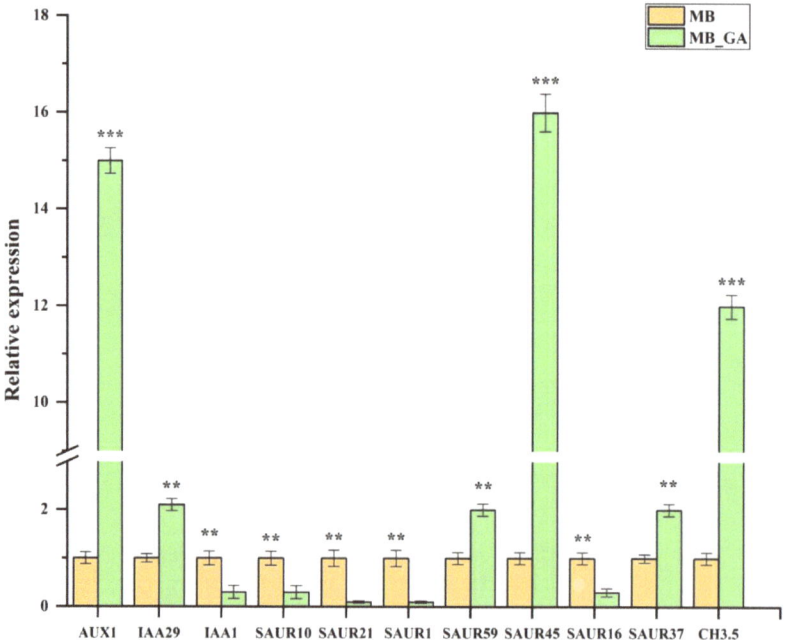

Figure 7. Expression profiles of 11 key differentially expressed genes (DEGs) in the MB and MB_GA. Error bars represent the standard error of the mean for three biological replicates. ** $p < 0.01$ and *** $p < 0.001$ determined by Student's *t*-test.

3.6. Exogenous GA₃ Altered Auxin Contents in the Rosette Shoot

The content of auxin hormones was determined in the rosette shoot of MB and MB_GA by HPLC (Figure 8). The auxin content significantly increased after spraying with GA$_3$. The results indicated that auxin may contribute to the development of rosette branch numbers.

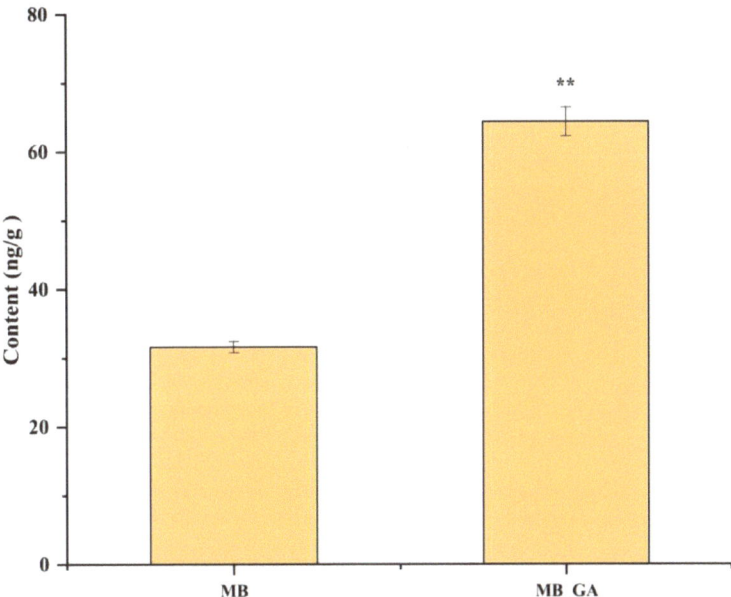

Figure 8. Determination of the content of auxin in MB and MB_GA. ** $p < 0.01$ determined by Student's *t*-test.

4. Discussion

The branching of the rosette shoot is crucial for increasing the yield of flowering Chinese cabbage. Previous studies have shown that the application of GA$_3$ reduced tillering in rice [43]. The overexpression of genes controlling GA$_3$ leads to reduced GA$_1$ levels producing increased branching phenotypes in pea [44], and treatment with higher concentrations of GA$_3$ inhibited the promotion of axillary bud (AB) outgrowth in apple trees [5]. Herein, we aimed to decipher the potential mechanism of exogenous GA$_3$ application to reduce the primary rosette branching in flowering Chinese cabbage. To this end, we firstly screened out a concentration of 600 mg/L GA$_3$ for spraying and found an increase in the length of the rosette shoots and a decrease in the primary rosette branching numbers. Following comparative transcriptome analysis and multiple phytohormone profiling by targeted metabolism, auxin has been found to be negatively regulated in rosette branching for flowering Chinese cabbage.

In this study, we detected 2183 genes that were differentially expressed between the multiple-branching Chinese flowering cabbage and GA treatment by RNA-seq, of which 1318 genes were upregulated and 865 genes were downregulated. The plant hormone signal transduction pathways were also significantly enriched by analyzing the functional annotation of the DEGs. The IAA signaling pathway had the highest number of DEGs, and transcriptome data indicated that the key genes in the IAA signaling pathway are all changing. Therefore, the genes involved in the IAA signaling pathway are of particular concern.

The degradation of *Aux/IAA* promoted by auxin is achieved by inhibiting the transcription of *ARFs*, and the degradation of *Aux/IAA* requires the participation of TIR1 [45–47].

The stability of the interaction between *Aux/IAA* and TIR1 is mainly mediated by the binding of auxin to the F-box of TIR1 [48–50]. The TIR1/AFB Aux/IAA pathway enhances the regulatory effect of *SAUR19*, and *SAUR19* is a very small family protein member that can be rapidly induced by auxin [51]. *Aux/IAAs* are believed to act as auxin-responsive proteins that mediate IAA signaling in the regulation of plant development [52]. Numerous studies have shown that *Aux/IAA* plays a significant role in regulating lateral branch development. The downregulation of *SlIAA2*, *SlIAA4*, *SlIAA7*, and *SlIAA9* has been found in transgenic tomatoes with multiple branches [53]. Furthermore, GA can also synergistically regulate axillary bud formation with IAA. The main influx and efflux of auxin rely on the polarity transport streams mediated by PIN and Aux/IAA proteins [54]. GA regulates *PIN* synthesis and accounts for auxin transport-dependent growth by the DELLA protein, which is a suppressor of GA biosynthesis and signal transduction [55]. It can also bind to the transcription factor SQUAMOSA PROMOTER BINDING PROTEIN-LIKE9 (SPL9) to facilitate axillary bud formation and regulate branching in *Arabidopsis* [19,56,57]. We identified DEGs of *AUX1*, *IAA29*, *SAUR1*, *SAUR10*, *SAUR16*, and *SAUR21* related to IAA metabolism and signal transduction between MB and MB_GA, which suggested that auxin may be involved in the formation of rosette branches in flowering Chinese cabbage. IAA concentration was increased and the IAA-related genes *IAA2*, *IAA11*, and *IAA29* were highly expressed in response to GA treatment to inhibit branching [58]. Similar to the above research result, the content of auxin determined by HPLC has a significant increase in MB_GA compared to MB. These results indicate that auxin may play an important role in flowering Chinese cabbage rosette branches.

Based on the results, we predicted a potential genetic mechanism of exogenous GA_3 application to reduce the primary rosette branching in flowering Chinese cabbage (Figure 9). It was speculated that exogenously sprayed GA_3 might upregulate the expression of *AUX1* influx carriers in specific areas of the rosette shoot, whereafter the *AUX1* might further induce the upregulated expression of the auxin receptor TIR1. TIR1 and AFB2 were shown to interact with *IAA29* by reducing the *IAA29* expression. The TIR1/AFB-Aux/IAA-ARF complex could directly or indirectly inhibit flowering Chinese cabbage rosette branches by suppressing the expression of *SAUR1*, *SAUR10*, *SAUR16*, and *SAUR21*.

The analysis of the hormone-responsive gene promoter region revealed various cis-regulatory elements like GA-responsive elements (GARE-motif, P-box) [59], auxin (TGA-element, AuxRR-core element and TATC-box) [60], ABA (ABRE element), ET (ERE element) [61], JA (MYC element, TGACG-motif) [62], SA (TCA-element) [63], etc., suggesting that they may have been involved in growth, development, and stress responses [64,65]. ARF24 combined with an auxin response region (AuxRR) affects kernel size in different maize haplotypes [66]. Two GARE-motif cis-acting elements are sufficient for gibberellin-upregulated proteinase expression in rice seeds [67]. In addition, the promoter of the gene in Figure 9 was analyzed (Table S11). The phytohormone-responsive cis-regulatory elements not only include GA-responsive elements (GARE-motif, P-box), but also many other hormone-responsive elements, including auxin (TGA-element, AuxRR-core element and TATC-box), ABA (ABRE element), JA (MYC element, TGACG-motif), SA (TCA-element), and ET (ERE element). This result indicates that auxin-related genes may respond to GA and various plant hormones to regulate branching development. Our results will be useful for elucidating the regulatory mechanisms of branching architecture in flowering Chinese cabbage.

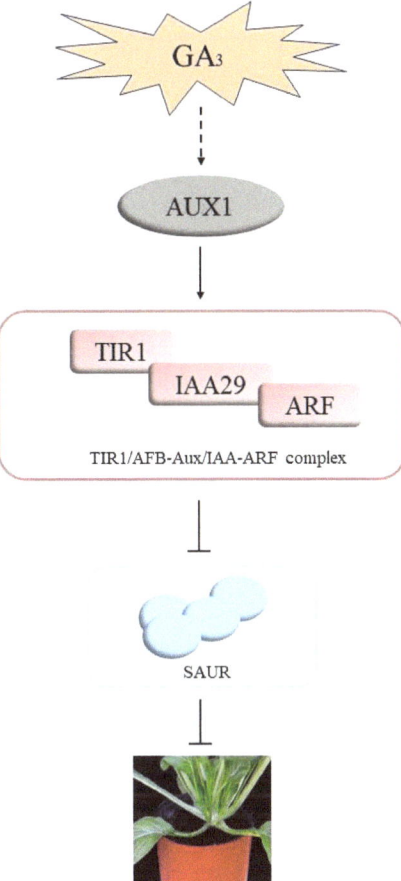

Figure 9. A hypothetical regulatory network of the rosette branching inhibited by exogenous GA$_3$ in flowering Chinese cabbage.

5. Conclusions

According to the comprehensive experimental results, GA$_3$ spraying can significantly decrease primary rosette branching in flowering Chinese cabbage, as well as influence the total and single plant yield. The transcriptome data indicate that GA$_3$ can interact with auxin-related genes to reduce the number of branches on the rosette stems in flowering Chinese cabbage. Using metabolic profiling, we identified that the content of auxin was significantly increased in MB_GA, which was highly associated with RNA-seq. Based on the above results, our results provide a new molecular mechanism for flowering Chinese cabbage branching, and can provide theoretical and practical guidance for its production.

Supplementary Materials: The following supporting information can be downloaded at: https://www.mdpi.com/article/10.3390/agronomy14040762/s1, Table S1. Summary of all expressed genes detected in the MB and MB_GA libraries; Table S2. Summary of clean data detected in the MB and MB_GA libraries; Table S3. Summary of GC content detected in the MB and MB_GA libraries; Table S4. List of all genes detected in the MB and MB_GA; Table S5. List of DEGs detected in the MB and MB_GA; Table S6. Significantly enriched GO terms identified in the MB compared with MB_GA; Table S7. Significantly enriched KEGG metabolic pathways in the MB compared with MB_GA; Table S8. DEGs involved in the plant hormone; Table S9. Transcription factors identified in the MB compared with MB_GA; Table S10. Primer sequences used for the qRT-PCR analysis; Table S11. Table S11 Cis-acting elements in auxin related genes.

Author Contributions: Experiment, Data curation, Figures, Writing—original draft, X.Q.; Funding acquisition, Writing—review, Y.Z. (Ying Zhao); Literature search, Data investigation, N.C.; Experimental design, Data collection, J.G.; Software, Data visualization, Z.L. (Zeji Liu); Methodology, Data analysis, Z.L. (Zhiyong Liu); Writing—review and editing, H.F.; Experimental design, Conceptualization, Funding acquisition, Writing—review and editing, Y.Z. (Yun Zhang). All authors have read and agreed to the published version of the manuscript.

Funding: This work was financially supported by Science and Technology Project of Liaoning-Applied Basic Research Plan (Youth Special Project), No.2023JH2/101600054 and National Training Program of Innovation and Entrepreneurship for Undergraduate (202310163039).

Data Availability Statement: The RNA-Seq data of all samples have been submitted to GenBank of the National Center for Biotechnology Information (https://www.ncbi.nlm.nih.gov/), and the Sequence Read Archive (SRA) accession number is PRJNA777406.

Conflicts of Interest: The authors declare no conflicts of interest.

References

1. Lin, L.Z.; Harnly, J.M. Phenolic component profiles of mustard greens, yu choy, and 15 other brassica vegetables. *J. Agric. Food Chem.* **2010**, *58*, 6850–6857. [CrossRef] [PubMed]
2. Teichmann, T.; Muhr, M. Shaping plant architecture. *Front. Plant Sci.* **2015**, *6*, 233. [CrossRef] [PubMed]
3. Wang, M.; Moigne, M.L.; Bertheloot, J.; Crespel, L.; Perez-Garcia, M.D.; Ogé, L.; Demotes-Mainard, S.; Hamama, L.; Davière, J.M.; Sakr, S. BRANCHED1: A Key Hub of Shoot Branching. *Front. Plant Sci.* **2019**, *10*, 76. [CrossRef] [PubMed]
4. Yang, Y.; Nicolas, M.; Zhang, J.; Yu, H.; Guo, D.; Yuan, R.; Zhang, T.; Yang, J.; Cubas, P.; Qin, G. The TIE1 transcriptional repressor controls shoot branching by directly repressing BRANCHED1 in Arabidopsis. *PLoS Genet.* **2018**, *14*, e1007296. [CrossRef] [PubMed]
5. Jung, H.; Lee, D.K.; Choi, Y.D.; Kim, J.K. OsIAA6, a member of the rice Aux/IAA gene family, is involved in drought tolerance and tiller outgrowth. *Plant Sci.* **2015**, *236*, 304–312. [CrossRef] [PubMed]
6. Bainbridge, K.; Guyomarc'h, S.; Bayer, E.; Swarup, R.; Bennett, M.; Mandel, T.; Kuhlemeier, C. Auxin influx carriers stabilize phyllotactic patterning. *Genes Dev.* **2008**, *22*, 810–823. [CrossRef] [PubMed]
7. Liu, Y.; Xu, J.; Ding, Y.; Wang, Q.; Li, G.; Wang, S. Auxin inhibits the outgrowth of tiller buds in rice (*Oryza sativa* L.) by downregulating OsIPT expression and cytokinin biosynthesis in nodes. *Am. J. Crop Sci.* **2011**, *5*, 169–174.
8. Reinhardt, D.; Pesce, E.R.; Stieger, P.; Mandel, T.; Baltensperger, K.; Bennett, M.; Traas, J.; Friml, J.; Kuhlemeier, C. Regulation of phyllotaxis by polar auxin transport. *Nature* **2003**, *426*, 255–260. [CrossRef]
9. Vernoux, T.; Besnard, F.; Traas, J. Auxin at the shoot apical meristem. *Cold Spring Harb. Perspect. Biol.* **2010**, *2*, a001487. [CrossRef]
10. Barbier, F.F.; Dun, E.A.; Kerr, S.C.; Chabikwa, T.G.; Beveridge, C.A. An Update on the Signals Controlling Shoot Branching. *Trends Plant Sci.* **2019**, *24*, 220–236. [CrossRef]
11. Dierck, R.; Keyser, E.D.; Riek, J.D.; Dhooghe, E.; Huylenbroeck, J.V.; Prinsen, E.; Straeten, D.V.D. Change in Auxin and Cytokinin Levels Coincides with Altered Expression of Branching Genes during Axillary Bud Outgrowth in Chrysanthemum. *PLoS ONE* **2016**, *11*, e0161732. [CrossRef] [PubMed]
12. Lin, Q.B.; Zhang, Z.; Wu, F.Q.; Feng, M.; Sun, Y.; Chen, W.W.; Cheng, Z.J.; Zhang, X.; Ren, Y.L.; Lei, C.l.; et al. The APC/C^{TE} E3 Ubiquitin Ligase Complex Mediates the Antagonistic Regulation of Root Growth and Tillering by ABA and GA. *Plant Cell* **2020**, *32*, 1973–1987. [CrossRef] [PubMed]
13. Wang, Y.; Sun, S.; Zhu, W.; Jia, K.; Yang, H.; Wang, X. Strigolactone/MAX2-induced degradation of brassinoste-roid transcriptional effector BES1 regulates shoot branching. *Dev. Cell* **2013**, *27*, 681–688. [CrossRef] [PubMed]
14. Yamaguchi, S. Gibberellin metabolism and its regulation. *Annu. Rev. Plant Biol.* **2008**, *59*, 225–251. [CrossRef] [PubMed]
15. Hedden, P.; Phillips, A.L. Gibberellin metabolism: New insights revealed by the genes. *Trends Plant Sci.* **2000**, *5*, 523–530. [CrossRef] [PubMed]
16. Eiichi, T.; Naohiko, Y.; Yoshio, M. Effect of Gibberellic Acid on Dwarf and Normal Pea Plants. *Physiol. Plant.* **1967**, *20*, 291–298. [CrossRef]
17. Tan, M.; Li, G.; Liu, X.; Cheng, F.; Ma, J.; Zhao, C.; Zhang, D.; Han, M. Exogenous application of GA3 inactively regulates axillary bud outgrowth by influencing of branching-inhibitors and bud-regulating hormones in apple (*Malus domestica* Borkh.). *Mol. Genet. Genomics* **2018**, *293*, 1547–1563. [CrossRef]
18. Tian, Y.T.; Wang, J.N.; Guo, H.P.; Qu, K.; Xu, D.; Hou, L.L.; Li, J.H. Transcriptome Analysis of Active Axillary Buds from Narrow-crown and Broad-crown Poplars Provides Insight into the Phytohormone Regulatory Network for Branching Angle. *Plant Mol. Biol. Report.* **2021**, *39*, 595–606. [CrossRef]
19. Zhang, Q.Q.; Wang, J.G.; Wang, L.Y.; Wang, J.F.; Wang, Q.; Yu, P.; Bai, M.Y.; Fan, M. Gibberellin repression of axillary bud formation in *Arabidopsis* by modulation of DELLA-SPL9 complex activity. *J. Integr. Plant Biol.* **2020**, *62*, 421–432. [CrossRef]

20. Lo, S.F.; Yang, S.Y.; Chen, K.T.; Hsing, Y.I.; Zeevaart, J.A.; Chen, L.J.; Yu, S.M. A novel class of gibberellin 2-oxidases control semidwarfism, tillering, and root development in rice. *Plant Cell* **2008**, *20*, 2603–2618. [CrossRef]
21. Liao, Z.; Yu, H.; Duan, J.; Yuan, K.; Yu, C.; Meng, X.; Kou, L.; Chen, M.; Jing, Y.; Liu, G.; et al. SLR1 inhibits MOC1 degradation to coordinate tiller number and plant height in rice. *Nat. Commun.* **2019**, *10*, 2738. [CrossRef]
22. Mauriat, M.; Sandberg, L.G.; Moritz, M. Proper gibberellin localization in vascular tissue is required to control auxin-dependent leaf development and bud outgrowth in hybrid aspen. *Plant J.* **2011**, *67*, 805–816. [CrossRef]
23. Agharkar, M.; Lomba, P.; Altpeter, F.; Zhang, H.; Kenworthy, K.L.T. Stable expression of AtGA2ox1 in a low-input turfgrass (*Paspalum notatum* Flugge) reduces bioactive gibberellin levels and improves turf quality under field conditions. *Plant Biotechnol. J.* **2007**, *5*, 791–801. [CrossRef]
24. Caruana, J.C.; Sittmann, J.W.; Wang, W.P.; Liu, Z.C. Suppressor of Runnerless encodes a DELLA protein that controls runner formation for asexual reproduction in strawberry. *Mol. Plant* **2018**, *11*, 230–233. [CrossRef]
25. O'Neill, D.P.; Ross, J.J. Auxin regulation of the gibberellin pathway in pea. *Plant Physiol.* **2000**, *130*, 1974–1982. [CrossRef]
26. Tenreira, T.; Lange, M.J.P.; Lange, T.; Bres, C.; Labadie, M.; Monfort, A.; Hernould, M.; Rothan, C.; Denoyes, B. A Specific Gibberellin 20-Oxidase Dictates the Flowering-Runnering Decision in Diploid Strawberry. *Plant Cell* **2017**, *29*, 2168–2182. [CrossRef]
27. Elfving, D.; Visser, D.; Henry, J. Gibberellins stimulate lateral branch development in young sweet cherry trees in the orchard. *Int. J. Fruit Sci.* **2011**, *11*, 41–54. [CrossRef]
28. Choubane, D.; Rabot, A.; Mortreau, E.; Legourrierec, J.; Péron, T.; Foucher, F.; Ahcène, Y.; Pelleschi-Trvier, S.; Leduc, N.; Hamama, L. Photocontrol of bud burst involves gibberellin biosynthesis in *Rosa* sp. *J. Plant Physiol.* **2012**, *169*, 1271–1280. [CrossRef]
29. Katyayini, N.U.; Rinne, P.L.H.; Tarkowská, D.; Strnad, M.; van der Schoot, C. Dual Role of Gibberellin in Perennial Shoot Branching: Inhibition and Activation. *Front. Plant Sci.* **2020**, *11*, 529247. [CrossRef] [PubMed]
30. Guan, J.; Li, J.; Yao, Q.; Liu, Z.; Feng, H.; Zhang, Y. Identification of two tandem genes associated with primary rosette branching in flowering Chinese cabbage. *Front. Plant Sci.* **2022**, *19*, 1083528. [CrossRef] [PubMed]
31. Gordon, S.P.; Tseng, E.; Salamov, A.; Zhang, J.; Meng, X.; Zhao, Z.; Wang, Z. Widespread Polycistronic Transcripts in Fungi Revealed by Single-Molecule mRNA Sequencing. *PLoS ONE* **2015**, *10*, e0132628. [CrossRef]
32. Mortazavi, A.; Williams, B.A.; McCue, K.; Schaeffer, L.; Wold, B. Mapping and quantifying mammalian transcriptomes by RNA-Seq. *Nat. Methods* **2008**, *5*, 621–628. [CrossRef]
33. Rosa, M.; Hilal, M.; González, J.A.; Prado, F.E. Low-temperature effect on enzyme activities involved in sucrose-starch partitioning in salt-stressed and salt-acclimated cotyledons of quinoa (*Chenopodium quinoa* Willd.) seedlings. *Plant Physiol. Biochem.* **2019**, *47*, 300–307. [CrossRef] [PubMed]
34. Livak, K.J.; Schmittgen, T.D. Analysis of relative gene expression data using real-time quantitative PCR. *Methods* **2002**, *25*, 402–408. [CrossRef] [PubMed]
35. Benjamini, Y.; Hochberg, Y. Controlling the False Discovery Rate: A Practical and Powerful Approach to Multiple Testing. *J. R. Stat. Soc. B* **1995**, *57*, 289–300. [CrossRef]
36. Benjamini, Y.; Yekutieli, D. The Control of the False Discovery Rate in Multiple Testing under Dependency. *Ann. Stat.* **2001**, *29*, 1165–1188. [CrossRef]
37. Abdi, H. Bonferroni and Šidák corrections for multiple comparisons. In *Encyclopedia of Measurement and Statistics*; Salkind, N., Ed.; CiteSeerX: Princeton, NJ, USA, 2007; pp. 103–107.
38. Yuan, C.; Xi, L.; Kou, Y.; Zhao, Y.; Zhao, L. Current perspectives on shoot branching regulation. *Front. Agric. Sci. Eng.* **2015**, *2*, 38–52. [CrossRef]
39. Lee, M.S.; An, J.H.; Cho, H.T. Biological and molecular functions of two EAR motifs of Arabidopsis IAA7. *J. Plant Biol.* **2016**, *59*, 24–32. [CrossRef]
40. Liu, Y.; Wang, Q.; Ding, Y.; Li, G.H.; Xu, J.X.; Wang, S.H. Effects of external ABA, GA_3 and NAA on the tiller bud outgrowth of rice is related to changes in endogenous hormones. *Plant Growth Regul.* **2011**, *65*, 247–254. [CrossRef]
41. Cui, D.; Neill, S.J.; Tang, Z.; Cai, W. Gibberellin-regulated XET is differentially induced by auxin in rice leaf sheath bases during gravitropic bending. *J. Exp. Bot.* **2005**, *56*, 1327–1334. [CrossRef]
42. Fukazawa, J.; Ito, T.; Kamiya, Y.; Yamaguchi, S.; Takahashi, Y. Binding of GID1 to DELLAs promotes dissociation of GAF1 from DELLA in GA dependent manner. *Plant Signal. Behav.* **2015**, *10*, e1052923. [CrossRef] [PubMed]
43. Tang, S.; Li, L.; Zhou, Q.Y.; Liu, W.Z.; Zhang, H.X.; Chen, W.Z.; Ding, Y.F. Expression of wheat gibberellins 2-oxidase gene induced dwarf or semi-dwarf phenotype in rice. *Cereal Res. Commun.* **2019**, *47*, 239–249. [CrossRef]
44. Nadeau, C.D.; Ozga, J.A.; Kurepin, L.V.; Jin, A.; Pharis, R.P.; Reinecke, D.M. Tissue-Specific Regulation of Gibberellin Biosynthesis in Developing Pea Seeds. *Plant Physiol.* **2011**, *156*, 897–912. [CrossRef] [PubMed]
45. Worley, C.K.; Zenser, N.; Ramos, J.; Rouse, D.; Leyser, O.; Theologis, A.; Callis, J. Degradation of AUX/IAA proteins is essential for normal auxinsignalling. *Plant J.* **2000**, *21*, 553–562. [CrossRef] [PubMed]
46. Tiwari, S.B.; Wang, X.J.; Hagen, G.; Guilfoyle, T.J. AUX/IAA proteins areactive repressors, and their stability and activity are modulated by auxin. *Plant Cell* **2001**, *13*, 2809–2822. [CrossRef] [PubMed]
47. Dharmasiri, N.; Dharmasiri, S.; Estelle, M. The F-box protein TIR1 is anauxin receptor. *Nature* **2005**, *435*, 441–445. [CrossRef] [PubMed]

48. Gray, W.M.; Kepinski, S.; Rouse, D.; Leyser, O.; Estelle, M. Auxin regulates SCF(TIR1)-dependent degradation of AUX/IAA proteins. *Nature* **2001**, *414*, 271–276. [CrossRef] [PubMed]
49. Jain, M.; Nijhawan, A.; Arora, R.; Agarwal, P.; Ray, S.; Sharma, P.; Kapoor, S.; Tyagi, A.K.; Khurana, J.P. F-box proteins in rice. Genome-wide analysis, classification, temporal and spatial gene expression during panicle and seeddevelopment, and regulation by light and abiotic stress. *Plant Physiol.* **2007**, *143*, 1467–1483. [CrossRef]
50. Trenner, J.; Poeschl, Y.; Grau, J.; Gogol-Döring, A.; Quint, M.; Delker, C. Auxin-induced expression divergence between Arabidopsis species may originate within the TIR1/AFB-AUX/IAA-ARF module. *J. Exp. Bot.* **2017**, *68*, 539–552. [CrossRef]
51. Spartz, A.K.; Lee, S.H.; Wenger, J.P.; Gonzalez, N.; Itoh, H.; Inzé, D.; Peer, W.A.; Murphy, A.S.; Overvoorde, P.J.; Gray, W.M. The *SAUR19* subfamily of *SMALL AUXIN UP RNA* genes promote cell expansion. *Plant J.* **2012**, *70*, 978–1068. [CrossRef]
52. Lee, M.S.; Choi, H.-S.; Cho, H.-T. Branching the auxin signaling; Multiple players and diverse interactions. *J. Plant Biol.* **2013**, *56*, 130–137. [CrossRef]
53. Pattison, R.J.; Catala, C. Evaluating auxin distribution in tomato (*Solanum lycopersicum*) through an analysis of the PIN and *AUX/LAX* gene families. *Plant J.* **2012**, *70*, 585–598. [CrossRef] [PubMed]
54. Vanneste, S.; Friml, J. Auxin: A trigger for change in plant development. *Cell* **2009**, *36*, 005–016. [CrossRef] [PubMed]
55. Salanenka, Y.; Verstraeten, I.; Löfke, C.; Tabata, K.; Naramoto, S.; Glanc, M.; Friml, J. Gibberellin DELLA signaling targets the retromer complex to redirect protein trafficking to the plasma membrane. *Proc. Natl. Acad. Sci. USA* **2018**, *115*, 3716–3721. [CrossRef] [PubMed]
56. Harrison, B.R.; Masson, P.H. ARL2, ARG1 and PIN3 define a gravity signal transduction pathway in root statocytes. *Plant J.* **2008**, *53*, 380–392. [CrossRef] [PubMed]
57. Gou, J.; Fu, C.; Liu, S.; Tang, C.; Debnath, S.; Flanagan, A.; Ge, Y.; Tang, Y.; Jiang, Q.; Larson, P.R.; et al. The miR156-SPL4 module predominantly regulates aerial axillary bud formation and controls shoot architecture. *New Phytol.* **2017**, *216*, 829–840. [CrossRef] [PubMed]
58. Jia, S.H.; Chang, S.; Wang, H.M.; Chu, Z.L.; Xi, C.; Liu, J.; Zhao, H.P.; Han, S.C.; Wang, Y.D. Transcriptomic analysis reveals the contribution of auxin on the differentially developed caryopses on primary and secondary branches in rice. *J. Plant Physiol.* **2021**, *256*, 153310. [CrossRef] [PubMed]
59. Yuan, G.; He, S.; Bian, S.; Han, X.; Liu, K.; Cong, P. Genome-wide identification and expression analysis of major latex protein (MLP) family genes in the apple (*Malus domestica* borkh.) genome. *Gene* **2020**, *733*, 144275. [CrossRef] [PubMed]
60. Chandler, J.W. Auxin response factors. *Plant Cell Environ.* **2016**, *39*, 1014–1028. [CrossRef]
61. Singh, K.B.; Foley, R.C.; Oñate-Sánchez, L. Transcription factors in plant defense and stress responses. *Curr. Opin. Plant Biol.* **2002**, *5*, 430–436. [CrossRef]
62. Boter, M.; Ruiz-Rivero, O.; Abdeen, A.; Prat, S. Conserved MYC transcription factors play a key role in jasmonate signaling both in tomato and Arabidopsis. *Genes* **2004**, *18*, 1577–1591. [CrossRef]
63. Wu, Q.; Bai, J.; Tao, X.; Mou, W.; Luo, Z.; Mao, L. Synergistic effect of abscisic acid and ethylene on color development in tomato (*Solanum lycopersicum* L.) fruit. *Sci. Hortic.* **2018**, *235*, 169–180. [CrossRef]
64. Yang, C.; Shi, G.; Li, Y.; Luo, M.; Wang, H.; Wang, J.; Yuan, L.; Wang, Y.; Li, Y. Genome-Wide Identification of SnRK1 Catalytic α Subunit and FLZ Proteins in Glycyrrhiza inflata Bat. Highlights Their Potential Roles in Licorice Growth and Abiotic Stress Responses. *Int. J. Mol. Sci.* **2023**, *24*, 121. [CrossRef]
65. Pan, W.; Zheng, P.; Zhang, C.; Wang, W.; Li, Y.; Fan, T. The effect of ABRE BINDING FACTOR 4-mediated FYVE1 on salt stress tolerance in arabidopsis. *Plant Sci.* **2020**, *296*, 110489. [CrossRef]
66. Gao, J.; Zhang, L.; Du, H.; Dong, Y.; Zhen, S.; Wang, C.; Wang, Q.; Yang, J.; Zhang, P.; Zheng, X.; et al. An ARF24-ZmArf2 module influences kernel size in different maize haplotypes. *J. Integr. Plant Biol.* **2023**, *65*, 1767–1781. [CrossRef]
67. Keita, S.; Daisuke, Y. Two cis-acting elements necessary and sufficient for gibberellin-upregulated proteinase expression in rice seeds. *Plant J.* **2003**, *34*, 635–645. [CrossRef]

Disclaimer/Publisher's Note: The statements, opinions and data contained in all publications are solely those of the individual author(s) and contributor(s) and not of MDPI and/or the editor(s). MDPI and/or the editor(s) disclaim responsibility for any injury to people or property resulting from any ideas, methods, instructions or products referred to in the content.

Article

The Role of the ADF Gene Family in Maize Response to Abiotic Stresses

Ruisi Yang [1,2,†], Fei Wang [2,†], Ping Luo [1,2], Zhennan Xu [2], Houwen Wang [2], Runze Zhang [1,2], Wenzhe Li [2], Ke Yang [1,2], Zhuanfang Hao [2,*] and Wenwei Gao [1,*]

1. College of Agriculture, Xinjiang Agricultural University, Urumqi 830052, China; yangruisixj@163.com (R.Y.); luoping987@126.com (P.L.); zrz04222022@163.com (R.Z.); yk864078904@163.com (K.Y.)
2. State Key Laboratory of Crop Gene Resources and Breeding, Institute of Crop Sciences, Chinese Academy of Agricultural Sciences, Beijing 100081, China; wangfei19990908@163.com (F.W.); xzn_caas@163.com (Z.X.); whw15797929108@163.com (H.W.); liwz16603632000@163.com (W.L.)
* Correspondence: haozhuanfang@163.com (Z.H.); gww0911@163.com (W.G.)
† These authors contributed equally to this work.

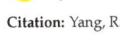

Citation: Yang, R.; Wang, F.; Luo, P.; Xu, Z.; Wang, H.; Zhang, R.; Li, W.; Yang, K.; Hao, Z.; Gao, W. The Role of the ADF Gene Family in Maize Response to Abiotic Stresses. *Agronomy* 2024, 14, 717. https://doi.org/10.3390/agronomy14040717

Academic Editor: Chenggen Chu

Received: 12 March 2024
Revised: 26 March 2024
Accepted: 27 March 2024
Published: 29 March 2024

Copyright: © 2024 by the authors. Licensee MDPI, Basel, Switzerland. This article is an open access article distributed under the terms and conditions of the Creative Commons Attribution (CC BY) license (https:// creativecommons.org/licenses/by/ 4.0/).

Abstract: The highly conserved actin depolymerizing factor (ADF) plays an important role in plant growth, development and responses to biotic and abiotic stresses. A total of 72 ADF genes in Arabidopsis, wheat, rice and sorghum can be divided into four groups. The multicollinearity analysis revealed that the maize ADF gene family exhibited more collinearity events with closely related gramineous plants. Fifteen ADF genes in maize were screened from the latest database, and bioinformatics analysis showed that these ADF genes were distributed across seven chromosomes in maize. The gene structure of the ADF gene family in maize exhibits significant conservation and cluster consistency. The promoter region contains rich regulatory elements that are involved in various regulations related to growth, development and adverse stresses. The drought-tolerant *ZmADF5* gene in maize was further studied, and it was found that the allelic variations in *ZmADF5* were mainly concentrated in its promoter region. A superior haplotype, with drought tolerance, was identified by candidate-gene association analysis of 115 inbred lines. By comparing the phenotypes of anthesis silking interval, grain yield and ear height, it was found that Hap2 performed better than Hap1 under drought stress. This study provides a theoretical reference for understanding the function of the ADF gene family and proposes further investigation into the role of *ZmADF5* in abiotic-stress tolerance.

Keywords: actin depolymerization factor; ADF gene family; candidate-gene association analysis; abiotic stresses

1. Introduction

Maize (*Zea mays* L.) is now the highest yielding cereal crop worldwide, and its planting area and production are increasing year by year [1]. As one of the main food crops in China, the majority of maize is processed into feed and biofuels, in addition to food [2–4]. Maize is a monoecious and cross-pollination crop, which is susceptible to various abiotic stresses during growth and development. Environmental stress at the flowering stage directly leads to serious yield reduction, with drought being one of the most serious abiotic stresses [5,6]. Therefore, it is of great practical significance to explore the abiotic stress-tolerant genes and study their functional mechanism to further improve the production or keep a stable level under abiotic stresses.

The actin depolymerizing factor (ADF/cofilin) is the main binding protein of microfilaments in the cytoskeleton and exists in all eukaryotic cells [7]. The first ADF in animals was isolated from chicken embryos and named cofilin, and the first ADF in plants was identified in lily [8–10]. In maize, plant biochemical characterization confirmed the conservative activity of ADF-binding F-actin and G-actin, observed for the first time in

ZmADF3 [11]. More ADF genes were identified in plants than in animals, including 11 ADF genes in Arabidopsis, 11 ADF genes in rice, 26 ADF genes in wheat, 15 ADF genes in maize, 9 ADF genes in sorghum and so on [12,13]. The number of plant ADF participate in important life activities, with more functional characteristics, such as cell movement, cell migration, cell division, cytoplasmic circulation, cell expansion, cell structure maintenance, intracellular material transport, polar growth and biotic- and abiotic-stress response [14–16]. It has been found that ADF genes have a non-negligible contribution to plant growth and development, especially in response to stress [17].

ADF belongs to the actin-binding protein (ABP) regulatory gene family, which participates in the dynamic regulation of actin cytoskeleton in cells by shearing and depolymerizing actin filaments [18]. Under abiotic stresses, ADF can regulate the rate of actin depolymerization by rapidly dissociating into small fragments and polymerizing into microfilaments to avoid plant damage [19]. Plant ADF genes are generally involved in the process of pollen tube germination, root hair growth and other polar growth processes related with microfilament skeleton rearrangement, where a large number of active G-actins concentrate [20]. Multiple rearrangements will happen when the growth apex and subapex continuously transform in the microfilament skeleton [21]. The microfilament skeleton is depolymerized at high speed, providing a carrier, power and anchor point for cytoplasmic circulation and vesicle transport [22].

In 2020, 13 ADF genes were publicly identified in maize [23]. Soon afterwards, a total of 15 maize ADF genes were updated based on the latest published maize genome database of the Zm-B73-REFERENCE-NAM-5.0 (B73 RefGen_v5) reference, along with the improvement of genome sequencing technology. Tandem and segmental duplication events are the main reasons driving the evolution of plant genomes [24].

In this study the new member of the maize ADF family *ZmADF14* undergoes segmental duplication with *ZmADF3* and tandem duplication with *ZmADF10*, respectively. In this study, the 15 ADF genes in maize were analyzed in detail by bioinformatics, including chromosome location, physical and chemical properties, functional clustering, collinearity, gene structure, conserved sequence, promoter elements, responses to abiotic stresses and so on. *ZmADF5*, which shows potential drought tolerance, was selected for candidate-gene association analysis. Combining with phenotypic data acquired from the field, it was proved that excellent variations from *ZmADF5* could help improving the drought tolerance in maize. This study is the first time to update the member information of maize ADF gene family and its related bioinformatics analysis after Huang which further proves that ADF gene family plays an important role in maize response to abiotic stresses. Secondly, a drought-tolerant candidate-gene *ZmADF5* and excellent variation was proposed, which laid an important foundation for further study on the function of the maize ADF gene.

2. Materials and Methods

2.1. Download and Arrangement of ADF Gene Family in Maize

The ZmADF family sequences (Zm-B73-REFERENCE-NAM-5.0, B73 RefGen_v5) were downloaded from the MaizeGDB (https://www.maizegdb.org/, accessed on 10 March 2023) database. ADF domains were used as queries to identify representative ZmADF proteins using TBtools (v1.120) (score $\geq$ 100 and e-value $\leq 1 \times 10^{-10}$) [25]. Non-redundant ZmADF protein sequences were then additionally screened with the NCBI Conserved Domain Database (https://www.ncbi.nlm.nih.gov/Structure/cdd/wrpsb.cgi, accessed on 10 March 2023) to confirm the presence of an ADF domain. The chromosomal locations of the 15 ZmADFs were determined according to maize reference genome B73 RefGen_v5 information. MapChart software (v 2.0) was used to plot their positions along the 7 chromosomes. ExPASy (https://web.expasy.org/, accessed on 10 March 2023) was used to analyze the biochemical parameters of ZmADFs family proteins, such as isoelectric point (pI), molecular weight (MW) and protein hydrophilic properties etc. [26].

2.2. Clustering, Collinearity, Gene Structure and Promoter Analysis of ADF Family

The species Arabidopsis, wheat, rice and sorghum sequences were downloaded from NCBI website (https://www.ncbi.nlm.nih.gov/, accessed on 10 March 2023). The aligned sequences were subjected to phylogenetic analysis by the maximum likelihood method using MEGA (v7.0) with 1,000 bootstrap replicates [27]. The online software tool iTOL (Interactive Tree of Life, v5) was applied to modify the phylogenetic tree [28]. Tandem and segmental duplication events for ZmADF genes were identified using MCScanX (https://github.com/wyp1125/MCScanX, accessed on 10 March 2023) and displayed through TBtools. Synteny analyses assessing the relationships among ADF genes encoded by maize and Arabidopsis, wheat, rice and sorghum bicolor were conducted with TBtools [29,30]. MEME (https://meme-suite.org/tools/meme, accessed on 10 March 2023) was applied to predict possible conserved motifs in the ZmADFs family, and the maximum number of motifs was set to 10. The intron and exon structures of the ZmADF gene were mapped by GSDS (http://GSDS.cbi.pku.edu.cn/, accessed on 10 March 2023). PlantCare online software (http://bioinformatics.psb.ugent.be/webtools/plancare/html/, accessed on 10 March 2023) was used to analyze the cis-acting elements in the promoter region of maize ADF genes.

2.3. Drought Stress Treatment in Seedling Stage of Maize

Fifty seeds of the inbred line Zheng 58 were soaked in 0.5% sodium hypochlorite for 10 min for surface disinfection, washed repeatedly with distilled water and soaked in saturated calcium persulfate for 12 h. The seeds were germinated on filter paper and then transferred to Hoagland nutrient solution for culture. The culture conditions were 28 °C, 16 h light and 8 h dark in a greenhouse. The treatment with 20% PEG6000 was applied to seedlings at the three-leaf stage. At each treatment point (0, 1, 3, 6 and 12 h), 3 seedlings were selected as 3 independent biological replicates.

2.4. Real-Time Fluorescence Quantitative PCR Detection

The total RNA of the sample was extracted with the FastPure Universal Plant Total RNA isolation kit (Nanjing Nuoweizan Biotechnology Co., Ltd., Nanjing, China), and the first-strand cDNA synthesis was performed using the FastQuant RT kit (Tiangen Biochemical Technology Co., Ltd., Beijing, China). The Real Master Mix (SYBR Green I) kit (Tiangen Biochemical Technology Co., Ltd.) was used for qRT-PCR experiments. The real-time PCR fluorescence quantitative reaction system was 20 µL: 2 × SuperReal PreMix Plus 10 µL, forward (reverse) primer (10 µL) 0.6 µL, cDNA template 2 µL, RNase-free ddH$_2$O 6.8 µL. Quantitative PCR was implemented using the SYBR Premix Ex Taq II (Takara) on an ABI 7500 real-time detection system (Applied Biosystems), and three independent RNA samples were prepared for each biological replicate. *ZmUBI* was used as the reference gene (Table 1). The relative expression of the gene and its standard deviation were analyzed according to the $2^{-\Delta\Delta CT}$ method [31].

Table 1. Real-time quantitative PCR primer sequence.

Name	Registration Number
ZmUBI-Forward primer	TGGTTGTGGCTTCGTTGGTT
ZmUBI-Reverse primer	GCTGCAGAAGAGTTTTGGGTACA
ZmADF5-Forward primer	CAGGGCCAAGATCCTGTACG
ZmADF5-Reverse primer	ATGACGTCGAAGCCCATCTC

2.5. Analysis of Transcription Level of Maize ADF Family under Abiotic Stresses

The maize transcriptome sequencing data published in the SRA database was downloaded and converted into Fastq data using Fastq-dump.2.11.0. Then, FastQC software (v0.11.5) (https://github.com/s-andrews/FastQC, accessed on 10 March 2023) was used to determine the quality of Fastq data [32]. Trimmomatic software (v0.33) was used to remove joints and low-quality sequences from the Fastq data, and ultimately to obtain filtered clean data [33]. The filtered data were compared with the maize B73 RefGen_v5 genome to generate a SAM file. SAM files were converted into sorted BAM files using SAMtools software (v1.17). StringTie software (v2.2.1) was used to estimate the expression data of each gene, and DESeq2 software (vR4.1.2) was used to analyze the differentially expressed genes [34].

2.6. Candidate-Gene Association Analysis in ZmADF5

In this study, we collected the phenotypic data of 188 maize inbred lines under drought stress (WS) and normal irrigation (WW) conditions in Hainan (HN) and Xinjiang (XJ) [35]. Subsequently, 115 maize NL inbred lines preserved in the laboratory were cultured to 1–2 leaf stage and genomic DNA was extracted by CTAB method. The plant genome database website was used to find the sequence of *ZmADF5* gene promoter region and gene region. The maize inbred line B73 genome was used as the reference sequence. Primer 5.0 software was used to design primers, including forward primer: 5′-TCTTCGGCAATCTCCAG-3′ and reverse primer: 5′-TCTACTCCACCCATCAACATC-3′. The PCR amplification reaction system was as follows: P520 mix 25 µL, upstream and downstream primers (10 µL) 1.5 µL, cDNA template 2 µL, RNase-free ddH$_2$O 20 µL. The PCR amplification program consisted of the following steps: 98 °C for 30 s, 35 cycles of 98 °C for 10 s, 60 °C for 5 s, 72 °C for 20 s, 72 °C for 1 min. Finally, the PCR product was sent to the biological company for sequencing.

3. Results

3.1. Chromosome Location and Physicochemical Properties of ADF Family in Maize

Based on the released genome information of maize (B73 RefGen_v5), 15 genes containing the ADF domain were screened by comparing them to the maize genome, and classified them as ADF genes. They were distributed on seven chromosomes in clusters or scattered conditions, with chromosome 1 containing the largest number of *ZmADFs* (Figure 1). Among them, *ZmADF7* and *ZmADF13*, *ZmADF1* and *ZmADF12*, *ZmADF6* and *ZmADF9*, *ZmADF5* and *ZmADF8*, *ZmADF14* and *ZmADF10* appeared as tandem duplications, respectively, and only *ZmADF3* and *ZmADF14* existed as segmental duplications. The protein sequence length of the maize ADF gene family ranged from 132 to 210 aa. The molecular weight ranged from 15,620.91 to 22,733.91 Da, with the isoelectric point values ranging from 4.81 to 9.51. The highest theoretical isoelectric point value was 9.51 for ZmADF8. Among the 15 ADF members, nine proteins were acidic (ZmADF1, 2, 3, 6, 7, 10, 11, 12, 14) and the other six proteins were alkaline (ZmADF4, 5, 8, 9, 13, 15). The prediction results showed that, except for ZmADF14, which exhibited hydrophobic properties, the remaining 14 proteins were hydrophilic proteins. The unstable parameter was between 37.04 and 64.63, and the aliphatic amino acid index ranged from 62.45 to 84.56. Subcellular localization prediction indicated that 15 genes were mainly located in the cytoplasm (Table 2).

Table 2. Physicochemical properties of maize ADF family members.

Name	Registration Number	Molecular Weight (Da)	Chromosome	Gene Location	Number of Amino Acids (aa)	Isoelectric Point	GRAVY	Unstable Parameter	Aliphatic Amino Acid Index	Subcellular Localization
ZmADF1	Zm00001eb321460_T001	16,554.63	7	155289254 155290450	144	6.32	−0.586	55.68	64.38	Cytoplasmic
ZmADF2	Zm00001eb105010_T001	16,083.09	2	206083396 206084459	139	5.57	−0.642	45.49	63.17	Cytoplasmic
ZmADF3	Zm00001eb062600_T001	15,899.93	1	300195447 300197511	139	5.46	−0.480	49.66	74.39	Cytoplasmic
ZmADF4	Zm00001eb266570_T001	15,855.13	6	43778235 43779873	139	7.66	−0.369	51.09	77.88	Cytoplasmic
ZmADF5	Zm00001eb010370_T001	16,413.96	1	32840219 32842834	143	8.41	−0.287	47.01	73.64	Cytoplasmic
ZmADF6	Zm00001eb213630_T001	16,833.10	5	5554015 5557587	145	6.15	−0.471	53.27	70.62	Cytoplasmic
ZmADF7	Zm00001eb186790_T001	15,855.09	4	158650414 158652626	139	6.31	−0.316	49.98	67.34	Cytoplasmic
ZmADF8	Zm00001eb398870_T001	20,038.07	9	149321956 149324583	172	9.51	−0.505	64.63	70.45	Cytoplasmic
ZmADF9	Zm00001eb059710_T001	15,620.91	1	291011670 291012909	132	7.78	−0.448	37.04	81.97	Cytoplasmic
ZmADF10	Zm00001eb211740_T001	15,913.96	5	2677695 2679728	139	5.47	−0.485	48.80	74.39	Cytoplasmic
ZmADF11	Zm00001eb021290_T002	22,733.91	1	80880988 80884908	210	5.64	−0.627	38.45	61.05	Cytoplasmic
ZmADF12	Zm00001eb074420_T001	15,981.05	2	20224012 20227127	139	5.27	−0.570	51.55	62.45	Cytoplasmic
ZmADF13	Zm00001eb249860_T001	15,893.23	5	197451938 197454933	139	7.56	−0.271	52.66	71.58	Cytoplasmic
ZmADF14	Zm00001eb062580_T001	15,698.58	1	300194049 300194984	149	4.81	0.019	38.49	84.56	Cytoplasmic
ZmADF15	Zm00001eb057310_T001	16,725.47	1	283186285 283187993	144	9.44	−0.274	59.98	75.90	Cytoplasmic

Note: Table 1 mainly shows gene location and the physicochemical characterization of the ZmADF genes family, in which the registration number is derived from (https://www.maizegdb.org/, accessed on 10 March 2023) based on the latest database of maize genome B73 RefGen_v5. Unstable parameter refers to chemical instability and physical instability. The isoelectric point, GRAVY value, aliphatic amino acid index and subcellular localization were all predicted by ExPASy software analysis.

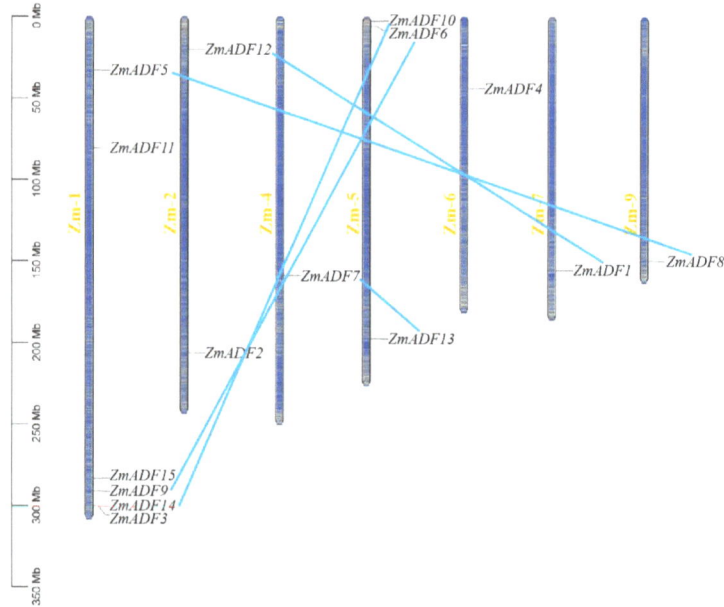

Figure 1. The distribution of ADF gene family on maize chromosomes. The figure shows 7 chromosomes of maize, namely chromosomes 1, 2, 4, 5, 6, 7 and 9, respectively. The left side of each chromosome is marked with yellow for chromosome number, and the right side is marked with red for chromosomal location distribution of 15 ZmADF genes. The blue line represents the gene pairs that undergo tandem duplication.

3.2. Phylogenetic Analysis of Maize ADF Family with Arabidopsis, Wheat, Rice and Sorghum

According to the classification results of the Arabidopsis ADF family, 72 ADF genes of these five species could be divided into four groups [36]. Group I contained the most genes, including *AtADF1*, *AtADF2*, *AtADF3* and *AtADF4*, which were constitutively expressed in all tissues except pollen. *AtADF7*, *AtADF8*, *AtADF10* and *AtADF11* were apical meristem-specific expression genes, which were expressed mainly in pollen and root. *ZmADF1*, *ZmADF2*, *ZmADF3*, *ZmADF4*, *ZmADF7*, *ZmADF10*, *ZmADF12*, *ZmADF13* and *ZmADF14* were found in the same group. Group II was a monocotyledon-specific ADF group, with the majority of its members belonging to ADF3. However, the monocotyledon plant *ZmADF3* was not clustered into this group. Protein sequence alignment analysis found that *ZmADF3* has low homology with *OsADF3*, *SbADF3* and *TaADF3*, with only 57.96% similarity, eventually leading to functional differentiation. According to the clustering results of *AtADF5* and *AtADF9*, group III was divided into several genes widely expressed in multiple tissues and organs, including meristem. *ZmADF5* and *ZmADF8* belong to this group. Represented by *ZmADF6*, group IV was speculated to be constitutively expressed in all tissues, including *ZmADF6*, *ZmADF9*, *ZmADF11* and *ZmADF15*, which exhibit high homology to each other (Figure 2).

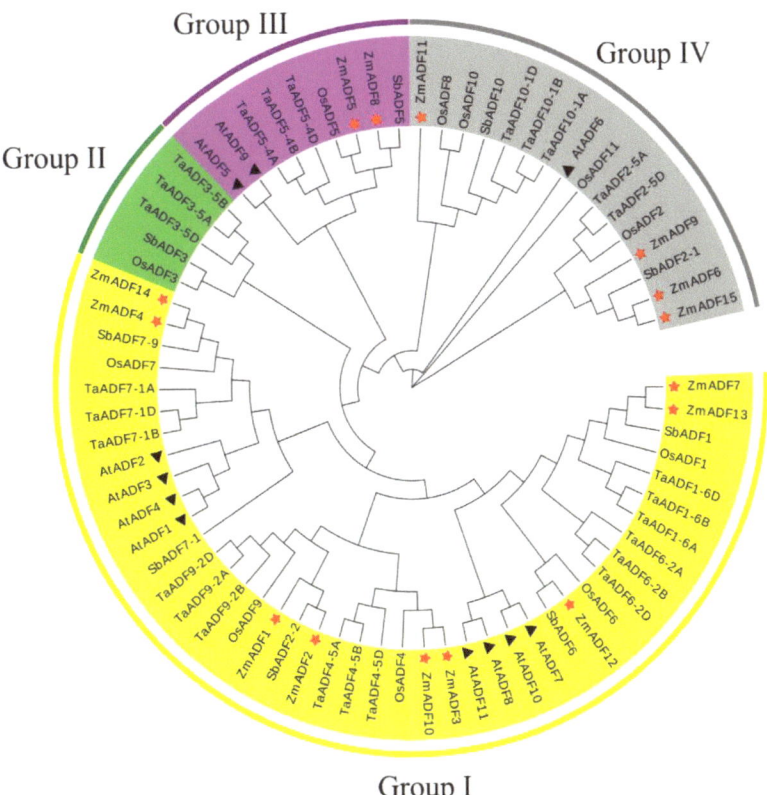

Figure 2. Phylogenetic analysis of ADF proteins from maize, Arabidopsis, wheat, rice and sorghum. The ADF genes were divided into four groups: yellow represents Group I, green represents Group II, purple represents Group III and gray represents Group IV. The four species are abbreviated as follows: maize (Zm), Arabidopsis (At), wheat (Ta), rice (Os) and sorghum (Sb). The black triangle marks the ADF genes of Arabidopsis, and the red pentagram marks the ADF genes in maize.

3.3. Collinearity Analysis of Maize ADF Gene Family with Arabidopsis, Wheat, Rice and Sorghum

Except for *ZmADF1*, the remaining 14 maize genes were collinear with Arabidopsis, wheat, rice and sorghum (Figure 3). *ZmADF4*, *ZmADF12* and *ZmADF13* had five collinear events with *AtADF1*, *AtADF6*, *AtADF7* and *AtADF11*. The collinearity logarithms of maize with wheat, rice and sorghum are 36, 15 and 10, respectively. Collinearity mainly exists between chromosomes 1, 2, 4, 5, 6, and 9. There are more collinearity events among wheat, rice, sorghum and maize, and fewer collinearity events with Arabidopsis. Combined with the results of cluster analysis, it was found that the genes with close genetic relationships within the same group were highly consistent with collinearity genes in other species, and the number of collinearity occurrences was also almost the same, exhibiting group-specific characteristics.

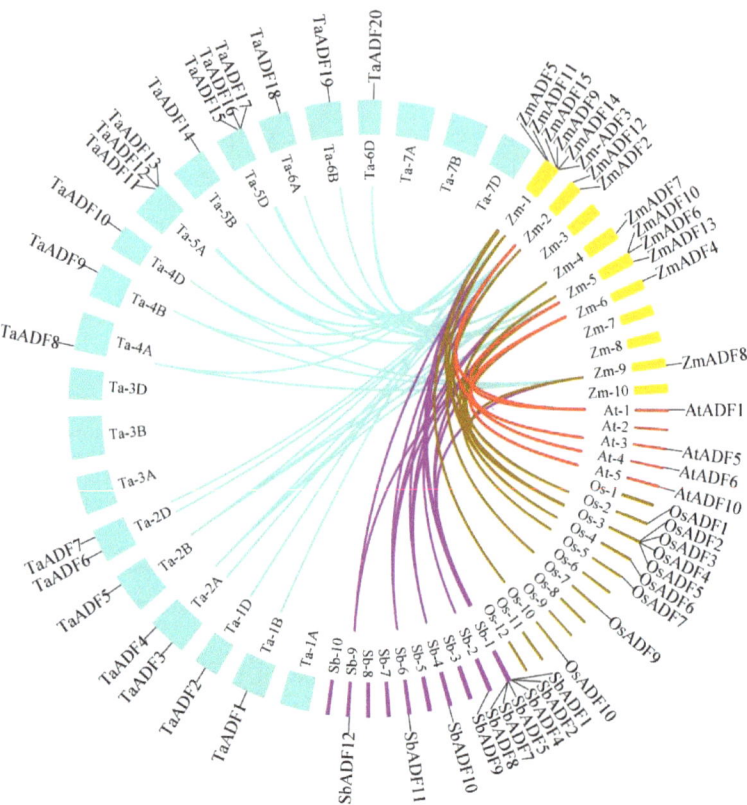

Figure 3. Collinearity analysis of maize ADF gene family in maize, Arabidopsis, wheat, rice and sorghum. The background of whole genome represents by gray, while the collinear gene pairs of ZmADF with different species are connected with different colors. Red represents Arabidopsis, yellow-green represents rice, purple represents sorghum, and blue-green represents wheat.

3.4. Gene Structure and Conserved Sequence of ADF Family in Maize

The 15 maize ADF gene families were individually clustered into four subfamilies, which was consistent with the clustering results in 2.2 (Figure 2). The conserved motifs of ADF proteins showed that the structures within the same group were similar each other. All 15 maize ADF gene families shared motif 3 (MAVADECKLKFVELKAKRSFRFIVFKIDE), which should be a highly conserved ADF domain (Figure 4A,C). The ADF gene family in maize exhibits relatively conserved characteristics, with a simple structure (Figure 4B).

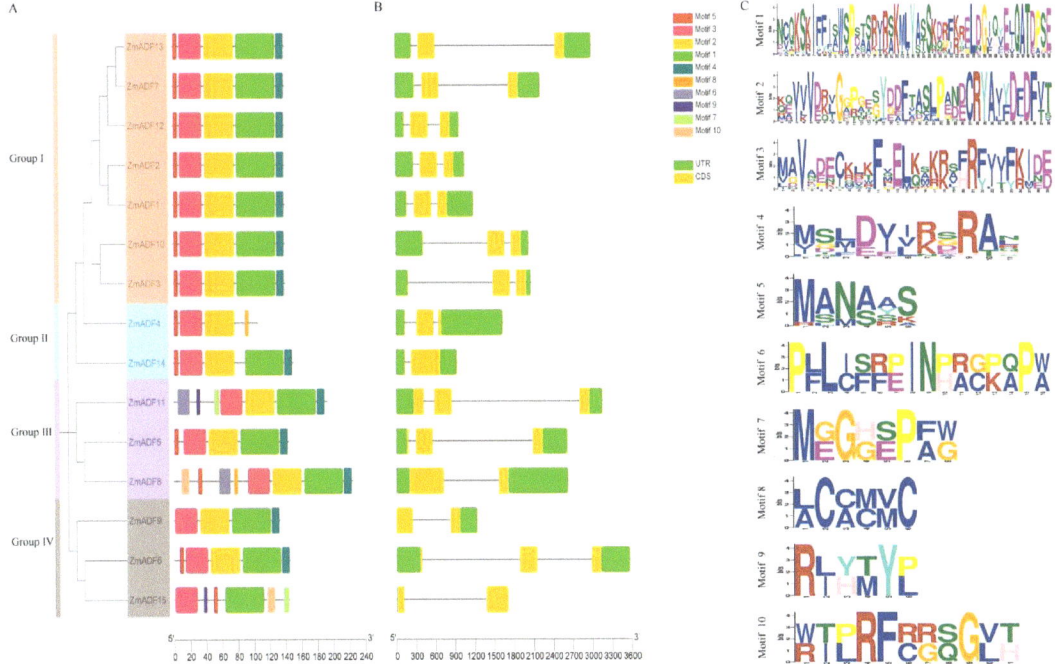

Figure 4. Exon-intron structures of ADF genes and a schematic diagram of the amino acid motifs of ADF proteins in maize. (**A**) The protein motif structure in ZmADFs. There are 10 main protein motifs displayed on the right side. (**B**) The gene structure in ZmADFs, including exons and introns. (**C**) Ten protein motif sequences are predicted in the ZmADFs protein.

3.5. Analysis of Cis-Acting Elements in the Promoter Regions of Maize ADF Family Genes

The promoter regions of the maize ADF gene family contained rich regulatory elements, of which 92 were light-responsive elements (LREs), 84 were jasmonic acid-responsive elements, 79 were ABA-responsive elements (ABRE) and 34 were anoxic inducibility elements (Figure 5A,B). On the whole, regulatory elements were divided into five categories according to their functions: hormone-responsive elements, abiotic stress-responsive elements, light-responsive elements, tissue-specific regulatory elements and other elements. Among them, the hormone-responsive elements constituted the highest proportion at 51.5%, while the proportion of abiotic stress-responsive elements was 25.2%. The light-responsive elements accounted for 16.16% (Figure 5C). Most elements play an important role in plant growth, development and responses to abiotic stresses. It has been reported that the ADF genes can participate in multiple functions at the same time, including growth and development [37].

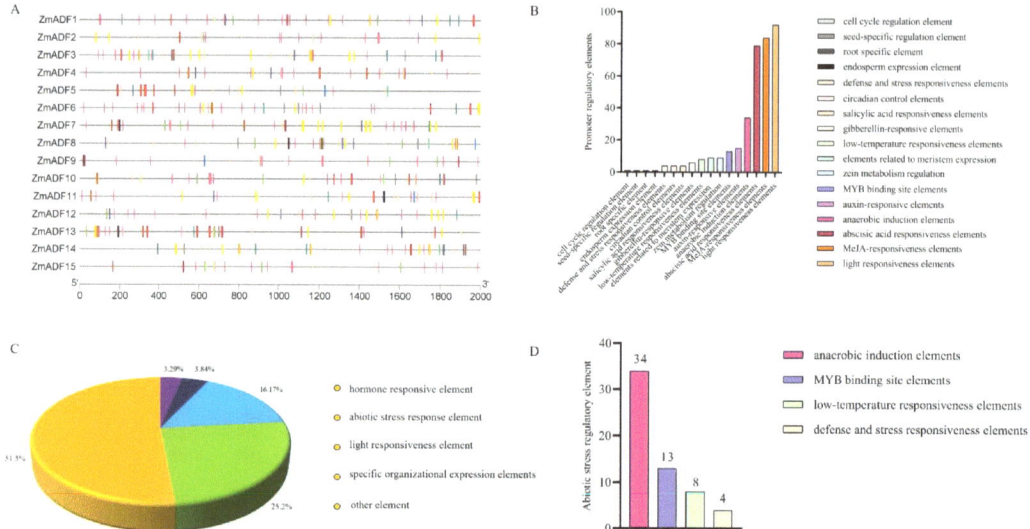

Figure 5. Cis-acting elements in the promoter regions of maize ADF family. (**A**) The distribution of cis-acting elements in the promoters of maize ADF gene family, where each color represents a cis acting element, and the same color in different genes represents the same cis acting element. (**B**) Quantitative statistics of different cis-acting elements in the promoters of maize ADF genes. (**C**) All elements can be divided into 5 categories, and the proportion of each type of component is showed. (**D**) Detailed information of 4 elements involved in abiotic-stress regulation.

3.6. Transcriptome Analysis of Maize ADF Gene Family under Abiotic Stresses

Based on the transcriptome sequencing data on maize under abiotic stress sourced from public databases and the transcriptome data on maize under drought stress in our laboratory, the expression heat maps of maize ADF gene families under four abiotic stresses, such as low temperature, drought, salt stress and high temperature, were drawn (Figure 6). Among them, *ZmADF1*, *ZmADF2*, *ZmADF7*, *ZmADF12* and *ZmADF13* are responsive to both drought and high temperature; *ZmADF3*, *ZmADF4*, *ZmADF5*, *ZmADF6*, *ZmADF10* and *ZmADF11* are mainly responsive to drought stress; and *ZmADF15* responds to low temperature. *ZmADF9* responds to low temperature and salt stress, *ZmADF8* responds to drought and high temperature, and *ZmADF14* responds to salt and high temperature (Figure 6A). The results showed that the up-regulated expression of 15 *ZmADF* genes after drought stress included *ZmADF3*, *ZmADF4*, *ZmADF5*, *ZmADF8*, *ZmADF11* and *ZmADF13* (Figure 6B). A comparison of both transcriptome results showed that most expression patterns of the ZmADF gene were the same, except for *ZmADF6*, *ZmADF8* and *ZmADF10*, which exhibited some differences in response to drought stress. From the FPKM value, the expression level of *ZmADF5* after drought was significantly different from that of the control (Figure 6C). *ZmADF5* had a strong response to drought stress both in the public database and the transcriptome results (Figure 6D).

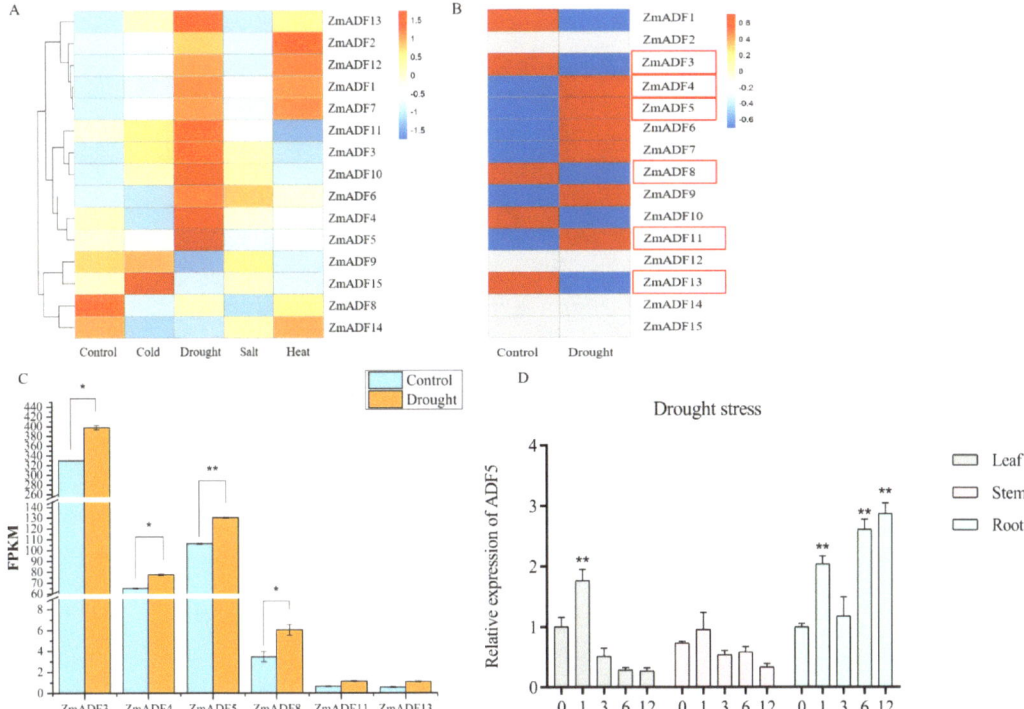

Figure 6. Transcriptional analysis of maize ADF gene under abiotic stresses. (**A**) Transcriptome data analysis on low temperature, drought, high temperature and salt stress from public databases. (**B**) Heat map analysis of ADF gene family expression in transcriptional data after drought treatment, sourced from RNA-seq, the red frame indicates genes that significantly respond to drought stress (**C**) FPKM value analysis of 6 drought resistance genes from transcriptome results. (**D**) Analysis of *ZmADF5* expression in leaves, stems and roots at 0, 1, 3, 6 and 12 h after drought treatment verified the response of *ZmADF5* to drought stress. Significant difference analysis in the figure: * represents $p < 0.05$, ** represents $p < 0.01$.

3.7. Association Analysis of ZmADF5 as a Candidate Gene for Drought Tolerance

To determine the allelic variations related to drought tolerance in *ZmADF5*, a total of 4521 bp length in the *ZmADF5* gene was sequenced, including the promoter, coding and non-coding regions, across 115 maize inbred lines (Figure 7). After candidate-gene association analysis, two indels and one single nucleotide polymorphism (SNP) variation loci in the promoter region of *ZmADF5* were identified that were significantly related to drought tolerance. These loci are located at positions ADF5-Indel-1511, ADF5-Indel-1435 and ADF5-SNP-206 (the first base of the start codon is marked as position 1, with positions being numbered negatively before it and positively after it) (Figure 7A). These three loci were in a complete linkage disequilibrium state and were significantly associated with ear height, grain yield and anthesis silking interval under drought stress. Based on these three loci, 115 inbred lines were divided into two haplotypes (Figure 7B). The anthesis silking interval of Hap2 was significantly shorter than that of Hap1, while the ear height and grain yield of Hap2 were significantly higher than those of Hap1 (Figure 7C).

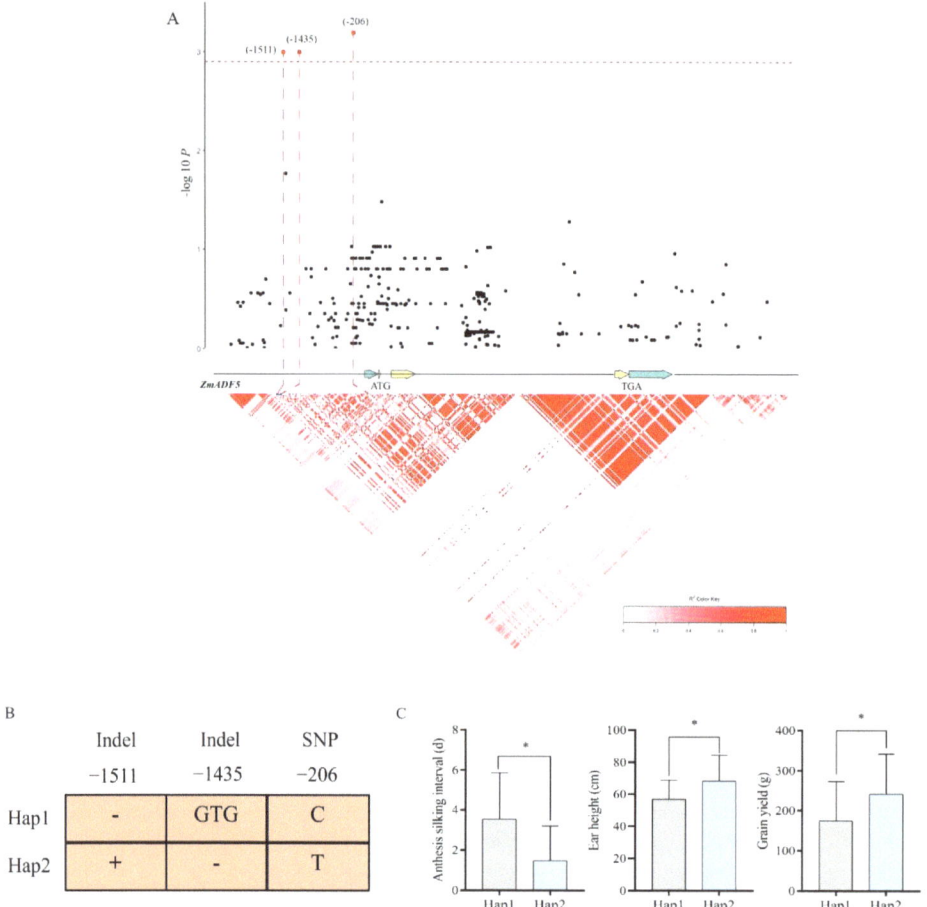

Figure 7. Candidate-gene association analysis in *ZmADF5*. (**A**) Linkage disequilibrium map for *ZmADF5* drawn based on phenotypic date of drought-tolerant-related traits by candidate-gene association analysis. The red dash line represents the screening threshold for achieving significant differences. The green arrow in the figure represents the 5′-UTR and 3′-UTR of the *ZmADF5* gene, and the yellow arrow represents the 3 exon regions. (**B**) Based on 2 Indel (−1511, −1435) and 1 SNP (−206), 115 inbred lines were divided into Hap1 and Hap2 haplotypes; "+" represents containing the corresponding mutation site, "−" represents not containing the corresponding mutation site. (**C**) Statistical analysis of anthesis silking interval, ear height and grain yield under Hap1 and Hap2 drought stress. Hap1 is represented in gray, Hap2 is represented in light blue. Significant difference analysis in the figure: * represents $p < 0.05$.

4. Discussion

4.1. Gene Duplication in ADF Gene Family and Its Possible Function in the Evolutionary Process of Genes

Tandem duplication and segmental duplication in genes are the main driving forces for the expansion and evolution of gene families [38]. The original genes provide raw materials for the formation of new genes, and the new genes promote functional diversification [39]. The ADF gene family has strong evolutionary conservation and relatively few gene duplication events. Only two genes have segmental duplication (*ZmADF3* and

ZmADF14), and five pairs of genes have tandem duplication (ZmADF7 and ZmADF13, ZmADF1 and ZmADF12, ZmADF6 and ZmADF9, ZmADF5 and ZmADF8, ZmADF14 and ZmADF10) (Figure 1). It was found that gene duplication events were clustered in the same group and had similar protein and gene structures (Figures 2 and 4). Therefore, it is speculated that the biological functions of genes with gene duplication are similar or complementary. For example, it was found that AtADF7 and AtADF10 showed tandem replication in Arabidopsis (Figure S1). Both genes showed distinct intracellular localizations during pollen germination. However, they could cooperate with nonequivalent functions in promoting pollen cells to achieve exquisite control of the turnover of different actin structures, thereby meeting different cellular needs [40]. Recently, ZmADF1 has been confirmed to negatively regulate pollen development [41]. AtADF5 has been confirmed to be involved in drought and low temperature stress in Arabidopsis [42,43]. ZmADF5 can improve the drought resistance of maize [44]. The function of ZmADF8 has not been reported, but with a homology of 97.3% to ZmADF5, it is speculated that ZmADF8 and ZmADF5 may have similar functions or be complementary to each other. The newly identified gene, ZmADF14, is likely produced by the segmental duplication of ZmADF3 and the tandem duplication of ZmADF10. ZmADF15 exhibited high homology with ZmADF6 and ZmADF9, but no gene duplication events occurred. Currently, there are relatively few studies on the maize ADF gene. According to evolutionary conservation and gene duplication within the ADF gene family, both the ADF gene of Arabidopsis and the reported maize ADF gene can be used as a reliable references for studying the biological function of maize ADF genes.

4.2. The Characterization of ADF Gene Family and Its Regulation in Response to Abiotic Stresses

The structure of the maize ADF gene family is relatively simple, but the promoter region contains rich regulatory elements, with the most abundant being hormone response elements. Hormone response can participate in plant growth and development and can also participate in plant stress response. The quantity of abiotic-stress response elements in the promoter region of the maize ADF gene family is also high, including drought, low temperature, stress response and so on. It can be seen that the maize ADF gene family has great potential in coping with abiotic stresses. For example, ADF1 in Group I can improve the heat tolerance of Arabidopsis and Chinese cabbage, as well as enhance the low-temperature tolerance of Arabidopsis [45–47]; AtADF4 can respond to osmotic stress and drought [48]; AtADF7 not only participates in pollen tube development but also positively regulates osmotic stress [49]. OsADF3, in Group II, can positively regulate drought tolerance in rice [19]. In Group III, AtADF5 can promote stomatal closure and improve drought tolerance in Arabidopsis by regulating ABA and actin cytoskeleton remodeling under drought stress. It can also respond to low temperature stress by regulating the stomata, while PeADF5 in Populus euphoretic is mainly responsive to drought [42,43,50]. To sum up, it can be seen that the ADF gene has multiple biological functions, which are associated with their numerous regulatory elements in the promoter region, as discussed in Section 2.5 (Figure 5). According to the evolutionary conservation of ADF genes, it is speculated that the maize ADF gene family might participate in a variety of abiotic-stress responses.

4.3. Excellent Allelic Variations Associated with Drought Tolerance in ZmADF5

ZmADF5 is a drought-tolerant gene identified by genome-wide association analysis, which has been verified through overexpression in maize. The main three excellent variation sites are concentrated in its promoter region. The promoter region of ZmADF5 contains a large number of hormone-responsive elements (MeJA, GA, ABA), followed by growth regulatory elements (light response, diurnal regulation) and stress-responsive elements (drought, low temperature and anoxic conditions) (Figure 5). According to the promoter elements, it is speculated that ZmADF5 might participate in plant growth, development and stress response. Through candidate-gene association analysis of ZmADF5, 115 inbred lines were divided into two haplotypes, of which Hap2 was significantly better compared to Hap1, with a grain yield 18.77% higher than that of Hap1. As a screened and confirmed drought-tolerant candidate gene, understanding the mechanism behind these variations in drought tolerance will help us to study the biological function of the maize ADF gene family, indirectly promoting the process of marker-assisted breeding in maize.

5. Conclusions

In this study, 15 ADF family genes were identified and characterized in the whole maize genome, updating it with new members of *ZmADF14* and *ZmADF15*. These ADF genes were distributed across seven chromosomes and phylogenetically divided into four groups, with conservation and gene duplication trajectory captured in their gene structure. Nonetheless, the rich regulatory elements in their promoter region endow them with multiple biological functions, especially in response to abiotic stresses. Candidate-gene association analysis revealed that the promoter region in *ZmADF5* contained three excellent variations associated with drought resistance. Overall, ADF genes are expected to participate in various abiotic stresses with high potential, and their excellent variations related to drought resistance could be used for marker-assisted breeding in the future.

Supplementary Materials: The following supporting information can be downloaded at: https://www.mdpi.com/article/10.3390/agronomy14040717/s1, Figure S1 The chromosome distribution and gene duplication map of the ADF gene family in Arabidopsis, which show five chromosomes, namely chromosomes 1, 2, 3, 4, and 5. The left side of each chromosome is marked with black chromosome number, and the right side is marked with red chromosome location distribution of 11 AtADF genes. The red line represents gene pairs undergoing tandem duplication.

Author Contributions: R.Y. and F.W.: methodology, software, investigation, writing—original draft preparation, writing—review and editing. W.G. and Z.H.: conceptualization, methodology, writing—review and editing, project administration. P.L. and Z.X. revised the manuscript and modified the language. H.W., R.Z., W.L. and K.Y. proofread the data and formatted the article. All authors have read and agreed to the published version of the manuscript.

Funding: This work was supported by the National Natural Science Foundation of China (32272049, 32261143757), Sustainable Development International Cooperation Program from Bill & Melinda Gates Foundation (2022YFAG1002), Key Research and Development Program of Xinjiang Uygur Autonomous Region (2022B02001-4), Theearmarked Fund for XJARS-02.

Data Availability Statement: All data supporting the findings of this study are included in this article.

Conflicts of Interest: The authors declare no conflicts of interest.

References

1. Tigchelaar, M.; Battisti, D.S.; Naylor, R.L.; Ray, D.K. Future warming increases probability of globally synchronized maize production shocks. *Proc. Natl. Acad. Sci. USA* **2018**, *115*, 6644–6649. [CrossRef]
2. Carvalho-Estrada, P.A.; de Andrade, P.A.M.; Paziani, S.F.; Nussio, L.G.; Quecine, M.C. Rehydration of dry maize preserves the desirable bacterial community during ensiling. *FEMS Microbiol. Lett.* **2020**, *367*, 139. [CrossRef]
3. Liu, S.; Meng, J.; Lan, Y.; Cheng, X.E.Y.; Liu, Z.; Chen, W. Effect of maize straw biochar on maize straw composting by affecting effective bacterial community. *Prep. Biochem. Biotechnol.* **2021**, *51*, 792–802. [CrossRef]
4. Wang, M.; Qiao, J.; Sheng, Y.; Wei, J.; Cui, H.; Li, X.; Yue, G. Bioconversion of maize fiber to bioethanol: Status and perspectives. *Waste Manag.* **2023**, *15*, 256–268. [CrossRef]
5. Bray, E.A. Plant responses to water deficit. *Trends Plant Sci.* **1997**, *2*, 48–54. [CrossRef]
6. Wang, B.; Liu, C.; Zhang, D.; He, C.; Zhang, J.; Li, Z. Effects of maize organ-specific drought stress response on yields from transcriptome analysis. *BMC Plant Biol.* **2019**, *19*, 335. [CrossRef]
7. Roy-Zokan, E.M.; Dyer, K.A.; Meagher, R.B. Phylogenetic patterns of codon evolution in the actin-depolymerizing factor/cofilin (ADF/CFL) gene family. *PLoS ONE* **2015**, *10*, 0145917. [CrossRef] [PubMed]
8. Bamburg, J.R.; Harris, H.E.; Weeds, A.G. Partial purification and characterization of an actin depolymerizing factor from brain. *FEBS Lett.* **1980**, *121*, 178–182. [CrossRef]
9. Nishida, E.; Maekawa, S.; Sakai, H. Cofilin, a protein in porcine brain that binds to actin filaments and inhibits their interactions with myosin and tropomyosin. *Biochemistry* **1984**, *23*, 5307–5313. [CrossRef]
10. Kim, S.R.; Kim, Y.; An, G. Molecular cloning and characterization of anther-preferential cDNA encoding a putative actin-depolymerizing factor. *Plant Mol. Biol.* **1993**, *21*, 39–45. [CrossRef]

11. Jiang, C.J.; Weeds, A.G.; Hussey, P.J. The maize actin-depolymerizing factor, *ZmADF3*, redistributes to the growing tip of elongating root hairs and can be induced to translocate into the nucleus with actin. *Plant J.* **1997**, *12*, 1035–1043. [CrossRef]
12. Nishida, E.; Iida, K.; Yonezawa, N.; Koyasu, S.; Yahara, I.; Sakai, H. Cofilin is a component of intranuclear and cytoplasmic actin rods induced in cultured cells. *Proc. Natl. Acad. Sci. USA* **1987**, *84*, 5262–5266. [CrossRef]
13. Xu, K.; Zhao, Y.; Zhao, S.H.; Liu, H.D.; Wang, W.W.; Zhang, S.H.; Yang, X.J. Genome-wide identification and low temperature responsive pattern of actin depolymerizing factor (ADF) gene family in wheat (*Triticum aestivum* L.). *Front. Plant Sci.* **2021**, *12*, 618984. [CrossRef]
14. Menand, B.; Calder, G.; Dolan, L. Both chloronemal and caulonemal cells expand by tip growth in the moss physcomitrella patens. *J. Exp. Bot.* **2007**, *58*, 1843–1849. [CrossRef]
15. Bou Daher, F.; van Oostende, C.; Geitmann, A. Spatial and temporal expression of actin depolymerizing factors ADF7 and ADF10 during male gametophyte development in *Arabidopsis thaliana*. *Plant Cell Physiol.* **2011**, *52*, 1177–1192. [CrossRef]
16. Niu, Y.; Qian, D.; Liu, B.; Ma, J.; Wan, D.; Wang, X.; He, W.; Xiang, Y. ALA6, a P4-type ATPase is involved in heat stress responses in *Arabidopsis thaliana*. *Front. Plant Sci.* **2017**, *8*, 01732. [CrossRef]
17. Sengupta, S.; Mangu, V.; Sanchez, L.; Bedre, R.; Joshi, R.; Rajasekaran, K.; Baisakh, N. An actin-depolymerizing factor from the halophyte smooth cordgrass, *Spartina alterniflora* (*SaADF2*), is superior to its rice homolog (*OsADF2*) in conferring drought and salt tolerance when constitutively overexpressed in rice. *Plant Biotechnol. J.* **2019**, *17*, 188–205. [CrossRef]
18. Inada, N. Plant actin depolymerizing factor: Actin microfilament disassembly and more. *J. Plant Res.* **2017**, *130*, 227–238. [CrossRef]
19. Huang, Y.C.; Huang, W.L.; Hong, C.Y.; Lur, H.S.; Chang, M.C. Comprehensive analysis of differentially expressed rice actin depolymerizing factor gene family and heterologous overexpression of *OsADF3* confers *Arabidopsis thaliana* drought tolerance. *Rice* **2012**, *5*, 33. [CrossRef] [PubMed]
20. Zheng, Y.; Xie, Y.; Jiang, Y.; Qu, X.; Huang, S. Arabidopsis actin-depolymerizing factor7 severs actin filaments and regulates actin cable turnover to promote normal pollen tube growth. *Plant Cell* **2013**, *25*, 3405–3423. [CrossRef] [PubMed]
21. Allard, A.; Bouzid, M.; Betz, T.; Simon, C.; Abou-Ghali, M.; Lemière, J.; Valentino, F.; Manzi, J.; Brochard-Wyart, F.; Guevorkian, K.; et al. Actin modulates shape and mechanics of tubular membranes. *Sci. Adv.* **2020**, *22*, 3050. [CrossRef]
22. Zhu, J.; Nan, Q.; Qin, T.; Qian, D.; Mao, T.L.; Yuan, S.J.; Wu, X.R.; Niu, Y.; Bai, Q.F.; An, L.Z.; et al. Higher-ordered actin structures remodeled by Arabidopsis actin-depolymerizing factor 5 are important for pollen germination and pollen tube growth. *Mol. Plant* **2017**, *10*, 1065–1081. [CrossRef]
23. Huang, J.; Sun, W.; Ren, J.; Yang, R.; Fan, J.; Li, Y.; Wang, X.; Joseph, S.; Deng, W.; Zhai, L. Genome-Wide Identification and Characterization of Actin-Depolymerizing Factor (ADF) Family Genes and Expression Analysis of Responses to Various Stresses in *Zea mays* L. *Int. J. Mol. Sci.* **2020**, *21*, 1751. [CrossRef]
24. Zhu, Y.; Wu, N.N.; Song, W.L.; Yin, G.J.; Qin, Y.J.; Yan, Y.M.; Hu, Y.K. Soybean (*Glycine max*) expansin gene superfamily origins: Segmental and tandem duplication events followed by divergent selection among subfamilies. *BMC Plant Biol.* **2014**, *14*, 93. [CrossRef] [PubMed]
25. Chen, C.J.; Chen, H.; Zhang, Y.; Thomas, H.R.; Frank, M.H.; He, Y.H.; Xia, R. TBtools: An integrative toolkit developed for interactive analyses of big biological data. *Mol. Plant* **2020**, *13*, 1194–1202. [CrossRef]
26. Duvaud, S.; Gabella, C.; Lisacek, F.; Stockinger, H.; Ioannidis, V.; Durinx, C. Expasy, the Swiss bioinformatics resource portal, as designed by its users. *Nucleic Acids Res.* **2021**, *49*, W216–W227. [CrossRef] [PubMed]
27. Kumar, S.; Stecher, G.; Tamura, K. MEGA7: Molecular evolutionary genetics analysis version 7.0 for bigger datasets. *Mol. Biol. Evol.* **2016**, *33*, 1870–1874. [CrossRef] [PubMed]
28. Letunic, I.; Bork, P. Interactive tree of life (iTOL) v4: Recent updates and new developments. *Nucleic Acids Res.* **2019**, *47*, W256–W259. [CrossRef]
29. Wang, Y.; Tang, H.; Debarry, J.D.; Tan, X.; Li, J.; Wang, X.; Lee, T.-h.; Jin, H.; Marler, B.; Guo, H. MCScanX: A toolkit for detection and evolutionary analysis of gene synteny and collinearity. *Nucleic Acids Res.* **2012**, *40*, e49. [CrossRef]
30. Krzywinski, M.; Schein, J.; Birol, I.; Connors, J.; Gascoyne, R.; Horsman, D.; Jones, S.J.; Marra, M.A. Circos: An information aesthetic for comparative genomics. *Genome Res.* **2009**, *19*, 1639–1645. [CrossRef]
31. Livak, K.J.; Schmittgen, T.D. Analysis of relative gene expression data using real-time quantitative PCR and the $2^{-\Delta\Delta Ct}$ Method. *Methods.* **2001**, *25*, 402–408. [CrossRef]
32. Brown, J.; Pirrung, M.; McCue, L.A. FQC dashboard: Integrates FastQC results into a web-based, interactive, and extensible FASTQ quality control tool. *Bioinformatics* **2017**, *33*, 3137–3139. [CrossRef] [PubMed]
33. Bolger, A.M.; Lohse, M.; Usadel, B. Trimmomatic: A flexible trimmer for illumina sequence data. *Bioinformatics* **2014**, *30*, 2114–2120. [CrossRef] [PubMed]
34. Varet, H.; Brillet-Guéguen, L.; Coppée, J.Y.; Dillies, M.A. SARTools: A DESeq2-and edgeR-based R pipeline for comprehensive differential analysis of RNA-Seq data. *PLoS ONE* **2016**, *11*, e0157022. [CrossRef]
35. Wang, N.; Wang, Z.P.; Liang, X.I.; Weng, J.F.; Lv, X.L.; Zhang, D.G.; Yang, J.; Yong, H.J.; Li, M.S.; Li, F.h.; et al. Identification of loci contributing to maize drought tolerance in a genome-wide association study. *Euphytica* **2016**, *210*, 165–179. [CrossRef]
36. Ruzicka, D.R.; Kandasamy, M.K.; McKinney, E.C.; Burgos-Rivera, B.; Meagher, R.B. The ancient subclasses of Arabidopsis actin depolymerizing factor genes exhibit novel and differential expression. *Plant J.* **2007**, *52*, 460–472. [CrossRef] [PubMed]

37. Sun, Y.; Shi, M.; Wang, D.; Gong, Y.; Sha, Q.; Lv, P.; Yang, J.; Chu, P.; Guo, S. Research progress on the roles of actin-depolymerizing factor in plant stress responses. *Front. Plant Sci.* **2023**, *16*, 1278311. [CrossRef] [PubMed]
38. Moore, R.C.; Purugganan, M.D. The early stages of duplicate gene evolution. *Proc. Natl. Acad. Sci. USA* **2003**, *100*, 15682–15687. [CrossRef] [PubMed]
39. Kong, H.; Landherr, L.L.; Frohlich, M.W.; Leebens-Mack, J.; Ma, H.; DePamphilis, C.W. Patterns of gene duplication in the plant SKP1 gene family in angiosperms: Evidence for multiple mechanisms of rapid gene birth. *Plant J.* **2007**, *50*, 873–885. [CrossRef]
40. Jiang, Y.; Lu, Q.; Huang, S. Functional non-equivalence of pollen ADF isovariants in Arabidopsis. *Plant J.* **2022**, *110*, 1068–1081. [CrossRef]
41. Lv, G.H.; Li, Y.F.; Wu, Z.X.; Zhang, Y.H.; Li, X.N.; Wang, T.Z.; Ren, W.C.; Liu, L.; Chen, J.J. Maize actin depolymerizing factor 1 (*ZmADF1*) negatively regulates pollen development. *Biochem. Biophys. Res. Commun.* **2024**, *703*, 149637. [CrossRef] [PubMed]
42. Zhang, P.; Qian, D.; Luo, C.X.; Niu, Y.Z.; Li, T.; Li, C.Y.; Xiang, Y.; Wang, X.Y.; Niu, Y. Arabidopsis ADF5 acts as a downstream target gene of CBFs in response to low-temperature stress. *Front. Cell Dev. Biol.* **2021**, *9*, 635533. [CrossRef] [PubMed]
43. Qian, D.; Zhang, Z.; He, J.X.; Zhang, P.; Ou, X.B.; Li, T.; Niu, L.P.; Nan, Q.; Niu, Y.; He, W.L.; et al. Arabidopsis ADF5 promotes stomatal closure by regulating actin cytoskeleton remodeling in response to ABA and drought stress. *J. Exp. Bot.* **2019**, *70*, 435–446. [CrossRef] [PubMed]
44. Liu, B.J.; Wang, N.; Yang, R.S.; Wang, X.N.; Luo, P.; Chen, Y.; Wang, F.; Li, M.S.; Weng, J.F.; Zhang, D.G.; et al. ZmADF5, a maize actin-depolymerizing factor conferring enhanced drought tolerance in maize. *Plants* **2024**, *13*, 619. [CrossRef] [PubMed]
45. Wang, L.; Cheng, J.; Bi, S.; Wang, J.; Cheng, X.; Liu, S.; Gao, Y.; Lan, Q.K.; Shi, X.W.; Wang, Y.; et al. Actin depolymerization factor ADF1 regulated by MYB30 plays an important role in plant thermal adaptation. *Int. J. Mol. Sci.* **2023**, *24*, 5675. [CrossRef] [PubMed]
46. Wang, B.; Zou, M.; Pan, Q.; Li, J. Analysis of actin array rearrangement during the plant response to bacterial stimuli. *Methods Mol. Biol.* **2023**, *2604*, 263–270. [PubMed]
47. Wang, L.; Qiu, T.Q.; Yue, J.R.; Guo, N.N.; He, Y.J.; Han, X.P.; Wang, Q.Y.; Jia, P.F.; Wang, H.D.; Li, M.Z.; et al. Arabidopsis ADF1 is regulated by MYB73 and is involved in response to salt stress affecting actin filament organization. *Plant Cell Physiol.* **2021**, *62*, 1387–1395. [CrossRef] [PubMed]
48. Yao, H.; Li, X.; Peng, L.; Hua, X.Y.; Zhang, Q.; Li, K.X.; Huang, Y.L.; Ji, H.; Wu, X.B.; Chen, Y.H.; et al. Binding of 14-3-3κ to ADF4 is involved in the regulation of hypocotyl growth and response to osmotic stress in Arabidopsis. *Plant Sci.* **2022**, *320*, 111261. [CrossRef] [PubMed]
49. Bi, S.T.; Li, M.Y.; Liu, C.Y.; Liu, X.Y.; Cheng, J.N.; Wang, L.; Wang, J.S.; Lv, Y.L.; He, M.; Cheng, X.; et al. Actin depolymerizing factor ADF7 inhibits actin bundling protein VILLIN1 to regulate root hair formation in response to osmotic stress in Arabidopsis. *PloS Genet.* **2022**, *18*, e1010338. [CrossRef]
50. Yang, Y.L.; Li, H.G.; Wang, J.; Wang, H.L.; He, F.; Su, Y.Y.; Zhang, Y.; Feng, C.H.; Niu, M.X.; Li, Z.H.; et al. ABF3 enhances drought tolerance via promoting ABA-induced stomatal closure by directly regulating ADF5 in *Populus euphratica*. *J. Exp. Bot.* **2020**, *71*, 7270–7285. [CrossRef]

Disclaimer/Publisher's Note: The statements, opinions and data contained in all publications are solely those of the individual author(s) and contributor(s) and not of MDPI and/or the editor(s). MDPI and/or the editor(s) disclaim responsibility for any injury to people or property resulting from any ideas, methods, instructions or products referred to in the content.

Article

Fine Mapping and Functional Verification of the *Brdt1* Gene Controlling Determinate Inflorescence in *Brassica rapa* L.

Cuiping Chen [1,†], Xuebing Zhu [1,2,†], Zhi Zhao [1,2], Dezhi Du [1,2] and Kaixiang Li [1,2,*]

1. Academy of Agricultural and Forestry Sciences of Qinghai University, Xining 810016, China; chencuiyang@126.com (C.C.); 15297190798@163.com (X.Z.); zhaozhi918@sohu.com (Z.Z.); qhurape@126.com (D.D.)
2. Key Laboratory of Spring Rape Genetic Improvement of Qinghai Province, Rapeseed Research and Development Center of Qinghai Province, Xining 810016, China
* Correspondence: 18997174190@163.com; Tel.: +86-0971-5366-520
† These authors contributed equally to this work.

Abstract: *Brassica rapa*, a major oilseed crop in high-altitude areas, is well known for its indeterminate inflorescences. However, this experiment revealed an intriguing anomaly within the plot: a variant displaying a determinate growth habit (520). Determinate inflorescences have been recognized for their role in the genetic enhancement of crops. In this study, a genetic analysis in a determinate genotype (520) and an indeterminate genotype (515) revealed that two independently inherited recessive genes (*Brdt1* and *Brdt2*) are responsible for the determinate trait. BSA-seq and SSR markers were employed to successfully locate the *Brdt1* gene, which is localized within an approximate region 72.7 kb between 15,712.9 kb and 15,785.6 kb on A10. A BLAST analysis of these candidate intervals revealed that Bra009508 (*BraA10.TFL1*) shares homology with the *A. thaliana TFL1* gene. Then, *BraA10.TFL1* (gene from the indeterminate phenotype) and *BraA10.tfl1* (gene from the determinate phenotype) were cloned and sequenced, and the results indicated that the open reading frame of the alleles comprises 537 bp. Using qRT-PCR, it was determined that *BraA10.TFL1* expression levels in shoot apexes were significantly higher in NIL-520 compared to 520. To verify the function of *BraA10.TFL1*, the gene was introduced into the determinate *A. thaliana tfl1* mutant, resulting in the restoration of indeterminate traits. These findings demonstrate that *BraA10.tfl1* is a gene that controls the determinate inflorescence trait. Overall, the results of this study provide a theoretical foundation for the further investigation of determinate inflorescence.

Keywords: *Brassica rapa*; *Brdt1*; determinate inflorescence; morphological observation; paraffin sectioning; mapping; quantitative real-time PCR; transformation

Citation: Chen, C.; Zhu, X.; Zhao, Z.; Du, D.; Li, K. Fine Mapping and Functional Verification of the *Brdt1* Gene Controlling Determinate Inflorescence in *Brassica rapa* L. *Agronomy* 2024, 14, 281. https://doi.org/10.3390/agronomy14020281

Academic Editor: Ferdinando Branca

Received: 25 December 2023
Revised: 21 January 2024
Accepted: 23 January 2024
Published: 27 January 2024

Copyright: © 2024 by the authors. Licensee MDPI, Basel, Switzerland. This article is an open access article distributed under the terms and conditions of the Creative Commons Attribution (CC BY) license (https://creativecommons.org/licenses/by/4.0/).

1. Introduction

Oil crops belonging to the genus Brassica hold significant importance in global agriculture. This genus comprises a diverse range of plant species, including the species *Brassica rapa* L., *Brassica napus* L., and *Brassica juncea* L. [1,2]. *Brassica rapa* is a diploid species (2n = 20, AA) and has a worldwide distribution [3]. It includes vegetable crops and oil seed crops [4]. Oil seed crops have excellent characteristics, such as barren, drought, and cold resistance [5,6]. Currently, the rapeseed cultivation area in China is approximately seven million hectares, with *B. rapa* accounting for about 15% of this area [7]. *Brassica rapa* plays an irreplaceable role in the provinces of the Yangtze River basin and the Northwest Plateau due to its short growth period, particularly in the rotation of rice and rapeseed. All *Brassica* crops possess an indeterminate growth habit, which is influenced by competition for resources within the plant canopy, both within and between plants [8]. This competition often leads to incomplete seed filling, immature pods, and sterility at the tip of the plant at maturity [9]. Additionally, certain varieties with indeterminate inflorescence growth habits have drawbacks that negatively impact yield, such as taller plants, an increased vulnerability to lodging, longer growth periods, and inconsistent ripening periods [10]. Transforming plants from indeterminate to determinate inflorescence offers a new approach for breeders. Thus, it is necessary to research the molecular mechanisms underlying the determinate inflorescence traits in *B. rapa* to enhance rapeseed genetics.

There are many research reports on determinate inflorescence [8,11–13]. Kaur and Banga [8] identified the determinate gene *Sdt1*, which was responsible for regulating determinate inflorescence on the B5 chromosome of *Brassica juncea*. Li et al. [9,14] discovered a double haploid (DH) line 4769 that exhibited a determinate inflorescence trait. Further investigation revealed that this trait was controlled by two independently inherited recessive genes (*Bnsdt1* and *Bnsdt2*). Wan et al. [15] employed BSA-Seq technology to identify two QTL loci associated with determinate inflorescence in a mutant of *B. napus*. These loci were found on the C02 and C06 chromosomes, respectively. Chen et al. [10] discovered that the determinate inflorescence natural mutant 6138 material of *B. napus* is controlled by a single recessive gene, *BnDM1*. Furthermore, they successfully identified the location of the gene on the C02 chromosome.

The *TERMINAL FLOWER 1* (*TFL1*) gene and its homolog play an important role in determinate inflorescence. The genetic mechanism of the *TFL1* gene in *A. thaliana* has been extensively studied [16–18]. The *TFL1* gene is predominantly expressed in the central region of the apical meristem of *Arabidopsis* [19]. Its interaction with genes, such as *LEAFY* (*LFY*), *APETALA1* (*AP1*), and *FLOWERING LOCUS T* (*FT*), influences the development of the stem tip of *A. thaliana*, leading to the formation of distinct inflorescences [20,21]. *TFL1*, being the main inhibitor of flower development, interferes with the development of flowers by binding to the transcription factor *FLOWERING LOCUS D* (*FD*) and inhibiting *FT*. This inhibition is crucial in regulating the expression of downstream flowering integrators *AP1* and *LFY* [22].

Additionally, there are many reports about *TFL1* in other plants. In rice, the *TFL1* homologous gene *RCN* initiates its regulatory pathway by binding to 14-3-3 proteins, followed by its interaction with *FD* to regulate flower recognition genes [23]. However, *CsTFL1* in cucumbers does not directly interact with *CsFD*, *CsFDP*, or *Cs14-3-3*. Instead, it interacts with *CsNOT2a*, binding *CsFD* or *CsFDP* and affecting the transcription of *CsAP1* and *CsLFY* [24]. Therefore, it can be concluded that the regulatory pathway of *TFL1* in tall plants is a relatively complex process.

Although there has been extensive research on determinate inflorescence in other plants [25–27], the genetic patterns and molecular basis of determinate inflorescence remain unclear in *B. rapa*. In order to investigate the genetic inheritance patterns controlling determinate inflorescence in *B. rapa*, this study focused on the determinate inflorescence mutant 520 of *B. rapa*. Two genes (*Brdt1* and *Brdt2*) associated with determinate inflorescence were identified and further subjected to the fine mapping and functional verification of *Brdt1*. These findings enhance our theoretical understanding of determinate inflorescence and lay the groundwork for future investigations.

2. Materials and Methods

2.1. Plant Material and Population Construction

Lines 520 and 515 of *B. rapa* were used as materials in the present study. The inflorescence of 520 was determinate (a natural mutant) (Figure 1a), while the inflorescence of 515 was indeterminate (Figure 1b). At maturity, they will exhibit different morphological structures (Figure 1c,d). Initially, F_1 generation was obtained through a hybrid cross between lines 515 and 520, completed in 2018. Subsequently, they were selfed to produce the F_2 generation, and the F_1 generation was backcrossed with the recessive parent line 520 to generate the BC_1F_1 generation (2019). The F_2 and BC_1F_1 populations were used for the genetic analysis of inflorescence traits. The inflorescence traits were investigated during the flowering period in 2020.

Figure 1. Two IM development phenotypes in *B. rapa*. (**a**) The determinate phenotype (520). (**b**) The indeterminate phenotype (515). (**c**) The line 520 phenotype at maturity. (**d**) The line 515 phenotype at maturity.

The *Brdt1* gene was identified using the BC_1F_1 population. From this population, 12 indeterminate individuals were selected for backcrossing with line 520. The resulting BC_2F_1 isolate line was then planted at the Xining Experimental Base in Qinghai Province, China, in 2020. Subsequently, the inflorescence traits of each line were investigated and it was found that the indeterminate to determinate separation ratio was 1:1 in 9 out of 12 lines. The dominant *BrDT1* gene was identified in an isolated line, with an indeterminate to determinate ratio of 1:1. To establish the BC_3F_1 population, 10 indeterminate individuals were selected for backcrossing with individuals from line 520, resulting in a population of 847 individuals (2021). From this BC_3F_1 population, 18 indeterminate individuals were chosen and backcrossed with individuals from line 520 to create the BC_4F_1 population, which consisted of 1267 individuals (2022). This BC_4F_1 population was used for the fine mapping of the *Brdt1* gene (Figure S1). Additionally, near-isogenic lines were constructed for gene expression analysis, including line 520 (determinate) and NIL-520 (indeterminate).

2.2. Morphological Observation and Paraffin Sectioning

The inflorescence variations were observed under the fluorescent microscope (4× magnification), based on stem apex meristem development in accordance to the method described by Kobayashi et al. [28]. In this study, a stereoscopic fluorescence microscope (Nikon SMZ25, Tokyo, Japan) was used to observe the inflorescence SAM in homozygous plants at different stages, including the 2-, 4-, 6-, and 8-leaf stages. The aim was to investigate the differences in apical SAM. The methods for paraffin sectioning plant SAMs were followed as described in previous studies [29]. The samples were treated with 70% formalin-acetic acid-alcohol (FAA), followed by Safranin O staining, rinsing, dehydration, clearing, infiltration, embedding, slicing, sealing, and examination using a Nikon microscope (Nikon, Tokyo, Japan).

2.3. DNA Extraction and BSA-Seq

DNA was extracted individually from fresh leaves using the cetyltrimethylammonium bromide (CTAB) method [30]. To map the *Brdt1* gene, 20 plants with indeterminate inflorescence and 20 plants with determinate inflorescence were selected from the BC_3F_1 population. Separate indeterminate and determinate bulks were created. The two parents and bulks were subjected to bulked segregant analysis (BSA) at Novogene Biological Company in Beijing, China. The Illumina HiSeq TM PE150 sequencing method was employed. Subsequently, the Burrows–Wheeler alignment (BWA) tool aligned the whole-genome sequencing (WGS) reads to the reference genome of 'chiifu' v1.5 (BRAD (http://brassicadb.org/brad/)). Single-nucleotide polymorphisms (SNPs) were detected using the Haplotype Caller of Genome Analysis Toolkit (GATK, version 3.7). The candidate region was determined based on the SNP index.

2.4. Development of SSR Marker

The initial localization of the determinate inflorescence gene *Brdt1* and the location of its homologous gene in *B. napus* chromosomes helped determine the approximate range of *Brdt1* on the A10 chromosome. The sequence segment was downloaded from the BRAD (http://brassicadb.org/brad/) database, and SSR loci were detected using SSRHunter 1.3. Using the Primer 3 software (Premier Biosoft International, Palo Alto, CA, USA) to design the SSR primers. SSR amplification was performed following the method described by Lowe [31]. The co-dominance of SSR markers was detected using 6% polypropylene gel electrophoresis.

2.5. Mapping

The *Brdt1* gene was mapped using the BC_3F_1 populations (847 individuals) and BC_4F_1 populations (1267 individuals). The SSR markers and individual phenotypes were analyzed using the JoinMap 4/MapDraw program, resulting in the construction of a partial linkage map for the chromosome region containing the *Brdt1* gene.

2.6. Cloning and Sequence Analysis of the Candidate Gene

The *Brdt1* gene was amplified from the gDNA of parents 515 and 520. Primer 3.0 software was used to design the specific primers (Bra.TFL1-orf-F/Bra.TFL1-orf-R) of the gene (Table S1). The amplification process was as follows: pre-denaturation at 95 °C for 3 min, followed by 35 cycles of denaturation at 95 °C for 30 s, annealing at 57 °C for 30 s, extension at 72.0 °C for 60 s, and terminal extension at 72 °C for 10 min. The amplified sequences were cloned using the PMD19-T vector and *E. coli* DH5α methods. The positive clones were verified using M13-specific primers (Tsingke Biotech, Beijing, China). The gDNA sequences were analyzed using DNAMAN8.0.

2.7. Plasmid Construction and Plant Transformation

To assess the functionality of the candidate gene, a primer was designed based on the *Bra009508* (*TFL1* homologous) sequences obtained from NCBI (National Center for Biotechnology Information (https://www.ncbi.nlm.nih.gov/)). The primers were modified to include the *EcoRI* and *PstI* enzyme restriction sites, as well as 15 bp sequences from both ends of the pCAMIBA2300 vector. The amplification of the gDNA fragment, which consisted of the upstream, full-length gene, and downstream sequences, was performed using Phusion Hot Start High Fidelity DNA Polymerase (NEB, Ipswich, MA, USA) and the recombinant primers (CEBra.TFL1-F/CEBra.TFL1-R, Table S1). The indeterminate inflorescence line 515 served as the source of DNA for this amplification. To create the complementation plasmid *pBraA10. TFL1: BraA10. TFL1*, a genomic fragment was sequenced and digested with *EcoRI* and *PstI*. The digested fragment was then ligated into the pCAMIBA2300 vector using a One Step Cloning Kit (Vazyme, Nanjing, China). Finally, the confirmed recombinant plasmid was introduced into GV3101 (*Agrobacterium tumefaciens*).

The inflorescence impregnation method was implemented to transform *Arabidopsis thaliana* (L.) Heynh. Specifically, the fused construct was introduced into the *A. thaliana tfl1-2* mutant. PCR was conducted to verify and identify the positive transgenic plants (PT-Bra.TFL1-F/PT-Bra.TFL1-R, Table S1).

2.8. RNA Extraction and qRT-PCR

Total RNA was extracted from 520 and NIL-520 materials at different developmental stages, including the 2-leaf seedlings, budding stage, bolting stage, and root, stem, and leaf tissues during the bolting stage. The extraction was performed using a TaKaRa MiniBEST Universal RNA Extraction Kit (TaKaRa, Dalian, China) with three biological replicates. The extracted RNA samples were frozen in liquid nitrogen and stored at −80 °C. First-strand cDNA was synthesized using the PrimeScriptTM RT Reagent Kit (TaKaRa, Dalian, China) following the manufacturer's protocol. The primers (qRT-braA10-1F/qRT-braA10-1R, Table S1) for qRT-PCR were designed using Primer-BLAST in NCBI (Primer designing tool (https://www.ncbi.nlm.nih.gov/tools/primer-blast/, accessed on 22 January 2024)). The actin gene was selected as the reference gene for the relative quantification of the candidate gene (Actin-F/Actin-R, Table S1). qRT-PCR was conducted using a CFX Opus 96 instrument (Bio-Rad, Hercules, CA, USA). The reaction system consisted of 25 µL, including 2 µL of gene-specific primer (10 ng µL^{-1}) (Table S1), 2 µL of cDNA (50 ng µL^{-1}), 12.5 µL of TB Green Premix Ex Taq II (TliRNaseH Plus) (TaKaRa, Dalian, China), and 8.5 µL of sterile water. The PCR conditions were 95 °C for 30 s, followed by 45 cycles of 95 °C for 5 s and 60 °C for 30 s. The data were processed using the $2^{-\Delta\Delta Ct}$ method.

3. Results

3.1. Observations of the SAM Apex in B. rapa

The formation of the SAM was observed using a stereoscopic fluorescence microscope and paraffin section (Figure 2). The aim was to determine the point at which the apices of indeterminate inflorescence and determinate inflorescence exhibit morphological differences during the growth process. This observation showed that *B. rapa* exhibits normal inflorescence differentiation at the SAM of indeterminate and determinate inflorescences during the two-to-six-leaf stage (Figure 2a–c,e–g). However, at the eight-leaf stage, indeterminate inflorescences continued to undergo normal inflorescence differentiation (Figure 2d), while the SAM of determinate inflorescences exhibited variation and could not differentiate inflorescence tissue normally (Figure 2h). The results of paraffin sections also indicated that during the two-to-six-leaf stage, the indeterminate and determinate shoot apex meristem displayed the same shapes (Figure 2i–k,m–o). However, in the eight-leaf stage, the indeterminate normal process continued (Figure 2i), and the determinate apices had already exhibited variation and could not differentiate into normal inflorescence meristems (Figure 2p). These findings provide morphological evidence for the formation period of determinate inflorescences and offer insights for observing determinate inflorescences in *B. rapa*.

3.2. Genetic Analysis

This study investigated and analyzed the growth habits of the F_2 and BC_1F_1 populations. It was observed that all F_1 individuals exhibited a complete indeterminate phenotype, indicating the dominance of indeterminate growth over determinate growth. In 515 × 520 F_2 plants, 297 plants displayed indeterminate growth while 27 plants showed determinate growth, consistent with a segregation ratio of 15:1. The BC_1 plants exhibited a segregation ratio of approximately 3:1 (indeterminate-to-determinate = 168:50) (Table 1). These findings suggest that determinate growth habits are controlled by two independently inherited recessive genes, and determinate inflorescence genes were tentatively designated *Brdt1* and *Brdt2*.

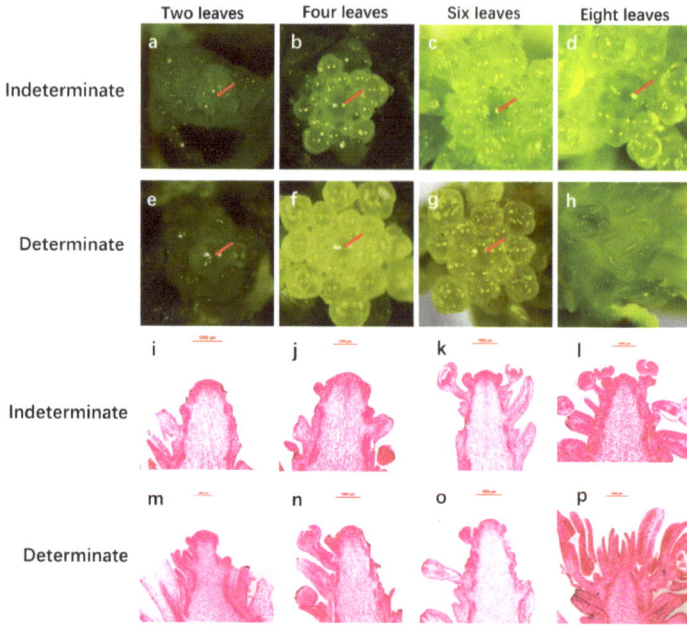

Figure 2. Microscope observation and paraffin section of the shoot apical meristem (SAM) of the indeterminate (**a–d,i–l**) and determinate inflorescences (**e–h,m–p**). (**a,e,i,m**) SAMs of indeterminate and determinate inflorescences at the two-leaf stage. (**b,f,j,n**) SAMs of indeterminate and determinate inflorescences at the four-leaf stage. (**c,g,k,o**) SAMs of indeterminate and determinate inflorescences at the six-leaf stage. (**d,h,i,p**) SAMs of indeterminate and determinate inflorescences at the eight-leaf stage. Red arrows indicate growth points. In the eight-leaf stage, the determinate inflorescence was beginning to appear.

Table 1. Segregation of inflorescence traits in the F_2 and BC_1 populations.

Combination	F1/RF1	Population	No. of INDT. Plants	No. of DT. Plants	Expected Ratio	X^2 Value
515 × 520	indeterminate	F_2	297	27	15:1	2.40
		BC_1F_1	168	50	3:1	0.49

3.3. Primary Mapping of the Determinate Gene Brdt1 Using BSA-Seq

The BC_3F_1 population obtained from a cross between 515 and 520 was used for BSA resequencing [32] to determine the position of the *Brdt1* gene. After sequencing the two bulks and their respective parents, a total of 54.429 Gb clean reads were obtained after quality control filtering. Specifically, the 515 parent, the 520 parent, the indeterminate bulk, and the determinate bulk accounted for 10.454 Gb, 10.731 Gb, 17.035 Gb, and 16.209 Gb, respectively. The sequencing data showed that more than 90.15% of the bases in both pools and parents had a quality score of more than 30 (Q30), and more than 96.13% had a quality score of more than 20 (Q20). Additionally, the average GC content was between 38.03% and 39.31%. The sequencing data were analyzed, and a total of 659,356 SNPs were identified between 515 and 520 by aligning with the 'chiifu' v1.5 reference genome. Out of these SNPs, 604,936 homozygous SNPs were found between the two parents, and these were used to calculate the SNP index for the two descendants. A graph of the ΔSNP index was plotted against the genomic regions (Figure 3a), revealing a significant peak in the 1.55 Mb region from 14.76 Mb to 16.31 Mb on chromosome A10 (Figure 3b). This finding suggests that the *Brdt1* gene, located in this specific region, could be a potential candidate locus.

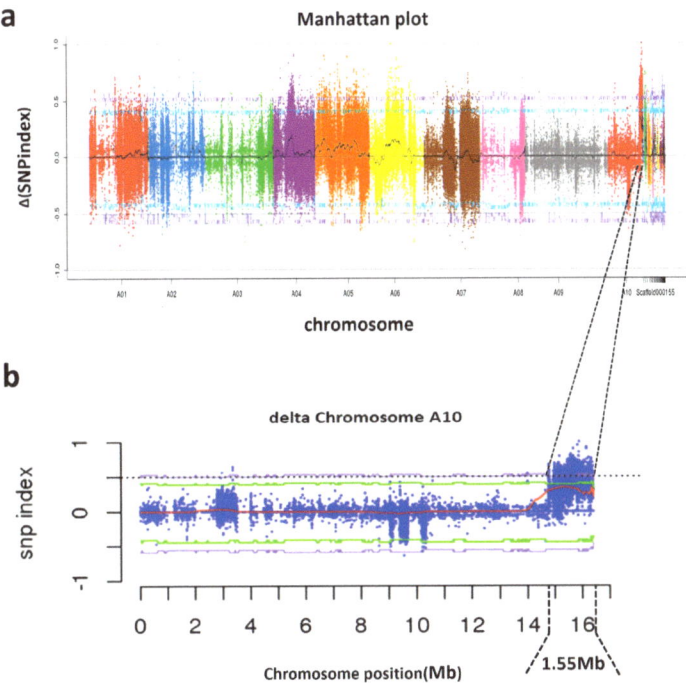

Figure 3. ΔSNP index Manhattan plot graphs. (**a**) The blue line indicates the 95% threshold value, and the purple line indicates the 99% threshold value. Different colors represent different chromosomes. (**b**) SNP index of delta chromosome A10. The red line is the SNP-index mean line, the green line is the 95% threshold line, and the purple line is the 99% threshold line.

3.4. Fine Mapping of the Brdt1 Gene

Based on the findings derived from the BSA-Seq analysis, it is likely that the *Brdt1* gene is located in the 14.76–16.31 Mb regions of A10 of *B. rapa*. Subsequently, the region sequence was downloaded from BRAD (http://brassicadb.org/brad/) and 60 SSR markers were developed. Among these, nine markers with polymorphism were identified and named BrSSR1 to BrSS9 (Table 2). A total of 847 BC_3F_1 individuals were screened using polymorphic SSR markers to assess the genetic distance between the *Brdt1* gene and the SSR markers and determine the order of these markers. Subsequently, the recombinants for each marker were recorded. The genetic distance was computed, and the genetic linkage map was constructed using JoinMap 4.0/MapDraw software V2.1. The results indicate that BrSSR1–BrSSR3 were located on one side of the *Brdt1* gene, BrSSR5–BrSSR9 were located on the other side of the *Brdt1* gene, and BrSSR4 co-segregated with the *Brdt1* gene (Figure 4a). Among the markers flanking the *Brdt1* gene, BrSSR3 and BrSSR5 displayed the closest linkage, with distances of 0.3 cM and 0.2 cM from the *Brdt1* gene, respectively (Table 2). These closely linked markers were subjected to BLAST analysis against BRAD (http://brassicadb.org/brad/). All markers were mapped to A10 of the 'chiifu' v1.5 reference genome (Table 2). Furthermore, the order of these markers on the map perfectly aligned with their counterparts on A10 of *B. rapa*. Based on this established order, the genomic region harboring the *Brdt1* gene was precisely delimited within an approximate interval of 232.3 kb, from 15,580.0 to 15,812.3 kb on A10 (Figure 4b).

Table 2. Information about the markers that were closely linked to *Brdt1*.

Type of Marker	Name	Size of Marker	Physical Position (kb)	Chromosome of 'chiifu' v1.5
SSR	BrSSR1	174	15,476,191	A10
SSR	BrSSR2	158	15,505,412	A10
SSR	BrSSR3	185	15,580,043	A10
SSR	BrSSR4	166	15,712,914	A10
SSR	BrSSR5	175	15,812,313	A10
SSR	BrSSR6	215	15,859,041	A10
SSR	BrSSR7	182	15,872,725	A10
SSR	BrSSR8	238	15,912,798	A10
SSR	BrSSR9	234	15,918,222	A10
SSR	BrSSR10	174	15,731,259	A10
SSR	BrSSR11	151	15,785,574	A10
SSR	BrSSR12	182	15,793,126	A10

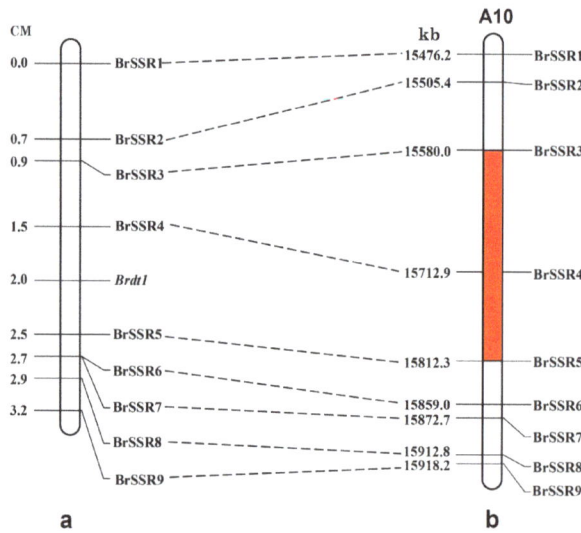

Figure 4. Mapping of *Brdt1* gene. (**a**) A partial genetic linkage map around the *Brdt1* gene. (**b**) A partial physical map of linkage markers around the *Brdt1* gene. The red region indicates candidate intervals, corresponding to 15,580.0–15,812.3 kb, identified through bulk segregant sequencing and SSR markers.

To narrow down the target region of *Brdt1*, a total of 1267 individuals from the BC_4F_1 population were used. Three SSR markers (BrSSR10–BrSSR12) were identified within the interval of 155,800 to 158,123 kb on A10 (Table 2). We utilized SSR markers including the previously used markers (BrSSR3–BrSSR5) and these SSR markers to conduct a screening of BC_4F_1 individuals. Our findings revealed that the *Brdt1* gene was positioned between BrSSR4 and BrSSR11 (Figure 5a). Using BLAST analysis against BRAD (http://brassicadb.org/brad/), the *Brdt1* gene was further narrowed down to an interval of approximately 72.7 kb, specifically between 15,712.9 kb and 15,785.6 kb on A10 of *B. rapa* (Figure 5b).

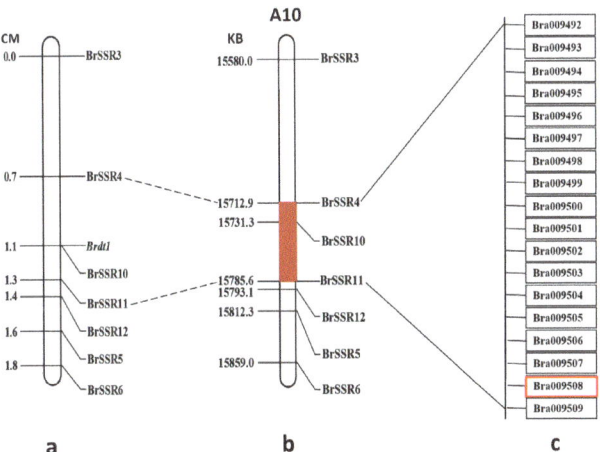

Figure 5. Fine mapping of *Brdt1* gene. (**a**) A partial genetic linkage map around the *Brdt1* gene. (**b**) A partial physical map of linkage markers around the *Brdt1* gene. The red region indicates candidate intervals corresponding to 15,712.9–15,785.6 kb. (**c**) Results of a BLAST analysis using sequences from candidate intervals against the 'chiifu' v1.5 genome.

3.5. Dissection of the Brdt1 Target Region

The candidate intervals were then submitted to the BRAD (brassicadb.cn) and TAIR (https://www.arabidopsis.org/) databases for BLAST analysis. The analysis revealed that the candidate region included 18 predicted genes (Figure 5c) from the *B. rapa* reference genome and showed homology to 17 genes on the *A. thaliana* chromosome (Table 3). Notably, according to the TAIR database of the gene annotation of these genes, the *Bra009508* (*BraA10.TFL1*) gene is homologous to the *AT5G03840* gene, which is known as the *TERMINAL FLOWER 1* (*TFL1*) gene. The *AT5G03840* gene encodes a phosphatidylethanolamine-binding protein (*PEBP*). In *Arabidopsis TFL1* mutants, determinate inflorescence can be formed at the SAM of the inflorescence. Based on this information, it was inferred that the *Bra009508* (*BraA10.TFL1*) gene was the most promising candidate gene for *Brdt1* and selected for further study.

3.6. Expression of BraA10.TFL1 in Different Tissues of B. rapa

A qRT-PCR analysis was conducted to investigate the expression levels of the *BraA10.TFL1* gene in different stages and tissues of *B. rapa* growth. Specifically, the gene was quantified in the shoot apex during the two-leaf stage, budding stage, and bolting stage of both indeterminate and determinate inflorescences. Additionally, gene expression was measured in the root, stem, and leaf tissues during the bolting stage. The findings revealed that the expression level of the *BraA10.TFL1* gene remained relatively consistent in the root, stem, and leaf tissues during the bolting stage. However, significant differences were observed in the expression of the *BraA10.TFL1* gene in the shoot apex between the 520 and NIL-520 lines, encompassing the two-leaf stage, budding stage, and bolting stage. Furthermore, the expression of *BraA10.TFL1* in the NIL-520 line was notably higher than in the 520 line, particularly during the P2 and P3 phases. These results emphasize the crucial role of the *BraA10.TFL1* gene in the development of distinct inflorescences (Figure 6). Therefore, it is reasonable to postulate that *BraA10.TFL1* is a potential candidate gene for the inflorescence trait.

Table 3. Results of BLASTN searches using the candidate interval gene.

Gene of B. rapa	Homologous Gene in A. thaliana	Putative Function
Bra009492	AT5G04030	unknown
Bra009493	AT5G04020	Calmodulin-binding
Bra009494	AT5G04010	F-box family protein
Bra009495	AT5G03990	FK506-binding-like protein
Bra009496	AT5G03980	SGNH hydrolase-type esterase superfamily protein
Bra009497	AT5G00970	F-box family protein
Bra009498	AT5G03960	IQ-domain 12
Bra009499	AT5G03940	Chloroplast signal recognition particle 54 KDa subunit protein
Bra009500	AT5G03910	ABC2 homolog 12
Bra009501	AT5G03905	Iron-sulfur cluster biosynthesis family protein
Bra009502	AT5G03900	Iron-sulfur cluster biosynthesis family protein
Bra009503	AT5G03900	Iron-sulfur cluster biosynthesis family protein
Bra009504	AT5G03890	unknown
Bra009505	AT5G03880	Thioredoxin family protein
Bra009506	AT5G00893	unknown
Bra009507	AT5G03850	Nucleic acid-binding, OB-fold-like protein s28
Bra009508	AT5G03840	TFL1 (TERMINAL FLOWER 1); PEBP (phosphatidylethanolamine binding protein) family protein
Bra009509	AT5G03795	Exostosin family protein

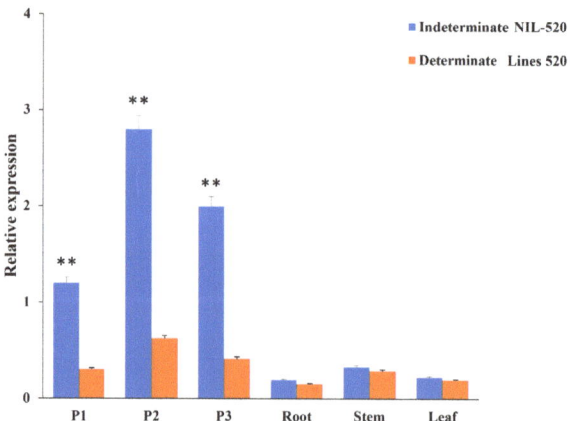

Figure 6. Expression analysis of the *BraA10.TFL1* gene among different stages and tissues of the NIL-520 and 520 lines in *B. rapa*. P1, P2, and P3 are two-leaf stage, budding stage, and bolting stage, respectively. The expression levels in the root, stem, and leaf tissues were consistent during the bolting stage. ** indicates significant differences at $p < 0.01$.

3.7. Cloning and Sequencing Analysis of the BraA10.TFL1/BraA10.tfl1 Gene

The sequences of gDNA and CDS of *BraA10.TFL1/BraA10.tfl1* were amplified from the indeterminate inflorescence line 515 and the determinate inflorescence line 520. DNA-MAN 8.0 software was used to analyze the gDNA and CDS sequences of the determinate and indeterminate inflorescences. The results revealed the presence of a gDNA sequence measuring 1066 bp and a 537 bp cDNA sequence in inflorescence lines 515 and 520, respectively. These sequences contained four exons and three introns. The sequence analysis of *BraA10.TFL1* (BraA10.TFL1_DNA) and *BraA10.tfl1* (BraA10.tfl1_DNA) in *B. rapa* revealed two SNP mutations (G 434 T and C 569 T) in the intron region (Figure 7a). Amino acid sequence prediction and the analysis of the *BraA10.TFL1/BraA10.tfl1* genes were performed using Premier 5. The results showed no differences in amino acid sequences (Figure 7b).

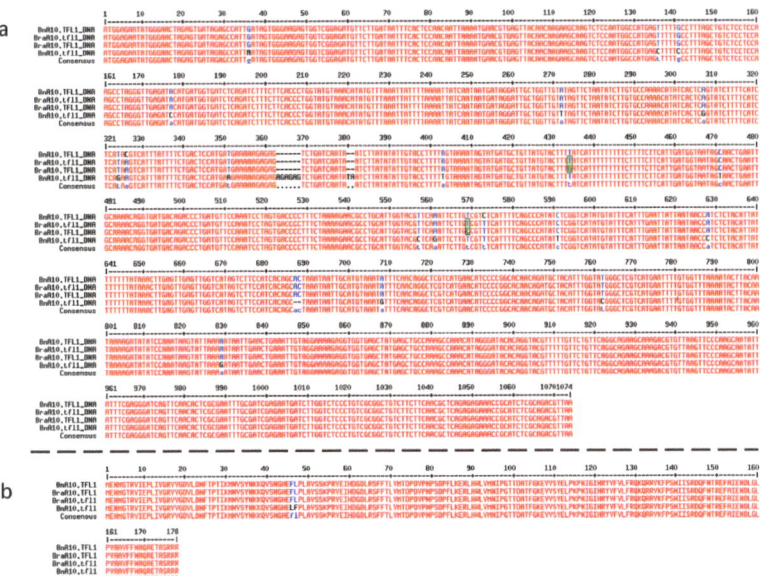

Figure 7. Sequence analysis of *BraA10.TFL1*. (**a**) The gDNA sequences of *BraA10.TFL1* (indeterminate) from the 515 line and *BraA10.tfl1* (determinate) from the 520 line were aligned with *BnA10.TFL1* (indeterminate) and *BnA10.tfl1* (indeterminate) from *B. napus*. The green box indicates the position of SNP differences between indeterminate 515 and determinate 520. Blue indicates the bases that are different in the sequence. (**b**) Amino acid sequence alignment between *BraA10.TFL1* (indeterminate) and *BraA10.tfl1* (determinate) with *BnA10.TFL1* (indeterminate) and *BnA10.tfl1* (indeterminate) from *B. napus*. Blue indicates amino acids with differences in the sequence.

Previous studies have reported that the gene *BnA10.tfl1*, responsible for determinate inflorescence, is located on the A10 chromosome of *B. napus*. A comparison of the sequences of the *BnA10.tfl1* and the *BnA10.TFL1* gene, which controls indeterminate inflorescence, revealed 22 differences [33]. Among these differences, two SNPs resulted in a change in two amino acids (Phe to Leu and Leu to Phe), potentially leading to a transition from indeterminate to determinate inflorescence. To further investigate this, the gene and amino acid sequences of *BraA10.TFL1* and *BraA10.tfl1* from *B. rapa* were compared with the *BnA10.TFL1* and *BnA10.tfl* genes from *B. napus*. The results indicated that *BraA10.tfl1* and *BnA10.TFL1* had the same size, consisting of 1066 bp, and showed high homology with only two differences out of 1066 bases. However, there were 20 sequence differences between *BraA10.tfl1* and *BnA10.tfl1*, resulting in changes in two amino acids (Phe to Leu and Leu to Phe). These unexpected findings highlight the importance of conducting functional validation to confirm the accuracy of the predicted genes.

3.8. BraA10.TFL1 Rescues the tfl1-2 Mutant Phenotype in A. thaliana

The complete genomic DNA sequence of *BraA10.TFL1* from the 515 line was amplified and utilized to construct *pBraA10.TFL1: BraA10.TFL1* in the PCAMBIA2300 vector. This vector contained a 3437 bp genomic fragment that included the *BraA10.TFL1* gene, with

1842 bp upstream, 1066 bp coding region, and 529 bp downstream (Figure S2). After the transformation process, we introduced this construct into the *A. thaliana tfl1-2* mutant (determinate inflorescence) (Figure 8b). Both the transgenic *A. thaliana* lines and *tfl1-2* mutants exhibited the expected outcomes (Figure 8b–d). The transgenic lines showed indeterminate inflorescence stem growth, similar to wild-type *A. thaliana* (Figure 8a). A total of 12 positive transgenic plants were obtained. The terminal flowers of *tfl1-2* mutants suppressed the differentiation of the SAM, resulting in the development of determinate inflorescence. It is important to note that the T1 *A. thaliana* plants displayed multiple buds, indicating an indeterminate phenotype (Figure 8c,d). In conclusion, these findings provide evidence supporting the functional similarities between *BraA10.TFL1* and *TFL1*.

Figure 8. Architecture of terminal racemes in transgenic *A. thaliana*. (**a**) Wild-type *A. thaliana* (WT). (**b**) The *A. thaliana tfl1-2* mutant. (**c**,**d**) T1 *A. thaliana* transgenic plants returned to an indeterminate phenotype.

4. Discussion

Inflorescences significantly impact the yield of *B. rapa* [34–36]. In previous studies, the formation of determinate inflorescences changed plant height and the flowering and maturity stages. This not only improved the plant type, but also facilitated mechanized harvesting. Among the determinate mutant plants currently studied, *A. thaliana* [37], *B. juncea* [8], and *Vigna radiata* [38], it has been observed that the determinant inflorescence is regulated by a recessive gene. Zhang et al. [39] examined a natural mutant strain, FM8, of *B. napus* with a determinant inflorescence, and genetic analysis revealed that the inheritance of this inflorescence type is controlled by the interaction of two recessive genes and one recessive epistasis suppressor gene. Li et al. [14] found that the determinate inflorescence strain of *B. napus* was controlled by two independently inherited recessive genes (*Bnsdt1* and *Bnsdt2*). However, the determinate inflorescence line 520 of *B. rapa* was used as material in this study, and it was found that it is controlled by two independent inherited recessive genes, *Brdt1* and *Brdt2*. This result may be due to the different mutant types of determinate inflorescence. Therefore, further in-depth research is necessary to understand the genetic mechanisms governing determinate inflorescence genes in rapeseeds.

This study utilized stereoscopic fluorescence microscopy and paraffin sectioning to observe the growth stages of the shoot apical meristem (SAM) in both determinate and indeterminate inflorescences. The results showed that the SAMs of determinate inflorescences (Figure 2e–g,m–o) were structurally similar to those of indeterminate inflorescences (Figure 2a–c,i–k) at the two-, four-, and six-leaf stages. However, the SAMs of determinate inflorescences (Figure 2h,p) showed differences compared to indeterminate inflorescences (Figure 2d,l) in the eight-leaf stage. The apex of the determinate SAM exhibited variation similar to the shape of a floral organ, which may cause determinate growth. Other studies have reported the observation of determinate SAMs forming terminal flowers, either singularly or in multiples [40,41]. While the phenotypes exhibited variation among different plants with determinate inflorescences, they had one fundamental characteristic in common: the shoot apical meristem (SAM) maintained its differentiation, producing fresh floral tissue, thereby causing a transition from an indeterminate to a determinate growth pattern.

In previous studies, the genes responsible for controlling quality traits were identified using bulk segregant sequencing (BSA-Seq) and map-based cloning approaches [42–46].

This study employed a method to map the determinate trait of *B. rapa*. Initially, the genomic sequence information of 'chiifu' v1.5 (*B. rapa*) was used as a reference to successfully map the *Brdt1* gene to the A10 chromosome. The *Brdt1* gene was delimited to an interval of approximately 72.7 kb, ranging from 15,712.9 kb to 15,785.6 kb on A10 of *B. rapa*. Furthermore, a gene annotation analysis within this interval identified a highly similar gene, *Bra009508* (*BraA10.TFL1*), which is homologous with the *TFL1* gene in *A. thaliana*. A sequence analysis of *BraA10.TFL1* revealed two SNP differences (G 434 T; C 569 T) in the intron region between indeterminate and determinate sequences. Subsequently, an expression pattern analysis of *BraA10.TFL1* was conducted, which demonstrated its specific expression in the shoot apex. Genetic transformation experiments in *A. thaliana* further confirmed the functionality of the *BraA10.TFL1* gene. When introduced into the *A. thaliana tfl1-2* mutant, the T1 *A. thaliana* plants reverted to an indeterminate state. Jia et al. [33] conducted a study in which they transferred the *TFL1* homologous gene *BnA10.TFL1* from *B. napus* into a determinate *Arabidopsis* mutant. The researchers observed that this transfer restored determinate expression in *Arabidopsis*, resulting in indeterminate inflorescence. These findings provide evidence that *BraA10.TFL1* shares a similar function to *TFL1*.

In this study, it is noteworthy that the sequence alignment results indicated no difference in the amino acid sequences encoded by the determinate inflorescence and indeterminate inflorescence sequences. However, the genetic transformation results demonstrated that the *BraA10.TFL1* gene from indeterminate inflorescence was successfully introduced into the *A. thaliana tfl1-2* mutant (regenerating plants from *B. rapa* through genetic transformation is challenging). As a result, the mutant's phenotype could be restored to indeterminate inflorescence. Based on these findings, it is hypothesized that the promoter region of the gene may contain a functional region that contributes to determinate inflorescence formation. However, additional evidence is necessary to substantiate this hypothesis.

5. Conclusions

As a novel combination of materials was used in this study, genetic analysis revealed that the determinate inflorescence is controlled by two independent recessive genes (*Brdt1* and *Brdt2*). Morphological observations indicated that the 520 strain with determinate inflorescence exhibited the characteristic at the eight-leaf stage. One of the genes, referred to as *Brdt1*, was mapped using BSA sequencing and SSR marker development. The gene was fine-mapped to the 15,712.9–15,785.6 kb interval on chromosome A10. Within this interval, it was found that the gene *Bra009508* (*BraA10.TFL1*) is homologous to the *AT5G03840* gene in *Arabidopsis*, which is annotated as the *TFL1* gene. Moreover, when the *BraA10.TFL1* gene from an indeterminate inflorescence plant was transferred to the *A. thaliana tfl1-2* mutant, the determinate traits became indeterminate. These findings suggest that *BraA10.TFL1* may play a role in controlling the determinate inflorescence trait. Overall, this research provides novel insights into the molecular mechanism of oilseed breeding for determinate inflorescences.

Supplementary Materials: The following supporting information can be downloaded at: https://www.mdpi.com/article/10.3390/agronomy14020281/s1, Figure S1: Population construction; Figure S2: The 3437 bp genomic fragment incorporating the *BraA10.TLF1* gene. Table S1: The primer sequences used in this study.

Author Contributions: C.C. and X.Z. performed the research and wrote the manuscript; Z.Z. conducted the data analysis. K.L. and D.D. designed the research and revised the article. All authors have read and agreed to the published version of the manuscript.

Funding: This research was financially supported by the Qinghai Provincial Natural Science Foundation of China (2022-ZJ-975Q).

Data Availability Statement: No new data were created.

Acknowledgments: We are grateful to Yongping Jia, Xutao Zhao, Lingxiong Zan, Liren Xie, and the Oil Crop Research Institute of the Chinese Academy of Agricultural Sciences for their help in purchasing *A. thaliana* mutants.

Conflicts of Interest: The authors declare no conflicts of interest.

References

1. Gupta, M.; Atri, C.; Banga, S.S. Cytogenetic stability and genome size variations in newly developed derived *Brassica juncea* allopolyploid lines. *J. Oilseed Brassica* **2014**, *5*, 9.
2. Wang, X.; Zheng, M.; Liu, H.; Zhang, L.; Chen, F.; Zhang, W.; Fan, S.; Peng, M.; Hu, M.; Wang, H.; et al. Fine-mapping and transcriptome analysis of a candidate gene controlling plant height in *Brassica napus* L. *Biotechnol. Biofuels* **2020**, *13*, 42. [CrossRef] [PubMed]
3. Akter, A.; Kakizaki, T.; Itabashi, E.; Kunita, K.; Shimizu, M.; Akter, M.A.; Mehraj, H.; Okazaki, K.; Dennis, E.S.; Fujimoto, R. Characterization of FLOWERING LOCUS C 5 in *Brassica rapa* L. *Mol. Breed.* **2023**, *43*, 58. [CrossRef] [PubMed]
4. Li, N.; Yang, R.; Shen, S.; Zhao, J. Molecular Mechanism of Flowering Time Regulation in Brassica rapa: Similarities and Differences with Arabidopsis. *Hortic. Plant J.* **2024**. [CrossRef]
5. Ma, L.; Coulter, J.A.; Liu, L.; Zhao, Y.; Chang, Y.; Pu, Y.; Zeng, X.; Xu, Y.; Wu, J.; Fang, Y.; et al. Transcriptome Analysis Reveals Key Cold-Stress-Responsive Genes in Winter Rapeseed (*Brassica rapa* L.). *Int. J. Mol. Sci.* **2019**, *20*, 1071. [CrossRef]
6. Raza, A.; Su, W.; Hussain, M.A.; Mehmood, S.S.; Zhang, X.; Cheng, Y.; Zou, X.; Lv, Y. Integrated Analysis of Metabolome and Transcriptome Reveals Insights for Cold Tolerance in Rapeseed (*Brassica napus* L.). *Front. Plant Sci.* **2021**, *12*, 721681. [CrossRef]
7. He, Y.-T.; Tu, J.-X.; Fu, T.-D.; Li, D.-R.; Chen, B.-Y. Genetic Diversity of Germplasm Resources of *Brassica campestris* L. in China by RAPD Markers. *Acta Agron. Sin.* **2002**, *28*, 7.
8. Kaur, H.; Banga, S.S. Discovery and mapping of *Brassica juncea* Sdt 1 gene associated with determinate plant growth habit. *Theor. Appl. Genet.* **2015**, *128*, 235–245. [CrossRef]
9. Li, K.; Yao, Y.; Xiao, L.; Zhao, Z.; Guo, S.; Fu, Z.; Du, D. Fine mapping of the *Brassica napus* Bnsdt1 gene associated with determinate growth habit. *Theor. Appl. Genet.* **2018**, *131*, 193–208. [CrossRef]
10. Chen, J.; Zhang, S.; Li, B.; Zhuo, C.; Hu, K.; Wen, J.; Yi, B.; Ma, C.; Shen, J.; Fu, T.; et al. Fine mapping of BnDM1-the gene regulating indeterminate inflorescence in *Brassica napus*. *Theor. Appl. Genet.* **2023**, *136*, 151. [CrossRef]
11. Alvarez, J.; Guli, C.L.; Yu, X.H.; Smyth, D.R. terminal flower: A gene affecting inflorescence development in *Arabidopsis thaliana*. *Plant J.* **2005**, *2*, 103–116. [CrossRef]
12. Nurmansyah; Alghamdi, S.S.; Migdadi, H.M.; Farooq, M. Novel inflorescence architecture in gamma radiation-induced faba bean mutant populations. *Int. J. Radiat. Biol.* **2019**, *95*, 1744–1751. [CrossRef] [PubMed]
13. Xu, C.; Park, S.J.; Van Eck, J.; Lippman, Z.B. Control of inflorescence architecture in tomato by BTB/POZ transcriptional regulators. *Genes. Dev.* **2016**, *30*, 2048–2061. [CrossRef] [PubMed]
14. Li, K.; Xu, L.; Jia, Y.; Chen, C.; Yao, Y.; Liu, H.; Du, D. A novel locus (Bnsdt2) in a TFL1 homologue sustaining determinate growth in *Brassica napus*. *BMC Plant Biol.* **2021**, *21*, 568. [CrossRef] [PubMed]
15. Wan, W.; Zhao, H.; Yu, K.; Xiang, Y.; Dai, W.; Du, C.; Tian, E. Exploration into Natural Variation for Genes Associated with Determinate and Capitulum-like Inflorescence in *Brassica napus*. *Int. J. Mol. Sci.* **2023**, *24*, 12902. [CrossRef] [PubMed]
16. Hanano, S.; Goto, K. Arabidopsis TERMINAL FLOWER1 is involved in the regulation of flowering time and inflorescence development through transcriptional repression. *Plant Cell* **2011**, *23*, 3172–3184. [CrossRef] [PubMed]
17. Repinski, S.L.; Kwak, M.; Gepts, P. The common bean growth habit gene PvTFL1y is a functional homolog of *Arabidopsis TFL1*. *Theor. Appl. Genet.* **2012**, *124*, 1539–1547. [CrossRef]
18. Fernandez-Nohales, P.; Domenech, M.J.; Martinez de Alba, A.E.; Micol, J.L.; Ponce, M.R.; Madueno, F. AGO1 controls arabidopsis inflorescence architecture possibly by regulating TFL1 expression. *Ann. Bot.* **2014**, *114*, 1471–1481. [CrossRef]
19. Wang, R.; Albani, M.C.; Vincent, C.; Bergonzi, S.; Luan, M.; Bai, Y.; Kiefer, C.; Castillo, R.; Coupland, G. Aa TFL1 confers an age-dependent response to vernalization in perennial *Arabis alpina*. *Plant Cell* **2011**, *23*, 1307–1321. [CrossRef]
20. Mundermann, L.; Erasmus, Y.; Lane, B.; Coen, E.; Prusinkiewicz, P. Quantitative modeling of Arabidopsis development. *Plant Physiol.* **2005**, *139*, 960–968. [CrossRef]
21. Jaeger, K.E.; Pullen, N.; Lamzin, S.; Morris, R.J.; Wigge, P.A. Interlocking feedback loops govern the dynamic behavior of the floral transition in Arabidopsis. *Plant Cell* **2013**, *25*, 820–833. [CrossRef]
22. Do, V.G.; Lee, Y.; Kim, S.; Kweon, H.; Do, G. Antisense Expression of Apple TFL1-like Gene (MdTFL1) Promotes Early Flowering and Causes Phenotypic Changes in Tobacco. *Int. J. Mol. Sci.* **2022**, *23*, 6006. [CrossRef]
23. Kaneko-Suzuki, M.; Kurihara-Ishikawa, R.; Okushita-Terakawa, C.; Kojima, C.; Nagano-Fujiwara, M.; Ohki, I.; Tsuji, H.; Shimamoto, K.; Taoka, K.I. TFL1-like Proteins in Rice Antagonize Rice FT-Like Protein in Inflorescence Development by Competition for Complex Formation with 14-3-3 and FD. *Plant Cell Physiol.* **2018**, *59*, 458–468. [CrossRef]
24. Wen, C.; Zhao, W.; Liu, W.; Yang, L.; Wang, Y.; Liu, X.; Xu, Y.; Ren, H.; Guo, Y.; Li, C.; et al. CsTFL1 inhibits determinate growth and terminal flower formation through interaction with CsNOT2a in cucumber. *Development* **2019**, *146*, dev180166. [CrossRef]
25. Jiang, Y.; Wu, C.; Zhang, L.; Hu, P.; Hou, W.; Zu, W.; Han, T. Long-day effects on the terminal inflorescence development of a photoperiod-sensitive soybean [*Glycine max* (L.) Merr.] variety. *Plant Sci.* **2011**, *180*, 504–510. [CrossRef]

26. Yang, J.; Bertolini, E.; Braud, M.; Preciado, J.; Chepote, A.; Jiang, H.; Eveland, A.L. The SvFUL2 transcription factor is required for inflorescence determinacy and timely flowering in *Setaria viridis*. *Plant Physiol.* **2021**, *187*, 1202–1220. [CrossRef]
27. Zhong, J.; van Esse, G.W.; Bi, X.; Lan, T.; Walla, A.; Sang, Q.; Franzen, R.; von Korff, M. *INTERMEDIUM-M* encodes an *HvAP2L-H5* ortholog and is required for inflorescence indeterminacy and spikelet determinacy in barley. *Proc. Natl. Acad. Sci. USA* **2021**, *118*, e2011779118. [CrossRef]
28. Kobayashi, K.; Yasuno, N.; Sato, Y.; Yoda, M.; Yamazaki, R.; Kimizu, M.; Yoshida, H.; Nagamura, Y.; Kyozuka, J. Inflorescence meristem identity in rice is specified by overlapping functions of three AP1/FUL-like MADS box genes and PAP2, a SEPALLATA MADS box gene. *Plant Cell* **2012**, *24*, 1848–1859. [CrossRef]
29. Chen, C.; Xiao, L.; Li, X.; Du, D. Comparative Mapping Combined with Map-Based Cloning of the *Brassica juncea* Genome Reveals a Candidate Gene for Multilocular Rapeseed. *Front. Plant Sci.* **2018**, *9*, 1744. [CrossRef]
30. Fulton, T.M.; Chunwongse, J.; Tanksley, S.D. Microprep protocol for extraction of DNA from tomato and other herbaceous plants. *Plant Mol. Biol. Report.* **1995**, *13*, 207–209. [CrossRef]
31. Lowe, A.J.; Jones, A.E.; Raybould, A.F.; Trick, M.; Moule, C.L.; Edwards, K.J. Transferability and genome specificity of a new set of microsatellite primers among *Brassica* species of the U triangle. *Mol. Ecol. Notes* **2002**, *2*, 7–11. [CrossRef]
32. Chen, S.; Yuan, H.; Yang, X.; Chen, L.; Chen, J.; Liu, Y.; Wu, L.; Hu, Y.; Huang, W.; Yao, Y.; et al. Identification and Analysis of Flax Resistance Genes to *Septoria linicola* (Speg.) Garassini. *J. Nat. Fibers* **2023**, *20*, 2163331. [CrossRef]
33. Jia, Y.; Li, K.; Liu, H.; Zan, L.; Du, D. Characterization of the BnA10.tfl1 Gene Controls Determinate Inflorescence Trait in *Brassica napus* L. *Agronomy* **2019**, *9*, 722. [CrossRef]
34. Kellogg, E.A. Genetic control of branching patterns in grass inflorescences. *Plant Cell* **2022**, *34*, 2518–2533. [CrossRef]
35. Chen, Z.; Li, W.; Gaines, C.; Buck, A.; Galli, M.; Gallavotti, A. Structural variation at the maize WUSCHEL1 locus alters stem cell organization in inflorescences. *Nat. Commun.* **2021**, *12*, 2378. [CrossRef]
36. Chen, Z.; Gallavotti, A. Improving architectural traits of maize inflorescences. *Mol. Breed.* **2021**, *41*, 21. [CrossRef] [PubMed]
37. Meeks-Wagner, S.S.O.R. A mutation in the *Arabidopsis TFL1* gene affects inforescence meristem development. *Plant Cell* **1991**, *3*, 92.
38. Isemura, T.; Kaga, A.; Tabata, S.; Somta, P.; Srinives, P.; Shimizu, T.; Jo, U.; Vaughan, D.A.; Tomooka, N. Construction of a genetic linkage map and genetic analysis of domestication related traits in mungbean (*Vigna radiata*). *PLoS ONE* **2012**, *7*, e41304. [CrossRef] [PubMed]
39. Zhang, Y.F.; Zhang, D.Q.; Yu, H.S.; Lin, B.G.; Hua, S.J.; Ding, H.D.; Fu, Y. Location and Mapping of the Determinate Growth Habit of *Brassica napus* by Bulked Segregant Analysis (BSA) Using Whole Genome Re-Sequencing. *Sci. Agric. Sin.* **2018**, *51*, 10. [CrossRef]
40. Liu, Y.; Gao, Y.; Gao, Y.; Zhang, Q. Targeted deletion of floral development genes in Arabidopsis with CRISPR/Cas9 using the RNA endoribonuclease Csy4 processing system. *Hortic. Res.* **2019**, *6*, 99. [CrossRef] [PubMed]
41. Balanza, V.; Martinez-Fernandez, I.; Sato, S.; Yanofsky, M.F.; Ferrandiz, C. Inflorescence Meristem Fate Is Dependent on Seed Development and FRUITFULL in *Arabidopsis thaliana*. *Front. Plant Sci.* **2019**, *10*, 1622. [CrossRef] [PubMed]
42. Azam, M.; Zhang, S.; Huai, Y.; Abdelghany, A.M.; Shaibu, A.S.; Qi, J.; Feng, Y.; Liu, Y.; Li, J.; Qiu, L.; et al. Identification of genes for seed isoflavones based on bulk segregant analysis sequencing in soybean natural population. *Theor. Appl. Genet.* **2023**, *136*, 13. [CrossRef] [PubMed]
43. Gao, J.; Dai, G.; Zhou, W.; Liang, H.; Huang, J.; Qing, D.; Chen, W.; Wu, H.; Yang, X.; Li, D.; et al. Mapping and Identifying a Candidate Gene Plr4, a Recessive Gene Regulating Purple Leaf in Rice, by Using Bulked Segregant and Transcriptome Analysis with Next-Generation Sequencing. *Int. J. Mol. Sci.* **2019**, *20*, 4335. [CrossRef]
44. Gao, Y.; Du, L.; Ma, Q.; Yuan, Y.; Liu, J.; Song, H.; Feng, B. Conjunctive Analyses of Bulk Segregant Analysis Sequencing and Bulk Segregant RNA Sequencing to Identify Candidate Genes Controlling Spikelet Sterility of Foxtail Millet. *Front. Plant Sci.* **2022**, *13*, 842336. [CrossRef] [PubMed]
45. Li, Y.; Zheng, L.; Corke, F.; Smith, C.; Bevan, M.W. Control of final seed and organ size by the DA1 gene family in *Arabidopsis thaliana*. *Genes Dev.* **2008**, *22*, 1331–1336. [CrossRef]
46. Wang, H.; Zhang, Y.; Sun, L.; Xu, P.; Tu, R.; Meng, S.; Wu, W.; Anis, G.B.; Hussain, K.; Riaz, A.; et al. WB1, a Regulator of Endosperm Development in Rice, Is Identified by a Modified MutMap Method. *Int. J. Mol. Sci.* **2018**, *19*, 2159. [CrossRef]

Disclaimer/Publisher's Note: The statements, opinions and data contained in all publications are solely those of the individual author(s) and contributor(s) and not of MDPI and/or the editor(s). MDPI and/or the editor(s) disclaim responsibility for any injury to people or property resulting from any ideas, methods, instructions or products referred to in the content.

Article

Agro-Morphological Variability of Wild *Vigna* Species Collected in Senegal

Demba Dramé [1,2,*], Amy Bodian [1], Daniel Fonceka [1,3], Hodo-Abalo Tossim [1], Mouhamadou Moussa Diangar [4], Joel Romaric Nguepjop [1,3], Diarietou Sambakhe [1], Mamadou Sidybe [2] and Diaga Diouf [2]

1. Centre d'Etude Régional pour l'Amélioration de l'Adaptation à la Sécheresse (CERAAS)/Institut Sénégalais de Recherches Agricoles (ISRA), Thiès BP 3320, Senegal
2. Laboratoire Campus de Biotechnologies Végétales, Département de Biologie Végétale, Faculté des Sciences et Techniques, Université Cheikh Anta Diop (UCAD), 10700, Dakar Fann, Dakar BP 5005, Senegal; diaga.diouf@ucad.edu.sn (D.D.)
3. Centre de Coopération Internationale en Recherche Agronyomique pour le Développement (Cirad), UMR AGAP, 34398 Montpellier, France
4. Centre de Recherches Agronomiques de Bambey, Institut Sénégalais de Recherches Agricoles (ISRA), Bambey BP 53, Senegal
* Correspondence: dembadrame10@hotmail.fr

Citation: Dramé, D.; Bodian, A.; Fonceka, D.; Tossim, H.-A.; Diangar, M.M.; Nguepjop, J.R.; Sambakhe, D.; Sidybe, M.; Diouf, D. Agro-Morphological Variability of Wild *Vigna* Species Collected in Senegal. *Agronomy* **2023**, *13*, 2761. https://doi.org/10.3390/agronomy13112761

Academic Editors: Jiezheng Ying, Zhiyong Li and Chaolei Liu

Received: 7 September 2023
Revised: 11 October 2023
Accepted: 17 October 2023
Published: 2 November 2023

Copyright: © 2023 by the authors. Licensee MDPI, Basel, Switzerland. This article is an open access article distributed under the terms and conditions of the Creative Commons Attribution (CC BY) license (https://creativecommons.org/licenses/by/4.0/).

Abstract: The domesticated *Vigna* species still need some of the beneficial characters that exist in the wild *Vigna* species, despite the improvements obtained so far. This study was carried out to enhance our understanding of the Senegalese wild *Vigna* diversity by exploring the agro-morphological characteristics of some accessions using 22 traits. The phenotyping was carried out in a shaded house for two consecutive rainy seasons (2021 and 2022) using the alpha-lattice experimental design with 55 accessions. Multiple correspondence analysis was carried out based on the qualitative traits, which showed considerable variability for the wild species (*Vigna unguiculata* var. *spontanea*, *Vigna racemosa*, *Vigna radiata* and the unidentified accession). The quantitative traits were subjected to statistical analysis using descriptive statistics and ANOVA. Our results revealed that ninety-five percent (95%) pod maturity ranged from 74.2 to 125.8 days in accession 3 of *V. unguiculata* and in accession 92 (*V. racemosa*), respectively. In addition, accession 14 of *V. radiata* recorded the highest weight for 100 seeds with a value of 4.8 g, while accession 18 of *V. unguiculata* had the lowest (1.48 g). The ANOVA showed significant differences for the accessions during each season ($p \leq 0.05$). Seasonal effects (accession × season) were observed for some quantitative traits, such as the terminal leaflet length and width, time to 50% flowering and 95% pod maturity, pod length and 100-seed weight. Principal component analysis showed that reproductive traits, such as the time to 50% flowering, number of locules per pod, pod length, pod width and 100-seed weight, were the major traits that accounted for the variations among the wild *Vigna* accessions. The genetic relationship based on qualitative and quantitative traits showed three clusters among the wild *Vigna* accessions. Indeed, the diversity observed in this study could be used to select parents for breeding to improve the cultivated species of *Vigna*.

Keywords: agro-morphological traits; wild *Vigna* species; cluster; breeding; Senegal

1. Introduction

The self-pollinating diploid crop belonging to the family of Fabaceae, cowpea (*Vigna unguiculata* (L.) Walp.), is one of the most important grain legumes growing in the tropical and subtropical regions [1]. It is cultivated worldwide for food and/or a cash crop [2]. According to Xiong et al. [3], the cowpea origin has been linked to Africa, based on the important diversity existing among the germplasm lines and the preponderance of wild relatives distributed in several parts of this continent. Cowpea has a major socio-economic impact in Sahelian countries including Senegal, where the crop is growing on

289,895 hectares with an annual production over 253,897 tons in 2021 [4]. In addition, being a legume, cowpea has the ability to fix nitrogen in association with Bradyrhizobia and to grow in low-fertility soil [5,6]. It is tolerant to high temperatures and drought compared to other legume crops [7]. For the wild species, each accession has unique traits that can be useful, but, in the *Vigna* genus, only a few of them are domesticated, such as (*V. radiata* (L.) Wilczeck), (*V. mungo* (L.) Hepper), (*V. aconitifolia* (Jacq.) Maréchal), (*V. angularis* (Willd.) Ohwi and Ohashi), (*V. umbellata* (Thunb.) Ohwi and Ohashi), (*V. subterranea* (L.) Verdc.) and (*V. unguiculata* (L.) Walp.) [8]. The wild *Vigna* species are reported to have important nutritional elements [9] and adaptability to unusual edaphic conditions, such as acidic soils, deserts and wetlands [10]. On top of that, it has also been reported that some wild species are used as supplements in traditional diets, fodder feeds or medicine. Through the domestication process, cowpea underwent many phenotypic changes compared to its wild progenitor (*Vigna unguiculata* var. *spontanea*). For many cultivated species, the loss of genetic diversity, in part due to the breeding programs associated with modern agricultural practices, has been dramatic [11], which is unlike their wild ancestors that present a very wide range of variability both in terms of important agronomic characteristics and genetic diversity. This statement is in accordance with the conclusions of Benjamin et al. [12] who reported a great agro-morphological variability among the wild accessions collected from Nigeria. Popoola et al. [13] again reported, using morphometric analysis of some species in the genus *Vigna,* considerable variabilities of the accessions in their growth habits, vegetative traits and flowering and reproductive attributes. In addition to this, Joshua et al. [14], in their work, evaluated cowpea landraces and their wild relative for growth and yield attributes in Bauchi, Northern Guinea. They reported a wide variability among the genotypes evaluated with respect to the traits studied, such as the number of seeds per pod and one hundred seed weight. Furthermore, the narrow genetic basis of the elite germplasm has increased their vulnerability to biotic and abiotic stress. Based on previous studies, cultivated cowpea has a narrow genetic basis, resulting from a single domestication event [15–18] that has limited its ability to respond to climate change or pathogen attacks [19]. Therefore, the conservation and characterization of a plant's genetic resources are a prerequisite for breeding programs in order to improve yield, pest and disease tolerance and resistance to other biotic and abiotic constraints [20]. Morphological traits are used in plant breeding for the description, classification of genetic material and the selection of intended genotypes [21]. Those traits are used in order to estimate diversity and select parental lines for crossing [22]. For years, these traits have formed a tool for breeders to attempt to check and capture the phenotypic differences among several crops worldwide [22,23] but also in cowpea [24]. According to Krichen et al. [21], morphological and agronomic traits remain imperative for breeders plants despite the extensive use of molecular markers in diversity studies [25]. Some of the wild *Vigna* species exhibit significant agro-morphological diversity, which could be utilized in crop improvement and domestication efforts. Thus, some authors have suggested to broaden the genetic basis of cultivated cowpea by using interspecific hybridization [26]. For this purpose, it would be very interesting to explore the agro-morphological diversity of the wild species for identifying the traits that can be used to improve the cultivated cowpea. Therefore, this study aimed to enhance our understanding of the genetic diversity of the Senegalese wild *Vigna* for identifying the relevant traits that can be used in cowpea breeding programs.

2. Materials and Methods
2.1. Plant Material

The plant material consisted of 55 accessions that were collected in different regions of Senegal between September and December 2016. The collection included 43 accessions of *Vigna unguiculata* var. *spontanea*, 1 hybrid accession of *V. unguiculata*, 9 accessions of *V. racemosa*, 1 accession of *V. radiata* and 1 unidentified accession. The *Vigna unguiculata* var. *spontanea* accessions were provided by the national wild germplasm, carried out by Sarr et al. [27] (Supplementary Table S1).

2.2. Site of Study and Meteorological Conditions

The trials were conducted in the Centre d'Etude Régional pour l'Amélioration de l'Adaptation à la Sécheresse (CERAAS) during two consecutive rainy seasons (2021 and 2022) from July to November. CERAAS is located in Thiès (latitude of 14°45'57'' N and longitude of 16°53'31'' W), Senegal. The meteorological characteristics (monthly rainfall) during the two rainy seasons were recorded. The total rainfall of the study site was 538 and 546.5 mm during the rainy seasons of 2021 and 2022, respectively.

2.3. Experimental Design and Sowing Process

Fifty-five (55) wild *Vigna* accessions were sown using the alpha-lattice design with 3 replications and 11 blocks per replication for two consecutive rainy seasons (2021 and 2022). The blocks contained 10 rows of 1.5 m each. The inter- and intra-row spacing were 50 cm. Each accession was sown in two rows. Two seeds were sown per pocket, and three weeks after sowing, they were thinned in one plant per pocket. Each accession was represented by 8 plants. For each accession, scarification was performed using a pair of scissors by cutting the seed coat. The scarified seeds were placed in Petri dishes containing soaked paper and left at room temperature. Radicle emergence was noticed 24 h after sowing.

2.4. Data Collection and Analysis

Twenty-two (22) traits, including 14 qualitative and 8 quantitative characters, were recorded during two consecutive rainy seasons (2021 and 2022) using both International Board for Plant Genetic Resources [28] and International Plant Genetic Resource Institute descriptors [29] (Table 1). These traits were chosen because of their usefulness for *Vigna* morphological description. The qualitative traits were measured based on visual scoring, while the quantitative characters were evaluated using a metric ruler or weighing balance. Twelve plants were analyzed for each accession in each rainy season. All the collected data were analyzed using R software version 4.1.2 [30]. Multiple correspondence analysis (MCA) was performed to reveal the most discriminant characters and relationships between the qualitative traits. To examine the relationships between accessions, the hierarchical cluster analysis (HCA) was carried out based on those traits. The quantitative traits were subjected to statistical analysis using descriptive statistics based on the mean of the two seasons. Pearson coefficient correlation was performed among the various quantitative traits using R software version 4.1.2. The quantitative data normality was checked with the Shapiro test for normality. To test for differences between accessions in order to indicate the accession effect, replicate effect and block effect, an analysis of variance (ANOVA) was calculated for each of these quantitative traits. In order to establish the interactions between the accessions and the seasons (years), a two-way ANOVA was performed. A principal component analysis (PCA) that was based on the quantitative traits was conducted to identify the relationships between variables and similarities between accessions. The construction of the plots and graphics was performed using R software version 4.1.2.

Table 1 provides information on the 8 quantitative and 14 qualitative traits related to the vegetative characteristics, flowers, pods and seeds used for the characterization of the wild *Vigna* accessions studied during the two rainy seasons based on IBPGR [28] and IPGRI [29] descriptors.

Table 1. List of the 22 morphological characters studied in the wild *Vigna* accessions.

Parameters	Descriptors
Qualitative traits	
Hypocotyl color	Green; Purple; Others
Leaf color	Pale green; Intermediate green; Dark green
Leaf texture	Cariaceous; Intermediate; Membranous
Growth habit	Erect; Intermediate; Prostrate; Climbing
Terminal leaflet shape	Globose; Sub-globose; Sub-hastate; Hastate; Others
Plant pigmentation	None; Moderate
Plant hairiness	Glabrescent; Short appressed hairs
Flower color	Violet; Dark blue; Yellow
Pod dehiscence	No shattering; Pods opened and twisted
Pod texture	Smooth; Rough
Seed texture	Smooth; Rough
Seed coat color	Grey; Marbled; brown; red; green; black
Eye color	White; Black
Seed shape	Kidney; Ovoid; Crowder; Globose; Rhomboid; Others

Parameters	Code	Unit
Quantitative traits		
Terminal leaflet length	Tlfl	cm
Terminal leaflet width	Tlfw	cm
Time to 50% flowering	T50%fw	day
Time to 95% pod maturity	T95%Rp	day
Pod length	Pdl	cm
Pod width	Pdw	cm
Number of locules per pod	Nlpd	-
100-seed weight	HSdw	g

3. Results

3.1. Qualitative Traits of the Wild Vigna Species

3.1.1. Morphological Variability of the Whole Set of *Vigna* Accessions

A wide range of variability was noticed among the morphological characters in the accessions used during this experiment (Figure 1), depending on their developmental stage. Figure 1 gives a pictorial description of some distinctive morphological traits of the wild *Vigna* accessions, studied on the basis of phenotypic observations made during the developmental stage and after harvest.

All accessions of *V. racemosa* as well as the unidentified accession showed a purple hypocotyl color. The *Vigna unguiculata* (var *spontanea* and hybrid) and *V. radiata* accessions showed a green hypocotyl color. The leaf color was dark green for all the *V. unguiculata* accessions, intermediate green for *V. racemosa* and *V. radiata* and pale green for the unidentified accession. The leaf texture was cariaceous for the *V. unguiculata* and *V. radiata* accessions, while it was membranous for *V. racemosa* and the unidentified accession (Supplementary Table S2). Among the *V. unguiculata* accessions, the terminal leaflet shapes registered were globose, sub-globose, hastate, sub-hastate and lanceolate. It was oval in *V. racemosa*, deltoid in the *V. radiata* accession and elliptical in the unidentified accession (Figure 1). All the *V. unguiculata* accessions and the unidentified accession showed a purple pigmentation on the stem and petiole, unlike those of *V. racemosa* and *V. radiata* that were unpigmented. For the plant hairiness, all the *V. racemosa*, *V. radiata* and unidentified accessions presented moderate pubescence. The *V. unguiculata* accessions were glabrous (Supplementary Table S2).

Figure 1. Qualitative trait variation in the wild *Vigna* species.

All the *V. unguiculata* accessions had violet flowers. The *V. radiata* and the unidentified accessions had yellow flowers. In *V. racemosa,* all accessions had dark blue flowers. The seed shapes observed were rhomboid in *V. unguiculata* and *V. racemosa*, globose in *V. radiata* and cylindrical in the unidentified accession. The seed eye color was white for all wild *Vigna* species except the unidentified accession that showed a dark color (Figure 1). All the *Vigna* accessions had twining growth habits and dehiscent pods. The unidentified accession had not dehiscent pods (Supplementary Table S2).

3.1.2. Multiple Correspondence Analysis

The first three axes of the multiple correspondence analysis (MCA) were the most appropriate for interpreting the observed variance. These axes explained 67.66% of the total variance (Table 2). Nine qualitative traits that were correlated with the first three axes contributed strongly to the description of the observed variation. Most qualitative traits were correlated with these axes.

Table 2. Multiple correspondence analysis for 14 qualitative traits.

Factor Axes	Dim 1	Dim 2	Dim 3
Variance	0.712	0.517	0.259
% of var.	32.385	23.497	11.773
Cumulative % of var.	32.385	55.882	67.665

The multiple correspondence analysis showed that the qualitative traits, which accounted for more variability in the Dim1, Dim2 and Dim3, were leaf color, terminal leaflet shape, hypocotyl color, plant hairiness, pigmentation on the stem and petiole, flower color, leaf texture, seed shapes and dehiscent pod (Figure 2).

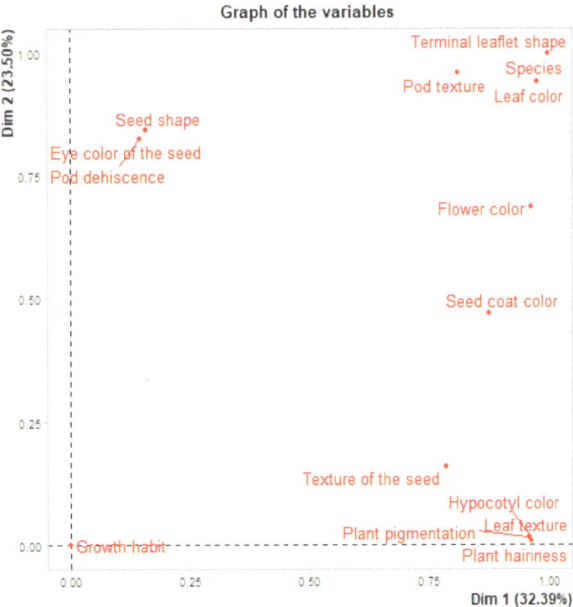

Figure 2. Multiple correspondence analysis on the basis of qualitative traits using R software (version 4.1.2).

3.1.3. Relationship between the Wild *Vigna* Accessions

The dendrogram, based on the qualitative traits including hypocotyl color, leaf color and texture, growth habit, terminal leaflet shape, plant pigmentation and hairiness, flower color, seed texture, coat color, shape and eye color, showed that 55 *Vigna* accessions were clustered into four major classes (Figure 3). Cluster 1 was formed with only the *V. radiata* accession. Cluster 2 gathered the hybrid accession of *V. unguiculata* (90) and all the *V. unguiculata* var *spontanea* accessions. Cluster 3 was formed by the unidentified accession, and cluster 4 consisted of the nine accessions of *V. racemosa* (Figure 3).

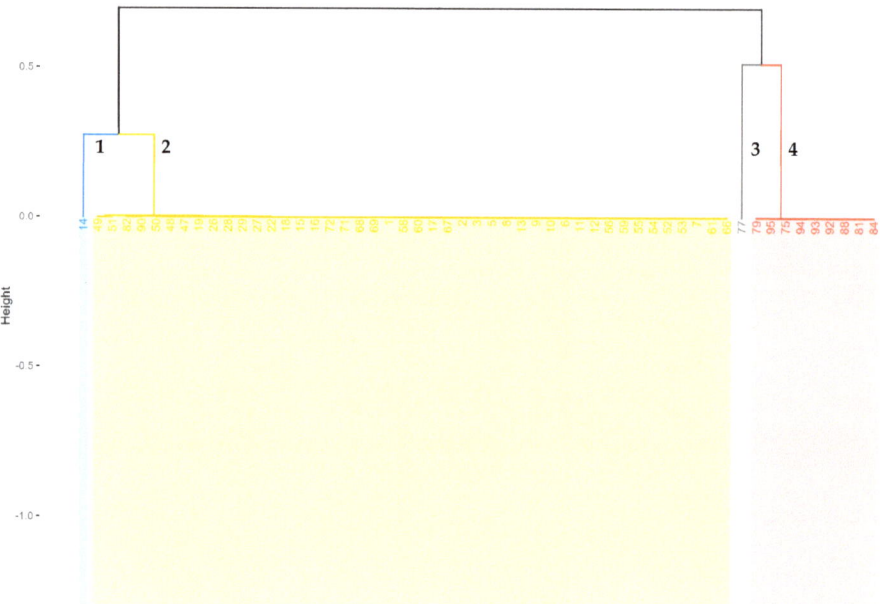

Figure 3. Dendrogram of the wild *Vigna* accessions based on qualitative traits with R software (version 4.1.2).

3.2. Diversity Description Based on Quantitative Traits

3.2.1. Descriptive Statistics Analysis

The terminal leaflet length ranged from 4.16 cm in the unidentified accession to 12.52 cm in the *V. unguiculata* var *spontanea* accession 7 (Figure 4A). Among the *V. unguiculata* accessions (hybrid and var *spontanea*), this trait ranged from 7.77 cm (accession 18) to 12.52 cm (accession 7). Among the *V. racemosa* accessions, the values of the terminal leaflet length ranged between 6.26 cm for accession 92 and 7.68 cm in accession 94. The terminal leaflet width ranged from 2.33 cm in the unidentified accession to 6.56 cm in the accession hybrid of *V. unguiculata* (number 90). Among *V. unguiculata*, it ranged from 2.95 cm in accession 56 to 6.56 cm (hybrid accession, 90). Among *V. racemosa*, the terminal leaflet width variations range from 3.3 cm for accession 94 to 4.28 cm in accession 93 (Figure 4B).

Fifty (50%) percent flowering varied widely from 52.2 days in accession 13 of *V. unguiculata* var *spontanea* to 112.3 days for *V. racemosa* accession 79 (Figure 4C). For 95% pod maturity, it ranged from 74.2 in accession 3 of *V. unguiculata* var *spontanea* to 125.8 days in *V. racemosa* (accession 92) (Figure 4D). The pod length varied from 2.36 cm in the unidentified accession to 10.4 cm in the hybrid accession of *V. unguiculata* (number 90) (Figure 4E). There was also variation in the 100-seed weight; accession 14 (*V. radiata*) recorded the highest value (4.8 g), while accession 18 of *V. unguiculata* var *spontanea* had the lowest value (1.48 g) (Figure 4F).

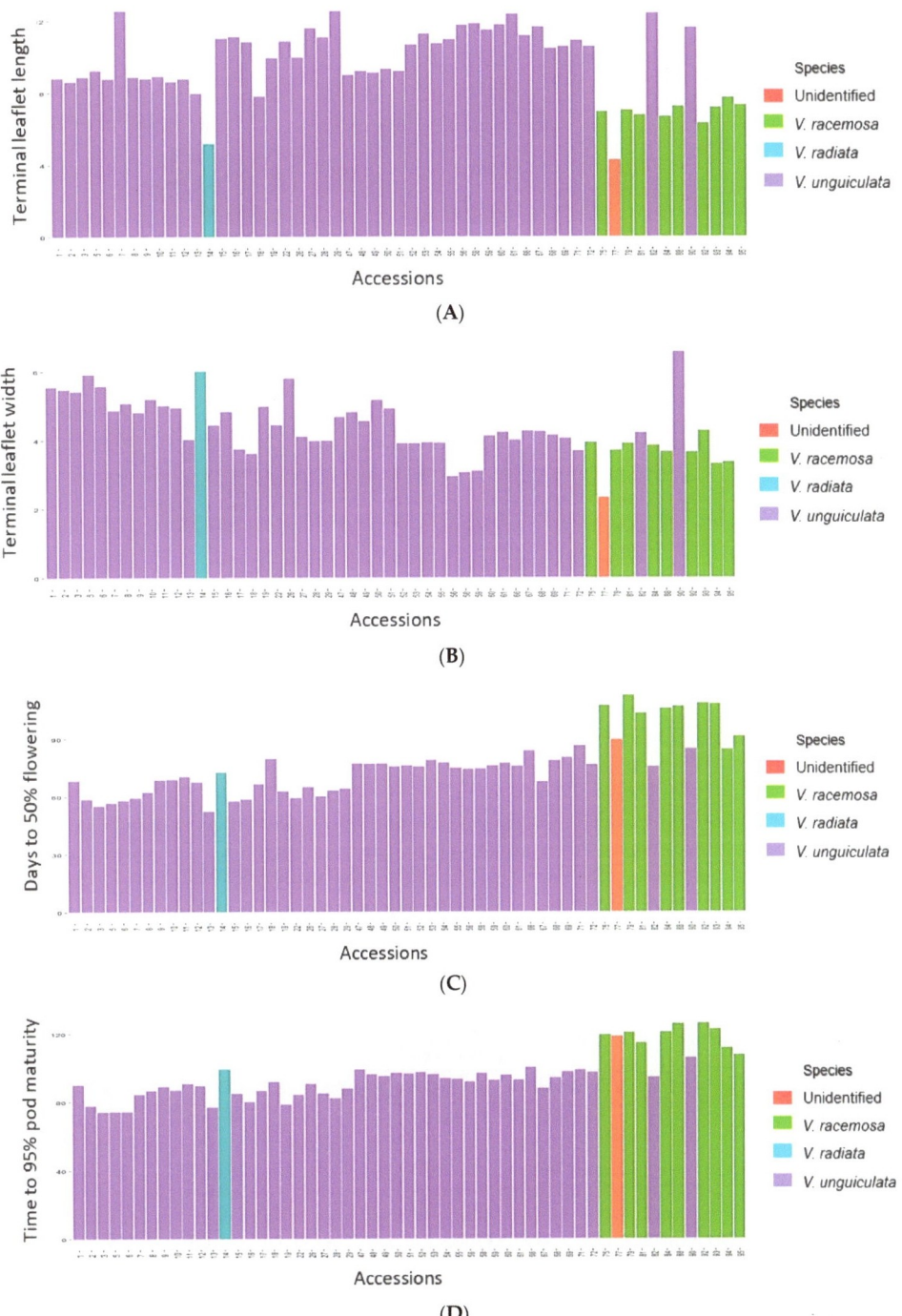

Figure 4. Cont.

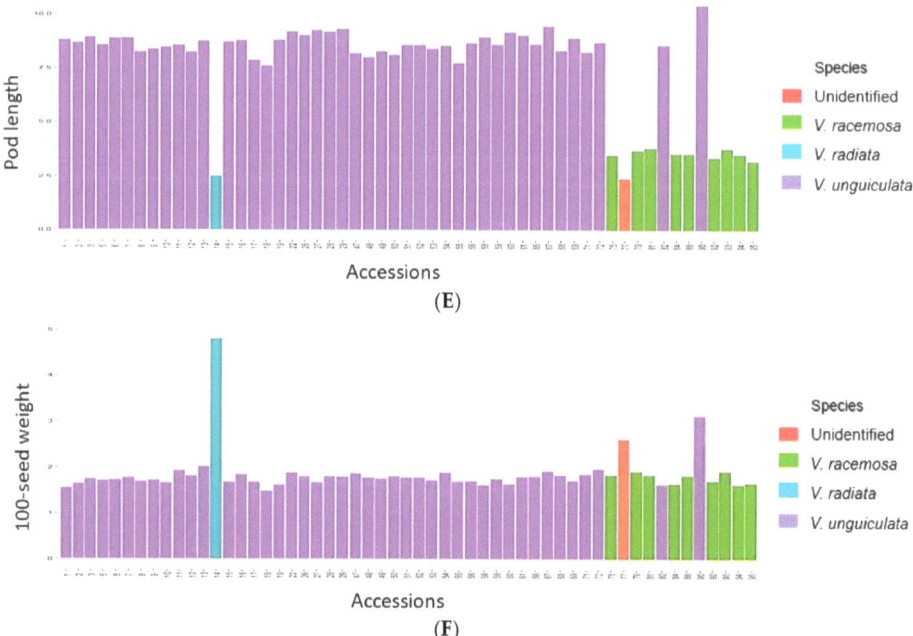

Figure 4. Descriptive statistics of some quantitative traits studied during the two seasons. Graphs generated with R software (version 4.1.2). Figures (**A–F**) showed the means of some quantitative traits during the two study seasons. (**A**): terminal leaflet length; (**B**): terminal leaflet width; (**C**): time to 50% flowering; (**D**): time to 95% pod maturity; (**E**): pod length and (**F**): 100-seed weight. Each species is represented by a color.

3.2.2. Analysis of Variance

Table 3a,b indicate the accession effect, block effect and replicate effect based on the analysis of variance (ANOVA). The results of the ANOVA indicate the existence of a significant difference ($p < 0.05$) between the accessions for all the measured traits from one season to the next. Replicate effects were only found ($p < 0.05$) for the time to 95% pod maturity, terminal leaflet length and width in 2021 (Table 3a).

Table 3. Analysis of variance for 8 quantitative traits in 2021 and 2022.

a. Analysis of Variance for 8 Quantitative Traits in 2021						
2021	Accessions Effect		Repetition Effect		Block Effect	
Traits	Mean Sq	Pr(>F)	Mean Sq	Pr(>F)	Mean Sq	Pr(>F)
T50%fw	760.100	$<2 \times 10^{-16}$	29.000	0.063	11.100	0.369
T95% Rp	675.800	$<2 \times 10^{-16}$	25.500	0.026	8.900	0.157
Tlfl	8.446	$<2 \times 10^{-16}$	3.319	1.4×10^{-6}	0.192	0.577
Tlfw	1.615	$<2 \times 10^{-16}$	0.233	0.029	0.098	0.064
Pdl	14.050	$<2 \times 10^{-16}$	0.123	0.384	0.085	0.889
Pdw	0.005	$<2 \times 10^{-16}$	0.000	0.239	0.000	0.617
Nlpd	34.890	$<2 \times 10^{-16}$	0.990	0.127	0.390	0.722
HSdw	1.051	$<2 \times 10^{-16}$	0.020	0.310	0.015	0.604

Table 3. Cont.

2022	b. Analysis of Variance for 8 Quantitative Traits in 2022					
	Accessions Effect		Repetition Effect		Block Effect	
Traits	Mean Sq	Pr(>F)	Mean Sq	Pr(>F)	Mean Sq	Pr(>F)
T50%fw	656.800	$<2 \times 10^{-16}$	177.600	7.3×10^{-6}	9.600	0.816
T95% Rp	252.570	$<2 \times 10^{-16}$	115.040	8.1×10^{-6}	7.870	0.509
Tlfl	15.773	$<2 \times 10^{-16}$	0.535	0.138	0.140	0.973
Tlfw	2.844	$<2 \times 10^{-16}$	0.204	0.114	0.079	0.653
Pdl	15.197	$<2 \times 10^{-16}$	0.134	0.041	0.029	0.828
Pdw	0.005	$<2 \times 10^{-16}$	0.000	0.225	0.000	0.442
Nlpd	42.740	$<2 \times 10^{-16}$	0.420	0.056	0.150	0.362
HSdw	0.479	$<2 \times 10^{-16}$	0.012	0.577	0.027	0.211

The results during the 2022 rainy season show that there were replicate effects ($p < 0.05$) for the time to 50% flowering, time to 95% pod maturity and pod length (Table 3b). Any block effect and significant difference have been obtained during the two rainy seasons ($p > 0.05$).

The quantitative traits studied showed significant differences for all the species during the two rainy seasons (Table 4). Among these, only six (time to 50% flowering, time to 95% pod maturity, terminal leaflet length and width, pod length and 100-seed weight) were affected by the season (accession × season).

Table 4. Analysis of variance for the interactions due to the season for the studied quantitative traits.

		p-Values for Season Effects		
Code	Trait	Season	Accession	Season × Accession
T50%fw	Time to 50% flowering	0.160	$<2 \times 10^{-16}$ ***	2.77×10^{-13} ***
T95% Rp	Time to 95% pod maturity	0.188	$<2 \times 10^{-16}$ ***	$<2 \times 10^{-16}$ ***
Tlfl	Terminal leaflet length	2×10^{-16} ***	$<2 \times 10^{-16}$ ***	2.33×10^{-14} ***
Tlfw	Terminal leaflet width	$<2 \times 10^{-16}$ ***	$<2 \times 10^{-16}$ ***	8.24×10^{-10} ***
Pdl	Pod length	0.0001 ***	$<2 \times 10^{-16}$ ***	1.3×10^{-12} ***
Pdw	Pod width	0.332	$<2 \times 10^{-16}$ ***	0.401
Nlpd	Number of locules per pod	0.0003 ***	$<2 \times 10^{-16}$ ***	2.98×10^{-7} ***
HSdw	Weight of 100 seeds	0.002 **	$<2 \times 10^{-16}$ ***	$<2 \times 10^{-16}$ ***

***, ** Significant at the 0.001 and 0.01 levels, respectively.

3.2.3. Principal Component Analysis

The two axes of the principal component analysis (PCA) explained 73.02% of the total variation and were retained to analyze the agro-morphological variability among the wild *Vigna* accessions (Figure 5). The time to 50% flowering (T50%fw), terminal leaflet length (Tlfl), number of locules per pod (Nlpd) and pod length (Pdl) were the traits with the highest contribution to the PC1. These traits explained 46.86% of the total variation to the PC1. Agronomic traits, such as pod width (Pdw), 100-seed weight (HSdw) and terminal leaflet width (Tlfw), were the most important traits that contributed to the variation in PC2.

A significant positive correlation ($p = 0.000$) was observed between the 100-seed weight (HSdw) and pod width (Pdw). The number of locules per pod (Nlpd) was significantly correlated with pod length (Pdl) ($p = 0.000$). The 100-seed weight was negatively correlated with pod length ($p = 0.023$) and number of locules per pod ($p = 0.010$) (Figures 5 and 6 and Table 5).

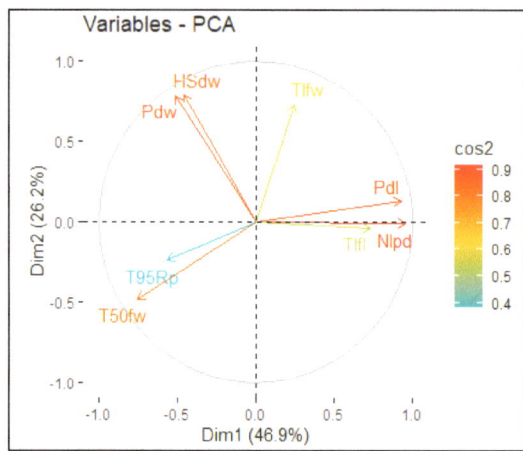

Figure 5. Principal component analysis of wild *Vigna* accessions using 8 quantitative traits under R (version 4.1.2). Variables with high cos2 values are colored orange. Variables with low cos2 values are colored blue.

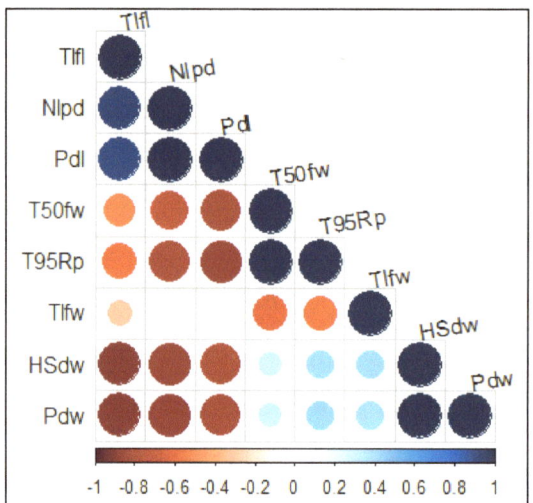

Figure 6. Correlation of the quantitative traits studied. Graph generated using 8 quantitative traits under R (version 4.1.2).

Table 5. Correlation between quantitative traits.

Traits	T50fw	T95Rp	HSdw	Nlpd	Pdl	Pdw	Tlfl	Tlfw
T50fw								
T95Rp	0.000							
HSdw	0.152	0.100						
Nlpd	0.000	0.000	0.010					
Pdl	0.000	0.000	0.023	0.000				
Pdw	0.205	0.140	0.000	0.013	0.029			
Tlfl	0.000	0.000	0.010	0.000	0.000	0.013		
Tlfw	0.003	0.005	0.711	0.056	0.032	0.685	0.138	

Tlfl: terminal leaflet length; Tlfw: terminal leaflet width; T50%fw: time to 50% flowering; Pdl: pod length; Pdw: pod width; T95%Rp: time to 95% pod maturity; Nlpd: number of locules per pod; HSdw: 100-seed weight.

3.2.4. Cluster Analysis Showing the Grouping of the Accessions Based on the Quantitative Traits

Three clusters were generated using the quantitative traits recorded on the wild *Vigna* accessions. Cluster 1 contained only the *V. radiata* accession. Cluster 2 encompassed the *V. racemosa* accessions and the unidentified accession (77). Cluster 3 was the largest and included only the *V. unguiculata* accessions (Figure 7). Cluster 3 included the hybrid accession of *V. unguiculata* (90) and all the *V. unguiculata* var *spontanea* accessions.

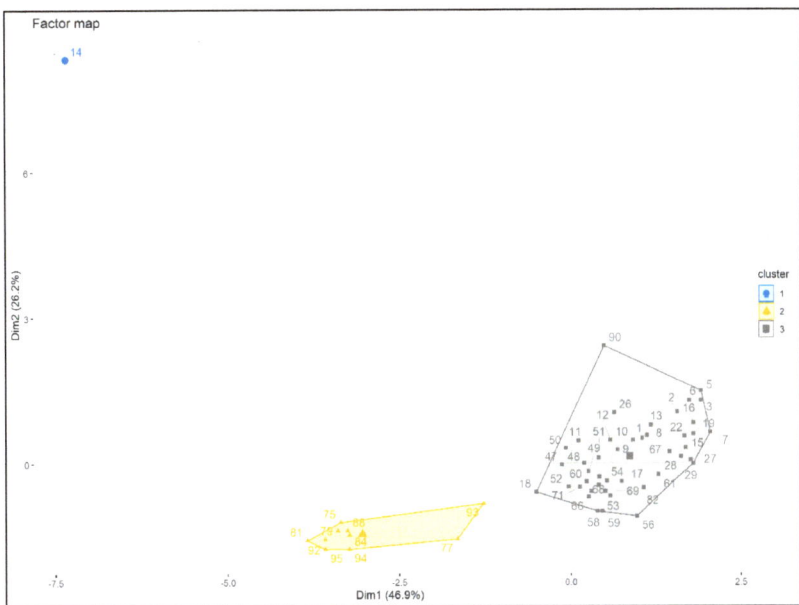

Figure 7. Grouping of the wild *Vigna* species based on 8 quantitative traits. Graph generated with R version 4.1.2.

Based on the quantitative traits, the wild *Vigna* were divided into three clusters. Cluster 1 included the *V. radiata* accession (14) with a high value of the 100-seed weight and a large pod compared to the overall average. Similarly, cluster 2 included the accessions with a low number of locules per pod and that were late flowering and late maturing compared to the overall average for each of these traits. Cluster 3 encompassed the accessions with long pods and a high number of locules per pod and that flowered earlier and matured earlier compared to the overall average for each of these measured traits (Table 6).

Table 6. Description of each cluster by quantitative variables.

	Cluster 1				Cluster 2				Cluster 3		
Traits	Mean in Category	Overall Mean	*p*-Value	Traits	Mean in Category	Overall Mean	*p*-Value	Traits	Mean in Category	Overall Mean	*p*-Value
Pdw	0.700	0.405	0	T50fw	101.500	75.701	0	Pdl	8.495	7.625	0
HSdw	4.795	1.853	0	T95Rp	97.190	91.249	0.013	Nlpd	14.508	13.133	0
Tlfw	5.998	4.365	0.05	Tlfw	3.592	4.365	0.001	Tlfl	10.285	9.542	0
Tlfl	5.161	9.542	0.025	Tlfl	6.710	9.542	0	Tlfw	4.503	4.365	0.015
Pdl	2.485	7.625	0.009	Nlpd	8.214	13.133	0	Pdw	0.400	0.405	0.046
Nlpd	2	13.133	0	Pdl	4.310	7.625	0	T95Rp	89.719	91.249	0.007
								T50fw	69.913	75.701	0

4. Discussion

The characterization and evaluation of cultivated and wild accessions are one of the most important activities for plant genetic resource management, leading to the discovery of relevant traits that can be used for crop adaptation. In the cowpea, the wild species remain the main reservoir of traits of interest that are usable in crop improvement due to their adaptation to a wide range of environments and their resistance to several diseases. So, their exploration in terms of genetic and agro-morphological study is essential for understanding the process of domestication. The present investigation of the agro-morphological variability of the wild *Vigna* accessions that was based on fourteen qualitative and eight quantitative traits, which included vegetative, floral, pod and seed traits, revealed that there is variation within the same species and among the *Vigna* accessions. These results are in agreement with the findings of Popoola et al. [13] and Joshua et al. [14]. Based on the traits examined, we discovered that there is more agro-morphological variability in *Vigna unguiculata* compared to that in the other *Vigna* species. In addition, some traits described in the *V. racemosa* accessions, such as the hairiness, that disappeared during the domestication process are desirable characters that could be also used in a breeding program. These traits could have potential use, since they are thought to be responsible for some benefits, such as the resistance to pests and diseases [31,32]. Similar findings were reported in previous studies [33–35] in the *Vigna unguiculata* var *spontanea* wild relatives and [36] in *Cajanus cajan* and *Sphenostylis stenocarpa*. In this work, sub-hastate, globose, hastate and sub-globose were the most occurring terminal leaflet shapes, unlike the lanceolate form, which was the least frequent. These morphological differences in terminal leaflet shape have been observed in *V. unguiculata* var *spontanea*. However, the terminal leaflet shapes oval (*V. racemosa*), deltoid (*V. radiata*) and elliptic (unidentified accession) were also observed in our study. Nicotra et al. [37] reported that the oval and elliptic terminal leaflets may help capture sunlight and carry out photosynthesis efficiently. Multiple correspondence analysis (MCA) suggests that morphological characters, such as leaf color, terminal leaflet shape, hypocotyl color, plant hairiness, pigmentation on the stem and petiole, flower color, leaf texture, seed shapes and dehiscent pod, have the potential to discriminate the wild *Vigna* accessions. The MCA, based on the qualitative phenotypic traits, was confirmed with an analysis of the Hierarchical Ascending Classification, which was used to construct a dendrogram. The dendrogram showed variation among the accessions for the traits studied. It also revealed the importance of morphological markers to differentiate and classify the wild *Vigna* species. Wild accessions with similar traits were found in the same group. This information would help breeders to make the right decision and choice in a cowpea breeding program. This study showed significant differences among the accessions for most of the agro-morphological characters evaluated. These results are in agreement with the report of Moalafi et al. [38]. The *V. unguiculata* var *spontanea* accession 13 reaches 50% flowering at the earliest time (52.2 days after sowing). The significant correlation among some traits, such as days to 50% flowering and time to 95% pod maturity, could be exploited in breeding to improve cultivated cowpea. Furthermore, the accessions with longer pods (accession 90) possessed a higher number of locules per pod and are potential raw materials to boost seed production of cowpea. However, for most of the *V. racemosa* accessions (such as accessions 79 and 92), we observed a late time to 50% flowering (approximately 112.3 days after sowing) during the two seasons; this could be due to a response of these accessions to photoperiod. Indeed, depending on photoperiod, flowering can be delayed in some genotypes [39]. During this study, the *V. radiata* accession (14) had the highest 100-seed weight, 4.8 g. The seed weight is a yield indicator [13,40] and, therefore, useful to increase the production. The analysis of variance (ANOVA) in the first and second season showed accession effects with significant differences for all the analyzed quantitative traits. For the same trait, the accessions differed significantly over one year. Similar results were described by Adewale et al. [41] in their studies of eleven cowpea genotypes. The *Vigna* accessions involved in this study revealed different phenotypic and probably genetic characteristics. Most of the quantitative traits, such as terminal leaflet length and width, pod length, number of locules per pod and 100-seed weight, were affected by the season (year). These variations in some quantitative traits for the same accession from season to season are likely due to the heavier precipitation in the 2022 season compared to that in the 2021 season and others environmental factors. The repetition effect observed could be due to specific factors such as soil characteristics. The significant effect of the season (accession × year effect) on some quantitative traits examined, such as the days to 50% flowering, time to 95% pod maturity and 100-seed weight, has also been reported by Adewale et al. [41] in their study of genotypic variability and stability of some grain yield components of cowpea. The principal component analysis (PCA), based on the quantitative traits, confirmed a high agro-morphological relationship among the wild *Vigna* accessions (Figures 5 and 6). High diversity among related accession groups is a good factor for the genetic improvement of species in the group. The cluster analysis, based on the eight quantitative traits studied, classify the accessions into three distinct groups. The cluster analysis clearly separated the accessions based on yield components, mainly those related to pods and seeds, as well as flowering, pod maturity and some vegetative traits. These quantitative traits are important in explaining the diversity of the *Vigna* accessions. Cluster 3 encompasses the largest number of accessions, while cluster 2 contains some accessions, and cluster 1 contains one accession (Figure 7), showing wide inter-cluster divergence, which is desirable for future hybridization programs. These are clear indications that these wild *Vigna* accessions could be made useful, and pre-breeding material (for example, a marker-assisted backcross) must be developed and evaluated in order to identify some useful traits. Thus, the combination of this agro-morphological information with the molecular information obtained with SSR markers by Sarr et al. [27] may provide added value for the use of some of these wild *Vigna* accessions in breeding programs. According to our study, despite the limited number of samples, the Senegalese wild *Vigna* accessions have a high level of morphological diversity, corroborating a previous genetic analysis [27], which showed that the wild forms remain more diverse than the cultivated.

5. Conclusions

This study revealed a high level of agro-morphological diversity between and within the wild *Vigna* species. Some particular traits of agronomic importance, including hairiness, a large length and width of the terminal leaflet, early maturity, a longer pod, a high number of locules per pod and a high weight of 100 seeds, can be targeted for subsequent hybridization for cowpea improvement. The high agro-morphological variability observed in this study suggests that these wild *Vigna* accessions constitute a true reservoir. Despite the influence of environmental factors (on some traits), the agro-morphological traits used in this study allowed for the characterization of the wild *Vigna* accessions. This study sets the basis for the genetic improvement of cultivated cowpea using wild relatives, since the observed diversity can be exploited in breeding and varietal improvement programs. However, further characterization that focuses on the study of the photoperiod in the wild *Vigna* species may be of great value. This would provide more information for characterizing diversity.

Supplementary Materials: The following supporting information can be downloaded at: https://www.mdpi.com/article/10.3390/agronomy13112761/s1. Table S1: List of the wild *Vigna* accessions used in this study and their region of provenance. Table S2: Agro-morphological characteristics of 55 wild *Vigna* accessions.

Author Contributions: D.F., A.B., D.D. (Demba Dramé), H.-A.T., M.M.D., D.D. (Diaga Diouf) and J.R.N. conceived, designed and performed the experiments; D.D. (Demba Dramé), M.S. and D.S. collected and analyzed the data and made the first draft of this manuscript; D.D. (Diaga Diouf) supervised the research; D.D. (Demba Dramé), D.D. (Diaga Diouf), D.F. and A.B. made the final internal review and revised the final draft of the manuscript. All authors have read and agreed to the published version of the manuscript.

Funding: This work constitutes part of doctoral research studies. This study is granted by the support of the American People provided to the Feed the Future Innovation Lab for Crop Improvement through the United States Agency for International Development (USAID). The APC was funded by the Centre d'étude Régional pour l'Amélioration de l'Adaptation à la sécheresse (CERAAS), Institut Sénégalais de Recherches Agricoles (ISRA). The scholarship was funded by the German academic exchange service (DAAD).

Acknowledgments: This study is granted by the support of the American People provided to the Feed the Future Innovation Lab for Crop Improvement through the United States Agency for International Development (USAID). The contents are the sole responsibility of the authors and do not necessarily reflect the views of USAID or the United States Government. Program activities are funded by the United States Agency for International Development (USAID) under Cooperative Agreement No. 7200AA-19LE-00005.

Conflicts of Interest: The authors declare no conflict of interest.

References

1. Oikeh, S.O.; Niang, A.; Abaidoo, R.; Houngnandan, P.; Futakuchi, K.; Koné, B.; Touré, A. Enhancing Rice Productivity and Soil Nitrogen Using Dual-Purpose Cowpea-NERICA®Rice Sequence in Degraded Savanna. *J. Life Sci.* **2012**, *6*, 1237–1250.
2. Lonardi, S.; Muñoz-Amatriaín, M.; Liang, Q.; Shu, S.; Wanamaker, S.I.; Lo, S.; Tanskanen, J.; Schulman, A.H.; Zhu, T.; Luo, M.-C.; et al. The genome of cowpea (*Vigna unguiculata* [L.] Walp). *Plant J.* **2019**, *98*, 767–782. [CrossRef] [PubMed]
3. Xiong, H.; Shi, A.; Mou, B.; Qin, J.; Motes, D.; Lu, W.; Ma, J.; Weng, Y.; Yang, W.; Wu, D. Genetic diversity and population structure of cowpea (*Vigna unguiculata* L. Walp). *PLoS ONE* **2016**, *11*, e0160941. [CrossRef] [PubMed]
4. ANSD. Bulletin Mensuel des Statistiques Économiques d'Octobre 2020. Available online: https://www.ansd.sn (accessed on 18 October 2023).
5. Ehlers, J.D.; Hall, A.E. Genotypic Classification of Cowpea Based on Responses to Heat and Photoperiod. *Crop. Sci.* **1996**, *36*, 673–679. [CrossRef]
6. Elowad, H.O.; Hall, A.E. Influences of early and late nitrogen fertilization on yield and nitrogen fixation of cowpea under well-watered and dry field conditions. *Field Crop. Res.* **1987**, *15*, 229–244. [CrossRef]
7. Hall, A.E. Breeding for adaptation to drought and heat in cowpea. *Eur. J. Agron.* **2004**, *21*, 447–454. [CrossRef]
8. Ba, F.S.; Pasquet, R.S.; Gepts, P. Genetic diversity in cowpea [*Vigna unguiculata* (L.) Walp as revealed by RAPD markers. *Genet. Resour. Crop. E* **2004**, *51*, 539–550. [CrossRef]
9. Difo, V.H.; Onyike, E.; Ameh, D.A.; Njoku, G.C.; Ndidi, U.S. Changes in nutrient and antinutrient composition of *Vigna racemosa* flour in open and controlled fermentation. *J. Food Sci. Technol.* **2015**, *52*, 6043–6048. [CrossRef]
10. Tomooka, N.; Kaga, A.; Isemura, T.; Vaughan, D. Vigna. In *Wild Crop Relatives: Genomic and Breeding Resources: Legume Crops and Forages*; Kole, C., Ed.; Springer: Berlin/Heidelberg, Germany, 2010; pp. 291–311. [CrossRef]
11. Wilkes, G.; Williams, J.T. Current status of crop plant germplasm. *Crit. Rev. Plant Sci.* **1983**, *1*, 133–181. [CrossRef]
12. Benjamin, U.; Olamide, F.; Oladipupo, D.; Yusuf, A.; Abdulhakeem, A. GSC Biological and Pharmaceutical Sciences Phenotypic variability studies in selected accessions of Nigerian wild cowpea (*Vigna unguiculata* L. Walp). *GSC Biol. Pharm. Sci.* **2018**, *3*, 19–27.
13. Popoola, J.O.; Aremu, B.R.; Daramola, F.Y.; Ejoh, A.S.; Adegbite, A.E. Morphometric Analysis of some Species in the *Genus Vigna* (L.) Walp: Implication for Utilization for Genetic Improvement. *J. Biol. Sci.* **2015**, *15*, 156–166. [CrossRef]
14. Joshua, N.N.; Namo, O.A.T. Agronomic evaluation of some landrace cowpeas (*Vigna unguiculata* (L.) Walp) and their wild relative (*dekindtiana* var. *pubescens*) for incorporation into cowpea breeding programme. *Eur. J. Agric. For. Res.* **2019**, *7*, 13–23.
15. Li, C.; Fatokun, C.A.; Ubi, B.; Singh, B.B.; Scoles, G.J. Determining Genetic Similarities and Relationships among Cowpea Breeding Lines and Cultivars by Microsatellite Markers. *Crop. Sci.* **2001**, *41*, 189–197. [CrossRef]
16. Kouakou, C.K.; Roy-Macauley, H.; Coudou, M.; Otto, M.C.; Rami, J.-F.; Cissé, N. Diversité génétique des variétés traditionnelles de niébé [*Vigna unguiculata* (L.) Walp] au Sénégal: étude préliminaire. *Plant Genet. Resour. Newsl.* **2007**, *152*, 33–44.
17. Badiane, F.A.; Gowda, B.S.; Cissé, N.; Diouf, D.; Sadio, O.; Timko, M.P. Genetic relationship of cowpea (*Vigna unguiculata*) varieties from Senegal based on SSR markers. *Evolution* **2012**, *11*, 292–304. [CrossRef] [PubMed]
18. Asare, A.T.; Gowda, B.S.; Galyuon, I.K.A.; Aboagye, L.L.; Takrama, J.F.; Timko, M.P. Assessment of the genetic diversity in cowpea (*Vigna unguiculata* L. Walp) germplasm from Ghana using simple sequence repeat markers. *Plant Genet. Resour. Charact. Util.* **2010**, *8*, 142–150. [CrossRef]
19. Manifesto, M.M.; Schlatter, A.R.; Hopp, H.E.; Suárez, E.Y.; Dubcovsky, J. Quantitative Evaluation of Genetic Diversity in Wheat Germplasm Using Molecular Markers. *Crop. Sci.* **2001**, *41*, 682–690. [CrossRef]
20. Kandel, B.P.; Shrestha, J. Characterization of rice (*Oryza sativa* L.) germplasm in Nepal: A mini review. *Farming Manag.* **2018**, *3*, 153–159. [CrossRef]
21. Krichen, L.; Audergon, J.M.; Trifi-Farah, N. Relative efficiency of morphological characters and molecular markers in the establishment of an apricot core collection. *Hereditas* **2012**, *149*, 163–172. [CrossRef]
22. Lee, O.N.; Park, H.Y. Assessment of genetic diversity in cultivated radishes (*Raphanus sativus*) by agronomic traits and SSR markers. *Sci. Hortic.* **2017**, *223*, 19–30. [CrossRef]

23. Arteaga, S.; Yabor, L.; Torres, J.; Solbes, E.; Muñoz, E.; Díez, M.J.; Vicente, O.; Boscaiu, M. Morphological and agronomic characterization of Spanish landraces of *Phaseolus vulgaris* L. *Agriculture* **2019**, *9*, 149. [CrossRef]
24. Menssen, M.; Linde, M.; Omondi, E.O.; Abukutsa-Onyango, M.; Dinssa, F.F.; Winkelmann, T. Genetic and morphological diversity of cowpea (*Vigna unguiculata* (L.) Walp) entries from East Africa. *Sci. Hortic.* **2017**, *226*, 268–276. [CrossRef]
25. Tanhuanpää, P.; Manninen, O. High SSR diversity but little differentiation between accessions of Nordic timothy (*Phleum pratense* L.). *Hereditas* **2012**, *149*, 114–127. [CrossRef] [PubMed]
26. Badiane, F.A.; Diouf, M.; Diouf, D. Cowpea. In *Broadening the Genetic Base of Grain Legumes*; Singh, M., Bisht, I.S., Dutta, M., Eds.; Springer: New Delhi, India, 2014; pp. 95–114. [CrossRef]
27. Sarr, A.; Bodian, A.; Gbedevi, K.M.; Ndir, K.N.; Ajewole, O.O.; Gueye, B.; Foncéka, D.; Diop, E.A.; Diop, B.M.; Cissé, N.; et al. Genetic Diversity and Population Structure Analyses of Wild Relatives and Cultivated Cowpea (*Vigna unguiculata* (L.) Walp) from Senegal Using Simple Sequence Repeat Markers. *Plant Mol. Biol. Rep.* **2021**, *39*, 112–124. [CrossRef]
28. IBPGR. Descriptors For Cowpea 377. 1983. Available online: https://www.scribd.com/document/556492902/Descriptors-for-Cowpea-377 (accessed on 18 October 2023).
29. IPGRI. *Key Characterization and Evaluation Descriptors: Methodologies for the Assessment of 22 Crops*; Bioversity International: Rome, Italy, 2011; 602p.
30. R Software. The Comprehensive R Archive Network. Available online: https://cran.r-project.org (accessed on 1 April 2021).
31. Oyatomi, O.; Fatokun, C.; Boukar, O.; Abberton, M.; Ilori, C.; Maxted, N.; Dulloo, M.E.; Ford-Lloyd, B.V. Screening wild *Vigna* species and cowpea (*Vigna unguiculata*) landraces for sources of resistance to Striga gesnerioides. In *Enhancing Crop Genepool Use: Capturing Wild Relatives and Landrace Diversity for Crop Improvement*; Maxted, N., Dulloo, M.E., Ford-Lloyd, B.V., Eds.; CABI: Boston, MA, USA, 2016; pp. 27–31. [CrossRef]
32. Popoola, J.O.; Adebambo, A.; Ejoh, S.; Agre, P.; Adegbite, A.E.; Omonhinmin, C.A. Morphological diversity and cytological studies in some accessions of *Vigna vexillata* (L.) A. Richard. *Annu. Res. Rev. Biol.* **2017**, *19*, 1–12. [CrossRef]
33. Padulosi, S.; Ng, N.Q. Origin, taxonomy, and morphology of *Vigna unguiculata* (L.) Walp. In *Advances in Cowpea Research*; Singh, B.B., Dashiell, K.E., Jackai, L.E.N., Eds.; IITA-JIRCAS: Ibadan, Nigeria, 1997; Volumes 1–12.
34. Damayanti, F.; Lawn, R.J.; Bielig, L.M. Genetic compatibility among domesticated and wild accessions of the tropical tuberous legume *Vigna vexillata* (L.) A. Rich. *Crop. Pasture Sci.* **2010**, *61*, 785–797. [CrossRef]
35. Marubodee, R.; Ogiso-Tanaka, E.; Isemura, T.; Chankaew, S.; Kaga, A.; Naito, K.; Ehara, H.; Tomooka, N. Construction of an SSR and RAD-Marker Based Molecular Linkage Map of *Vigna vexillata* (L.) A. Rich. *PLoS ONE* **2015**, *10*, e0138942. [CrossRef] [PubMed]
36. Popoola, J.O.; Adegbite, A.E.; Obembe, O.O. Cytological studies on some accessions of African Yam Bean (AYB) (*Sphenostylis stenocarpa* Hochst. Ex. A. Rich. Harms). *Int. Res. J. Plant Sci.* **2011**, *2*, 249–253.
37. Nicotra, A.B.; Leigh, A.; Boyce, C.K.; Jones, C.S.; Niklas, K.J.; Royer, D.L.; Tsukaya, H. The evolution and functional significance of leaf shape in the angiosperms. *Funct. Plant Biol.* **2011**, *38*, 535–552. [CrossRef]
38. Moalafi, S.; Sanka, G.; Apuyor, B. Genetic diversity in cultivated cowpea (*Vigna unguiculata* L.). *Afr. J. Agric. Sci.* **2010**, *32*, 841–850.
39. Timko, M.P.; Singh, B.B. *Cowpea, a Multifunctional Legume, in Genomics of Tropical Crop Plants*, 1st ed.; Moore, P.H., Ming, R., Eds.; Genomics of Tropical Crop Plants. Plant Genetics and Genomics: Crops and Models; Springer: New York, NY, USA, 2008; Volume 1, pp. 227–258. [CrossRef]
40. Burger, J.C.; Chapman, M.A.; Burke, J.M. Molecular insights into the evolution of crop plants. *Am. J. Bot.* **2008**, *95*, 113–122. [CrossRef]
41. Adewale, B.D.; Okonji, C.; Oyekanmi, A.A.; Akintobi, D.A.C.; Aremu, C.O. Genotypic variability and stability of some grain yield components of Cowpea. *Afr. J. Agric. Res.* **2010**, *5*, 874–880.

Disclaimer/Publisher's Note: The statements, opinions and data contained in all publications are solely those of the individual author(s) and contributor(s) and not of MDPI and/or the editor(s). MDPI and/or the editor(s) disclaim responsibility for any injury to people or property resulting from any ideas, methods, instructions or products referred to in the content.

Article

Mapping of the Waxy Gene in *Brassica napus* L. via Bulked Segregant Analysis (BSA) and Whole-Genome Resequencing

Junying Zhang [†], Jifeng Zhu [†], Liyong Yang, Yanli Li, Weirong Wang, Xirong Zhou * and Jianxia Jiang *

Key Laboratory of Germplasm Innovation and Genetic Improvement of Grain and Oil Crops (Co-Construction by Ministry and Province), Ministry of Agriculture and Rural Affairs, Crop Breeding and Cultivation Research Institute, Shanghai Academy of Agricultural Sciences, Shanghai 201403, China; zhangjunying@saas.sh.cn (J.Z.); zhujifeng0224@163.com (J.Z.); yangliyong@saas.sh.cn (L.Y.); 15921009648@163.com (Y.L.); wangwr71@sina.com (W.W.)

* Correspondence: zhouxr63@mail.sh.cn (X.Z.); xiajianjiang321@163.com (J.J.); Tel.: +86-213-719-5613 (J.J.)
[†] These authors contributed equally to this work.

Citation: Zhang, J.; Zhu, J.; Yang, L.; Li, Y.; Wang, W.; Zhou, X.; Jiang, J. Mapping of the Waxy Gene in *Brassica napus* L. via Bulked Segregant Analysis (BSA) and Whole-Genome Resequencing. *Agronomy* 2023, *13*, 2611. https://doi.org/10.3390/agronomy13102611

Academic Editors: Jiezheng Ying, Zhiyong Li and Chaolei Liu

Received: 25 September 2023
Revised: 8 October 2023
Accepted: 11 October 2023
Published: 13 October 2023

Copyright: © 2023 by the authors. Licensee MDPI, Basel, Switzerland. This article is an open access article distributed under the terms and conditions of the Creative Commons Attribution (CC BY) license (https:// creativecommons.org/licenses/by/ 4.0/).

Abstract: Plant cuticular wax is the covering of the outer layer of the plant. It forms a protective barrier on the epidermis of plants and plays a vital role like a safeguard from abiotic and biotic stresses. In the present study, *Brassica napus* L. materials with and without wax powder were observed. Genetic analysis showed that the separation ratio of waxy plants to waxless plants was 15:1 in the F_2 population, which indicated that the wax powder formation was controlled by two pairs of genes. In order to identify the candidate genes associated with the wax powder trait of *B. napus* L., bulked segregant analysis (BSA) was performed. The homozygous waxy plants, the homozygous waxless plants, and plants from three parents were selected for establishing five DNA pools for genome-wide resequencing. The results of the resequencing showed that the site associated with wax powder trait was located in the region of 590,663–1,657,546 bp on chromosome A08. And 48 single nucleotide polymorphisms (SNPs) were found between the DNA sequences of waxy plants and waxless plants in this region. These SNPs were distributed across 16 gene loci. qRT-PCR analysis was conducted for the 16 candidate genes and three genes (*BnaA08g01070D*, *BnaA08g02130D*, and *BnaA08g00890D*) showed significantly differential expression between waxy and waxless parents. *BnaA08g01070D* and *BnaA08g02130D* were significantly down-regulated in the waxless parent, while *BnaA08g00890D* was significantly up-regulated in the waxless parent. Gene Ontology (GO) and Kyoto Encyclopedia of Genes and Genomes (KEGG) pathway enrichment analyses revealed that the *BnaA08g02130D* gene was enriched in lipid biosynthetic or metabolic processes. All the results in our study would provide valuable clues for exploring the genes involved in wax powder development.

Keywords: *Brassica napus* L; bulked segregant analysis; whole-genome resequencing; waxy; waxless

1. Introduction

A plant cuticle constitutes the interface between the leaf and the external environment, and is an important barrier for plants to cope with biotic and abiotic stresses. The cuticle, comprising wax, is highly hydrophobic and impermeable to water and other solutes. The cuticle is composed of an insoluble membrane impregnated by and covered with soluble waxes such as cutin [1], cutan [2], and other epicuticular waxes. The functions of cuticular wax include sealing any aerial organ's surface to protect the plant against uncontrolled water loss or water stress [3–5], preventing UV damage [6–9], resisting extreme temperatures [10], maintaining a clean and water-repellent surface [11], and protecting plants from the attack of pests and diseases [12,13]. Moreover, the cuticular wax content in *Arabidopsis thaliana* and *Brassica* plants was also closely related to pollen fertility [14,15]. The cuticle also plays a very important role in growth and development.

Plant cuticular wax is composed of very-long-chain fatty acids (VLCFAs) with 20 to 34 carbon atoms and their derivatives (including alkanes, alcohols, aldehydes, ketones, and esters), of which primary alcohols and esters are mainly produced from VLCFAs via the acyl reduction pathway, while alkanes, secondary alcohols, and ketones are mainly formed from VLCFAs through the decarbonylation pathway [16]. The cuticular wax content of a plant is closely related to its developmental status and the surrounding environment. The composition and total amount of epidermal wax differ significantly between different tissues and organs of a plant in the same or different species [17]. The regulation of the formation of cuticular wax involves the coordination of multiple genes [18]. Several genes related to the synthesis, transport, and regulation of wax have been identified and cloned in the model plant *A. thaliana*. The *Arabidopsis lacs1* and *lacs2* mutants showed a significantly reduced cutin content and increased wax content, and the *lacs1* and *lacs2* double mutant displayed an organ fusion phenotype [19]. The *KCS* gene family plays an important role in the catalytic synthesis of VLCFAs. *KCS6* encodes a VLCFA-condensing enzyme. The wax content in the stem of the *kcs6* mutant plant was reduced to only 6% to 7% of that of the wild type plant [16,20]. In *Arabidopsis*, the *CER2* gene encoding a soluble protein was the first cloned wax synthesis-associated regulatory gene [21,22]. *CER2* was located in the nuclei and expressed at the highest level in the early stage of plant development, with tissue-specificity [23]. *CER6* mutation caused defective VLCFAs on the surface of pollens and stems, wax content reduction, and sterile pollen. The over-expression of *CER6* increased the wax content in the epidermis of plants [20]. Moreover, the ethylene response factor (ERF), *WAX INDUCER 1* (*WIN1*) gene was cloned and its over-expression in *A. thaliana* significantly increased the wax content in leaves and stems [24]. The over-expression of the ERF transcription factor genes, *WAXPRODUCTION 1* (*WXP1*) and *WAXPRODUCTION 2* (*WXP2*), cloned from *Medicago sativa*, in *Arabidopsis* resulted in wax accumulation in leaf epidermis [25].

Rapeseed is an oil crop widely planted across the world. It is the largest oil crop in China. Rapeseed plants have a typical cuticle, and the epidermis of leaves and stems is usually covered with wax powder. In a novel dominant glossy mutant (*BnaA.GL*) in *Brassica napus*, the expression of the *BnCER1* gene and other waxy synthesis genes were down-regulated [26]. Over-expression of the *Brassica campestris* lipid transfer protein gene *BraLTP1* in *B. napus* reduced wax deposition in leaves and affected cell division and flower development [27]. The candidate genes controlling the waxy phenotype of cabbage (*Brassica oleracea*) were located on chromosome C08. Genome-wide analysis of the coding and non-coding RNAs in the cuticular wax synthesis process in cabbage revealed that *CER1* and *CER4* genes were down-regulated in the waxless mutant [28]. The waxless trait has been used as a morphological marker in crossbreeding. The presence or absence of wax powder on the surface of plants can be used as a morphological marker in hybrid seed production [29]. Therefore, the wax powder on the surface of plants is of great importance. Bulked segregant analysis (BSA) is a fast and simple method for mapping target traits. The method has the advantage that it does not require the investigation of individual progenies, and the progenies of two opposite traits are pooled. This method was established and a downy mildew resistance gene was mapped in lettuce in a 25 cM chromosomal region [30]. Early BSA applications were based on molecular markers such as RFLP, RAPD, and SSR [31,32]. With the wide application of high-throughput sequencing technology, a large number of SNPs and Indel markers generated by chips, RNAseq, and genome-wide resequencing facilitated the use of BSA in rapidly mapping and identifying important crop traits and genes, such as the peanut branching habit [33], citrus polyembryonic loci [34], Asian pear skin color [35], and watermelon dwarfism gene [36].

In the present study, *Brassica napus* L. materials without wax powder were found. Waxless mutant plants showed significant reduction or absence of wax powder on leaves and stems. To find out whether the wax powder trait is dominant or recessive, and whether the trait is controlled by a single gene or multiple genes, we performed genetic analyses. Moreover, in order to map the genes related to wax powder development in *B. napus*, BSA-based genome-wide resequencing was performed. The resequencing data was analyzed and the candidate regions associated with wax powder trait were localized to chromosomes.

The SNPs were found between the DNA sequences of waxy plants and waxless plants and the corresponding candidate genes were found. qRT-PCR analysis was conducted to explore the expression patterns of these candidate genes. Meanwhile, GO and KEGG pathway enrichment analyses were performed for gene function prediction. The results would provide insights for exploring the genes involved in wax powder development.

2. Materials and Methods

2.1. Plant Materials

In the current study, B. napus male sterile line P202001 was used as the maternal parent and the temporary maintainer line P202002 was used as the male parent, and their hybrid offspring was male sterile line (Figure 1). The male sterile line was further hybridized with the waxless restorer line P202003 to obtain F_1. F_1 plants were self-inbred to obtain F_2. In the F_2 population, the ratio of waxy plants to waxless plants was 15:1, and several waxy plants were selected and self-inbred to obtain F_3. According to the genetic laws and the phenotypic situation, four kinds of groups appeared in F_3. In group I, all the plants were waxy plants. In group II, the ratio of waxy plants to waxless plants was 15:1. In group III, the ratio of waxy plants to waxless plants was 3:1. In group IV, all the plants were waxless plants. Several waxy plants from group III were selected and self-inbred to obtain F_4. In the F_4 population, the ratio of waxy plants to waxless plants was 3:1. Several waxy plants were selected and self-inbred to obtain F_5. Based on the genetic laws and the phenotypic situation, F_5 was divided into two groups. In group I, all the plants were waxy plants. In group II, the ratio of waxy plants to waxless plants was 3:1. Several waxy plants from group II were selected and self-inbred to obtain F_6. In the F_6 population, two kinds of groups occurred. Group I was numbered '17-4107', and all plants in group I had wax powder. Group II was numbered '17-4105', which contained waxy plants and waxless plants with a ratio of 3:1. All these plant materials were planted in the experimental farm of Zhuanghang comprehensive experimental station of Shanghai Academy of Agricultural Sciences. And all the crossings/selfings were conducted in the farm. During seedling stage, leaves of 50 waxless plants were, respectively, harvested from the '17-4105' group for the construction of DNA pool of waxless plants. Meanwhile, leaves of 50 waxy plants were, respectively, harvested from the '17-4107' group for the construction of DNA pool of waxy plants. In addition, leaves of 20 plants were, respectively, harvested from parents P202001, P202002, and P202003 for the construction of three parental DNA pools. All the leaf samples were quickly frozen in liquid nitrogen and stored at $-80\ °C$.

2.2. Observation of Wax Powder Trait on Waxy Plants and Waxless Plants

The wax power trait of waxy plants and waxless plants was observed in the flowering and carob stages. Fresh leaves from the waxy plant and waxless plants were fixed on specimen holders with glue. The samples were transferred to a preparation chamber under vacuum for coating (SD-900, BoYuan, Beijing, China). Then, the wax powder layer on the surface of the leaves was scanned and photographed using scanning electron microscope (TM4000plus, Hitachi, Tokyo, Japan).

2.3. Sample Pooling and Genotyping

Take 0.2 g of leaf tissue from each plant for DNA extraction. The leaf DNAs of 50 waxy plants were mixed in equal amounts to construct a DNA pool of waxy plants. The leaf DNAs of 50 waxless plants were also mixed in equal amounts to construct a DNA pool of waxless plants. In addition, three parental DNA pools were constructed. A TruSeq DNA LT Sample Prep kit was used for the library construction of the two extreme DNA pools and three parent DNA pools. Genome-wide resequencing at $30\times$ coverage was performed using Illumina Hiseq Xten (Shanghai OE Biotech Co., Ltd, Shanghai, China). Quality control of raw sequencing data was performed using Trimmomatic v0.36 [37]. Clean reads were aligned and compared to Brassica napus genome (http://www.genoscope.cns.fr/brassicanapus/data/Brassica_napus_v4.1.chromosomes.fa.gz, accessed on 29 June 2018) using BWA v0.7.17 [38]; then, SAMtools v1.6 [39] was used to convert the format of the obtained data. Subsequently, the PCR repeats were removed using the Picard module of GATK v4.0.2.1 (GATK). Based on the alignment of the sample sequence with the reference genome, the SNP and InDel sites in the samples were detected using the SAMtools v1.6 mpileup module with default parameters. SNPs and InDels were annotated using snpEff v4.1 [40].

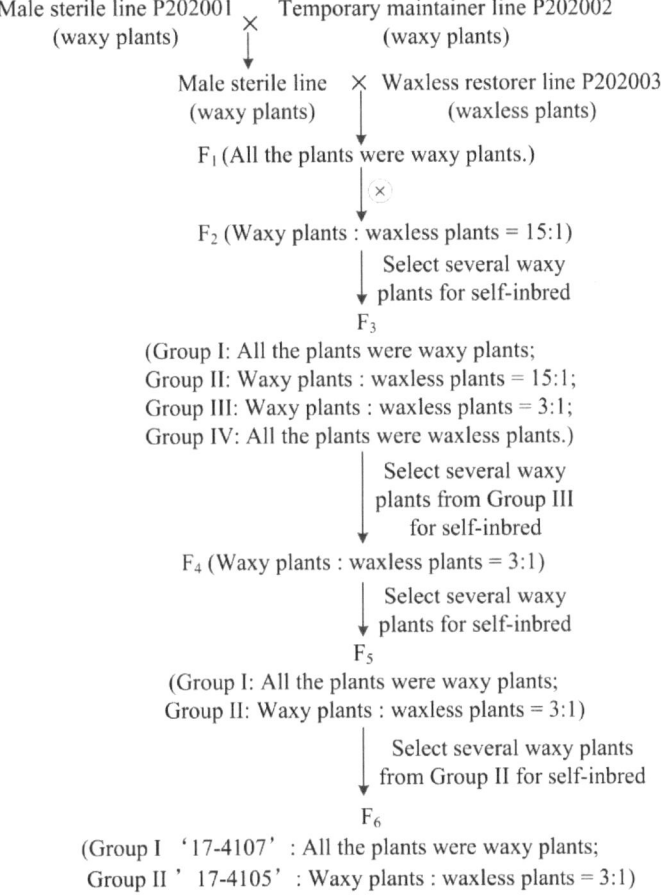

Figure 1. The genetic analysis chart of wax powder trait.

2.4. Localization and Prediction of Candidate Genes Associated with Wax Powder Development

Based on the genotyping results, the homozygous polymorphic SNPs between parents were screened. The parents, P202001, P202002, and P202003, were used as references. The frequency of the SNPs (SNP-index) at each polymorphic site of the two progeny pools was calculated according to the method reported by Takagi et al. [41]. For SNPs in progeny, the reads of mutant-type parental origin were determined based on parental genotypes, and the SNP-indexes were calculated. The average distribution of the SNP-index was estimated with 1-Mb window size and 10-kb step and was plotted to generate an SNP-index curve. The peak in the linkage region became distinct after the noise was partially eliminated. The SNP-index plot was generated for the two progeny pools using the absolute value of the ΔSNP-index that was the difference in the SNP-index between the two pools at the segregation position. The 95% and 99% confidence levels were chosen as screening thresholds, and areas above the threshold were possible phenotypic linkage regions. By comparing the type of the SNPs in the candidate region with that of the corresponding position of the reference genome, the mutant type of SNPs in parental lines and waxless and waxy pools was estimated. Candidate genes with the same type of SNPs in a parent and the progeny pool with a similar phenotype were selected. The candidate genes were screened primarily through homology-based annotation using NR (Non-Redundant Protein Sequence), GO (Gene Ontology), and KEGG (Kyoto Encyclopedia of Genes and Genomes) databases. The BRAD database (http://brassicadb.org/brad/, accessed on 1 September 2023) and TAIR database (https://www.arabidopsis.org, accessed on 1 September 2023) were used to analyze and predict the functions of candidate genes.

2.5. Expression Analysis of Candidate Genes

At the 12-leaf stage of rapeseed seedlings, the amount of wax powder on the waxy plants was large and the wax powder trait was very easy to be observed, so the waxy and waxless plants could be clearly distinguished. At the 12-leaf stage, the leaves of P202001 and P202003 were sampled and continued on day 7, 14, and 21 after the first sampling. And total RNAs of these leaf samples were extracted using RNeasy Plant Mini Kit (Qiagen). The concentration and OD260/OD280 were measured using a NanoDrop 2000 spectrophotometer (Thermo Scientific, Waltham, MA, USA). The RNA integrity was analyzed via agarose gel electrophoresis. The RNA was reverse transcribed into cDNA using HiScript II Q RT SuperMix for qPCR (+gDNA wiper) (Vazyme, R223-01, Nanjing, China). The reverse transcription reaction procedure included 2 steps. Step 1: A reaction including 0.5 µg total RNA, 2 µL 4× gDNA wiper mix, and 8 µL nuclease-free H_2O was prepared and placed at 42 °C for 2 min. Step 2: In this reaction, 2 µL 5 × HiScript II Q RT SuperMix II^a was added, and the reaction was incubated at 25 °C for 10 min, 50 °C for 30 min, and 85 °C for 5 min. The total volume of the reaction was 10 µL. After reverse transcription, 90 µL nuclease-free H_2O was added, and the solution was stored in a −20 °C freezer. The relative expression levels of the selected candidate genes in the leaves of *B. napus* were analyzed using qRT-PCR. The PCR primers (Table S1) were designed using the software Primer Premier 5 and were synthesized by Shanghai Sunny Biotech Co., Ltd. (Shanghai, China). The amplification reaction was prepared by adding 0.5 µL (about 100 ng) cDNA, 10 µL SYBR Premix Ex Taq II (2×), 0.8 µL 10 µmol·L^{-1} each of the forward and reverse primers, and ddH_2O to a final volume of 20 µL. The amplification was conducted using the real-time PCR thermo cycler Bio-Rad CFX96. The thermo cycles were pre-denaturing at 95 °C for 2 min, followed by 40 cycles of 95 °C for 10 s and 60 °C for 30 s. The qRT-PCR results were analyzed using the Bio-Rad CFX96 Manager, and the threshold line was automatically set using the software. The *B. napus* gene *BnActin7* (GenBank accession number EV086936) was used as an internal reference. The relative expression levels of the target genes were quantified by using the $2^{-\Delta\Delta Ct}$ method [42]. Each experiment was conducted in three biological replicates, and the same sample was performed in three technical replicates. The Student's *t*-test was used to test the probability of significant differences between samples.

3. Results

3.1. Genetic Analysis and Morphological Observation of the Waxless Trait of Rapeseed

In this study, genetic analysis showed that the separation ratio of waxy plants to waxless plants was 15:1 in the F_2 population. And in F_3, when a pair of genes was homozygous, the separation ratio of waxy powder plants and waxless powder plants was 3:1. The whole genetic analysis indicated that the wax powder formation was controlled by two pairs of genes and the waxy phenotype was dominant over the waxless phenotype (Figure 1). In contrast to waxy plants, waxless plants had no wax on the surface of stalks, leaves, floral buds, and siliques (Figure 2A,B). The waxless trait was distinctly visible throughout the growth period. The results of the scanning electron microscopy showed there were abundant and intact wax crystals on the leaf blade of waxy parent P202001.

However, the leaf epidermis of waxless parent P202003 was smooth, without wax crystals (Figure 2C,D).

Figure 2. Plant appearance of waxless P202003, waxy P202001, and the corresponding micro-structure of leaf blade. (**A**) Plants at the beginning of flowering, waxless P202003 (left) and waxy P202001 (right). (**B**) Plants in the carob stage, waxless P202003 (left) and waxy P202001 (right). (**C**) Microstructure of waxless P202003 blade ventral. (**D**) Microstructure of P202001 blade ventral, wax crystals were dense, with high proportion of tubular-like wax crystals under high magnification (5000×), the red arrows indicate the wax powder crystal structures, and all similar structures in the figure are wax powder crystal structures. Scale bar = 10 μm.

3.2. RNA-Seq Analysis

To rapidly map the candidate genes associated with the wax powder trait, BSA-Seq analysis was performed. Five DNA pools were constructed, including three parental DNA pools (P202001, P202002 and P202003), waxless plants DNA pool (waxless plants from '17-4105' group), and waxy plants DNA pool (waxy plants from '17-4107' group). These five DNA pools were subjected to whole-genome resequencing. As shown in Table 1, after filtering the raw data, the clean data of each DNA pool ranged from 28,936 to 56,780 million bp. High-quality sequencing data (Q20 $\geq$ 97.33% and Q30 $\geq$ 91.97%) were obtained with the GC $\geq$ 36.37%. The numbers of clean reads ranged from 198.2 to 383.2 million. The clean reads were then mapped to the *B. napus* reference genome (http://brassicadb.cn/, accessed on 29 June 2018) with a mapping rate of 94.9% to 99.1%. The mean coverage ranged from 29.2× to 51.3×. The mean mapping quality ranged from 42.2 to 43.3. Compared to the *B. napus* reference genome, a total of 21,826,671 SNPs and 2,630,037 InDel loci were detected in the five pools. The numbers of SNPs and InDels in the upstream and downstream regions of genes and the intergenic regions were higher than those in other regions of chromosomes. The total number of SNPs and InDels detected in different pools was similar (Table 2).

Table 1. The alignment statistics results with the reference genome for all samples.

Sequencing Information	17-4105	17-4107	P202001	P202003	P202002
HQ clean data (bp)	33,302,724,403	34,577,724,967	28,936,047,349	37,904,376,134	56,780,469,639
Q20 (%)	97.33%	97.33%	97.56%	98.70%	98.74%
Q30 (%)	92.01%	91.97%	92.53%	96.00%	96.10%
GC content (%)	36.37%	36.47%	37.20%	37.82%	37.64%
HQ clean reads number and percent (%)	228,591,684 (87.97%)	237,268,442 (88.16%)	198,207,190 (89.67%)	255,719,126 (93.6%)	383,242,398 (93.83%)
Total mapped reads and mapped rate (%)	202,613,330 (98.69%)	210,587,454 (99.12%)	171,778,467 (96.74%)	202,899,761 (94.94%)	296,825,463 (96.27%)
Mean coverage	34.4175×	35.7912×	29.2245×	35.0443×	51.3133×
Mean mapping quality	43.316	43.2226	43.2545	42.2296	43.0776

3.3. Screening and Analysis of Candidate Genes

The distribution of the SNP-index on the chromosomes of the waxy and waxless DNA pools was estimated. The absolute value of the difference in the number of SNP-indexes between the two pools, which was |ΔSNP-index|, was calculated. By setting 95% and 99% confidence thresholds, a candidate region between the 590,663 and 1,657,546 bp on chromosome A08 was identified (Figure 3). There were 48 SNPs distributed across 16 gene loci in this region.

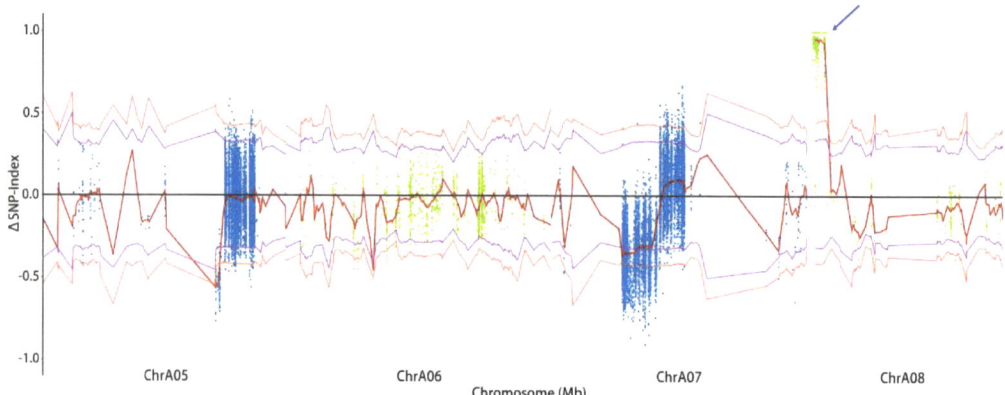

Figure 3. Map of the ΔSNP-index of waxless and waxy plants after screening with parental homozygous polymorphic loci. Note: The blue arrow indicates the candidate region (590,663–1,657,546 bp) on chromosome A08 with an association threshold higher than 1%. (1) Green and blue dots both represent the SNP-index or ΔSNP-index. (2) The red line represents the fitted line of the SNP-index or ΔSNP-index after window sliding. (3) The purple lines represent the 95% confidence threshold. (4) The orange lines represent the 99% confidence threshold.

To predict the biological functions of the 16 potential candidate genes, GO and KEGG pathway enrichment analyses were performed. The annotated genes were divided into three major functional categories, namely, biological processes (BP), cellular components (CC), and molecular functions (MF, Figure 4). The results showed that *BnaA08g01250D*, *BnaA08g01020D*, and *BnaA08g02130D* were related to "metabolic process (GO:0008152)"; *BnaA08g02130D*, *BnaA08g013300D*, and *BnaA08g01270D* were related to "ATP binding, glucose binding and Ca^{2+} binding activity (GO:0005488)", respectively; *BnaA08g00820D* and *BnaA08g02130D* were associated with biological regulation (GO:0065007)"; *BnaA08g01020D* and *BnaA08g1000D* were annotated as "catalytic activity (GO:0003824)" and *BnaA08g00820D* was annotated as "regulation of biological process (GO:0050789)".

Table 2. Statistics of SNPs and Indels.

Sample	Type	Stopgain	Stoploss	Missense	Splicing	CDS	Synonymous	Upstream	Downstream	Intronic	Intergenic	Other	Total
17-4105	SNPs	12,872	10,720	316,979	41,969	608,670	264,950	1,871,755	787,613	102,305	1,087,082	3149	4,499,394
	Indels	336	250	/	10,141	19,694	/	305,119	109,734	18,138	81,927	72	544,753
17-4107	SNPs	12,818	10,706	315,030	41,634	605,295	263,636	1,858,996	781,415	101,499	1,072,178	3105	4,461,017
	Indels	339	247	/	10,106	19,657	/	303,705	109,008	17,942	80,938	71	541,356
P202001	SNPs	11,996	9940	293,070	39,015	561,866	244,017	1,733,726	730,785	95,315	1,028,954	2843	4,189,661
	Indels	328	238	/	9132	18,051	/	280,020	100,232	16,417	76,408	61	500,260
P202002	SNPs	12,619	10,618	306,490	40,551	588,424	255,703	1,808,653	761,545	96,946	1,037,526	2994	4,333,645
	Indels	340	235	/	9870	19,384	/	295,891	105,899	17,387	78,665	65	527,096
P202003	SNPs	12,358	10,418	303,309	40,212	583,115	254,081	1,797,291	758,743	97,673	1,065,920	2949	4,342,954
	Indels	344	250	/	9706	18,749	/	287,494	104,172	17,266	79,185	65	516,572

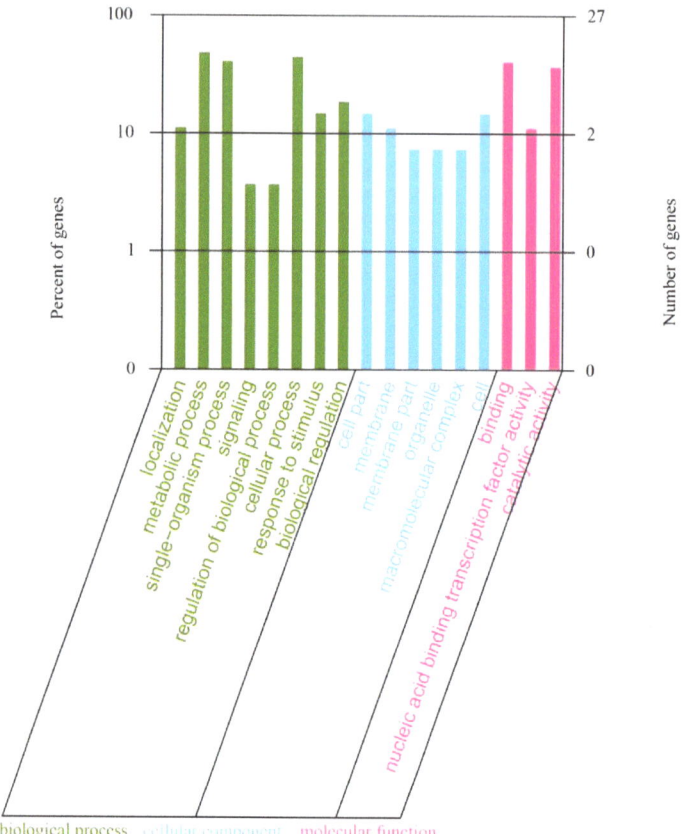

Figure 4. GO annotation of identified 16 candidate genes possibly involved in wax powder development. X-axis shows GO terms; y-axis shows gene number and percentage in log scale.

KEGG pathway analysis was conducted to investigate whether these candidate genes were involved in special pathways. For these candidate genes, the most significantly enriched pathway were metabolic pathways, proteasome, pyruvate metabolism, biosynthesis of secondary metabolites, and citrate cycle.

The homologous genes of these 16 candidate genes in Arabidopsis were found and annotated based on the Brassicaceae Database (BRAD, http://brassicadb.cn/#/, accessed on 1 September 2023) and The Arabidopsis Information Resource (TAIR, https://www.arabidopsis.org, accessed on 1 September 2023) (Table 3). According to previous reports, *BnaA08g00820D*, which was homologous to *AT1G54200*, was likely to regulate the plant biomass, stress tolerance, seed weight, and yield. *BnaA08g00890D* was likely to regulate reactive oxygen species homeostasis and leaf senescence. *BnaA08g01000D* encoded alpha5 subunit of 20s proteosome involved in protein degradation and RNA degradation. *BnaA08g01010D* may refer to 26S proteasome AAA-ATPase subunit RPT1a. *BnaA08g01020D* was *SRF6a* gene, which might be involved in stress-related processes, including responses to light, heat, and plant immunity. *BnaA08g01030D* encoded Calcineurin-like metallophosphoesterase superfamily protein. *BnaA08g01070D* was *ETHE1* gene and encoded a sulfur dioxygenase essential for embryo and endosperm development. *BnaA08g01170D* was a member of the plant TLP family, with an unknown function. *BnaA08g01250D* was *mMDH1*, which encoded a mitochondrial malate dehydrogenase. It was involved in leaf respiration and altered photorespiration and plant growth. *BnaA08g01260D* was *TCP3*, which was involved in heterochronic regulation of leaf differentiation. *BnaA08g01270D* was the *ATNCL* gene, which encoded a Na^+/Ca^{2+} exchanger-like protein that participated in the maintenance of Ca^{2+} homeostasis. The previous study showed it was involved in salt stress in Arabidopsis. *BnaA08g01330D* encoded putative DUF616 with an unknown function. *BnaA08g01350D* encoded ubiquitin-conjugating enzyme E2. *BnaA08g02130D* was reported to be involved in glucose-ethylene crosstalk. *BnaA08g02150D* and *BnaA08g02160D* were homologous to *AT1G50420*. Their other names were *SCL-3* or *SCL3*, which refer to regulatory network in *Arabidopsis* roots.

Table 3. Information on candidate genes identified by BSA-seq.

Candidate Genes	Arabidopsis Genes	Functional Description	References
BnaA08g00820D	AT1G54200	Regulation of *Big Grain1* in rice and *Arabidopsis* increases plant biomass, stress tolerance, seed weight, and yield.	[43,44]
BnaA08g00890D	AT1G54115	CCX1, a putative cation/Ca^{2+} exchanger, participates in regulation of reactive oxygen species homeostasis and leaf senescence.	[45]
BnaA08g01000D	AT1G53850	Encodes alpha5 subunit of 20s proteosome involved in protein degradation and RNA degradation.	[46]
BnaA08g01010D	AT1G53750	26S proteasome AAA-ATPase subunit RPT1a (RPT1a) mRNA.	[46,47]
BnaA08g01020D	AT1G53730	SRF6. SRF6 has a role in the defense against pathogenic fungi. SRF6 gene may be involved in stress-related processes, including responses to light, and heat. The potato SRF family gene, StLRPK1, is involved in plant immunity.	[48,49]
BnaA08g01030D	AT1G53710	Calcineurin-like metallo-phosphoesterase superfamily protein, GPI, anchor biosynthetic process.	[50]
BnaA08g01070D	AT1G53470	Other name: ETHE1, ETHE1-LIKE. Arabidopsis ETHE1 encodes a sulfur dioxygenase that is essential for embryo and endosperm development. (Holdorf et al., 2012). Deficiency of the mitochondrial sulfide regulator ETHE1 disturbs cell growth, glutathione level and causes proteome alterations outside mitochondria (Sahebekhtiari et al., 2019).	[51,52]
BnaA08g01170D	AT1G53320	Member of plant TLP family. TLP7 is tethered to the PM but detaches upon stimulus and translocates to the nucleus.	[53]
BnaA08g01250D	AT1G53240	mMDH1 encodes a mitochrondrial malate dehydrogenase. It is expressed at higher levels than the other mitochrondrial isoform mMDH2 (At3G15020) according to transcript and proteomic analyses. It has copper-ion-binding and L-malate dehydrogenase activity. In *Arabidopsis*, mitochondrial malate dehydrogenase reduces leaf respiration and alters photorespiration and plant growth.	[54]
BnaA08g01260D	AT1G53230	Other name: TCP3. Encodes a member of a recently identified plant transcription factor family that includes Teosinte branched 1, Cycloidea 1, and proliferating cell nuclear antigen (PCNA) factors, PCF1 and 2. Regulated by miR319. Involved in heterochronic regulation of leaf differentiation.	[55]
BnaA08g01270D	AT1G53210	Other name: ATNCL. Encodes a Na^+/Ca^{2+} exchanger-like protein that participates in the maintenance of Ca^{2+} homeostasis. It is involved in salt stress in Arabidopsis. ATNCL may be a novel Ca^{2+} transporter in higher plants (Wang et al., 2012).	[56]
BnaA08g01330D	AT1G53040	tRNA (met) cytidine acetyltransferase, putative (DUF616). Protein of unknown function DUF616.	[57]
BnaA08g01350D	AT1G53020	Ubiquitin-conjugating enzyme E2.	[58]
BnaA08g02130D	AT1G50460	It has ATP binding, glucose-binding and hexokinase activities and participates in lipid biosynthesis and protein metabolism. Involved in glucose–ethylene crosstalk.	[59]
BnaA08g02150D *BnaA08g02160D*	AT1G50420	Other name: SCARECROW-LIKE 3; SCL-3 or SCL3. SCL3-induced regulatory network in Arabidopsis thaliana roots.	[60]

3.4. Specific Expression of Candidate Genes

To explore whether the 16 candidate genes were associated with wax powder development, qRT-PCR was used to detect their expression levels in the leaves of waxy parent P202001 and waxless parent P202003 (Figure 5). The qRT-PCR results showed that the expression levels of *BnaA08g01070D* and *BnaA08g02130D* in the waxy parent P202001 were significantly higher than those in the waxless parent P202003. The expression level of *BnaA08g00890D* in the waxless parent P202003 was significantly higher than that in the waxy parent P202001 (6.15-fold). The expression levels of these three genes were further analyzed at four time points when the wax powder was clearly visible (Figure 6). The expression levels of *BnaA08g01070D* and *BnaA08g02130D* in the leaves of the waxy parent P202001 were consistently higher than in the leaves of the waxless parents P202003. Their expression levels gradually increased, reaching their highest levels at day 14 and decreasing at day 21. *BnaA08g00890D* was up-regulated in the leaves of the waxy parent P202001 at day 0, day 7, and day 14. At day 21, its expression level was highest in the waxless parent, but the lowest in the waxy parent.

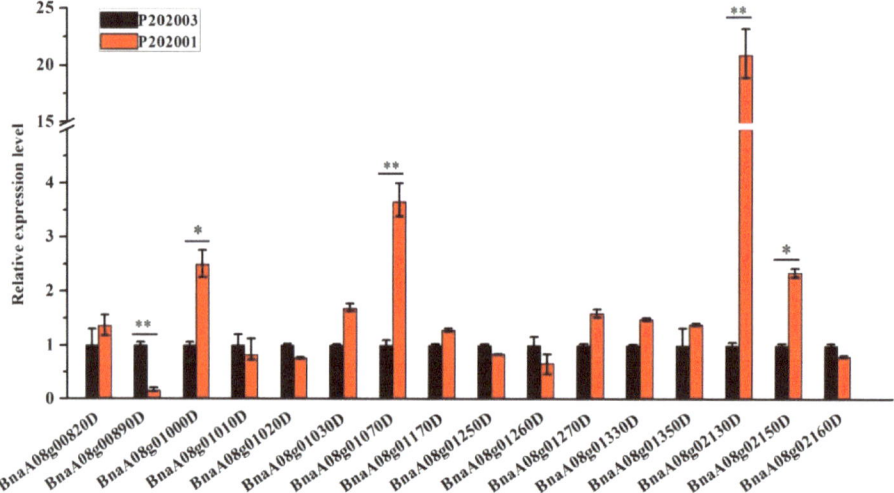

Figure 5. Expression profile of 16 candidate genes in the leaves of waxless parent P202003 and waxy parent P202001, revealed via qRT-PCR. Data represent means of three replicates ± SD. *, $p < 0.05$, **, $p < 0.01$. Student's *t*-test.

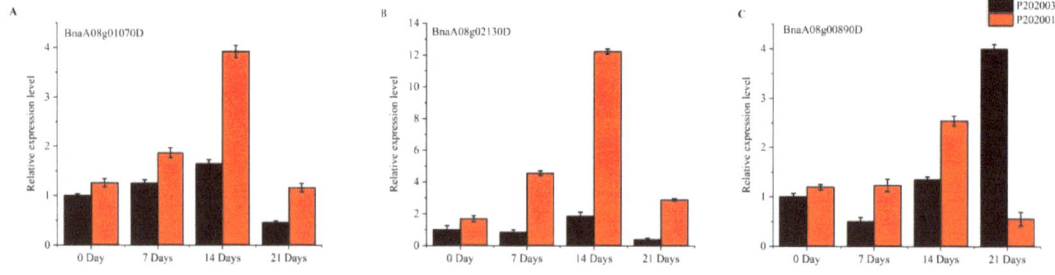

Figure 6. Expression patterns of three candidate genes in the leaves from the waxless and the waxy parents at four time points when the wax powder was clearly visible. P202001, the waxy parent. P202003, the waxless parent. The relative expression levels of *BnaA08g01070D* (**A**), *BnaA08g02130* (**B**), and *BnaA08g00890D* (**C**) by qRT-PCR.

4. Discussion

Wax powder is a mixture of organic compounds, which is widely distributed in the plant epidermis to prevent direct contact between plant tissues and the outside world, thereby preventing strong light radiation, high temperature burning, drought, frost damage, air pollution, mechanical damage, and other abiotic stress injuries [26,61–65]. Wax powder plays an important role in improving the stress resistance of plants. Drought increased the amount and content of cuticular waxes in rapeseed [26,64]. Cuticular wax is a major factor to defend crop plants from extreme ultraviolet (UV) radiation. It is reported that changes in cuticular wax production and gas exchange depend on high UV-B radiation in plants [63]. And cuticular waxes accumulations have an effect on the rate of gas exchange in the leaf surface of canola (*Brassica napus*) [61]. In addition, it was reported that cuticular wax could increases drought tolerance in *Brassica napus* [65]. Therefore, cuticular wax production and accumulation is helpful to improve the stress tolerance and indirectly improve crop quality and yield.

The wax powder trait is easy to observe, and waxy plants can be visually distinguished from waxless plants from the seedling stage, and the wax powder trait can be stably inherited under natural conditions, so the wax powder trait can be used as morphological marker in hybrid production. In rapeseed, the recessive sterility or fertility traits can be marked with wax powder trait, and the fertility of plants can be judged by observing the presence or absence of wax powder at the seedling stage, then 50% of the fertile plants can be removed. Using wax powder trait as morphological marker can quickly simplify the two-line hybrid breeding process.

Although wax powder synthesis pathway has been identified and the related genes have been cloned in *Arabidopsis*, the mechanism of wax powder formation is not well understood in *Brassica*. Most studies on *Brassica* have focused on the analysis of genetic models, and there were few in-depth studies [29,66]. Zhang et al. [67] closely mapped the *BrWax1* gene in a recessive waxless mutant of Chinese cabbage, which is located in an 86.4 kb region of the A1 linkage group. There were 15 annotated genes in this region, of which *Bra013809* was a homologous gene of *Arabidopsis CER2*, and an insertion mutation was found in its sequence. The expression level of *Bra013809* was significantly reduced in the waxless mutant. The candidate gene controlling the cabbage waxy phenotype was located on the C08 chromosome. Genome-wide analysis of the coding and non-coding RNAs in the process of cabbage wax synthesis showed that the *CER1* and *CER4* genes were down-regulated in the waxless materials [28]. BSA coupled with RNA-Seq and appropriate statistical procedures (bulked segregant RNA-Seq or BSR-Seq) were used to clone the *glossy3* (*gl3*) gene of maize, which was responsible for epicuticular wax accumulation on juvenile leaves [26]. Mutants of the glossy loci exhibit altered accumulation of epicuticular waxes on juvenile leaves. RNA sequencing performed on bulked RNA from blueberry progenies with and without wax power helped in the identification of genes related to the protective waxy coating on blueberry fruit [68].

In the current study, *B. napus* L. material without wax powder was found. In order to identify the candidate genes involved in wax power development, bulked segregant analysis (BSA) was performed. The results indicated that the 590,663 to 1,657,546 bp region on the chromosome A08 might be associated with wax power trait. There were 48 SNPs distributed across 16 gene loci in this region. Because waxy plants and waxless plants differed only in the presence or absence of wax powder, we reasoned that these genes might be the candidate genes associated with wax powder development. The qRT-PCR analysis revealed that the expression levels of the candidate genes, *BnaA08g01070D* and *BnaA08g02130D*, were significantly more up-regulated in the leaves of waxy parent P202001 than that in the leaves of waxless parent P202003, while the expression levels of candidate gene *BnaA08g00890D* were significantly down-regulated in the leaves of waxy parent P202001. Therefore, *BnaA08g01070D*, *BnaA08g02130D*, and *BnaA08g00890D* might be involved in wax power development.

5. Conclusions

In the present study, B. napus L. materials with and without wax powder were observed. Through genetic analysis, it was found that the wax powder trait was controlled by two pairs of genes. In order to identify the candidate genes involved in wax powder development in B. napus L., bulked segregant analysis (BSA) was performed. The resequencing results showed that the wax powder trait-associated loci might be located in the region of 590,663~1,657,546 bp on chromosome A08. There were 48 single-nucleotide polymorphisms (SNPs) found in this region between the DNA sequences of waxy plants and waxless plants. And these SNPs were distributed across 16 gene loci. qRT-PCR analysis was conducted for the 16 candidate genes and three genes (*BnaA08g01070D*, *BnaA08g02130D*, and *BnaA08g00890D*) showed significantly differential expression between the leaves from waxless and waxy parents. GO and KEGG pathway enrichment analysis revealed that the *BnaA08g02130D* gene was enriched in lipid biosynthetic or metabolic processes. All the results in our study would provide valuable clues for exploring the genes involved in wax powder development.

Supplementary Materials: The following supporting information can be downloaded at https://www.mdpi.com/article/10.3390/agronomy13102611/s1. Table S1 qRT-PCR primers in this study.

Author Contributions: J.Z. (Junying Zhang), J.Z. (Jifeng Zhu) and X.Z. designed the research and performed the experiments. J.Z. (Junying Zhang) and J.J. analyzed the sequencing data and finished the manuscript. L.Y., Y.L. and W.W. performed the field experiment. All authors have read and agreed to the published version of the manuscript.

Funding: This work was supported by grant from the Shanghai Agriculture Applied Technology Development Program, China (Grant No.T20210219).

Data Availability Statement: The BSA resequencing data is unavailable due to privacy. We may use the resequencing data for further study.

Conflicts of Interest: The authors declare no conflict of interest.

References

1. Holloway, P.J. The chemical constitution of plant cutins. In *The Plant Cuticule*; Culter, D.F., Alvin, K.L., Eds.; Academic Press: London, UK, 1982; pp. 45–85.
2. Tegelaar, E.W.; Leeuw, J.W.D.; Largeau, C.; Derenne, S.; Schulten, H.R.; Müller, R.; Boon, J.J.; Nip, M.; Sprenkels, J.C.M. Scope and limitations of several pyrolysis methods in the structural elucidation of a macromolecular plant constituent in the leaf cuticle of *Agave americana* L. *J. Anal. Appl. Pyrolysis.* **1989**, *15*, 29–54. [CrossRef]
3. Kosma, D.K.; Bourdenx, B.; Bernard, A.; Parsons, E.P.; Lu, S.; Joubes, J.; Jenks, M.A. The impact of water deficiency on leaf cuticle lipids of Arabidopsis. *Plant Physiol.* **2009**, *151*, 1918–1929. [CrossRef]
4. Schreiber, L. Transport barriers made of cutin, suberin and associated waxes. *Trends Plant Sci.* **2010**, *15*, 546–553. [CrossRef]
5. Chen, G.; Komatsuda, T.; Ma, J.F.; Li, C.; Yamaji, N.; Nevo, E. A functional cutin matrix is required for plant protection against water loss. *Plant Signal. Behav.* **2014**, *6*, 1297–1299. [CrossRef]
6. Kinnunen, H.; Huttunen, S.; Laakso, K. UV- absorbing compounds and waxes of Scots pine needles during a third growing season of supplemental UV-B. *Environ. Pollut.* **2001**, *112*, 215–220. [CrossRef]
7. Skorska, E.; Szwarc, W. Influence of UV-B radiation on young triticale plants with different wax cover. *Biol. Plantarum.* **2007**, *51*, 189–192. [CrossRef]
8. Wang, S.; Duan, L.; Eneji, A.E.; Li, Z. Variations in growth, photosynthesis and defense system among four weedspecies under increased UV-B radiation. *Acta Botanica Sinica.* **2010**, *49*, 621–627.
9. Ni, Y.; Xia, R.E.; Li, J.N. Changes of epicuticular wax induced by enhanced UV-B radiation impact on gas exchange in Brassica napus. *Acta Physiol. Plant.* **2014**, *36*, 2481–2490. [CrossRef]
10. Yeats, T.H.; Rose, J.K. The formation and function of plant cuticles. *Plant Physiol.* **2013**, *163*, 5–20. [CrossRef]
11. Neinhuis, C.; Barthlott, W. Characterization and distribution of water- repellent, self- cleaning plant surfaces. *Ann. Bot.* **1997**, *79*, 667–677. [CrossRef]
12. Castillo, L.; Díaz, M.; González-Coloma, A.; Alonso-Paz, E.; Bassagoda, M.J.; Rossini, C. Clytostoma callistegioides (Bignoniaceae) wax extract with activity on aphid settling. *Phytochemistry* **2010**, *71*, 2052–2057. [CrossRef] [PubMed]
13. Hansjakob, A.; Bischof, S.; Bringmann, G.; Riederer, M.; Hildebrandt, U. Very-longchain aldehydes promote in vitro prepenetration processes of Blumeria graminis in a dose- and chain length-dependent manner. *New Phytol.* **2010**, *188*, 1039–1054. [CrossRef] [PubMed]

14. Koch, K.; Ensikat, H.J. The hydrophobic coatings of plant surfaces: Epicuticular wax crystals and their morphologies, crystallinity and molecular self- assembly. *Micron* **2008**, *39*, 759–772. [CrossRef]
15. Koch, K.; Bhushan, B.; Barthlott, W. Multifunctional surface structures of plants: An inspiration for biomimetics. *Prog. Mater. Sci.* **2009**, *54*, 137–178. [CrossRef]
16. Millar, A.A.; Clemens, S.; Zachgo, S.; Giblin, E.M.; Taylor, D.C.; Kunst, L. CUT1, an *Arabidopsis* gene required for cuticular wax biosynthesis and pollen fertility, encodes a very-long-chain fatty acid condensing enzyme. *Plant Cell.* **1999**, *11*, 825–838. [CrossRef]
17. Reicosky, D.A.; Hanover, J.W. Physiological effects of surface waxes: I. Light reflectance for glaucous and nonglaucous picea pungens. *Plant Physiol.* **1978**, *62*, 101–104. [CrossRef]
18. Suh, M.C.; Samuels, A.L.; Jetter, R.; Kunst, L.; Pollard, M.; Ohlrogge, J.; Beisson, F. Cuticular lipid composition, surface structure, and gene expression in *Arabidopsis* stem epidermis. *Plant Physiol.* **2005**, *139*, 1649–1665. [CrossRef]
19. Weng, H.; Molina, I.; Shockey, J.; Browse, J. Organ fusion and defective cuticle function in a laes1 lacs2 double mutant of *Arabidopsis. Planta* **2010**, *231*, 1089–1100. [CrossRef]
20. Fiebig, A.; Mayfield, J.A.; Miley, N.L.; Chau, S.; Fischer, R.L.; Preuss, D. Alterations in CER6, a gene identical to CUT1, differentially affect long-chain lipid content on the surface of pollen and stems. *Plant Cell.* **2019**, *12*, 2001–2008. [CrossRef]
21. Negruk, V.; Yang, P.; Subramanian, M.; Mcnevin, J.P.; Lemieux, B. Molecular cloning and characterization of the *CER2* gene of *Arabidopsis thaliana. Plant J.* **2010**, *9*, 137–145. [CrossRef]
22. Xia, Y.; Nikolau, B.J.; Schnable, P.S. Cloning and characterization of *CER2*, an *Arabidopsis* gene that affects cuticular wax accumulation. *Plant Cell.* **1996**, *8*, 1291–1304. [PubMed]
23. Xia, Y.; Nikolau, B.J.; Schnable, N.P.S. Developmental and hormonal regulation of the *Arabidopsis CER2* gene that codes for a nuclear- localized protein required for the normal accumulation of cuticular waxes. *Plant Physiol.* **1997**, *115*, 925–937. [CrossRef] [PubMed]
24. Broun, P.; Poindexter, P.; Osborne, E.; Jiang, C.Z.; Riechmann, J.L. *WIN1*, a transcriptional activator of epidermal wax accumulation in *Arabidopsis. Proc. Natl. Acad. Sci. USA* **2004**, *101*, 4706–4711. [CrossRef] [PubMed]
25. Zhang, J.Y.; Broeckling, C.D.; Sumner, L.W.; Wang, Z.Y. Heterologous expression of two Medicago truncatula putative ERF transcription factor genes, WXP1 and WXP2, in Arabidopsis led to increased leaf wax accumulation and improved drought tolerance, but differential response in freezing tolerance. *Plant Mol. Biol.* **2007**, *64*, 265–278. [CrossRef]
26. Pu, Y.Y.; Gao, J.; Guo, Y.L.; Liu, T.T.; Zhu, L.X.; Xu, P.; Yi, B.; Wen, J.; Tu, J.X.; Ma, C.Z.; et al. A novel dominant glossy mutation causes suppression of wax biosynthesis pathway and deficiency of cuticular wax in *Brassica napus. BMC Plant Biol.* **2013**, *13*, 215. [CrossRef]
27. Liu, F.; Xiong, X.J.; Wu, L.; Fu, D.H.; Hayward, A.; Zeng, X.H.; Cao, Y.L.; Wu, Y.H.; Li, Y.J.; Wu, G. A lipid transfer protein gene involved in epicuticular wax deposition, cell proliferation and flower development in *Brassica napus. PLoS ONE* **2014**, *9*, e110272. [CrossRef]
28. Zhu, X.W.; Tai, X.; Ren, Y.Y.; Chen, J.X.; Bo, T.Y. Genome-Wide Analysis of Coding and Long Non-Coding RNAs Involved in Cuticular Wax Biosynthesis in Cabbage (*Brassica oleracea* L. var. capitata). *Int. J. Mol. Sci.* **2019**, *20*, 2820. [CrossRef]
29. Mo, J.G.; Li, W.Q.; Wang, J.H. Inheritance and agronomic performance of waxless character in *Brassica napus* L. *Plant Breeding.* **1992**, *108*, 256–259.
30. Michelmore, R.W.; Paran, I.; Kesseli, R.V. Identification of markers linked to disease-resistance genes by bulked segregant analysis: A rapid method to detect markers in specific genomic regions by using segregating populations. *Proc. Natl. Acad. Sci. USA* **1991**, *88*, 9828–9832. [CrossRef]
31. Van der Lee, T.; Robold, A.; Testa, A.; van't Klooster, J.W.; Govers, F. Mapping of avirulence genes in Phytophthora infestans with amplified fragment length polymorphism markers selected by bulked segregant analysis. *Genetics* **2001**, *157*, 949–956. [CrossRef]
32. Morlais, I.; Severson, D.W. Identification of a polymorphic mucin-like gene expressed in the midgut of the mosquito, *Aedes aegypti*, using an integrated bulked segregant and differential display analysis. *Genetics* **2001**, *158*, 1125–1136. [CrossRef] [PubMed]
33. Galya, K.; Yael, B.; Adi, F.D.; Abhinandan, P.; Ilan, H.; Ran, H. Fine-mapping the branching habit trait in cultivated peanut by combining bulked segregant analysis and high-throughput sequencing. *Front. Plant Sci.* **2017**, *8*, 1–11.
34. Wang, X.; Xu, Y.T.; Zhang, S.Q.; Cao, L.; Huang, Y.; Cheng, J.F.; Wu, G.Z.; Tian, S.L.; Chen, C.L.; Liu, Y.; et al. Genomic analyses of primitive, wild and cultivated citrus provide insights into asexual reproduction. *Nat. Genet.* **2017**, *49*, 765–772. [CrossRef] [PubMed]
35. Xue, H.B.; Shi, T.; Wang, F.F.; Zhou, H.K.; Yang, J.; Wang, L.; Wang, S.; Su, Y.L.; Zhang, Z.; Qiao, Y.S.; et al. Interval mapping for red/green skin color in Asian pears using a modified QTL-seq method. *Hortic. Res.* **2017**, *4*, 17053. [CrossRef]
36. Dong, W.; Wu, D.F.; Li, G.S.; Wu, D.W.; Wang, Z.C. Next-generation sequencing from bulked segregant analysis identifies a dwarfism gene in watermelon. *Sci. Rep.* **2018**, *8*, 2908. [CrossRef]
37. Bolger, A.M.; Lohse, M.; Usadel, B. Trimmomatic: A flexible trimmer for Illumina sequence data. *Bioinformatics* **2014**, *30*, 2114–2120. [CrossRef]
38. Li, H.; Durbin, R. Fast and accurate short read alignment with Burrows-Wheeler transform. *Bioinformatics* **2010**, *25*, 1754–1760. [CrossRef]
39. Li, H.; Handsaker, B.; Wysoker, A.; Fennell, T.; Ruan, J.; Homer, N.; Marth, G. The Sequence Alignment/Map Format and SAMtools. *Bioinformatics* **2009**, *25*, 2078–2079. [CrossRef]

40. Cingolani, P.; Platts, A.; Wang, L.L.; Coon, M.; Nguyen, T.; Wang, L.; Lang, S.J.; Lu, X.Y.; Ruden, D.M. A program for annotating and predicting the effects of single nucleotide polymorphisms, SnpEff: SNPs in the genome of Drosophila melanogaster strain w1118; iso-2; iso-3. *Fly* **2012**, *6*, 80–92. [CrossRef]
41. Takagi, H.; Abe, A.; Yoshida, K.; Kosugi, S.; Natsume, S.; Mitsuoka, C.; Uemura, A.; Utsushi, H.; Tamiru, M.; Takuno, S.; et al. QTL-seq: Rapid mapping of quantitative trait loci in rice by whole genome resequencing of DNA from two bulked populations. *Plant J.* **2013**, *74*, 174–183. [CrossRef]
42. Wu, G.; Zhang, L.; Wu, Y.H.; Cao, Y.L.; Lu, C.M. Comparison of five endogenous reference Genes for specific PCR detection and quantification of *Brassica napus*. *J Agric Food Chem.* **2010**, *58*, 2812–2817. [CrossRef] [PubMed]
43. Liu, L.; Tong, H.; Xiao, Y.; Che, R.; Xu, F.; Hu, B.; Liang, C.; Chu, J.; Li, J.; Chu, C. Activation of Big Grain1 significantly improves grain size by regulating auxin transport in rice. *Proc. Natl. Acad. Sci. USA* **2015**, *112*, 11102–11107. [CrossRef] [PubMed]
44. Lo, S.F.; Cheng, M.L.; Hsing, Y.C.; Chen, Y.S.; Lee, K.W.; Hong, Y.F.; Hsiao, Y.; Hsiao, A.S.; Chen, P.J.; Wong, L.I.; et al. Rice Big Grain 1 promotes cell division to enhance organ development, stress tolerance and grain yield. *Plant Biotechnol. J.* **2020**, *18*, 1969–1983. [CrossRef] [PubMed]
45. Li, Z.P.; Wang, X.L.; Chen, J.Y.; Gao, J.; Zhou, X.; Kuai, B. CCX1, a Putative Cation/Ca^{2+} Exchanger, Participates in Regulation of Reactive Oxygen Species Homeostasis and Leaf Senescence. *Plant Cell Physiol.* **2016**, *57*, 2611–2619. [CrossRef] [PubMed]
46. Dielen, A.S.; Sassaki, F.T.; Walter, J.; Michon, T.; Ménard, G.; Pagny, G.; Krause-Sakate, R.; Maia, I.D.G.; Badaoui, S.; Gall, O.L.; et al. The 20S proteasome α5 subunit of Arabidopsis thaliana carries an RNase activity and interacts in planta with the lettuce mosaic potyvirus HcPro protein. *Mol. Plant Pathol.* **2011**, *12*, 137–150. [CrossRef]
47. Voges, D.; Zwickl, P.; Baumeister, W. The 26S proteasome: A molecular machine designed for controlled proteolysis. *Annu. Rev. Biochem.* **1999**, *68*, 1015–1068. [CrossRef]
48. Eyüboglu, B.; Pfister, K.; Haberer, G.; Chevaller, D.; Fuchs, A.; Mayer, K.; Schneitz, K. Molecular characterisation of the STRUBBELIG-RECEPTOR FAMILY of genes encoding putative leucine-rich repeat receptor-like kinases in *Arabidopsis thaliana*. *BMC Plant Biol.* **2007**, *7*, 16. [CrossRef]
49. Wang, H.X.; Chen, Y.L.; Wu, X.T.; Long, Z.S.; Sun, C.L.; Wang, H.R.; Wang, S.M.; Birch, P.R.J.; Tian, A.D. A potato STRUBBELIG-RECEPTOR FAMILY member, StLRPK1, associates with StSERK3A/BAK1 and activates immunity. *J. Exp. Bot.* **2018**, *69*, 5573–5586. [CrossRef]
50. Blanco, F.; Garretón, V.; Frey, N.; Dominguez, C.; Pérez-Acle, T.; Van der Straeten, D.; Jordana, X.; Holuigue, L. Identification of NPR1-dependent and independent genes early induced by salicylic acid treatment in Arabidopsis. *Plant Mol. Biol.* **2005**, *59*, 927–944. [CrossRef]
51. Holdorf, M.M.; Owen, H.A.; Lieber, S.R.; Yuan, L.; Adams, N.; Dabney-Smith, C.; Makaroff, C.A. Arabidopsis ETHE1 encodes a sulfur dioxygenase that is essential for embryo and endosperm development. *Plant Physiol.* **2012**, *160*, 226–236. [CrossRef]
52. Sahebekhtiari, N.; Fernandez-Guerra, P.; Nochi, Z.; Carlsen, J.; Bross, P.; Palmfeldt, J. Deficiency of the mitochondrial sulfide regulator ETHE1 disturbs cell growth, glutathione level and causes proteome alterations outside mitochondria. *Biochim. Biophys. Acta Mol. Basis Dis.* **2019**, *1865*, 126–135. [CrossRef] [PubMed]
53. Wang, T.; Hu, J.; Ma, X.; Li, C.; Yang, Q.; Feng, S.; Li, M.; Li, N.; Song, X. Identification, evolution and expression analyses of whole genome-wide TLP gene family in Brassica napus. *BMC Genomics.* **2020**, *21*, 264. [CrossRef] [PubMed]
54. Tomaz, T.; Bagard, M.; Pracharoenwattana, I.; Lindén, P.; Lee, C.P.; Carroll, A.J.; Ströher, E.; Smith, S.M.; Gardeström, P.; Millar, A.H. Mitochondrial Malate Dehydrogenase Lowers Leaf Respiration and Alters Photorespiration and Plant Growth in Arabidopsis. *Plant Physiol.* **2010**, *154*, 1143–1157. [CrossRef] [PubMed]
55. Ballester, P.; Navarrete-Gómez, M.; Carbonero, P.; Oñate-Sánchez, L.; Ferrándiz, C. Leaf expansion in Arabidopsis is controlled by a TCP-NGA regulatory module likely conserved in distantly related species. *Physiol. Plant.* **2015**, *155*, 21–32. [CrossRef]
56. Wang, P.; Li, Z.W.; Wei, J.S.; Zhao, Z.L.; Sun, D.; Cui, S.J. A Na^+/Ca^{2+} exchanger-like protein (AtNCL) involved in salt stress in Arabidopsis. *J. Biol. Chem.* **2012**, *287*, 44062–44070. [CrossRef]
57. Rehrauer, H.; Aquino, C.; Gruissem, W.; Henz, S.R.; Hilson, P.; Laubinger, S.; Naouar, N.; Patrignani, A.; Rombauts, S.; Shu, H.; et al. AGRONOMICS1: A new resource for Arabidopsis transcriptome profiling. *Plant Physiol.* **2010**, *152*, 487–499. [CrossRef]
58. Kraft, E.; Stone, S.L.; Ma, L.; Su, N.; Gao, Y.; Lau, O.S.; Deng, X.W.; Callis, J. Genome analysis and functional characterization of the E2 and RING-type E3 ligase ubiquitination enzymes of Arabidopsis. *Plant Physiol.* **2005**, *139*, 1597–1611. [CrossRef]
59. Duncan, O.; Taylor, N.L.; Carrie, C.; Eubel, H.; Kubiszewski-Jakubiak, S.; Zhang, B.; Narsai, R.; Millar, A.H.; Whelan, J. Multiple Lines of Evidence Localize Signaling, Morphology, and Lipid Biosynthesis Machinery to the Mitochondrial Outer Membrane of Arabidopsis. *Plant Physiol.* **2011**, *157*, 1093–1113. [CrossRef]
60. Weng, C.Y.; Zhu, M.H.; Liu, Z.Q.; Zheng, Y.G. Integrated bioinformatics analyses identified SCL3-induced regulatory network in Arabidopsis thaliana roots. *Biotechnol. Lett.* **2020**, *42*, 1019–1033. [CrossRef]
61. Qaderi, M.; Reid, D.M. Growth and physiological responses of canola (*Brassica napus*) to UV-B and CO_2 under controlled environment conditions. *Physiol. Plant.* **2005**, *125*, 247–259. [CrossRef]
62. Tom, S.; Griffiths, D.W. The effects of stress on plant cuticular waxes. *New Phytol.* **2006**, *171*, 469–499.
63. Ni, Y.; Song, C.; Wang, X. Investigation on response mechanism of epicuticular wax on *Arabidopsis thaliana* under cold stress. *Sci. Agric. Sin.* **2014**, *47*, 252–261.
64. Tassone, E.E.; LIpka, A.E.; Tomasi, P.; Lohrey, G.T.; Qian, W.; Dyer, J.M.; Gore, M.A.; Jenks, M.A. Chemical variation for leaf cuticular waxes and their levels revealed in a diverse panel of *Brassica napus* L. *Ind. Crop. Prod.* **2016**, *79*, 77–83. [CrossRef]

65. Wang, Y.M.; Jin, S.R.; Xu, Y.; Li, S.; Zhang, S.J.; Yuan, Z.; Li, J.N.; Ni, Y. Overexpression of *BnKCS1-1*, *BnKCS1-2*, and *BnCER1-2* promotes cuticular wax production and increases drought tolerance in *Brassica napus*. *Crop J.* **2020**, *8*, 26–37. [CrossRef]
66. Farnham, M.W. Glossy and nonglossy near-isogenic lines USVL115-GL, USVL115-NG, USVL188-NG of broccoli. *Hortscience* **2010**, *45*, 660–662. [CrossRef]
67. Zhang, X.; Liu, Z.Y.; Wang, P.; Wang, Q.S.; Yang, S.; Feng, H. Fine mapping of *BrWax1*, a gene controlling cuticular wax biosynthesis in Chinese cabbage (*Brassica rapa* L. ssp. *Pekinensis*). *Mol. Breeding.* **2013**, *32*, 867–874. [CrossRef]
68. Qi, X.; Ogden, E.L.; Die, J.V.; Ehlenfeldt, M.K.; Polashock, J.J.; Darwish, O.; Alkharouf, N.; Rowland, L.J. Transcriptome analysis identifies genes related to the waxy coating on blueberry fruit in two northern-adapted rabbiteye breeding populations. *BMC Plant Biol.* **2019**, *19*, 460. [CrossRef]

Disclaimer/Publisher's Note: The statements, opinions and data contained in all publications are solely those of the individual author(s) and contributor(s) and not of MDPI and/or the editor(s). MDPI and/or the editor(s) disclaim responsibility for any injury to people or property resulting from any ideas, methods, instructions or products referred to in the content.

Article

Overexpression of the Peanut *AhDGAT3* Gene Increases the Oil Content in Soybean

Yang Xu [1,2,†], Fan Yan [1,†], Zhengwei Liang [2], Ying Wang [1], Jingwen Li [1], Lei Zhao [1], Xuguang Yang [1], Qingyu Wang [1,*] and Jingya Liu [1,*]

[1] College of Plant Science, Jilin University, Changchun 130062, China; xuyang@iga.ac.cn (Y.X.); fenfeiyongyuan@163.com (F.Y.); wangying2009@jlu.edu.cn (Y.W.); ljwk9@163.com (J.L.); zhaolei2010@jlu.edu.cn (L.Z.); xgyang@jlu.edu.cn (X.Y.)
[2] Northeast Institute of Geography and Agroecology, Chinese Academy of Sciences, Changchun 130102, China; liangzw@iga.ac.cn
* Correspondence: qywang@jlu.edu.cn (Q.W.); yj_liu@jlu.edu.cn (J.L.)
† Yang Xu and Fan Yan are co-first authors.

Citation: Xu, Y.; Yan, F.; Liang, Z.; Wang, Y.; Li, J.; Zhao, L.; Yang, X.; Wang, Q.; Liu, J. Overexpression of the Peanut *AhDGAT3* Gene Increases the Oil Content in Soybean. *Agronomy* **2023**, *13*, 2333. https://doi.org/10.3390/agronomy13092333

Academic Editors: Zhiyong Li, Chaolei Liu and Jiezheng Ying

Received: 14 August 2023
Revised: 30 August 2023
Accepted: 31 August 2023
Published: 7 September 2023

Copyright: © 2023 by the authors. Licensee MDPI, Basel, Switzerland. This article is an open access article distributed under the terms and conditions of the Creative Commons Attribution (CC BY) license (https://creativecommons.org/licenses/by/4.0/).

Abstract: Soybean (*Glycine max*) is the main oilseed crop that provides vegetable oil for human nutrition. The main objective of its breeding research is to increase the total oil content. In the Kennedy pathway, Diacylglycerol acyltransferase (DGAT) is a rate-limiting enzyme that converts diacylglycerol (DAG) to triacylglycerol (TAG). Here, the *AhDGAT3* gene was cloned from peanut and overexpressed in the wild-type (WT) *Arabidopsis*. The total fatty acid content in T$_3$ *AhDGAT3* transgenic *Arabidopsis* seeds was 1.1 times higher on average than that of the WT. Therefore, *AhDGAT3* was transferred into the WT (JACK), and four T$_3$ transgenic soybean lines were obtained, which proved to be positive using molecular biological detection. Specific T-DNA insertion region location information was also obtained via genome re-sequencing. The results of high-performance gas chromatography showed that the contents of oleic acid (18:1) composition and total fatty acids in transgenic soybean plants were significantly higher than that of the WT. However, linoleic acid (18:2) was much lower compared to the WT. The agronomic trait survey showed that the quantitative and yield traits of *AhDGAT3* transgenic soybean were better than those of the WT. These results suggest that fatty acids in transgenic soybeans, especially oleic acid and total fatty acid, are enhanced by the over-expression of *AhDGAT3*.

Keywords: *AhDGAT3* gene; gene expression; oil content; peanut; transgenic soybean; triacylglycerol (TAG)

1. Introduction

Soybean (*Glycine max*) and peanut (*Arachis hypogaea* L.) are the main sources of vegetable protein and oil, which are the most important bioenergy resources for humans to achieve sustainable development in the world. Peanut is recognized as one of the world's major oil crops and is widely planted and cultivated in China. The oil content of peanut seeds accounts for more than 50% of the seeds' dry weight. Among several common edible oil crops, the total oil content of peanut is second only to sesame (*Sesamum indicum* L.) but is higher than many other common oil crops, such as rapeseed (*Brassica campestris* L.), soybean, and cottonseed (*Gossypium* spp.) [1]. More than 80% of unsaturated fatty acids with high quality are stored in peanut seeds, which is approximately 2.5 times more than the two major oil crops, soybean and corn [2]. Therefore, it will be an innovative discovery if the functional genes that regulate the oil content in peanuts can be transformed into other common oil crops.

Soybean is a diploid plant that developed from an ancient tetraploid and contains a set of pathways to synthesize complex lipids. The oil synthesis pathways of soybean involve the process from synthesis to desaturation and the final formation of TAG. There are two methods for the synthesis of soybean fatty acids: one method involves their function as glycerin and phospholipid to form cell membranes, and the other is their storage in seeds, mainly to form TAG [3]. TAG is the main form of oil and fat stored in all kinds of living creatures and is very important for the formation of oil in plant seeds [4]. In addition, many enzymes are involved in the lipid synthesis pathway, including the expression and regulation of different genes. Therefore, it is much more complex to reveal the molecular regulation mechanism of changes in the content of total fatty acids and various fatty acid compositions [5].

DGAT is a rate-limiting enzyme in the Kennedy pathway and plays a key role in the synthesis and accumulation of TAG, which is mainly responsible for the conversion of DAG to TAG. There are four types of DGATs in the DGAT family, including DGAT1, DGAT2, DGAT3, and WS/DGAT [6]. DGAT1 is a member of the Acyl-CoA cholesterol acyltransferase family, which was first cloned in mice and then in Arabidopsis [7,8]. The DGAT1 gene was subsequently cloned in many other plants, including nasturtium, castor, maize, tobacco, and rape [9–13]. The DGAT2 gene was cloned in Mortierella ramanniana and Arabidopsis [14,15], castor [16], and tung tree [17]. The DGAT3 gene has only been identified in peanut [18] and Arabidopsis [19]. However, the DGAT3 protein sequence showed low homology with those of the DGAT1 and DGAT2 subfamilies. WS/DGAT was cloned in fungal microorganisms and has rarely been studied so far.

There are also four types of DGAT genes in peanuts. AhDGAT1 is involved in the lipid synthesis of yeast [20]. AhDGAT2 was found to significantly increase the content of fatty acids in Escherichia coli [21]. The AhDGAT3 gene is a soluble enzyme located in the cytoplasm with an unclear function [18]. However, the in-depth functional research on AhWSD/DGAT is still very limited. Peanut oil has a high nutritional value, and it is easily digested and absorbed by the human body from the composition and proportion of fatty acids in peanut seeds. Peanut oil is mainly composed of 12 fatty acid compositions, in which the sum of palmitic acid (16:0), oleic acid (18:1), and linoleic acid (18:2) accounts for about 90% of the total fatty acids. Kamisaka et al. found that DGAT was purified from the liposome in a lipid-producing fungus [21]. Lardizabal observed that overexpression of UrDGAT2A significantly increased the oil content in transgenic soybean seeds [22]. Chen et al. found that GmDGAT2D was overexpressed in hairy roots, increasing the 18:1 and 18:2 TAG content, whereas overexpression of GmDGAT1A increased the 18:3 TAG content. The overexpression of GmDGAT2D increased 18:1 TAG production and decreased 18:3 TAG in mutant seeds [23]. However, no study has reported that the AhDGAT3 gene is heterologously overexpressed in soybean or verified its further functions and regulation mechanisms so far.

In our study, the AhDGAT3 gene was cloned from peanut seeds, and the role of AhDGAT3 in regulating the total fatty acid content, fatty acid compositions, growth, and development was characterized in transgenic soybean by heterologously overexpressing AhDGAT3 to provide a theoretical basis for further verifying its function in soybean.

2. Materials and Methods

2.1. Plant Materials

Total RNAs were extracted from 30-day-old immature peanut seeds after flowering and used for cloning AhDGAT3 by RT-PCR. Mature seeds of WT soybean (JACK) were used as a negative control for subsequent experiments. JACK was a common soybean variety and receptor for soybean cotyledon node genetic transformation because of its high genetic transformation efficiency. T_3 AhDGAT3 transgenic soybean lines were selected and randomly sampled for specificity analyses of gene expression levels and fatty acid compositions. All soybean and peanut seeds were planted in the crop genetics and breeding station in May. Tissues including roots, stems, leaves, flowers, different period of pods after flowering, and mature seeds were sampled randomly for our experiments at particular time. Arabidopsis (Col-0) was used to transform the AhDGAT3 gene and measure the content of fatty acid compositions. The specific disinfection, vernalization, and culture methods of *Arabidopsis* referred to the relevant literature [24]. The mature seeds of the WT and 5 transgenic *Arabidopsis* lines after harvesting were selected to extract their DNA, RNA, and protein and measure the total fatty acid content and 5 key fatty acid compositions. All samples were selected and collected randomly from 3 different lines and stored in $-80\ °C$ freezer after freezing in liquid nitrogen.

2.2. Isolation and Sequence Analysis of AhDGAT Gene in Peanut

The complete coding sequence of the *AhDGAT3* gene was inquired, obtained, and accessed on 4 May 2020 using the NCBI Nucleotide Blast tool (http://blast.ncbi.nlm.nih.gov/Blast.cgi) and peanut genome database (http://www.peanutbase.org). The conserved sequences and regions of DGAT3 in different plants were aligned, analyzed and accessed on 5 May 2020 using the online software MEME 5.5.4 (https://meme-suite.org/meme/tools/meme). Total RNAs were sampled and extracted from 30-day-old immature peanut seeds after flowering. Then, the extracted RNAs were reverse-transcribed into cDNAs, which were used as a template for cloning the *AhDGAT3* using RT-PCR method. The specific PCR primers were designed according to the cDNA sequence using Primer Premier 5.0 software. All the RT-PCR primer pairs are provided in Table 1. The RT-PCR reaction contained the forward and reverse primers, cDNA, dNTP mixture, PCR reaction buffer, and double distilled water. After being sequenced in the Biosciences Company (Comate, Changchun, China) for accuracy, the PCR product containing target gene would be constructed to the cloning vector pMD18-T for further studies.

Table 1. Gene specific primer pairs and information used for PCR and molecular detection.

Purpose	Gene Name (Amplification Length)	Accession Number	Forward Primer (5'-3')	Reverse Primer (5'-3')
RT-PCR	*AhDGAT3* (1438 bp)	XM_016339125	AATAGAAATAGAAATGTGATAATGG	ACAAATCAGGCTCTGGAAGTT
qRT-PCR	*AhDGAT3* (140 bp)	XM_016339125	AGAATGGAACCGCTATGT	CTCTGCCCTTACTTGCTC
	β-Tubulin (185 bp)	GMU12286	GGAAGGCTTTCTTGCATTGGTA	AGTGGCATCCTGGTACTGC
	Actin (155 bp)	J01298	GTCCTTTCAGGAGGTACAACC	CCTTGAAGTATCCTATTGAGC
Detection	*Bar* (220 bp)		GTCTGCACCATCGTCAACCACTACA	AGACGTACACGGTCGACTCGGCCGT
	AhDGAT3 (304 bp)	XM_016339125	AGAATGGAACCGCTATGT	CTCTGCCCTTACTTGCTC

2.3. Arabidopsis and Soybean Transformation of AhDGAT3

The open reading frame of the *AhDGAT3* gene was constructed into an overexpression vector pTF101-35s using heat-shock method, and the DH5α competent cells were treated by $CaCl_2$, which needed to be prepared in advance. Then, the resultant plasmid pTF101-AhDGAT3 was transformed into *Agrobacterium* tumefaciens EHA105 using heat-shock method for the genetic transformation of *Arabidopsis* and soybean. Then, the resultant plasmid was transformed into *Arabidopsis* Col-0 using *Agrobacterium* transformation and floral-dip method [25] and transformed into soybean JACK (WT) using *A.* tumefaciens to infect the soybean cotyledon nodes. The transformed *Arabidopsis* and soybean were screened on the MS medium, which contained 5 mg L^{-1} glufosinate-ammonium. Then, T_0 transgenic *Arabidopsis* and soybeans were detected using PCR and *bar* strip, which is a simple and accurate method to identify the positive plants. All the positive *AhDGAT3* transgenic *Arabidopsis* and soybean lines were bred to T_3 generation for subsequent experiments.

2.4. Molecular Detection of AhDGAT3 Transgenic Soybean

The regenerated plants were transplanted and cultivated in the plant tissue incubator. Leaves of T_0 transgenic plants were sampled and identified using PCR and *bar* strip for positive detection. Transgenic plants were tested via RT-PCR, from which the DNA of *AhDGAT3* transgenic soybean leaves was extracted and used as the PCR template. All the specific primer pairs were provided in Table 1 and used to amplify the *AhDGAT3* gene (1438 bp) and *bar* gene (220 bp). Southern blot was performed to verify that the *AhDGAT3* gene has been transformed into soybean successfully at the DNA level. The *bar* and *AhDGAT3* genes were chosen as DNA probes, which referred to the DIG High Prime DNA Labeling and Detection Starter Kit II. Western blot was performed to verify the expression level of AhDGAT3 protein in transgenic soybeans on a qualitative level. Protein of JACK leaves was extracted as the negative control. The primary monoclonal antibody, which was required for the experiment, was Mouse Monoclonal with 1:5000 titers. The HRP Goat Anti-Mouse IgG was chosen to be the secondary antibody with 1:6000 titers. ELISA was performed to determine whether the PAT protein was expressed in transgenic soybeans on a quantitative level. The specific experiment methods all referred to the relevant literature [24].

2.5. Genome Re-Sequencing

Four T_3 *AhDGAT3* transgenic soybean strains were sent to Biotechnology company (Biomarker, Beijing, China) for genome re-sequencing analysis. After the Genomic DNA samples were qualified, the DNA was fragmented using ultrasonic mechanical interrupt method, and then the end of the DNA fragment was purified and repaired. Then, 3' and A ends of the DNA were sequenced, connected with joints, and then the DNA fragment was selected and sequenced. Finally, the genomic library was constructed, qualified, and sequenced with Illumina instrument. After the quality of the sequenced original reads (double-ended sequences) was evaluated, the clean reads were selected to be compared with the reference genome sequences, including the mutation detection and annotation of SNP. InDel and mutations were carried out on the basis of sequence alignment. The functional genes were finally discovered and annotated at DNA level.

The *AhDGAT3* transgenic soybeans (R01–R04), respectively, were aligned to the reference genome and exogenous T-DNA sequence data, according to the comparison results. The short end of two kinds of matching sequences was found, and the sample data were aligned to the reference genome and the exogenous insertion sequence, respectively. According to the result of alignment, two types of paired ends were found. The insertion sequence was aligned to the reference genome by blast for homology assessment. The assembled contig sequences were aligned to the reference genome by blast, and then the contig sequences were selected to be compared to the regions of chromosomes according to the alignment results. These regions were verified by IGV screenshots, and then the region location information where *AhDGAT3* was inserted was obtained.

2.6. Quantitative RT-PCR Analysis

Total RNAs were extracted from the WT soybean, peanut, and *AhDGAT3* transgenic soybean tissues, including vegetative growth (root, stem, and leaf) and reproductive growth (different period of pods after flowering and mature seeds). Then, RNAs were reverse-transcribed into cDNAs using RNAiso Plus and M-MLV Kit. All the qRT-PCR primer pairs of the *AhDGAT3* and internal reference genes including *β-Tubulin* and *Actin* genes, which were used to standardize the data [26–28], are provided in Table 1. The qRT-PCR conditions referred to the instructions of TAKARA Biotechnology Company. All the qRT-PCR samples followed the principles of three biological and technical replicates.

2.7. Measurement of Fatty Acid Content in Arabidopsis and Soybean

The total fatty acids of *AhDGAT3* transgenic *Arabidopsis* were extracted using the methylester method. The fatty acid samples were tested and analyzed using a flame ionization detector (Agilent 7890A GC system, Santa Clara, CA, USA) and referring to the methods described in the literature [29]. Five key fatty acid compositions (16:0, 18:0, 18:1, 18:2, and 18:3) were standardized as the control. The total fatty acids of mature seeds in

AhDGAT3 transgenic soybeans were extracted using the hydrolysis method (GB5009.168-2016). Two standard sample solutions of the two methods were formulated and used to measure different peak values at different time. And then, we plotted the standard curve in different concentration gradients. All the values were plugged into the standard curve for calculation. The specific experiment methods both referred to the relevant literature [24].

2.8. The Agronomic Trait Analysis of WT and Transgenic Soybeans

After JACK and T_3 AhDGAT3 transgenic soybean plants matured, the following agronomic traits were analyzed for each plant, including plant height, main stem node number, effective branch number, 100-seed weight, pod number per plant, seed number per plant, seed weight per plant, and podding height. We randomly selected 30 WT (JACK) and 40 T_3 transgenic soybeans (10 of each line). Each result was the average of one soybean line. The 10-day to 50-day pods after flowering of JACK and AhDGAT3 transgenic soybeans were randomly selected and assessed for pod length and width. The pods were measured 3 times. And then, the average length and width of a single soybean were calculated. The results of the agronomic traits were determined by tape measure, vernier calipers, grain counting machines, rulers, and scales.

2.9. Statistical Analysis

All statistical data were calculated and analyzed using SPSS 20.0 (IBM Corp, Armonk, NY, USA) software. Significant differences among the means of samples were compared at $p < 0.05$ (significant difference) or $p < 0.01$ (extremely significant difference), based on an independent-sample t-test.

3. Results

3.1. Isolation and Characterization of AhDGAT3 Gene

In order to verify the functions of AhDGAT3, all the AhDGAT family genes were searched, obtained, and accessed on 4 May 2020 from the plant comparative genome database, Phytozome (http://phytozome-next.jgi.doe.gov/), and the Peanut genome database, (http://www.peanutbase.org/), for homologous search. Four candidate genes were identified, named AhDGAT1 (XM_016346849), AhDGAT2 (XM_016095296), AhDGAT3 (XM_016339125), and AhWSD/DGAT (XM_016109460). The AhDGAT1, AhDGAT2, and AhDGAT3 genes were closely related but were far related to the AhWSD/DGAT. In the DGAT3 conserved domain, Motif1 contains the characteristic fatty-acid-binding protein sequence KSGSIALLQEFERVVGAEG, and Motif2 contains the CKCMGKCKSAPNVRIQNSTAD conserved sequences. Motif3 contains the speculated structural motif NPLCIGVGLEDVDAIVA, and Motif4 contains the DDLQGNLTWDAAEVLMKQLEQVRAEEKELKKKQKQEKKAKL conserved sequences. The subfamily also contains the KKRVLFDDL active sites of the acyltransferase family [15]. In addition, this subfamily also contains the thioesterase characteristic sequence TNPDCESSSSSSSSESESES (Figure 1).

3.2. Oil Content in Seeds of Different Genotype Arabidopsis Plants

To reveal and verify the functions of AhDGAT3 in the accumulation of oil in seeds, the AhDGAT3 gene was transformed into the WT Arabidopsis (Col-0) plants. Mature seeds from T_3 AhDGAT3 transgenic lines were sampled to measure the oil content. The results showed that the content of 18:1 composition in AhDGAT3 transgenic Arabidopsis seeds was significantly higher than that of the WT, reaching 1.3 times on average, respectively (Figure 2A). However, the contents of 16:0, 18:2, and 18:3 compositions in the transgenic lines were slightly lower than those in the WT. The total fatty acid content in four AhDGAT3 transgenic soybean lines was also enhanced a lot (Figure 2B).

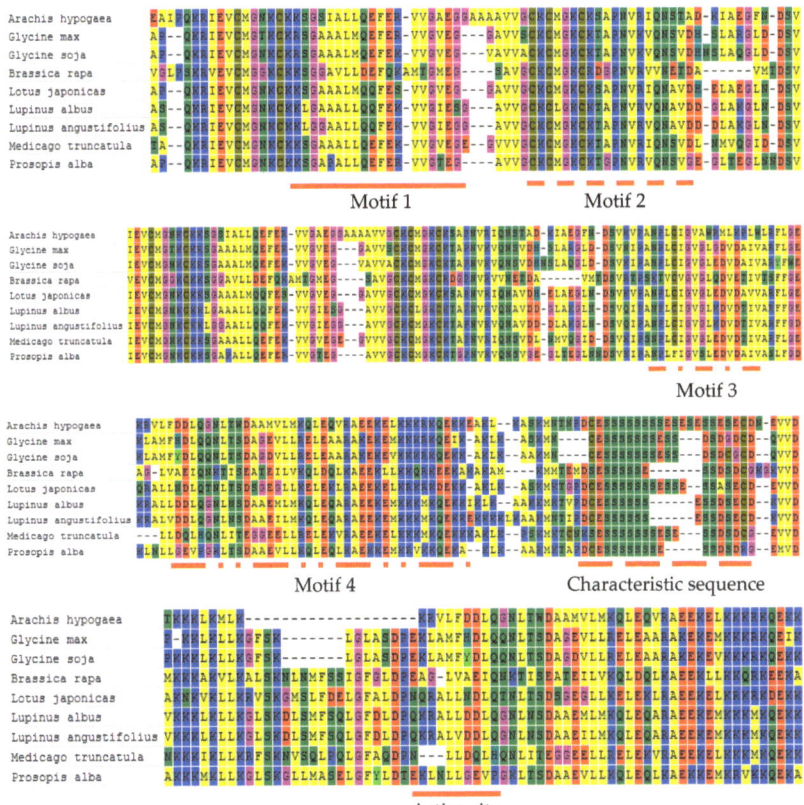

Figure 1. Partial alignment of DGAT3 deduced amino acids in different plants. Sequences were aligned using MEGA 7.0 software. Different shapes of lines represent four different special functional domains and two special sites. Accession numbers of 9 genes in GenBank are as follows: *Arachis hypogaea*, AAX62735.1; *Glycine max*, XP_003542403.1; *Glycine soja*, XP_028209585.1; *Brassica rapa*, RID49571.1; *Lotus japonicas*, AFK37850.1; *Lupinus angustifolius*, XP_019419028.1; *Medicago truncatula*, XP_003609890.1; *Prosopis alba*, XP_028760027.1.

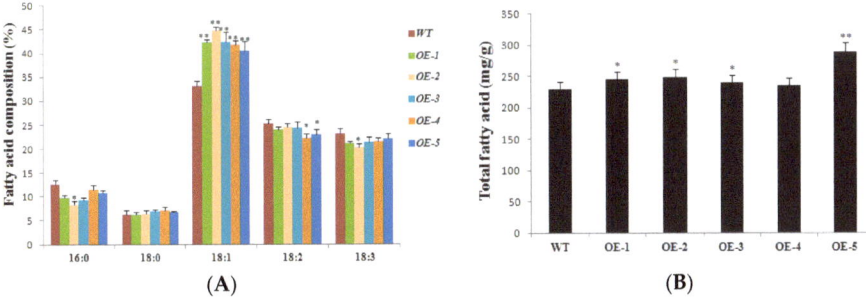

Figure 2. (**A**) Content of 5 fatty acid compositions in the WT and *AhDGAT3* transgenic *Arabidopsis* lines. (**B**) Total fatty acid content in the WT and *AhDGAT3* transgenic *Arabidopsis* seeds. The data represent the average of 3 independent repetitions of experiments. Error bars indicate the standard error. * $p < 0.05$ and ** $p < 0.01$ represent the significant difference. The peaks and peak time of each group were palmetto acid (16:0), 11.672 min; stearic acid (18:0), 16.011 min; oleic acid (18:1), 16.471 min; linoleic acid (18:2), 17.779 min; and linolenic acid (18:3), 20.345 min.

3.3. Overexpression AhDGAT3 Gene in Soybean

To verify the further functions of *AhDGAT3* in soybean, four T_3 *AhDGAT3* transgenic soybean lines were obtained and detected via a series of molecular biology methods, including the *bar* strip, Southern blot, and Western blot. The results showed that *AhDGAT3* was transformed into a JACK cultivar background (Figure 3). ELISA analysis revealed that the expression level in transgenic soybeans was higher than that of the WT (Table 2).

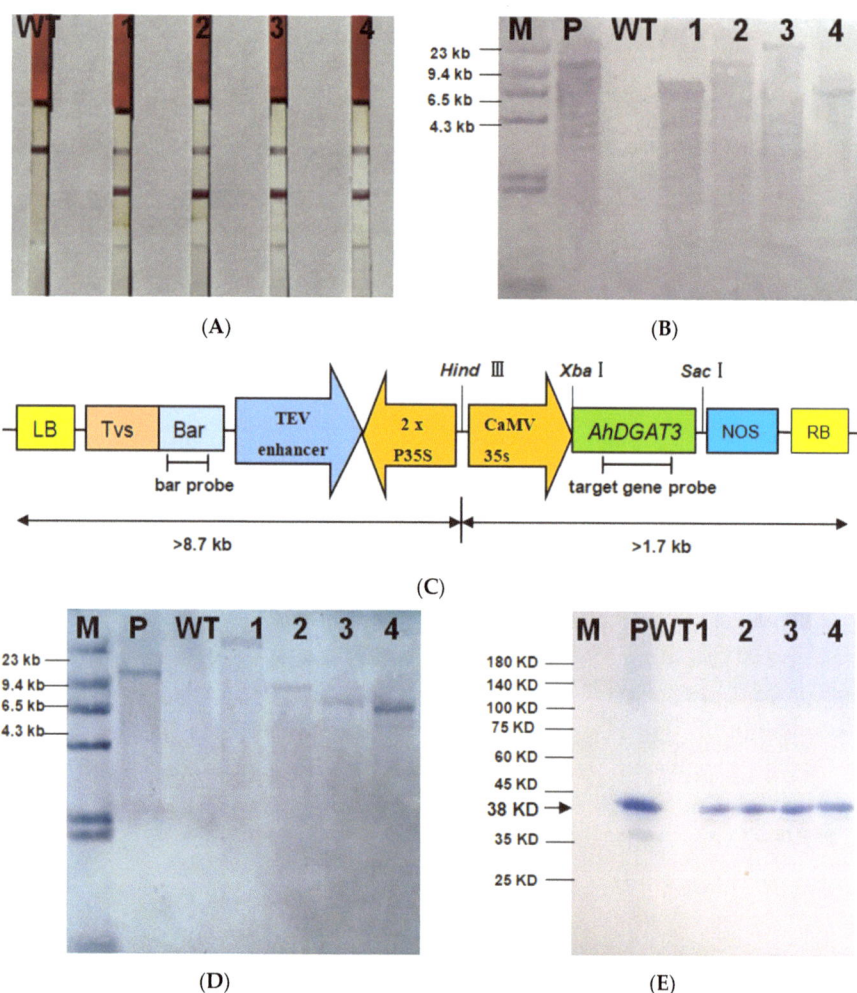

Figure 3. Positive detection of *AhDGAT3* transgenic soybean lines. (**A**) *Bar* strip analysis of 4 *AhDGAT3* transgenic soybeans. (**B**) Southern blot of T_3 transgenic soybeans using the *bar* gene fragment as a probe. M represents DNA marker; P represents recombinant plasmid (positive control); WT represents JACK (negative control). (**C**) Map of pTF101-*AhDGAT3* recombinant plasmid in *AhDGAT3* transgenic soybeans. (**D**) Southern blot of T_3 transgenic soybeans using the target gene fragment as a probe. M represents DNA marker; P represents recombinant plasmid (positive control); WT represents JACK (negative control). (**E**) Western blot of T_3 transgenic soybeans. M represents pre-stained protein marker; P represents AhDGAT3 protein (positive control); WT represents JACK (negative control).

Table 2. AhDGAT3 protein levels of different tissues in JACK and 4 transgenic lines (OE-1 to OE-4). Data represent the average values of three biological replicates. NA means that the value is not detected and calculated through the standard curve, and there were no detection results.

Material	JACK	OE-1 (ng/g)	OE-2 (ng/g)	OE-3 (ng/g)	OE-4 (ng/g)
Nodules	NA	196.2 ± 10.2	220.6 ± 14.3	200.5 ± 13.8	233.9 ± 14.9
Root	NA	236.1 ± 19.2	389.5 ± 8.7	248.2 ± 1.5	415.5 ± 20.4
Stem	NA	203.7 ± 14.9	208.7 ± 6.9	200.6 ± 7.9	317.1 ± 11.7
Leave	NA	483.7 ± 16.7	523.7 ± 14.4	511.3 ± 10.2	655.1 ± 11.4
Flower	NA	288.3 ± 27.2	385.1 ± 17.7	348.6 ± 11.5	417.6 ± 10.5
10-day Pod	NA	803.8 ± 14.8	841.7 ± 9.2	816.9 ± 3.8	869.5 ± 6.9
20-day Pod	NA	676.1 ± 22.9	706.6 ± 10.5	796.1 ± 4.6	787.1 ± 27.1
30-day Pod	NA	700.6 ± 15.9	804.1 ± 5.4	753.5 ± 32.3	883.7 ± 8.8
40-day Pod	NA	913.2 ± 5	985.4 ± 3.8	951.4 ± 20.3	1014.4 ± 11.2
Seed	NA	949.3 ± 17.9	1096.9 ± 11.1	1004.6 ± 16.9	1166.1 ± 22.1

The 1038 bp *AhDGAT3* gene was inserted into the pTF101-35s vector with the *Xba* I/*Sac* I restriction site (Figure 3C). The distances between the *bar* gene probe and the left/right boundaries of T-DNA were approximately 9 kb and 2 kb, respectively. The distances between the target gene and the left/right boundaries of T-DNA were approximately 9.6 kb and 1 kb, respectively (Figure 3C). The results of the Southern blot showed that only one hybridization band appeared in four *AhDGAT3* transgenic lines (OE-1 to OE-4), and no hybridization band was detected in the WT. This suggested that exogenous T-DNA was integrated in the form of single-copy DNA in four *AhDGAT3* transgenic lines (Figure 3B,D). Specific insertion sites were analyzed and studied via genome re-sequencing in transgenic soybean. The *AhDGAT3* gene was translated into the protein and synthesized monoclonal antibodies to identify the transgenic soybeans. The Western blot results showed that there was an apparent hybridization signal at 38 KD, indicating that the *AhDGAT3* gene was over-expressed at the protein level in transgenic plants (Figure 3E).

The ELISA results showed that the AhDGAT3 protein was not expressed in the WT (negative control), and no value was calculated using a standard curve. The AhDGAT3 protein of four transgenic lines was detected and calculated using a standard curve. AhDGAT3 expression levels in the vegetative growth stage were lower than those of the reproductive growth stage. AhDGAT3 level accumulation was highest in mature seeds as the most important organ for oil storage, which was approximately two to five times higher than that of the nutritious organs. This result can further indicate that AhDGAT3 proteins are expressed in four transgenic soybean lines (Table 2).

3.4. Genome Re-Sequencing of AhDGAT3 Transgenic Soybean

In this analysis, the genomic DNA from four T_3 transgenic soybean leaves was re-sequenced. Clean data with a total data volume of 92.59 Gbp were obtained, and Q30 reached 90.51%. The average contrast ratio of the sample to the W82 reference genome was 98.88%. The depth of the average coverage was 21× and the depth of the genome coverage was 98.28%, which made sure at least one base was covered.

The re-sequencing data of four transgenic soybeans (R01–R04) were compared with reference genome and exogenous T-DNA sequences, and all short sequences that could be compared with exogenous T-DNA sequences were selected for the analysis. The assembled contig sequences were blasted to the reference genome. And then, the contig sequences were selected to be compared to the regions of chromosomes according to the alignment results. These regions were verified by IGV screen shots to obtain the location information of exogenous insertion fragments. All the results of gene insertion information are shown in Table 3.

Table 3. The insertion region location information of four *AhDGAT3* transgenic strains (R01–R04).

Sample	Chromosome	Integration Sites	Integrated Way
R01	3	36,280,392	Single copy
R02	10	19,795,197	Single copy
R03	11	32,500,049	Single copy
R04	14	8,451,776	Single copy

3.5. Expression Analysis of AhDGAT3 in WT and Transgenic Soybean

Tissues in vegetative and reproductive growth stages from JACK and transgenic soybeans were collected, and the total RNAs were extracted to verify gene over-expression in the transgenic soybean using qRT-PCR. The results showed that *AhDGAT3* expression in transgenic soybeans was significantly ($p < 0.01$) higher than that of the WT in the leaves and 10- and 40-day-old pods after flowering. The *AhDGAT3* expression level in the flower and 30-day-old pods of transgenic soybeans was significantly higher ($p < 0.05$) than that in the WT (Figure 4).

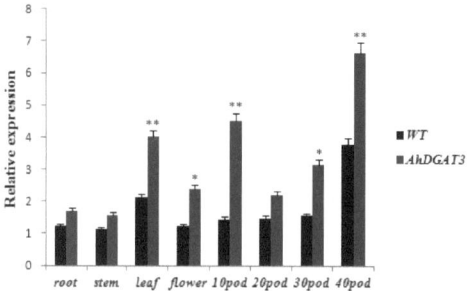

Figure 4. Expression levels of *AhDGAT3* in different tissues of soybean. The data represent the average of 3 independent experiments ± SD. Error bars indicate the standard error. * $p < 0.05$ and ** $p < 0.01$ represent the significant difference.

3.6. AhDGAT3 Over-Expression Enhances Fatty Acid Content in Transgenic Soybean

Mature soybean seeds (0.01 g) were collected and sampled with three biological replicates, including the WT (JACK) and *AhDGAT3* transgenic lines, which were used for oil determination and analysis. The results of high-performance gas chromatography (HPGC) showed no significant difference in the content of 16:0, 18:0, and 18:3 fatty acid compositions between *AhDGAT3* transgenic soybeans and JACK. However, the difference of the 18:1 content was significant, with the transgenic line being 27.3% higher than JACK, respectively. The content of the 18:2 fatty acid compositions was significantly lower than JACK. The total fatty acid content of the T_3 *AhDGAT3* transgenic soybean was up to 4% higher than that of JACK (Figure 5).

3.7. AhDGAT3 Overexpression Improved the Agronomic Traits of Transgenic Soybean

The analysis of the agronomic traits in the WT and *AhDGAT3* transgenic soybean seeds showed that plant height, effective branch number, pod number per plant, and seed weight per plant in the *AhDGAT3* transgenic soybean were significantly superior to the WT (Figure 6A and Table 4). The size of 10-day to 50-day pods after flowering of the *AhDGAT3* transgenic soybean was significantly bigger than the WT at different stages of reproductive growth (Figure 6B–D). The results showed that overexpression of the *AhDGAT3* gene may also affect the growth and development of soybean.

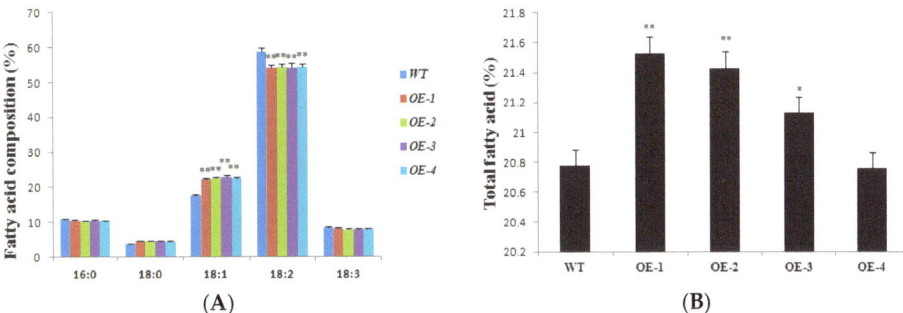

Figure 5. Content of different fatty acid compositions and total fatty acids in soybean seeds of different genotypes. (**A**) The content of different fatty acid compositions in JACK and 4 different *AhDGAT3* transgenic lines (OE-1 to 4) of soybean, measured by HPGC. (**B**) Total fatty acid of 4 *AhDGAT3* overexpressing soybean seeds. Error bars indicate the standard error. * $p < 0.05$ and ** $p < 0.01$ represent the significant difference.

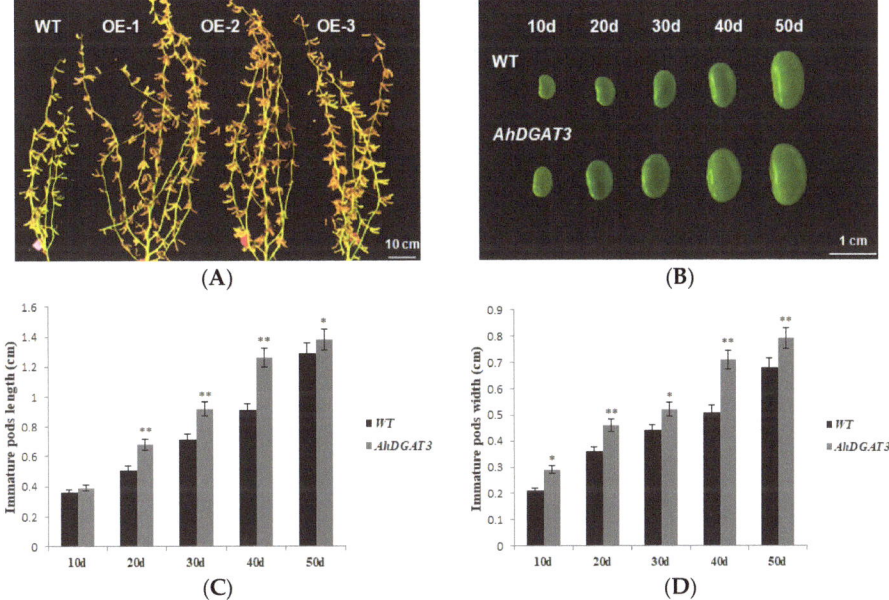

Figure 6. Phenotypic observation of *AhDGAT3* transgenic soybean and the WT. (**A**) Three *AhDGAT3* transgenic lines (OE-1 to OE-3) and the WT in mature period. (**B**) The size of 10-day to 50-day immature pods after flowering in the WT and *AhDGAT3* transgenic soybean. (**C**) The difference of 10-day to 50-day immature pods' length between the WT and *AhDGAT3* transgenic lines. (**D**) The difference of 10-day to 50-day immature pods' width between the WT and *AhDGAT3* transgenic lines. Error bars indicate the standard error. * $p < 0.05$ and ** $p < 0.01$ represent the significant difference.

Table 4. Agronomic traits in the WT and *AhDGAT3* transgenic soybean (OE-1 to 3). The data for sample mean ± standard error ($p < 0.01$). Significantly different results are indicated by different letters (a, b): a represents no significant difference; b represents significant difference.

Material	WT	OE-1	OE-2	OE-3
Plant height (cm)	101.4 ± 0.14 a	114.1 ± 0.17 b	109.2 ± 0.21 b	105.1 ± 0.27 a
Effective branch number	4 ± 0.21 a	8 ± 0.16 b	7 ± 0.28 b	6 ± 0.12 b
Pod number per plant	78 ± 0.14 a	188 ± 0.11 b	168 ± 0.18 b	161 ± 0.27 b
Seed number per plant	155 ± 0.16 a	316 ± 0.29 b	293 ± 0.19 b	289 ± 0.31 b
100-seed weight (g)	13.5 ± 0.21 a	16.3 ± 0.22 b	15.6 ± 0.24 a	16.9 ± 0.17 b
Seed weight per plant (g)	25.7 ± 0.17 a	57.2 ± 0.24 b	49.7 ± 0.26 b	45.7 ± 0.26 b
Main stem node number	18 ± 0.25 a	22 ± 0.26 b	21 ± 0.21 b	21 ± 0.24 b
Podding height (g)	8.4 ± 0.29 a	10.1 ± 0.31 a	9.9 ± 0.31 a	10.1 ± 0.27 a

4. Discussion

Soybean is an important food and oil crop in the world, accounting for a significant proportion [30]. How to effectively improve vegetable oil content and the composition ratio of unsaturated fatty acids is an important breeding goal for us. Heterologous gene expression is an important means to improve soybean oil content.

Recently, studies on how to improve the oil content of plant seeds have mainly focused on the lipid metabolism pathway. All kinds of functional enzyme genes and transcription factors, including *PDAT*, *FAD*, *SAD*, *LEC1*, *LEC2*, Dof, and ABI3, were cloned and studied, which proved to play an important role [31]. It is a major technical method to verify the coordinated expression of multiple genes in metabolic pathways via genetic transformation technology [32]. However, as an enzyme gene that plays a key catalytic role in the conversion process of DAG to TAG. *DGAT* plays an important role in regulating the accumulation of fatty acids and lipid synthesis in oil crops [33]. The oil content of seeds depends on quantitative heredity and is regulated by all kinds of factors, including the synthesis and accumulation of fatty acids and the development of seeds [34,35].

Compared to animal fatty acids, various unsaturated fatty acids in plants are very beneficial to our health. The demand for high-oleic-acid vegetable oil is also increasing with the improvement of our living standards. The results showed that *AhDGAT3* was over-expressed in *Arabidopsis* and soybean. The contents of 18:1 unsaturated fatty acid compositions in transgenic plant seeds were significantly increased. This result is consistent with recent research reports [23]. *AtDGAT3* was involved in a soluble cytosolic process in the circulation of linoleic and linolenic acid to TAG during the degradation and breakdown of seed oil [19]. However, no in-depth studies showed the functions of the *AhDGAT3* gene using heterologous transformation and genetic engineering technology between peanut and soybean. Most recent studies on *DGAT* genes mainly focus on functional analysis in *Arabidopsis*. The *UrDGAT2* gene of Umbelopsis ramanniana was successfully transferred into soybean and over-expressed in seeds. The *UrDGAT2* transgenic soybean showed a higher oil content [22]. TAG, as the main storage in oil crop seeds, provided enough C sources for seed germination and development. The up-regulation of the *DGAT1* gene was involved in the fatty acid mobilization and the catalyzation in the reverse reaction of TAG synthesis during seed germination and development [36,37]. The results of the agronomic traits showed that the phenotype of the *AhDGAT3* transgenic soybean was better than that of the WT soybean in terms of the size of different period pods, the number of effective branches, the number of seeds per plant, pod number per plant, main stem node number, and the weight of seeds per plant at harvest time. We speculated that *AhDGAT3* was also closely related to the growth and development of soybeans. Moreover, the next step is to study the internal regulation mechanism of *AhDGAT3*. We aimed to develop high-quality transgenic soybean lines with high oil content and yield, creating excellent soybean germplasm resources.

5. Conclusions

We characterized the major functions of *AhDGAT3* in *Arabidopsis* and soybean. The content of total fatty acids and 18:1 compositions in the *AhDGAT3*-overexpressed transgenic soybean was significantly higher than those of the WT (JACK). The re-sequencing of *AhDGAT3* transgenic soybean revealed that T-DNA was integrated into the soybean genome as a single copy. Overexpression of the *AhDGAT3* gene can affect the size of soybean immature pods and a series of agronomic traits. Therefore, we also speculate that this gene may also play a key role in soybean growth and development, which needs further study. Our results indicated that molecular-assisted breeding is a very efficient and fast way of increasing the oil content of soybean seeds, which can also be used in the modulation of other agronomic traits of soybean in the future.

Author Contributions: Conceptualization, Y.X., F.Y. and Q.W.; Data curation, F.Y. and Y.W.; Formal analysis, Y.X., Z.L., J.L. (Jingwen Li) and Y.W.; Funding acquisition, Q.W.; Investigation, Y.X. and X.Y.; Methodology, F.Y. and L.Z.; Project administration, Q.W.; Resources, J.L. (Jingwen Li); Software, J.L. (Jingya Liu); Supervision, Q.W.; Validation, Q.W.; Visualization, F.Y.; Writing—original draft, Y.X.; Writing—review and editing, Y.X. and F.Y. All authors have read and agreed to the published version of the manuscript.

Funding: This research was funded by the Science and Technology Development Plan Project of Jilin Province (No. 20220202009NC), the National Natural Science Foundation of China (No. 32101689), and the Science and Technology research project of Education Department of Jilin Province (No. JJKH20221034KJ).

Data Availability Statement: Not applicable.

Conflicts of Interest: The authors declare no conflict of interest.

References

1. Liao, B.S. Analysis on the competitiveness of peanut oil industry in China. *J. Peanut* **2003**, *32*, 11–15.
2. Li, X.D.; Cao, Y.L.; Hu, Y.P.; Xiao, L.; Wu, Y.; Wu, G.; Lu, C. Study on fatty acid accumulation patterns in peanut seed development. *Chin. J. Oil Crops* **2009**, *31*, 157–162.
3. Mekhedov, S.; De Ilárduya, O.M.; Ohlrogge, J. Toward a functional catalog of the plant genome. A survey of genes for lipid biosynthesis. *Plant Physiol.* **2000**, *122*, 389–402. [CrossRef] [PubMed]
4. Barthole, G.; Lepiniec, L.; Rogowsky, P.M.; Baud, S. Controlling lipid accumulation in cereal grains. *Plant Sci.* **2012**, *185–186*, 33–39. [CrossRef]
5. Liu, Z.J. Cloning of Genes Involved in Seed Oil Content and Transgenic Research of Cotton. Ph.D. Thesis, China Agricultural University, Beijing, China, 2013.
6. Yen, C.E.; Stone, S.J.; Koliwad, S.; Harris, C.; Farese, R.V., Jr. DGAT enzymes and triacylglyrol biosynthesis. *J. Lipid Res.* **2008**, *49*, 2283–2301. [CrossRef] [PubMed]
7. Cases, S.; Smith, S.J.; Zheng, Y.W. Identification of a gene encoding an acyl-Co A: Diacylglycerol acyltransferase, a key enzyme in triacylglycerol synthesis. *Proc. Natl. Acad. Sci. USA* **1998**, *95*, 13018–13023. [CrossRef]
8. Hobbs, H.D.; Chaofu, L.; Hills, M. Cloning of a c DNA encoding acyltransferase from *Arabidopsis thaliana* and its functional expression. *FEBS Lett.* **1999**, *452*, 145–149. [CrossRef]
9. Xu, J.; Francis, T.; Mietkiewska, E.; Giblin, E.M.; Barton, D.L.; Zhang, Y.; Zhang, M.; Taylor, D.C. Cloning and characterization of an acyl Co A-dependent diacylglycerol acyltransferase 1 (*DGAT1*) gene from Tropaeolummajus, and a study of the functional motifs of the DGAT protein using site-directed mutagenesis to modify enzyme activity and oil content. *Plant Biotechnol. J.* **2008**, *6*, 799–818. [CrossRef]
10. He, X.H.; Grace, Q.C.; Lin, J.T.; McKeon, T.A. Regulation of diacylglycerol acyltransferase in developing seeds of castor. *Lipids* **2004**, *39*, 865–871. [CrossRef] [PubMed]
11. Zheng, P.; Allen, W.B.; Roesler, K.; Williams, M.E.; Zhang, S.; Li, J.; Glassman, K.; Ranch, J.; Nubel, D.; Solawetz, W.; et al. A phenylalanine in *DGAT* is a key determinant of oil content and composition in maize. *Nat. Genet.* **2008**, *40*, 367–372. [CrossRef]
12. Bouvier-Navé, P.; Benveniste, P.; Oelkers, P.; Sturley, S.L.; Schaller, H. Expression in yeast and tobacco of plant c DNAs encoding acyl-CoA: Diacylglycerol acyltransferase. *Eur. J. Biochem.* **2000**, *267*, 85–96. [CrossRef]
13. Nykiforuk, C.L.; Laroche, A.; Weselake, R.J. Isolation and characterization of a c DNA encoding a second putative diacylglycerol acyltransferase from a microspore-derived cell suspension culture of *Brassica napus* L. cv Jet Neuf. *Plant Physiol.* **1999**, *121*, 1957–1959.
14. Salanoubat, M.; Lemcke, K.; Rieger, M.; Ansorge, W.; Unseld, M.; Fartmann, B.; Valle, G.; Blöcker, H.; Perez-Alonso, M.; Obermaier, B.; et al. Sequence and analysis of chromosome 3 of the plant *Arabidopsis thaliana*. *Nature* **2000**, *408*, 820–822.

15. Lardizabal, K.D.; Mai, J.T.; Wagner, N.W.; Wyrick, A.; Voelker, T.; Hawkins, D.J. DGAT2 is a new diacylglycerol acyltransferase gene family: Purification, cloning, and expression in insect cells of two polypeptides from *Mortierella ramanniana* with diacylglycerol acyltransferase activity. *J. Biol. Chem.* **2001**, *276*, 38862–38869. [CrossRef] [PubMed]
16. Kroon, J.T.; Wei, W.; Simon, W.J.; Slabas, A.R. Identification and functional expression of a type 2 acyl-Co A: Diacylglycerol acyltransferase (*DGAT2*) in developing castor bean seeds which has high homology to the major triglyceride biosynthetic enzyme of fungi and animals. *Phytochemistry* **2006**, *67*, 2541–2549. [CrossRef]
17. Shockey, J.M.; Gidda, S.K.; Chapital, D.C.; Kuan, J.-C.; Dhanoa, P.K.; Bland, J.M.; Rothstein, S.J.; Mullen, R.T.; Dyer, J.M. Tung tree *DGAT1* and *DGAT2* have nonredundant functions in triacylglycerol biosynthesis and are localized to different subdomains of the endoplasmic reticulum. *Plant Cell* **2006**, *18*, 2294–2313. [CrossRef]
18. Saha, S.; Enugutti, B.; Rajakumari, S.; Rajasekharan, R. Cytosolic triacylglycerol biosynthetic pathway in oilseeds. Molecular cloning and expression of peanut cytosolic diacylglycerol acyltransferase. *Plant Physiol.* **2006**, *141*, 1533–1543. [CrossRef]
19. Hernandez, M.L.; Whitehead, L.; He, Z.; Gazda, V.; Gilday, A.; Kozhevnikova, E. A cytosolic acyltransferase contributes to triacylglycerol synthesis in sucrose-rescued *Arabidopsis* seed oil catabolism mutants. *Plant Physiol.* **2012**, *160*, 215–225. [CrossRef]
20. Zheng, L.; Shockey, J.; Guo, F.; Shi, L.; Li, X.; Shan, L.; Wan, S.; Peng, Z. Discovery of a new mechanism for regulation of plant triacylglycerol metabolism: The peanut diacylglycerol acyltransferase-1 gene family transcriptome is highly enriched in alternative splicing variants. *J. Plant Physiol.* **2017**, *219*, 62–70. [CrossRef] [PubMed]
21. Kamisaka, Y.; Mishra, S.; Nakahara, T. Purification and characterization of diacylglycerol acyltransferase from the lipid body fraction of an oleagious fungus. *J. Biol. Chem.* **1997**, *121*, 1107–1114.
22. Lardizabal, K.; Effertz, R.; Levering, C.; Mai, J.; Pedroso, M.C.; Jury, T.; Aasen, E.; Gruys, K.; Bennett, K. Expression of Umbelopsis ramanniana *DGAT2A* in seed increases oil in soybean. *Plant Physiol.* **2008**, *148*, 89–96. [CrossRef] [PubMed]
23. Chen, B.B.; Wang, J.J.; Zhang, G.Y. Two types of soybean diacylglycerol acyltransferases are differentially involved in triacylglycerol biosynthesis and response to environmental stresses and hormones. *Sci. Rep.* **2016**, *6*, 28541. [CrossRef]
24. Xu, Y.; Yan, F.; Liu, Y.; Wang, Y.; Gao, H.; Zhao, S.; Wang, Q.; Li, J. Quantitative proteomic and lipidomics analyses of high oil content GmDGAT1-2 transgenic soybean, illustrates the regulatory mechanism of *lipoxygenase* and *oleosin*. *Plant Cell Rep.* **2021**, *40*, 2303–2323. [CrossRef] [PubMed]
25. Clough, S.J.; Bent, A.F. Floral dip: A simplified method for Agrobacterium-mediated transformation of *Arabidopsis thaliana*. *Plant J.* **1998**, *16*, 735–743. [CrossRef] [PubMed]
26. Bo, J.; Liu, B.; Bi, Y.; Hou, W.; Wu, C. Validation of Internal Control for Gene Expression Study in Soybean by Quantitative Real-Time PCR. *BMC Mol. Biol.* **2008**, *9*, 59.
27. Hu, R.; Fan, C.; Li, H.; Zhang, Q. Evaluation of Putative Reference Genes for Gene Expression Normalization in Soybean by Quantitative Real-Time RT-PCR. *BMC Mol. Biol.* **2009**, *10*, 93. [CrossRef]
28. Le, D.T.; Aldrich, D.L.; Valliyodan, B.; Watanabe, Y.; Ha, C.V.; Nishiyama, R.; Guttikonda, S.K.; Quach, T.N.; Gutierrez-Gonzalez, J.J.; Tran, L.S.P.; et al. Evaluation of Candidate Reference Genes for Normalization of Quantitative RT-PCR in Soybean Tissues under Various Abiotic Stress Conditions. Edite par Christian Schönbach. *PLoS ONE* **2012**, *7*, e46487. [CrossRef]
29. Browse, J.; McCourt, P.J.; Somerville, C.R. Fatty acid composition of leaf lipids determined after combined digestion and fatty acid methyl ester formation from fresh tissue. *Anal. Biochem.* **1986**, *152*, 141–145. [CrossRef]
30. Broun, P.; Gettner, S.; Somerville, C. Genetic engineering of plant lipids. *Annu. Rev. Nutr.* **1999**, *19*, 197–216. [CrossRef]
31. Li-Beisson, Y.; Shorrosh, B.; Beisson, F.; Andersson, M.X.; Arondel, V.; Bates, P.D.; Baud, S.; Bird, D.; Debono, A.; Durrett, T.P.; et al. Acyl-lipidmetabolism. *Arab. Book* **2010**, *8*, e0133. [CrossRef]
32. Sun, Q.X. Cloning and Characterization of Genes Involved in the Biosynthesis of Very Long Chain Polyunsaturated Fatty Acid and the Reconstitution of This Pathway in Crop Plants. Ph.D. Thesis, Shandong Agricultural University, Taian, China, 2011.
33. Jako, C.; Kumar, A.; Wei, Y.D.; Zou, J.; Barton, D.L.; Giblin, E.M.; Covello, P.S.; Taylor, D.C. Seed-specific over-expression of an *Arabidopsis* cDNA encoding a diacylglycerol acyltransferase enhances seed oil content and seed weight. *Plant Physiol.* **2001**, *126*, 861–874. [CrossRef] [PubMed]
34. Bao, X.; Ohlrogge, J. Supply of triacylglycerol in developing embryos. *Plant Physiol.* **1999**, *120*, 1057–1062. [CrossRef] [PubMed]
35. Yun, S.; Isleib, T.G. Genetic analysis of fatty acids in American peanuts. *Chin. J. Oil Crops* **2000**, *22*, 35–37.
36. Zou, J.; Wei, Y.D.; Jako, C.; Kumar, A.; Selvaraj, G.; Taylor, D.C. The *Arabidopsis thaliana* TAG1 mutant has a mutation in a diacylglycerol acyltransferase gene. *Plant J.* **1999**, *19*, 645–653. [CrossRef]
37. Feussner, I.; Kuhn, H.; Wasternack, C. Lipoxygenase-dependent degradation of storage lipids. *Trends Plant Sci.* **2001**, *6*, 268–273. [CrossRef] [PubMed]

Disclaimer/Publisher's Note: The statements, opinions and data contained in all publications are solely those of the individual author(s) and contributor(s) and not of MDPI and/or the editor(s). MDPI and/or the editor(s) disclaim responsibility for any injury to people or property resulting from any ideas, methods, instructions or products referred to in the content.

Editorial

Advances in Crop Molecular Breeding and Genetics

Wanning Liu, Guan Li, Jiezheng Ying and Zhiyong Li *

State Key Laboratory of Rice Biology and Breeding, China National Rice Research Institute, Hangzhou 311400, China; dearliuwanning@126.com (W.L.); liguan@caas.cn (G.L.); yingjiezheng@caas.cn (J.Y.)
* Correspondence: lizhiyong@caas.cn

Citation: Liu, W.; Li, G.; Ying, J.; Li, Z. Advances in Crop Molecular Breeding and Genetics. *Agronomy* 2023, *13*, 2311. https://doi.org/10.3390/agronomy13092311

Received: 14 August 2023
Accepted: 30 August 2023
Published: 1 September 2023

Copyright: © 2023 by the authors. Licensee MDPI, Basel, Switzerland. This article is an open access article distributed under the terms and conditions of the Creative Commons Attribution (CC BY) license (https://creativecommons.org/licenses/by/4.0/).

Selecting crop varieties with high and stable yields, as well as improving quality and economic benefits, has become a long-term topic while facing the continuous increasing population and the adverse effects of environmental changes. Crop breeding is an important way to ensure human food security all over the world. Over recent decades, the process of crop breeding has greatly accelerated due to the application of modern molecular biology technologies and the basis of plant genetics. However, scientists and crop breeders still have a long way to go. Recently, some researchers have improved rice yields or quality by controlling grain size [1], some have optimized crop growth by increasing nitrogen use efficiency [2], and some aim to make crops more resilient to cold, salt, and drought to ensure high yields [3–5]. These understandings of the molecular basis for improved product quality, nutrient utilization, and adaptation to stress highlight the need for higher crop yields.

This issue focuses on the latest fundamental discoveries in crop genetics, germplasm resources, crop adaptation to climate change, and their potential applications. These findings will play an important role in regulating important developmental processes, finding beneficial agronomic traits, and achieving high yields. This issue has published five articles on the topic "Crop Molecular Breeding and Genetics". Different technologies were used to explore the relevant genes of important traits that affect crop yield and quality, such as whole genome analysis, phylogenetic analysis, multi-omics technology, and molecular marker assistance technology. There are also different perspectives, including molecular mechanisms related to seed germination, grain development, rice quality, heat stress, and carotenoid regulation. It helps to understand the regulatory mechanisms of important agronomic traits and provides gene resources for crop genetic improvement.

Maize (*Zea mays* L.) is an important source of food, feed, and industrial raw materials. Currently, global maize production has exceeded that of rice and wheat [6]. Healthy seed germination is important for improving the yield and quality of maize, but the molecular mechanisms regulating maize seed germination are still unclear. Generally, gibberellin (GA) is considered a phytohormone which has the function of releasing seed dormancy and promoting seed germination [7]. A recent study by Han et al. investigated the molecular mechanism of GA-induced maize seed germination, using multi-omics analysis, including transcriptome, miRNA, and degradome sequencing. Multiple items were found to be closely related to the seed germination process. A newly discovered lipid metabolism-related gene *ZmSLP* has a negative regulatory effect on maize germination. Over-expression of this gene in *Arabidopsis* can lead to seed lipid metabolism disorders and inhibit seed germination and seedling growth. This study provided valuable information for molecular research on maize seed germination.

In high plants, HD-Zip transcription factors play an important role in plant growth and tolerance to environmental stress [8]. So far, the HD-Zip gene has been extensively and systematically studied in *Arabidopsis thaliana*, *Manihot esculenta*, and *Zea mays* [9–11]. Yin et al. analyzed the function of HD-Zip gene family related to heat stress and carotenoid accumulation in three genomes of Brassicaceae plants, *B. rapa*, *B. oleracea*, and *B. napus*.

They identified 93, 96, and 184 HD-Zip genes, respectively. They found that the expression level of *BraA09g011460.3C* was up-regulated after heat stress treatment, and significantly decreased in varieties with high carotenoid content, indicating that it has the potential for heat tolerance and regulating the level of carotenoid. This study provides important gene resources for the follow-up breeding of Chinese cabbage.

The NLR (nucleotide-binding site leucine-rich repeat receptor) gene family is large and diverse, and can activate ETI (effector-triggered immunity) in response to pathogen effectors and subsequently mediate immune signaling [12,13]. The unique composition of the NLR gene family in papaya (*Carica papaya* L.) has attracted researchers to study its characteristics, evolution, and function.

Papaya is a special plant with fewer genes than most flowering plant genomes, and the lack of disease resistance genes was reconfirmed in the latest genome release [14]. Wu et al. identified 59 NLR genes from the improved papaya genome via a customized RGAugury. They conducted a comprehensive analysis, including structural composition, sequence diversity, chromosome distribution, and phylogenetic analysis. The NLR family members were identified and classified more accurately than previous research on papaya NLR. Wu et al. showed that the NLR family of papaya is a simplified set of NLRs in typical Eudicots, making papaya a suitable plant model for studying basic disease resistance genes. This study provides a new perspective for the evolution of the NLR gene in papaya, which will help us better understand the complex and diverse disease resistance genes in Eudicots, and provides a basis for disease resistance breeding of crops.

Among various traits related to rice yield, grain filling is considered a limiting factor affecting rice yield and quality, and some reports support a close relationship between grain filling and starch metabolism in plants [15,16]. We still require further research to solve the problem of high yields and high grain-filling ratio in cultivated rice varieties using modern molecular biology technologies [17,18]. The new research findings reported by Lee et al. revealed several QTLs related to rice grain filling. They used a doubled haploid (DH) population along with Kompetitive allele-specific PCR (KASP) markers and Fluidigm markers to achieve this. Notably, *qFG3*, *qFG5-1*, and *qFG5-2* were significant in grain filling. The newly discovered *qFG3* has been detected in both early and normal cultivation environment and is considered a stable QTL that can serve as a useful gene source for breeding. *qFG3* carries genes related to cell division, elongation and differentiation, photosynthesis, and starch synthesis. This study provides target QTL regions for future breeding work, including fine mapping and functional characterization of candidate genes during rice filling.

In recent decades, the yield of rice has increased significantly, basically meeting the demand. However, there is a contradiction between high yield and high quality in crops generally. With the gradual improvement of people's living standard and consumption level, improving the quality of rice has become more and more important [19]. Gong et al. reviewed the genes that had beneficial effects on rice quality and their applications in breeding practices from four aspects: milling quality, appearance quality, edible and cooking quality, and nutritional quality. Significant progress has been made in the study of rice quality functional genomics, with the cloning of many important genes related to the regulation of rice quality traits. However, rice quality is composed of multiple traits, and there are always interactions between different quality traits and environmental factors. The application of new technologies such as GWAS, genetically modified organisms, and gene editing can accelerate the improvement of rice quality. Gong et al. also believe that different populations show diverse taste preferences under different environmental conditions. Therefore, developing fragrant rice varieties, strengthening the breeding of high-quality conventional indica rice, and developing functional rice with special nutritional value or specific needs of specific populations are important directions for future rice breeding. This review further deepens our understanding of rice quality regulation and breeding applications.

In summary, the studies presented in "Crop Molecular Breeding and Genetics" will help to improve the understanding of important traits and their molecular mechanisms and help to develop effective and efficient trait improvement strategies. The research of Han et al. revealed a new mechanism for regulating seed germination in maize. Wu et al. provided a new perspective for the evolution of the NLR gene in papaya and a basis for disease resistance breeding of crops. Lee et al. revealed several QTLs related to rice grain filling and provided a new target region for downstream breeding. Gong et al. reviewed excellent genes that affect rice quality and their applications in breeding. They will be beneficial for future work on genomic selection, QTL mapping, marker-assisted selection, gene editing, and breeding design in crops. It is expected that crop molecular breeding will have a broader development, from laboratory research to field selection, and new varieties with ideal traits will be cultivated using comprehensive methods.

Funding: This research was funded by grants from the National Natural Science Foundation of China (32201805), China Postdoctoral Science Foundation (2023T160701), and Zhejiang Provincial Natural Science Foundation of China (LQ21C130003).

Conflicts of Interest: The authors declare that the research was conducted in the absence of any commercial or financial relationships that could be construed as a potential conflict of interest.

References

1. Ren, D.; Ding, C.; Qian, Q. Molecular bases of rice grain size and quality for optimized productivity. *Sci. Bull.* **2023**, *68*, 314–350.
2. Hou, M.; Yu, M.; Li, Z.; Ai, Z.; Chen, J. Molecular Regulatory Networks for Improving Nitrogen Use Efficiency in Rice. *Int. J. Mol. Sci.* **2021**, *22*, 9040. [CrossRef] [PubMed]
3. Li, J.; Zhang, Z.; Chong, K.; Xu, Y. Chilling tolerance in rice: Past and present. *J. Plant Physiol.* **2022**, *268*, 153576.
4. Qin, H.; Li, Y.; Huang, R. Advances and Challenges in the Breeding of Salt-Tolerant Rice. *Int. J. Mol. Sci.* **2020**, *21*, 8385. [CrossRef]
5. Oladosu, Y.; Rafii, M.Y.; Samuel, C.; Fatai, A.; Magaji, U.; Kareem, I.; Kamarudin, Z.S.; Muhammad, I.; Kolapo, K. Drought Resistance in Rice from Conventional to Molecular Breeding: A Review. *Int. J. Mol. Sci.* **2019**, *20*, 3519. [CrossRef] [PubMed]
6. Cui, W.; Song, Q.; Zuo, B.; Han, Q.; Jia, Z. Effects of Gibberellin (GA_{4+7}) in Grain Filling, Hormonal Behavior, and Antioxidants in High-Density Maize (*Zea mays* L.). *Plants* **2020**, *9*, 978. [CrossRef] [PubMed]
7. Jin, Y.; Wang, B.; Tian, L.; Zhao, L.; Guo, S.; Zhang, H.; Xu, L.; Han, Z. Identification of miRNAs and their target genes associated with improved maize seed vigor induced by gibberellin. *Front. Plant Sci.* **2022**, *13*, 1008872. [CrossRef] [PubMed]
8. Sharif, R.; Raza, A.; Chen, P.; Li, Y.; El-Ballat, E.M.; Rauf, A.; Hano, C.; El-Esawi, M.A. HD-ZIP Gene Family: Potential Roles in Improving Plant Growth and Regulating Stress-Responsive Mechanisms in Plants. *Genes* **2021**, *12*, 1256. [CrossRef] [PubMed]
9. Kamata, N.; Okada, H.; Komeda, Y.; Takahashi, T. Mutations in epidermis-specific HD-ZIP IV genes affect floral organ identity in Arabidopsis thaliana. *Plant J.* **2013**, *75*, 430–440. [CrossRef] [PubMed]
10. Ding, Z.; Fu, L.; Yan, Y.; Tie, W.; Xia, Z.; Wang, W.; Peng, M.; Hu, W.; Zhang, J. Genome-wide characterization and expression profiling of HD-Zip gene family related to abiotic stress in cassava. *PLoS ONE* **2017**, *12*, e0173043. [CrossRef]
11. Vernoud, V.; Laigle, G.; Rozier, F.; Meeley, R.B.; Perez, P.; Rogowsky, P.M. The HD-ZIP IV transcription factor OCL4 is necessary for trichome patterning and anther development in maize. *Plant J.* **2009**, *59*, 883–894. [CrossRef] [PubMed]
12. Lolle, S.; Stevens, D.; Coaker, G. Plant NLR-triggered immunity: From receptor activation to downstream signaling. *Curr. Opin. Immunol.* **2020**, *62*, 99–105. [CrossRef] [PubMed]
13. Wang, J.; Song, W.; Chai, J. Structure, biochemical function, and signaling mechanism of plant NLRs. *Mol. Plant* **2023**, *16*, 75–95. [CrossRef] [PubMed]
14. Yue, J.; VanBuren, R.; Liu, J.; Fang, J.; Zhang, X.; Liao, Z.; Wai, C.M.; Xu, X.; Chen, S.; Zhang, S.; et al. SunUp and Sunset genomes revealed impact of particle bombardment mediated transformation and domestication history in papaya. *Nat. Genet.* **2022**, *54*, 715–724. [CrossRef] [PubMed]
15. Tang, T.; Xie, H.; Wang, Y.; Lü, B.; Liang, J. The effect of sucrose and abscisic acid interaction on sucrose synthase and its relationship to grain filling of rice (*Oryza sativa* L.). *J. Exp. Bot.* **2009**, *60*, 2641–2652. [CrossRef] [PubMed]
16. Jiang, Z.; Chen, Q.; Chen, L.; Yang, H.; Zhu, M.; Ding, Y.; Li, W.; Liu, Z.; Jiang, Y.; Li, G. Efficiency of Sucrose to Starch Metabolism Is Related to the Initiation of Inferior Grain Filling in Large Panicle Rice. *Front. Plant Sci.* **2021**, *12*, 732867. [CrossRef] [PubMed]
17. Zhang, W.; Cao, Z.; Zhou, Q.; Chen, J.; Xu, G.; Gu, J.; Liu, L.; Wang, Z.; Yang, J.; Zhang, H. Grain Filling Characteristics and Their Relations with Endogenous Hormones in Large- and Small-Grain Mutants of Rice. *PLoS ONE* **2016**, *11*, e0165321. [CrossRef] [PubMed]

18. Yang, J.; Zhang, J. Grain-filling problem in 'super' rice. *J. Exp. Bot.* **2010**, *61*, 1–5. [CrossRef]
19. Rao, Y.; Li, Y.; Qian, Q. Recent progress on molecular breeding of rice in China. *Plant Cell Rep.* **2014**, *33*, 551–564. [CrossRef] [PubMed]

Disclaimer/Publisher's Note: The statements, opinions and data contained in all publications are solely those of the individual author(s) and contributor(s) and not of MDPI and/or the editor(s). MDPI and/or the editor(s) disclaim responsibility for any injury to people or property resulting from any ideas, methods, instructions or products referred to in the content.

Article

Multi-Omics Revealed the Molecular Mechanism of Maize (*Zea mays* L.) Seed Germination Regulated by GA3

Zanping Han *,†, Yunqian Jin †, Bin Wang and Yiyang Guo

College of Agronomy, Henan University of Science and Technology, Luoyang 471003, China; jyq920422@163.com (Y.J.); cansong@163.com (B.W.); guoyiyang0830@163.com (Y.G.)
* Correspondence: hnlyhzp@163.com
† These authors contributed equally to this work.

Abstract: Maize is a valuable raw material for feed and food production. Healthy seed germination is important for improving the yield and quality of maize. However, the molecular mechanisms that regulate maize seed germination remain unclear. In this study, multi-omics was used to reveal the molecular mechanism of seed germination induced by gibberellin (GA) in maize. The results indicated that 25,603 genes were differentially expressed (DEGs) and annotated in the GO database, of which 2515 genes were annotated in the KEGG database. In addition, 791 mature miRNAs with different expression levels were identified, of which 437 were known in the miRbase database and 354 were novel miRNAs. Integrative analysis of DEGs and miRNAs suggested that carbohydrate, lipid, amino acid, and energy metabolisms are the primary metabolic pathways in maize seed germination. Interestingly, a lipid metabolism-related gene named *ZmSLP* was found to negatively regulate maize germination. We transformed this gene into *Arabidopsis thaliana* to verify its function. The results showed that the germination rate of transgenic *Arabidopsis* seeds was obviously decreased, and the growth of seedlings was weaker and slower than that of WT plants, suggesting that this gene plays an important role in promoting seed germination. These findings provide a valuable reference for further research on the mechanisms of maize seed germination.

Keywords: maize; seed germination; multi-omics; gibberellin; lipid metabolism

Citation: Han, Z.; Jin, Y.; Wang, B.; Guo, Y. Multi-Omics Revealed the Molecular Mechanism of Maize (*Zea mays* L.) Seed Germination Regulated by GA3. *Agronomy* **2023**, *13*, 1929. https://doi.org/10.3390/agronomy13071929

Academic Editors: Jiezheng Ying, Zhiyong Li and Chaolei Liu

Received: 29 June 2023
Revised: 18 July 2023
Accepted: 19 July 2023
Published: 21 July 2023

Copyright: © 2023 by the authors. Licensee MDPI, Basel, Switzerland. This article is an open access article distributed under the terms and conditions of the Creative Commons Attribution (CC BY) license (https://creativecommons.org/licenses/by/4.0/).

1. Introduction

Maize (*Zea mays* L.) is one of the most important cereal crops worldwide and is widely used as a terrestrial food, fodder, and industrial raw material [1,2]. More outstanding maize production is required to meet the growing demands of the ever-increasing world population [3]. Studies have shown that the global production of maize has recently exceeded that of rice and wheat [2]. Seed germination is a complicated process initiated by the uptake of water (imbibition) by dry seeds and ending with the emergence of the radicle through the seed coat. The following processes include the induction of translation, transcription, cell division, and energy metabolism, which involve a series of differentially expressed genes and their corresponding regulatory networks [4–7]. Therefore, the efficient and healthy germination of maize seeds, which is essential for maize production, directly affects maize yield and grain quality.

In recent decades, plant growth regulators (PGRs), including gibberellins (GA), have attracted the interest of agricultural scientists and are widely used in agronomic crops. Previous studies have reported that GAs, which are ubiquitous in higher plants, contain a large family of hormones that have long been known as endogenous GAs. GAs promote plant growth and developmental processes such as seed germination, cell division, stem elongation, dormancy, leaf expansion, flowering, and fruit development [2,8,9]. GAs and abscisic acid (ABA) are the two major phytohormones that regulate seed germination; GA promotes seed germination, whereas ABA induces seed dormancy. In addition, GA can break quiescence and promote seed germination by increasing the content of hydrolytic enzymes, soluble sugars, and amino acids [5,10,11]. In contrast, GA-deficient mutants exhibit stronger seed dormancy and fail to complete seed germination without applying exogenous gas [12,13]. Despite extensive research being published about the molecular mechanism of maize seed germination, it remains insufficient.

Increasingly advanced technologies and research methods are used to decode the molecular mechanisms of various plant life activities, including transcriptomics, miRNA, degradome, DNA methylation sequencing, metabolomics, and isobaric tags for relative and absolute quantitation (iTRAQ) proteomic approaches. Guo et al. utilized transcription-associated metabolomics to establish a model to reveal an ABA-dependent maize acclimation mechanism to the stress combination [14]. Studies have shown that two hub genes, auxin response factor 4 (*ARF4*) and amino acid permease 3 (*AAP3*), play central roles in the regulation of Cd-responsive genes using integration analysis of small RNAs and degradome and transcriptome sequencing in the hyperaccumulator *Sedum alfredii* [15]. High seed vigor and high-quality seed germination are important for agriculture. Therefore, to better understand the involvement and regulatory mechanism in the process, Gong et al. identified miRNAs and their targets associated with sweet corn seed vigor by combing small RNAs and degradome sequencing, finally obtaining 26 target genes cleaved by nine differentially expressed miRNAs that might play roles in the regulation of seed vigor [16]. However, reports on the molecular mechanisms regulating GAs using multi-omics research methods are rare.

In this study, two inbred maize lines, Yu537 and Yu82, were selected and treated with 400 mg/L GA3 to investigate the molecular mechanism of corn seed germination by multi-omics analysis, including transcriptome, miRNA, and degradome sequencing. This can provide valuable information for understanding the regulatory mechanism of corn seed germination.

2. Materials and Methods

2.1. Plant Material and Seed Germination

Yu82 and Yu537A, two maize (*Z. mays*) inbred lines, were selected for this study. They were obtained from the widely planted cultivar, Yuzong5. To eliminate the age and maternal effects on seeds, two inbred lines were planted in Sanya with the same management conditions. The seeds of Yu82 and Yu537A were harvested at the same developmental stage (mature stage, black layer formed). Seeds from two inbred lines were used for surface sterilization with 75% ethyl alcohol. GA3 (400 mg/L) was used to treat Yu537A seeds to promote germination. The conditions used for the germination of two seeds were below 25 °C, 14/10 h (light/dark), and an illumination intensity of 5000 lx. Purified water (20 mL) was added daily at a fixed time. Each treatment was repeated in triplicate.

2.2. RNA Extraction and the Library Construction of Transcriptome Sequencing

Seed embryos were cut with a sterile test blade. The fresh embryos of seeds were frozen in liquid nitrogen and stored at −80 °C. High-quality total RNA was isolated using the Tiangen RNAprep extraction kit. The concentration and quality of the RNA were checked using a Nanodrop 2000. The eligible RNAs were used for library construction and transcriptome sequencing.

2.3. Library Construction for Transcriptome Sequencing

Purified RNA was fragmented into short segments using a fragmentation buffer. Fragmented mRNA was used as the template to synthesize the first cDNA strand using random hexamers, and then the second was obtained using buffer solution, dNTPs, RNaseH, and DNA Polymerase I. T4 DNA polymerase and Klenow DNA polymerase were used to repair the sticky end of DNA into a flat end and add base A and adapters to the 3′ end. Finally, a sequencing library was constructed by PCR amplification. An eligible library was used for sequencing with the Illumina Hiseq4000 platform, and the sequencing read length was double-ended at 2 × 150 bp (PE150).

2.4. Filtration of Sequencing Data

The raw data file contained short sequences (reads) of approximately 150 bp, which could not be directly used for mRNA analysis. To ensure accurate and reliable results, raw data must be preprocessed, including removing sequencing connectors (introduced during database construction) and low-quality sequencing data (due to sequencer errors).

2.5. Comparative Analysis with the Reference Genome

Valid data were used to map the reference genome after filtering out invalid data using the Hisat program. Gene location information specified in the genome annotation gtf file was statistically analyzed as follows: (1) read statistics of sequencing data were compared with the reference genome; (2) a summary of the regional distribution of sequencing data was compared with the reference genome; and (3) the chromosome density distribution of sequencing data was compared with the reference genome. Information regarding the regions of the reference genome can be defined as comparisons to exons, introns, and intergenic regions. Under normal circumstances, the percentage of sequence localization in the exon region should be the highest. In contrast, the comparison of reads to intron and intergenic regions may be due to splicing events of precursor mRNA, incomplete genome annotation, DNA contamination, and background noise.

2.6. Expression Analysis of Differentially Expressed Genes

Gene expression level was calculated with the FPKM (Fragments Per Kilobase of exon model per Million mapped reads) value. Differentially expressed genes were obtained with R-language at p value < 0.05 and p value < 0.01. The exon model FPKM value was used for gene expression measurements. The number of genes was counted in different expression regions. Based on the differentially expressed genes, GO and KEGG enrichments were analyzed in the public database.

2.7. Association Analysis of Multi-Omics

The miRNA sequences used in the analysis of the degradation group were all miRNAs identified by small RNA sequencing, and the database to be compared was made up of sequences spliced by transcriptome sequencing. The relationship between miRNAs and their target genes was determined using degradation group sequencing. Based on the analysis of the degradation group, we integrated the expression profiles of miRNAs and target genes in the different comparison groups to obtain an overall table of miRNAs and target genes. In addition, information was extracted from the general table to find negative regulatory relationship pairs between miRNAs and target genes in different comparison groups, and network regulatory analysis of miRNAs and target genes was conducted.

GO and KEGG enrichment analyses were performed on target genes of miRNA in each comparison group. First, the number of genes corresponding to target genes of all selected miRNAs corresponding to each function or pathway annotation was statistically calculated. A hypergeometric test was then applied to calculate the p value for significant enrichment. The main biological functions of the miRNA-target gene relationship can be determined by functional significance enrichment analysis, with a p-value ≤ 0.05 as the threshold.

The formula for calculating the p-value of significance is as follows:

$$p = 1 - \sum_{i=0}^{s-1} \frac{\binom{B}{i}\binom{TB-B}{TS-i}}{\binom{TB}{TS}}$$

where S is the number of annotated genes with significant expression differences in a GO item; TS is the number of genes with significant expression differences; B is the number of genes in a GO item; and TB is the number of total genes.

2.8. Acquisition of Transgenic Arabidopsis Plants

Agrobacterium vectors were inoculated into LB liquid medium containing 25 mg·L^{-1} Rif and 50 mg·L^{-1} kanamycin to activate the culture for 2 d. The culture conditions were set at 28 °C and 200 rpm. According to the ratio of bacteria solution to LB medium of 1:50 (V/V), 1 mL of the above bacteria solution was absorbed and added into LB liquid medium containing the corresponding resistance at 28 °C, followed by overnight culture at 200 rpm until the OD600 value was between 0.6 and 1.0. That is, the color of the medium changed from clear brick red to orange-yellow. The Agrobacterium solution with the expected OD value was centrifuged at 5000 rpm for 8 min. The supernatant was discarded to collect the bacteria. Sterile water with a 1/2 volume of bacterial solution was used to suspend the bacteria, and an equal volume of 2× suspension buffer was added before the transformation. On the second day, the mixed transformation vector was placed in a 50 mL centrifuge tube, and the inflorescences of the white flower buds were immersed in the bacterial solution for approximately 30 s. The inflorescences were then dripped onto the flower buds using a sterile dropper. After transformation, the plants were shaded in black plastic bags and placed in a paper box with 24 h of moisture. The material was placed under normal growth conditions at the end of shading. The inflorescences were retreated using the above transformation method after one week. Finally, the material was placed under normal conditions until the seeds matured. Total DNA was extracted using the improved CTAB method [17], and PCR detection was performed with specific primers for screened marker genes (*NptII*); positive plants were used for verification tests.

3. Results

3.1. Overview of the Maize Seed Transcriptome

To investigate the molecular mechanisms of seed germination induced by GA$_3$ in maize, differentially expressed genes (DEGs) were explored using RNA-Seq technology. Six groups containing 18 samples were obtained with three replicates (Supplemental Table S1). In total, 139.54 GB of valid bases with a mean GC content of 53.08% were obtained. After data processing, the Q30 values ranged from 91.69% to 95.49% (Supplemental Tables S2 and S3). In total, 138,049 transcripts and 44,117 genes with different expression levels were identified. Among these, 25,603 genes were annotated in the gene ontology (GO) database, and 2515 genes were annotated in the KEGG (Kyoto Encyclopedia of Genes and Genomes) database. All valid reads were mapped to the reference genome (maize B73) (Table 1). The average mapped ratio was 87.53%, providing a reliable reference. According to the genomic region information of the reference genome, the mapping regions could be divided into exon, intron, and intergenic regions. The results indicated that the percentage of exon regions was the highest. The percentage of intron regions was the lowest (Supplemental Figure S1), which may correlate with the cleavage events of pre-mRNAs, incomplete genome annotation, DNA contamination, and background noise.

Table 1. Statistical analysis of mapped reads.

Sample	Valid Reads	Mapped Reads	Unique Mapped Reads	Multi Mapped Reads	PE Mapped Reads	Non-Splice Reads	Splice Reads
E_6h1	44,522,844	38,872,898 (87.31%)	25,118,045 (56.42%)	13,754,853 (30.89%)	35,328,974 (79.35%)	23,649,652 (53.12%)	1,049,1882 (23.57%)
E_6h2	52,779,658	45,489,407 (86.19%)	27,416,327 (51.94%)	18,073,080 (34.24%)	40,859,244 (77.41%)	26,677,569 (50.55%)	12,594,962 (23.86%)
E_6h3	48,573,036	42,144,424 (86.77%)	25,152,219 (51.78%)	16,992,205 (34.98%)	38,151,374 (78.54%)	24,484,992 (50.41%)	11,414,228 (23.50%)
S_6h1	56,094,828	47,116,337 (83.99%)	26,205,903 (46.72%)	20,910,434 (37.28%)	41,867,540 (74.64%)	26,352,750 (46.98%)	12,264,970 (21.86%)
S_6h2	56,605,766	47,223,659 (83.43%)	27,515,806 (48.61%)	19,707,853 (34.82%)	41,761,020 73.78%)	28,352,616 (50.09%)	12,250,405 (21.64%)
S_6h3	56,941,752	47,673,165 (83.72%)	28,036,288 (49.24%)	19,636,877 (34.49%)	42,222,202 (74.15%)	28,223,671 (49.57%)	12,889,242 (22.64%)
E_53h1	51,623,746	45,966,061 (89.04%)	29,463,816 (57.07%)	16,502,245 (31.97%)	42,035,450 (81.43%)	22,812,230 (44.19%)	18,799,134 (36.42%)
E_53h2	51,263,292	45,899,826 (89.54%)	31,031,019 (60.53%)	14,868,807 (29.00%)	42,108,328 (82.14%)	23,255,978 (45.37%)	19,879,897 (38.78%)
E_53h3	56,832,242	50,841,433 (89.46%)	32,269,169 (56.78%)	18,572,264 (32.68%)	46,704,020 (82.18%)	26,033,538 (45.81%)	19,548,995 (34.40%)
S_53h1	54,294,104	47,469,186 (87.43%)	30,395,012 (55.98%)	17,074,174 (31.45%)	43,186,082 (79.54%)	23,781,631 (43.80%)	19,888,369 (36.63%)
S_53h2	48,938,972	43,265,705 (88.41%)	27,044,065 (55.26%)	16,221,640 (33.15%)	39,668,690 (81.06%)	21,719,154 (44.38%)	16,712,438 (34.15%)
S_53h3	50,377,860	44,288,776 (87.91%)	28,249,543 (56.08%)	16,039,233 (31.84%)	40,480,912 (80.35%)	22,362,326 (44.39%)	18,059,268 (35.85%)
SGA_53h1	53,650,262	47,512,991 (88.56%)	30,986,382 (57.76%)	16,526,609 (30.80%)	43,394,170 (80.88%)	23,359,239 (43.54%)	20,725,004 (38.63%)
SGA_53h2	45,060,898	39,937,286 (88.63%)	24,497,940 (54.37%)	15,439,346 (34.26%)	36,729,340 (81.51%)	20,075,697 (44.55%)	15,951,613 (35.40%)
SGA_53h3	55,721,064	49,400,868 (88.66%)	30,618,696 (54.95%)	18,782,172 (33.71%)	45,147,652 (81.02%)	24,821,328 (44.55%)	20,102,437 (36.08%)
S_78h1	51,162,146	44,975,042 (87.91%)	28,480,225 (55.67%)	16,494,817 (32.24%)	41,083,328 (80.30%)	23,069,493 (45.09%)	18,061,692 (35.30%)
S_78h2	53,589,910	48,079,252 (89.72%)	31,263,293 (58.34%)	16,815,959 (31.38%)	44,274,240 (82.62%)	23,985,179 (44.76%)	21,043,894 (39.27%)
S_78h3	42,286,200	37,561,743 (88.83%)	23,876,043 (56.46%)	13,685,700 (32.36%)	34,416,830 (81.39%)	18,772,494 (44.39%)	16,135,439 (38.16%)

Note: Samples and sample names for sequencing. Valid reads represent clean data obtained after quality control. The mapped reads represent the number of reads matched to the genome. Uniquely Mapped reads represent the number of reads that can be uniquely matched to one location in the genome. Multi-mapped reads represent the number of reads that can be mapped to multiple locations in a genome. PE Mapped reads are paired-end sequencing reads mapped to the genome. Reads mapped to the sense strand represent the reads that were mapped to the sense strand. Reads mapped to the antisense strand represent reads that were mapped to the antisense strand. Non-splice reads represent reads that can be mapped to genomic regions by end-to-end alignment. Splice reads cannot be mapped to genomic regions by end-to-end alignment.

3.2. Analysis of DEGs Induced by GA3 in the Germination Process of Maize Seeds

The fragments per kilobase of transcript per million mapped reads (FPKM) value was used to represent the expression of each gene (Supplemental Figure S2), and significant DEGs were screened based on a p-value < 0.05. The number of significant DEGs in the different groups was counted (Figure 1), and the results showed noticeable expression changes. A total of 4312 significant DEGs were induced in the E_53h vs. E_6h group, of which 2593 DEGs were upregulated and 1719 DEGs were downregulated. A total of 5696 significant DEGs were identified in the S_53h vs. S_6h group, of which 3760 were upregulated and 1936 were downregulated. In the group of SGA_53h vs. S_6h, a total of 4815 DEGs were discovered, including 2104 upregulated DEGs and 2711 downregulated DEGs. Only 2704, 1086, and 835 DEGs were obtained from groups S_78h vs. S_6h, S_78h vs. SGA_53h, and SGA_53h vs. S_53h, respectively. Drastically reduced numbers of DEGs in the later stages of maize seed germination suggest that the regulation of the early stages of imbibition germination of maize seeds is more complex and requires more DEGs. After GA3 treatment, only 835 significant DEGs were induced in the SGA_53h sample compared to the S_53h sample, of which 322 were upregulated and 513 were downregulated, indicating that GA3 may induce other specific germination-related genes.

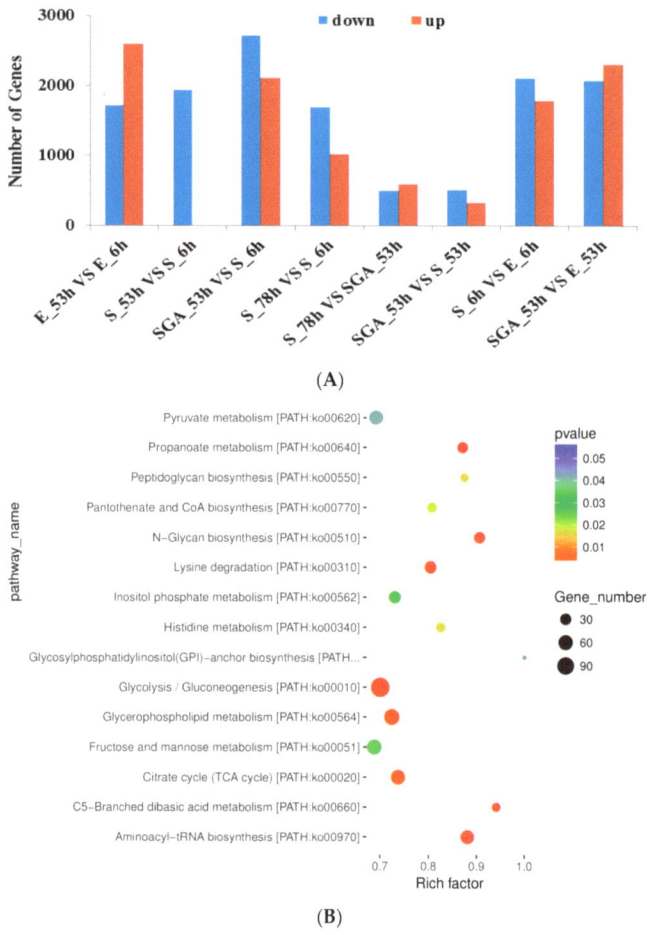

Figure 1. Number of significant differentially expressed genes (DEGs) in different groups and KEGG analysis of DEGs. (**A**) Number of significant DEGs in different groups. Red and blue indicate upregulation and downregulation, respectively. (**B**) KEGG analysis of significant DEGs. Different colors represent different p-values, and the size of the color dot represents the number of DEGs; the bigger the dot, the more enriched the genes.

GO analysis (including biological processes, cellular components, and molecular functions) was performed to analyze the functions of the DEGs. The results showed that oxidation-reduction (GO: 0055114), carbohydrate metabolic (GO: 0005975), integral components of the membrane (GO: 0016021), mitochondrion (GO: 0005739), ATP binding (GO: 0005524), protein binding (GO: 0051082), and catalytic activity (GO: 0003824) processes were the most enriched terms. KEGG analysis showed that the significant DEGs were mainly enriched in Gly/gluconeogenesis (ko00010), glycerophospholipid (ko00564), fructose and mannose (ko00051), and pyruvate metabolism (ko00620). To further determine the accuracy of the RNA sequencing results, 15 DEGs involved in the regulation of seed germination induced by GA3 were selected for RT-qPCR, and specific primers for these genes were designed using online NCBI primer software. RNA sequencing results showed that the expression levels of eight DEGs were significantly upregulated and seven DEGs were significantly downregulated. The RT-qPCR results showed that the expression levels of the selected DEGs were consistent with the RNA-seq data, which proved that the transcriptome sequencing data of corn seeds were reliable.

3.3. miRNAs Were Involved in the Germination Regulation of Maize Seeds

miRNAs have been reported to be involved in the regulation of corn seed vigor. Several miRNAs with specific functions have been reported in maize seeds [16,18,19]. To fully understand the regulatory mechanism of maize seeds, miRNA sequencing was performed. Significant DEGs in the germination process of maize seeds were explored. In total, 791 mature miRNAs with different expression levels were identified (Supplemental Figure S3). Of these, 437 were already known in the miRbase database, and 354 were novel miRNAs. As shown in Figure 2, compared with 6 HAI, 28 miRNAs with different expression levels were obtained at 53 HAI ($p < 0.05$) in Yu82, of which 15 miRNAs were upregulated and 13 were downregulated. Only six miRNAs were upregulated at 53 HAI in Yu537A ($p < 0.05$), indicating that the regulatory mechanisms in Yu82 were more complicated than those in Yu537A. After GA3 treatment, the number of differentially expressed miRNAs increased to 22 from six, suggesting that GA3 induced more DEGs involved in maize seed germination. Among the 22 significantly differentially expressed miRNAs, only one was downregulated, whereas the others were upregulated. Genome location clusters of pre-miRNAs showed that approximately 26.00% were located on chromosome 2, indicating that the miRNAs on chromosome 2 played an important role in maize seed germination.

Mature plant miRNAs generally interact with their targets through perfect or near-perfect complementarity, leading to the cleavage of target mRNAs [20–22], which is guided by the RNA-induced silencing complex (RISC) [23,24]. In our study, all the miRNA targets were identified using prediction and degradome sequencing. To identify the effective targets of the miRNAs, we investigated all the targets of significantly differentially expressed miRNAs in each comparison group. In the E_53h vs. E_6h group, 28 significantly differentially expressed miRNAs and 113 target genes were gained with negative expression. While only four significantly differentially expressed miRNAs with 22 target genes were observed to have negative expressions. However, after the GA3 treatment, the number of significantly differentially expressed miRNAs, and their target genes increased to 14 and 41, respectively. These results indicate that GA3 induced significantly more differentially expressed miRNAs to regulate germination-related genes and promote the germination of maize seeds.

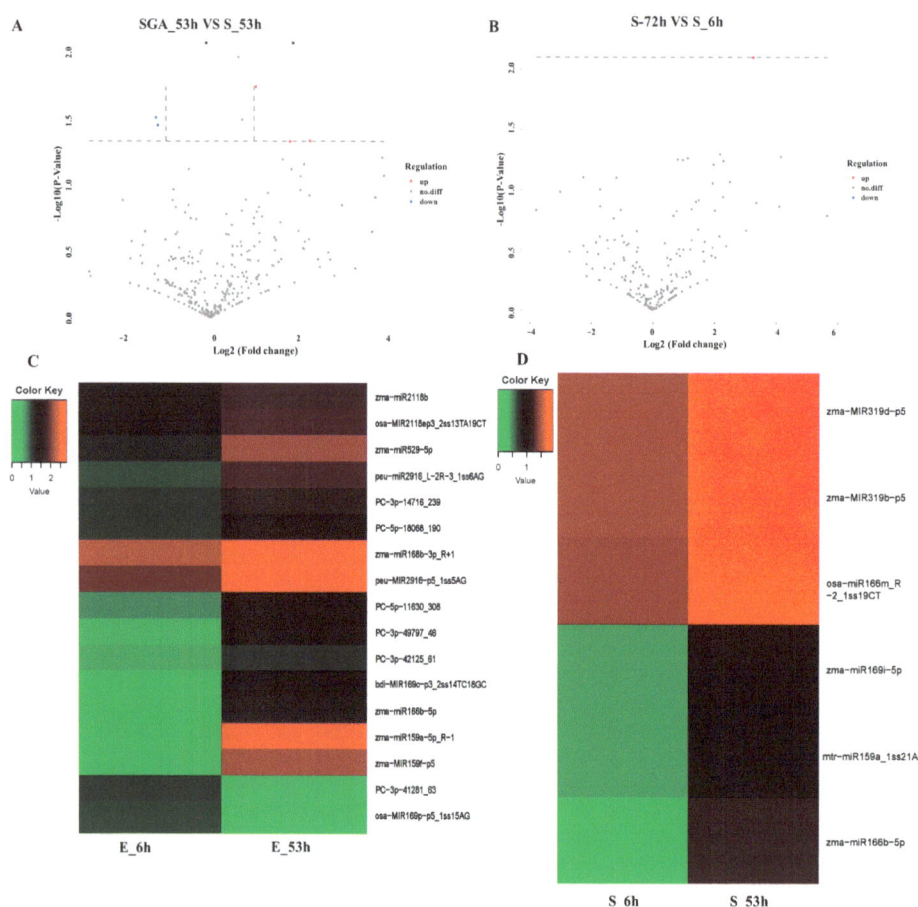

Figure 2. Differentially expressed miRNAs in two different germination stages in two maize inbreds. (**A**) Volcano of differentially expressed miRNAs in the SGA_53h vs. S_53h. (**B**) Volcano of differentially expressed miRNAs in the group of S-72h vs. S_6h. (**C**) Heatmap analysis of differentially expressed miRNAs in the E_53h vs. E_6h. (**D**) Heatmap analysis of differentially expressed miRNAs in the SGA_53h vs. S_53h.

3.4. Integrative Analysis of DEGs and miRNAs Involved in the Germination of Maize Seed Treated with GA3

To further understand the molecular mechanism of increased maize seed vigor induced by GA3, multi-omics, including transcriptome, miRNA, and degradome sequencing, were used to identify key genes related to maize seed vigor. The results showed that in the SGA_53h vs. S_53h group, 43 miRNA-mRNA pairs were identified at a significance level of $p \leq 0.05$. As shown in Figure 3A, GO enrichment analysis of the differentially expressed miRNAs and DEGs revealed that protein binding (GO: 0005515), ATP binding (GO: 0005524), and lipid metabolic processes (GO: 0006629) were the most enriched. In addition, 355 and 1781 miRNA-mRNA pairs were identified in S_53h vs. S_6h and E_53h vs. E_6h, respectively. GO enrichment analysis showed these differentially expressed miRNA-mRNA pairs were mainly enriched in items of oxidation-reduction (GO: 0055114), carbohydrate binding (GO: 0030246), hydrolase activity (GO: 0016787), zinc ion binding (GO: 0008270), porphyrin-containing compound biosynthesis (GO: 0006779), protein glycosylation (GO: 0006486), and ATP binding (GO: 0005524) processes (Figure 3A). Furthermore, ATP binding

(GO: 0005524) and lipid metabolic processes (GO: 0006629) were enriched in S_53h vs. S_6h and E_53h vs. E_6h, indicating DEGs in these items play a vital role in the energy supply during germination. KEGG enrichment analysis showed that carbohydrate, lipid, amino acid, and energy metabolisms were the main metabolic pathways involved in maize seed germination (Figure 3B).

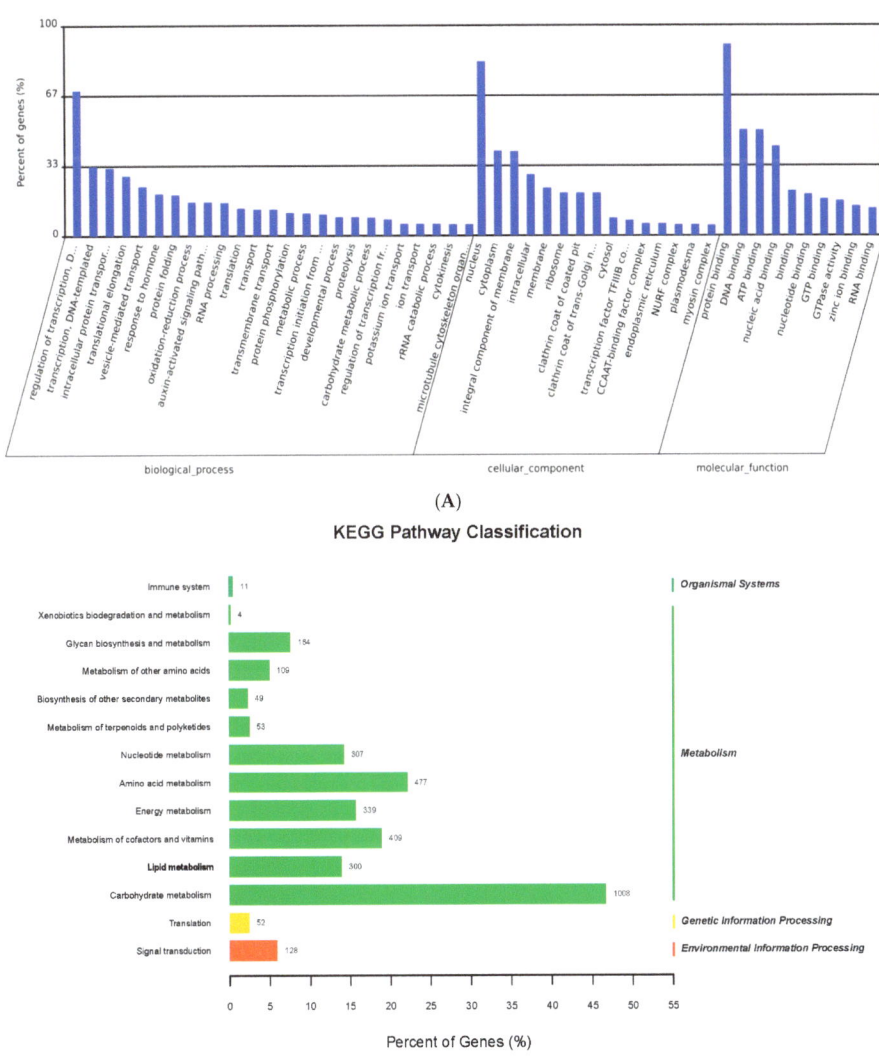

Figure 3. GO and KEGG enrichment analysis of miRNA-mRNA targets. (A) GO enrichment analysis of miRNA-mRNA targets. (B) KEGG enrichment analysis of miRNA-mRNA targets.

Therefore, it is speculated that the primary seed germination process mainly uses the energy stored in the seed body to activate the corresponding hydrolase and protease activities and redox reactions to hydrolyze sugar substances to release energy and promote germination. Maize is an important food and feed crop, and the hydrolysis of lipid metabolism negatively regulates the germination of maize seeds. Therefore, differentially expressed miRNAs and mRNAs in lipid metabolic processes are crucial for the healthy germination of maize seeds.

3.5. Function Verification of the Lipid Metabolism-Related Gene ZmSLP

To explore the molecular mechanism regulating maize seed vigor, expression profiles of differentially expressed miRNAs and mRNAs were analyzed, and finally the target gene LOC100282421 was obtained from the differential metabolic pathway K00561 (glycerolipid metabolism) and named *ZmSLP* (Figure 4A–D). A miRNA–mRNA pair, zma-MIR169c-3p–ZmSLP, was identified and is involved in seed germination regulation. Under normal conditions, low expression of the miRNA zma-MIR169c-3p induces high expression of the target gene *ZmSLP*. While at the germination stage, the miRNA zma-MIR169c-3p was upregulated to inhibit the expression of the target gene *ZmSLP*, promoting seed germination. To verify the function of *ZmSLP* in regulating maize seed vigor, an expression vector for *ZmSLP* was constructed and transfected into *Arabidopsis thaliana*. The results indicated that transgenic *A. thaliana* seeds germinated earlier and the seedlings were more robust than wild-type *A. thaliana* plants (Figure 4E,F), suggesting that *ZmSLP* significantly improved the germination ability of transgenic *A. thaliana* seeds.

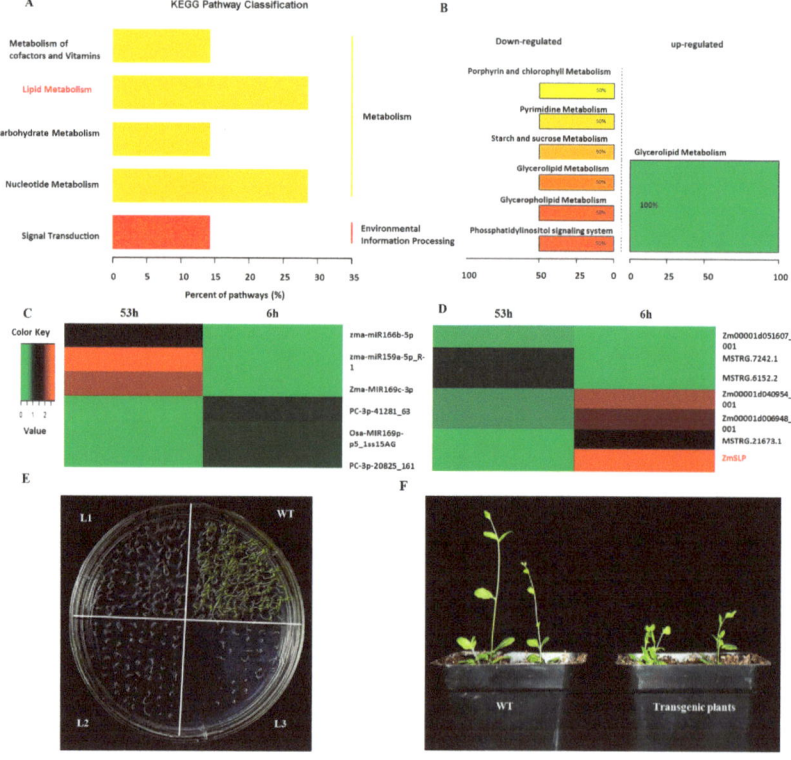

Figure 4. KEGG enrichment and expression analysis of target *ZmSLP*. (**A**) Classification of significantly enriched KEGG pathways. (**B**) Ratio of upregulated or downregulated genes in significantly enriched pathways. (**C**) Expression analysis of partial miRNAs. (**D**) Expression analysis of the target *ZmSLP* gene. (**E**) Comparison of Arabidopsis seed germination on solid medium. WT represents wild-type Arabidopsis seeds. L1, L2, and L3 represent three transgenic Arabidopsis lines. (**F**) Growth difference between WT and transgenic plants.

4. Discussion

Maize (*Zea mays* L.) is an excellent model plant for genetic studies and an important staple crop typically used as a raw material for human food, animal feed, and industrial production [25]. Healthy seed germination is an important event for high-yield and high-quality maize and involves a series of physiological and genetic regulations, such as miRNAs, an important class of non-coding RNAs [26]. GA is generally regarded as a phytohormone that releases seed dormancy and promotes seed germination [27]. However, ABA is another important phytohormone that regulates seed germination, and the GA-ABA ratio is a key factor in breaking and promoting seed germination [28]. Several thousand DEGs were identified in different comparison groups, and the early stages of imbibition germination of maize induced more DEGs. GO and KEGG enrichment analysis suggested that multiple items were closely correlated with the seed germination process, including the oxidation-reduction process (GO: 0055114), carbohydrate metabolic process (GO: 0005975), an integral component of membrane (GO: 0016021), mitochondrion (GO: 0005739), ATP binding (GO: 0005524), Gly/gluconeogenesis (ko00010), glycerophospholipid (ko00564), fructose and mannose (ko00051), and pyruvate metabolism (ko00620). Among these, using informatics analysis, the lipid metabolism-related gene *ZmSLP* was identified for its close correlation with maize seed germination.

Oil bodies are carriers of lipids in maize seeds and provide energy for seed germination and early seedling growth. Lipids exist in different species as triacylglycerols (TAG) [29–31]. As an excellent food crop with high nutritional value, maize is an important raw material source for feed for the animal husbandry, aquaculture, and fish industries. In this study, glycerophospholipid metabolism (ko00564) was significantly enriched, suggesting that lipid metabolism plays an important role in maize seed germination. Furthermore, the *ZmSLP* gene was identified and functionally verified in transgenic *Arabidopsis* plants, and the overexpression of this gene significantly reduced the germination quality. The main nutrients and energy used for seed germination are mostly derived from stored starch, because maize is also one of the most important food crops in our country due to its high content of starch. At the seed germination stage, starch provides energy and a carbon skeleton for seed germination, which is necessary for maize seed germination. Li et al. discovered a putative rice non-specific lipid transport protein named *OsLTPL23*, which plays an important role in the seed germination process [32]. Hydro-electro hybrid priming (HEHP), a new seed introduction technique, promotes carrot seed germination by activating lipid metabolism [33].

In the process of maize seed germination, free amino acids and soluble sugars in the embryo are first consumed, and the utilization of nutrients in the storage tissue begins only when the seeds absorb water fully and expand to a certain stage. Overexpression of *ZmSLP* in Arabidopsis would lead to a disorder of lipid metabolism, inhibiting the germination of maize seeds and strong growth at the seedling stage. In this study, we observed that lipid metabolism is important for seed germination. Therefore, we established a regulation model for maize seed germination, providing valuable information for molecular research on maize seed germination.

Supplementary Materials: The following supporting information can be downloaded at: https://www.mdpi.com/article/10.3390/agronomy13071929/s1, Figure S1: Distribution regions of mapped reads in each sample; Figure S2: Expression Profiles of genes obtained in each sample; Figure S3: Differentially expressed of partial miRNAs among different samples; Table S1: Information of each sample; Table S2: Quality of RNA sequencing reads of maize seeds; Table S3: Statistical results of the comparisons with reference genome.

Author Contributions: Z.H. and Y.J. designed the experiments and wrote the manuscript. B.W. and Y.G. helped with the experiments and manuscript revisions. All authors have read and agreed to the published version of the manuscript.

Funding: This study was supported by grants from the National Natural Science Foundation of China (No. U1504315) and the Science and Technology Project in Henan Province (No. 212102110244).

Institutional Review Board Statement: Not applicable.

Informed Consent Statement: Not applicable.

Data Availability Statement: The datasets generated in this study are available from the GEO repository (accession number: GSE196738). All analysis results generated during this study are included in this article and the Supplementary Information.

Acknowledgments: We thank Hangzhou LC-Bio Technology Co., Ltd. for their assistance with sequencing and bioinformatics analysis.

Conflicts of Interest: The authors declare no conflict of interest.

References

1. Achard, P.; Renou, J.P.; Berthome, R.; Harberd, N.P.; Genschik, P. Plant DELLAs restrain growth and promote survival of adversity by reducing the levels of reactive oxygen species. *Curr. Biol. CB* **2008**, *18*, 656–660. [CrossRef] [PubMed]
2. Cui, W.; Song, Q.; Zuo, B.; Han, Q.; Jia, Z. Effects of Gibberellin (GA4+7) in Grain Filling, Hormonal Behavior, and Antioxidants in High-Density Maize (*Zea mays* L.). *Plants* **2020**, *9*, 978. [CrossRef]
3. Zhang, Y.; Wang, Y.; Ye, D.; Xing, J.; Duan, L.; Li, Z.; Zhang, M. Ethephon-regulated maize internode elongation associated with modulating auxin and gibberellin signal to alter cell wall biosynthesis and modification. *Plant Sci. Int. J. Exp. Plant Biol.* **2020**, *290*, 110196. [CrossRef]
4. White, C.N.; Proebsting, W.M.; Hedden, P.; Rivin, C.J. Gibberellins and seed development in maize. I. Evidence that gibberellin/abscisic acid balance governs germination versus maturation pathways. *Plant Physiol.* **2000**, *122*, 1081–1088. [CrossRef] [PubMed]
5. Han, Z.; Wang, B.; Tian, L.; Wang, S.; Zhang, J.; Guo, S.; Zhang, H.; Xu, L.; Chen, Y. Comprehensive dynamic transcriptome analysis at two seed germination stages in maize (*Zea mays* L.). *Physiol. Plant.* **2020**, *168*, 205–217. [CrossRef]
6. Weitbrecht, K.; Muller, K.; Leubner-Metzger, G. First off the mark: Early seed germination. *J. Exp. Bot.* **2011**, *62*, 3289–3309. [CrossRef] [PubMed]
7. Rajjou, L.; Duval, M.; Gallardo, K.; Catusse, J.; Bally, J.; Job, C.; Job, D. Seed germination and vigor. *Annu. Rev. Plant Biol.* **2012**, *63*, 507–533. [CrossRef]
8. Ahmad, I.; Kamran, M.; Ali, S.; Cai, T.; Bilegjargal, B.; Liu, T.; Han, Q. Seed filling in maize and hormones crosstalk regulated by exogenous application of uniconazole in semiarid regions. *Environ. Sci. Pollut. Res. Int.* **2018**, *25*, 33225–33239. [CrossRef]
9. Hedden, P. Gibberellin Metabolism and Its Regulation. *J. Plant Growth Regul.* **2001**, *20*, 317–318. [CrossRef]
10. Shu, K.; Liu, X.D.; Xie, Q.; He, Z.H. Two Faces of One Seed: Hormonal Regulation of Dormancy and Germination. *Mol. Plant* **2016**, *9*, 34–45. [CrossRef]
11. Holdsworth, M.J.; Bentsink, L.; Soppe, W.J.J. Molecular networks regulating Arabidopsis seed maturation, after-ripening, dormancy and germination. *New Phytol.* **2008**, *179*, 33–54. [CrossRef] [PubMed]
12. Lee, S.C.; Cheng, H.; King, K.E.; Wang, W.F.; He, Y.W.; Hussain, A.; Lo, J.; Harberd, N.P.; Peng, J.R. Gibberellin regulates Arabidopsis seed germination via RGL2, a GAI/RGA-like gene whose expression is up-regulated following imbibition. *Genes Dev.* **2002**, *16*, 646–658. [CrossRef] [PubMed]
13. Shu, K.; Zhang, H.W.; Wang, S.F.; Chen, M.L.; Wu, Y.R.; Tang, S.Y.; Liu, C.Y.; Feng, Y.Q.; Cao, X.F.; Xie, Q. ABI4 Regulates Primary Seed Dormancy by Regulating the Biogenesis of Abscisic Acid and Gibberellins in Arabidopsis. *PLoS Genet.* **2013**, *9*, e1003577. [CrossRef]
14. Guo, Q.; Li, X.; Niu, L.; Jameson, P.E.; Zhou, W. Transcription-associated metabolomic adjustments in maize occur during combined drought and cold stress. *Plant Physiol.* **2021**, *186*, 677–695. [CrossRef]
15. Han, X.; Yin, H.; Song, X.; Zhang, Y.; Liu, M.; Sang, J.; Jiang, J.; Li, J.; Zhuo, R. Integration of small RNAs, degradome and transcriptome sequencing in hyperaccumulator Sedum alfredii uncovers a complex regulatory network and provides insights into cadmium phytoremediation. *Plant Biotechnol. J.* **2016**, *14*, 1470–1483. [CrossRef]
16. Gong, S.; Ding, Y.; Huang, S.; Zhu, C. Identification of miRNAs and Their Target Genes Associated with Sweet Corn Seed Vigor by Combined Small RNA and Degradome Sequencing. *J. Agric. Food Chem.* **2015**, *63*, 5485–5491. [CrossRef] [PubMed]
17. Springer, N.M. Isolation of plant DNA for PCR and genotyping using organic extraction and CTAB. *Cold Spring Harb. Protoc.* **2010**, *2010*, pdb.prot5515. [CrossRef]
18. Kang, M.M.; Zhao, Q.; Zhu, D.Y.; Yu, J.J. Characterization of microRNAs expression during maize seed development. *BMC Genom.* **2012**, *13*, 360. [CrossRef]
19. Li, D.; Wang, L.; Liu, X.; Cui, D.; Chen, T.; Zhang, H.; Jiang, C.; Xu, C.; Li, P.; Li, S.; et al. Deep sequencing of maize small RNAs reveals a diverse set of microRNA in dry and imbibed seeds. *PLoS ONE* **2013**, *8*, e55107. [CrossRef]
20. Aukerman, M.J.; Sakai, H. Regulation of flowering time and floral organ identity by a MicroRNA and its APETALA2-like target genes. *Plant Cell* **2003**, *15*, 2730–2741. [CrossRef]
21. Mallory, A.C.; Vaucheret, H. Functions of microRNAs and related small RNAs in plants. *Nat. Genet.* **2006**, *38*, S31–S36. [CrossRef] [PubMed]
22. Voinnet, O. Origin, biogenesis, and activity of plant microRNAs. *Cell* **2009**, *136*, 669–687. [CrossRef] [PubMed]

23. Bonnet, E.; Van de Peer, Y.; Rouze, P. The small RNA world of plants. *New Phytol.* **2006**, *171*, 451–468. [CrossRef] [PubMed]
24. Baumberger, N.; Baulcombe, D.C. Arabidopsis ARGONAUTE1 is an RNA Slicer that selectively recruits rnicroRNAs and short interfering RNAs. *Proc. Natl. Acad. Sci. USA* **2005**, *102*, 11928–11933. [CrossRef]
25. Liu, J.; Guo, X.; Zhai, T.; Shu, A.; Zhao, L.; Liu, Z.; Zhang, S. Genome-wide identification and characterization of microRNAs responding to ABA and GA in maize embryos during seed germination. *Plant Biol.* **2020**, *22*, 949–957. [CrossRef]
26. Bewley, J.D. Seed germination and dormancy. *Plant Cell* **1997**, *9*, 1055–1066. [CrossRef] [PubMed]
27. Jin, Y.Q.; Wang, B.; Tian, L.; Zhao, L.X.; Guo, S.L.; Zhang, H.C.; Xu, L.R.; Han, Z.P. Identification of miRNAs and their target genes associated with improved maize seed vigor induced by gibberellin. *Front. Plant Sci.* **2022**, *13*, 8872. [CrossRef]
28. Gazzarrini, S.; Tsai, A.Y.L. Hormone cross-talk during seed germination. *Essays Biochem.* **2015**, *58*, 151–164.
29. Hsieh, K.; Huang, A.H.C. Endoplasmic reticulum, oleosins, and oils in seeds and tapetum cells. *Plant Physiol.* **2004**, *136*, 3427–3434. [CrossRef]
30. Schmidt, M.A.; Herman, E.M. Suppression of soybean oleosin produces micro-oil bodies that aggregate into oil body/ER complexes. *Mol. Plant* **2008**, *1*, 910–924. [CrossRef]
31. Murphy, S.; Martin, S.; Parton, R.G. Lipid droplet-organelle interactions; sharing the fats. *Biochim. Biophys. Acta* **2009**, *1791*, 441–447. [CrossRef] [PubMed]
32. Li, Q.; Zhai, W.; Wei, J.; Jia, Y. Rice lipid transfer protein, *OsLTPL23*, controls seed germination by regulating starch-sugar conversion and ABA homeostasis. *Front. Genet.* **2023**, *14*, 1111318. [CrossRef] [PubMed]
33. Zhao, S.; Garcia, D.; Zhao, Y.; Huang, D. Hydro-Electro Hybrid Priming Promotes Carrot (*Daucus carota* L.) Seed Germination by Activating Lipid Utilization and Respiratory Metabolism. *Int. J. Mol. Sci.* **2021**, *22*, 11090. [CrossRef] [PubMed]

Disclaimer/Publisher's Note: The statements, opinions and data contained in all publications are solely those of the individual author(s) and contributor(s) and not of MDPI and/or the editor(s). MDPI and/or the editor(s) disclaim responsibility for any injury to people or property resulting from any ideas, methods, instructions or products referred to in the content.

Review

Genetic Improvements in Rice Grain Quality: A Review of Elite Genes and Their Applications in Molecular Breeding

Diankai Gong [1], Xue Zhang [1], Fei He [2], Ying Chen [1], Rui Li [1], Jipan Yao [1], Manli Zhang [1], Wenjing Zheng [1,*] and Guangxing Yu [1,*]

1. Liaoning Rice Research Institute, Shenyang 110115, China; gdkrice0709@126.com (D.G.)
2. Rural Revitalization Research Institute, Tianjin Agricultural University, Tianjin 300392, China
* Correspondence: zwj27@126.com (W.Z.); xuanyuan2gege@126.com (G.Y.); Tel.: +86-150-4020-6835 (G.Y.)

Abstract: High yield and superior quality are the main objectives of rice breeding and research. While innovations in rice breeding have increased production to meet growing demand, the universal issue of balancing high yield and superior quality has led to a lack of focus on improving rice quality. With rising living standards, improving rice quality has become increasingly important. Rice grain quality is a complex trait influenced by both genetic and environmental factors, with four primary aspects: milling quality, appearance quality, eating and cooking quality, and nutritional quality. While different populations have varying demands for rice quality, the core traits that contribute to rice quality include grain shape and chalkiness in terms of appearance, as well as endosperm composition that influences cooking and sensory evaluation. Researchers have made substantial advancements in discovering genes/QTLs associated with critical traits including appearance, aroma, texture, and nutritional properties. Markers derived from these genetic discoveries have provided an efficient tool for marker-assisted selection to improve rice quality. Thus, this review focuses on elite genes and their applications in breeding practices to quickly develop superior quality rice varieties that meet various market demands.

Keywords: rice; grain quality; gene cloning; marker-assisted selection; genetic improvement

Citation: Gong, D.; Zhang, X.; He, F.; Chen, Y.; Li, R.; Yao, J.; Zhang, M.; Zheng, W.; Yu, G. Genetic Improvements in Rice Grain Quality: A Review of Elite Genes and Their Applications in Molecular Breeding. *Agronomy* **2023**, *13*, 1375. https://doi.org/10.3390/agronomy13051375

Academic Editor: Antonio Lupini

Received: 14 April 2023
Revised: 7 May 2023
Accepted: 13 May 2023
Published: 15 May 2023

Copyright: © 2023 by the authors. Licensee MDPI, Basel, Switzerland. This article is an open access article distributed under the terms and conditions of the Creative Commons Attribution (CC BY) license (https://creativecommons.org/licenses/by/4.0/).

1. Introduction

Rice (*Oryza sativa* L.) is one of the most important staple crops in the world. High yield and superior quality have always been the main goals of rice breeding and basic research. In recent decades, through technological innovations such as dwarf breeding, heterosis utilization, and super rice breeding, rice production has increased significantly, basically meeting the growing demand for rice consumption. However, the contradiction between high yield and superior quality of crops is a universal scientific problem, and there has been insufficient attention given to rice quality in the actual rice variety breeding process. With the gradual improvement in people's living conditions and consumption levels, improving the quality of rice is becoming increasingly important [1]. The primary focus of current basic research and genetic improvement in rice quality is to develop new superior quality rice varieties that meet diverse market demands and facilitate rapid improvement of rice quality.

Rice quality is a comprehensive trait that encompasses various fundamental characteristics of the product from rice production to consumption after processing. Rice quality is a complex trait influenced by genetic and environmental factors. Although different populations have different focuses on rice quality, rice breeders, producers, and consumers generally have a consistent definition of rice quality, which generally includes four aspects: milling quality, appearance quality, eating and cooking quality, and nutritional quality [2–4]. Each aspect of quality can be described by a series of corresponding quantifiable indicators. Milling quality is an important aspect that refers to the ability of rice to withstand the processing procedure, characterized by brown rice recovery (BR) and head rice recovery (HR), whereas appearance quality is primarily concerned with grain length (GL), grain width (GW), the length/width ratio, chalkiness, and grain translucency. Eating and cooking quality (ECQ) mainly reflects the properties and taste of cooked rice and is associated with three starch-related traits, namely, amylose content (AC), gel consistency (GC), and gelatinization temperature (GT). Meanwhile, rice nutritional quality (NQ) is determined by the content and quality of proteins, lipids, minerals, and other beneficial substances that can contribute to human health. While this article does not focus on discussing individual rice quality indicators, several excellent reviews or chapters in books published in recent years can help people gain a more comprehensive understanding of rice quality [5–8].

Different populations show diverse demands for rice quality. Milling quality is defined as the percentage of the marketable product obtained during the milling process, which begins with brown rice and ends with unbroken white rice. Obviously, milling quality is of greater concern to rice producers. BR and HR are important indicators of rice quality evaluation standards in different countries and are of interest to breeders [6]. The properties exhibited by rice grains following processing are known as the appearance quality, which includes grain shape, chalkiness, and translucency, and directly affects the commercial value of rice. Rice with a uniform shape, less chalkiness, good translucency, and a glossy appearance is more attractive to consumers [9]. Eating and cooking quality refers to all the physical and chemical characteristics and sensory properties of rice during the cooking process and eating. Similar to appearance quality, ECQ is a fundamental indicator of rice quality that consumers can directly perceive. While consumers in various regions may have distinct preferences for rice ECQ, superior quality rice generally exhibits intact grains that are clean and fragrant, soft and elastic but not sticky, palatable, and can maintain their texture when cold [6,10]. Rice is mainly composed of starch, with very low amounts of other nutrients including protein, amino acids, fat, vitamins, and mineral elements. However, additional nutrients in rice have received increasing attention due to their potential importance for maintaining human health.

In general, rice quality traits are determined by a combination of genetic and environmental factors. Core traits that contribute to rice quality include grain shape and chalkiness in terms of appearance, as well as the starch composition that influences cooking and sensory evaluation. The quality of rice grain is closely correlated with rice seed characteristics, such as the pericarp, seed coat, aleurone, starchy endosperm, and embryo [11]. Recently, many quantitative trait loci (QTLs) and critical genes responsible for rice quality traits have been identified, and useful molecular markers have been developed to improve rice quality. Notably, researchers have made significant progress in identifying and characterizing the genes responsible for crucial traits such as appearance (shape and chalk), aroma, texture, and nutritional properties. Here, we mainly discuss some elite genes and their applications in breeding practice.

2. Genetics of Grain Shape

Rice grain appearance is an essential and extensively researched quality trait that is usually determined by grain shape, chalkiness, and transparency. Grain shape is is defined by its length, width, and length/width ratio, and it is a crucial target trait for yield, appearance, domestication, and breeding [12–17]. In Indica rice, slender grains are preferred, while Japonica rice is characterized by short and round grains. Over the past few decades, geneticists have identified numerous QTLs related to grain shape in rice, but only a few have been cloned (Figure 1).

2.1. Grain Length

The first QTL that has been cloned for grain length in rice was *GS3* (*Grain Size on Chromosome 3*). *GS3* was identified from a NIL population of Minghui 63 (large grain) and Chuan 7 (small grain) and it negatively regulates grain length. By comparing the *GS3* sequences from rice varieties with different grain lengths, it was found that the long-grain varieties had a cysteine codon (TGC) to a termination codon mutation in the second exon of the *GS3* gene, which resulted in premature termination and a 178-aa truncation of the C-terminus of the protein and led to larger grains [18]. This mutation was strongly selected in both Japonica and Indica rice varieties, providing an explanation for the prevalence of long grain in Indica rice genotypes [19]. *GS3* has four functional domains (OSR in the N terminus, a transmembrane domain, TNFR/NGFR, and VWFC in the C terminus), and different alleles of *GS3* formed due to natural variation [20]. The GS3 gene is a key regulator of grain size in rice, with different alleles exhibiting distinct effects on grain length. The wild-type *GS3* allele (*GS3-1*) has a medium-length grain shape, which is the genotype for the majority of medium-grain Indica varieties. In most temperate Japonica varieties, such as Nipponbare, the *GS3-2* allele has minimal impact on grain length due to a single amino acid difference at the C-terminus compared to that of *GS3-1*. In contrast, the *GS3-3* allele in the Minghui 63 cultivar results in longer grains by eliminating OSR function. Similarly, the *GS3-4* allele in the Chuan 7 cultivar leads to extremely short grains by the loss of functional domains at the C-terminus. It was later discovered that *GS3* is a G protein γ subunit gene that regulates grain length through the G protein pathway [21,22]. *GS3* negatively regulates rice grain length by competitively binding to G protein β subunits with two other γ subunits, DEP1 and GGC2, which inhibits the function of DEP1 and GGC2 and leads to shorter grains [22].

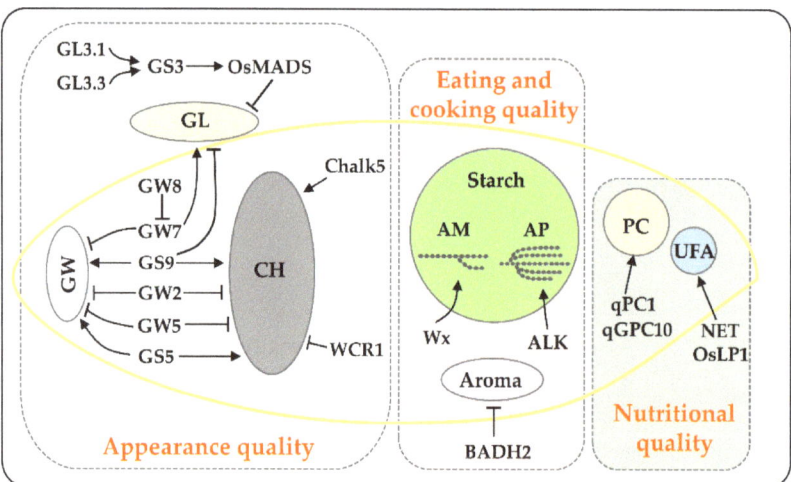

Figure 1. The major genes controlling rice grain quality. This figure shows some extensively studied genes/QTLs related to the core quality traits including appearance quality, eating and cooking quality, and nutritional quality. Genetic dissection of these genes has enabled the development of functional markers to accelerate breeding programs for improving rice quality. GL, grain length; GW, grain width; CH, chalkiness; AM, amylose; AP, amylopectin; PC, protein content; UFA, unsaturated fatty acids.

The *GL3.1/qGL3* locus encodes a protein phosphatase kelch (PPKL) family member and functions by negatively modulating the longitudinal cell number in grain glumes. This locus is closely linked to *GS3* and acts as an enhancer gene, enhancing the effect of *GS3* on grain length [23–25]. Furthermore, it has been discovered that the *GS3* locus displays epistatic interaction with *GL3.3*, a gene encoding a kinase similar to GSK3/SHAGGY. Plants carrying both *gs3* and *gl3.3* genotypes exhibit significantly longer grains compared to those carrying only one of the genotypes [26].

2.2. Grain Width

GW2 (*Grain Width on Chromosome 2*) was the first QTL cloned for rice grain width, which encodes a RING-type E3 ubiquitin ligase [27]. In exon 4 of the *GW2* gene, a 1 bp deletion introduces a premature stop codon, leading to a truncated protein and a large-grain phenotype. GW2 negatively regulates cell division by directing its substrate to proteasomes for regulated proteolysis. The loss function of *gw2* results in increased cell numbers in the spikelet hull, thereby, enhancing grain width, weight, and yield.

qSW5 (*Seed Width on Chromosome 5*)/*GW5* (*Grain Width on Chromosome 5*) is a major QTL that controls rice grain width and has been mapped to a genomic region of 2263 bp and 21 kb, respectively [28,29]. The former two studies identified a 1212 bp deletion in this locus as a determinant of wide-grain phenotype. Further investigations revealed that the true gene of *GW5* encodes a membrane-localized protein with an IQ calmodulin-binding motif domain and it is located downstream of the 1212 bp deletion [30,31]. GW5 participates in the brassinosteroid pathway by regulating grain growth through inhibition of the kinase activity of GSK2 on OsBZR1 and DLT [31]. Natural variation of *GW5* has been found in different haplotypes, including a 1212 bp deletion in most Japonica rice varieties, a 950 bp deletion in most wide-grain Indica varieties, and no deletion in most narrow-grain Indica varieties [30]. This InDel marker can be used to identify functional alleles of *GW5*/*qSW5* for wide and narrow grains. A recent functional analysis revealed that the transcription factor *OsSPL12* promoted the expression of *GW5* by directly binding to the 1212 bp region in most Indica varieties. This suggests that *OsSPL12* may have co-evolved with *GW5* to determine the differences in grain shape between Indica and Japonica rice [32].

GS5 (*Grain Size on Chromosome 5*) is another major QTL in rice that controls grain size by positively regulating grain width, filling, and weight, and it has no other significant effect. It encodes a putative serine carboxypeptidase, and higher expression of *GS5* is correlated with larger grain size. The promoter region of *GS5* shows the natural variation that is linked to grain width, thereby, contributing to the diverse grain size of rice [33]. Through the identification of two SNPs located between the wide-grain allele *GS5-1* and the narrow-grain allele *GS5-2* in the upstream region of the gene, differential expression of *GS5* during the development of young panicles is attributed to these SNPs. Moreover, enhanced expression of GS5 competitively inhibits the interaction between OsBAK1-7 and OsMSBP1 by occupying the extracellular leucine-rich repeat (LRR) domain of OsBAK1-7. This, in turn, prevents OsBAK1-7 from undergoing endocytosis by interacting with OsMSBP1 and explains the positive association between grain size and *GS5* expression [34].

2.3. Grain Length/Width

Grain length and width are negatively correlated, and this requires a mechanism to balance these two traits. Indeed, *GW8* and *GW7* loci have been proposed to interact and form a module for regulating grain shape. *GW8* (*Grain Width on Chromosome 8*) is a positive regulator of grain width which encodes the OsSPL16 protein that is responsible for cell proliferation [35]. A 10 bp deletion in the *OsSPL16* promoter in Basmati rice is associated with the formation of a slenderer grain. GW8 directly binds to the *GW7* (*Grain Width on Chromosome 7*) promoter and represses its expression [36]. *GW7*, a locus also known as *GL7* (*Grain Length on Chromosome 7*)/*SLG7* (*Slender Grain on Chromosome 7*), is a positive regulator for the grain length/width ratio encoding a TONNEAU1-recruitment motif (TRM) protein homologous to *Arabidopsis* LONGIFOLIA proteins [36–38]. Copy number variations (CNVs) at the *GW7* locus lead to the upregulation of *GW7*, inducing slender grains with higher grain appearance without loss in yield. Thus, the *GW8*-*GW7* module represents a promising strategy for improving rice grain quality. In addition, *GS9* (*Grain Shape Gene on Chromosome 9*) negatively regulates the length/width ratio of rice grains by regulating horizontal cell division and vertical cell elongation [39]. Introducing the *gs9* null allele into elite rice cultivars produces significantly more slender rice grains than those with normal *GS9* allele and with no changes in grain thickness or weight, suggesting its potential application in the breeding of rice varieties with optimized grain shape [39,40].

2.4. The Utilization of Grain Shape Genes in Genetic Improvement of Rice Grain Quality

The rapid progress in rice genetics and functional genomics has facilitated the development of numerous molecular markers. Marker-assisted selection (MAS) has become a widely used approach in conventional plant breeding programs. MAS is widely used in various breeding techniques such as backcross breeding, forward selection, reverse selection, gene pyramiding, and background screening. By incorporating genes related to grain shape, rice yields and appearance quality can be significantly improved [41] (Table 1).

Sequencing and association analyses of natural variant populations have confirmed that *GS3* is the most effective and crucial gene that controls rice grain length. The premature termination of the *gs3* mutation is advantageous for designing functional CAPS molecular markers to detect the long- and short-grain genotypes of *GS3* [42,43]. Utilizing the five molecular markers developed for *GS3*, a chromosome segment harboring the recessive *gs3* allele has been successfully introgressed into a leading cultivar, Kongyu 131, resulting in the development of new varieties with longer grains (a 12.05% increase in grain length) [44].

Table 1. Elite alleles in the genetic improvements of rice grain quality by MAS.

Trait	Allele	Variation	Function Description	Reference
Grain shape	*gs3*	Loss of function mutation in exon 2	Increased grain length to produce slender grains.	[35,36,44–46]
	GW7^{TFA}	SNPs and Indels in the promoter	Pyramiding of *GW7^{TFA}* and *gs3* produced much longer grains and increased grain yield.	[36]
	gw8	10-bp deletion in the promoter	Pyramiding of *gw8* and *gs3* increased grain length with no yield loss.	[35]
	OsMADS^{lgy3}	An insertion–deletion polymorphism in the splice site of the intron 7/exon 8 junction	Pyramiding of *lgy3* and *gs3* increased grain length and yield.	[45]
	dep1	A 625 bp deletion in exon 5	Increased grain length in Japonica rice.	[46]
	gl3.1	C-A at 1092 bp and C-T at 1495 bp	Increased grain length and yield in hybrid rice.	[24]
	GSE5^{ZJB}	No deletion	Produced longer but smaller grains with lower chalkiness.	[47]
	gs9	3 bp Insertion in exon 1 and 4 SNPs in exon 4; 7 kb insertion at 311 bp	Increased grain length.	[39,40]
Chalkiness	*chalk5*	2 SNPs in the promoter	Reduced grain chalkiness in Indica rice.	[48]
	WCR1^A	A functional SNP (A/G) at −1696 bp in the promoter	Reduced grain chalkiness in Japonica rice.	[49]
Eating and cooking quality	*Wx^{mq}*	G-A in exon 4	Decreased the AC to 10~15% to produce semi-glutinous rice.	[50]
	Wx^b	G-T at splicing site in intron 1	Decreased the AC to 10~15%.	[51–53]
	Wx^{mw}	A-C in exon 6	Decreased the AC to 14% and improved endosperm transparency.	[54]
	ALK^b	3 SNPs in exon 8	Decreased the AC and GT.	[55]
	badh2-E7	8 bp Deletion and 3 SNPs in exon 7	Produced rice variety with aroma.	[55]
Protein content	*qPC-1^{Habataki}*	Not clear	Decreased the protein content to enhance ECQ.	[56,57]

The combination of genes involved in grain shape is a valuable approach for producing rice varieties with high yield and superior quality. The large number of genes that regulate grain shape, as well as their wide genetic variation, allow breeders to create diverse rice grain shapes. For example, the pyramiding of *gw8* and *gs3* into a short-grain variety (HJX74) has produced long-grain shaped rice similar to Basmati 385. Similarly, the development of a new elite Indica variety, Huabiao1, with elongated grains, was achieved by pyramiding these two genes, which substantially improved grain quality [35]. Furthermore, the CNVs and InDel in the *GW7* locus can be used to create functional molecular markers without any significant adverse effects, allowing for the design of superior-quality rice varieties. The new hybrid Indica rice varieties, Taifengyou55 and Taifengyou208, developed through QTL pyramiding of the $GW7^{TFA}$ (an allelic variant of the *GW7* from a cytoplasmic male sterility line TaifengA) and *gs3* alleles, have substantially improved grain quality [36]. Pyramiding of $OsMADS1^{lgy3}$ (a key effector downstream of *GS3* to regulate grain length) and *gs3* alleles in the Liangyoupeijiu genetic background has been shown to enhance overall grain yield with improved grain quality [45].

Indica rice is known for having longer grains and better appearance quality compared to Japonica rice. In recent years, improving the grain shape and appearance quality of Japonica rice has been a major focus. The development of the Jiahe series serves as an exemplary case of improving Japonica rice grain shape [17]. The Jiahe series of rice varieties are characterized by enhanced appearance quality and elongated grain shape, which can be attributed to the incorporation of *dep1* and *gs3* into the breeding program. DEP1 (*Dense and Erect Panicle 1*) encodes another atypical Gγ subunit that is similar to *GS3* [46]. The successful integration of *gs3*, $GW7^{TFA}$, and *dep1* led to the development of Jiahe 218, a novel long-grain Japonica rice variety with improved grain shape and quality. Derivatives of Jiahe, such as Zhongjia 8 and Jiafengyou 2, also carry *gs3* and *dep1* and have both good grain shape and quality.

In addition, some other beneficial alleles that control grain shape have been utilized. The *qgl3* allele has been applied in breeding elite rice varieties, and crossing between 9311 and NIL-*qgl3* with three commercial photo-thermosensitive male sterile lines has been shown to increase grain length and yield without negatively impacting grain quality [24]. The introgression of the *gs9* allele, which has slender grains due to the loss of function, has also been used to improve the appearance quality of the Japonica variety Wuyungeng 27 which carries *dep1-1* [39,40]. Moreover, pyramiding multiple favorable genes has proven effective in significantly enhancing the appearance quality of rice grains. Rational allelic combinations of the $GW5^{NIP}$-$GS3^{9311}$-$GW7^{9311}$-$GW8^{NIP}$-$LGY3^{9311}$ module simultaneously improve grain yield and quality [58], offering promising implications for rice improvement in the future. A recent genetic analysis of grain shape-related genes in the Guangdong Simiao rice varieties, a widely cultivated series of rice varieties in southern China, has identified *GS3*, *GW5*, *GL7*, and *GS5* as crucial genes. The study found that the combination of $GS3^{allele3}$-$GW5^{allele2}$-$GL7^{allele2}$-$GS5^{allele2}$ is the most predominant allelic combination associated with grain shape in this rice variety series [59].

3. Genetics of Grain Chalkiness

Grain chalkiness is one of the most important traits in rice grain quality since it results in an opaque endosperm phenotype caused by the loose and irregular distribution of starch granules and protein bodies [60]. This undesirable trait affects the appearance, milling, cooking, and nutritional quality of rice, thereby, decreasing its marketability and commercial value [10]. Since the formation of chalkiness is highly susceptible to environmental conditions, especially temperature, and the complexity of measuring the chalkiness trait, the majority of major QTLs that impact this trait are challenging to identify. For naturally occurring chalkiness, only two QTLs have been successfully cloned so far (Figure 1).

Chalk5 (*Chalk on Chromosome 5*) is the first major QTL that has been cloned in rice and is responsible for positively regulating grain chalkiness [48]. It was identified from a double-haploid population created from a cross between two Indica varieties, H94 (with low chalkiness) and Zhenshan 97 (with high chalkiness). *Chalk5* encodes a vacuolar H^+-translocating pyrophosphatase, and its increased expression leads to an elevation in vacuole H^+ concentration, which disturbs endomembrane pH homeostasis and protein formation, causing the formation of air gaps. This abnormal spatial arrangement of storage substances eventually leads to an increase in grain chalkiness [48]. Two conserved SNPs in the *Chalk5* promoter result in the high chalkiness of Zhenshan 97, which carries a naturally highly expressed allele of *Chalk5*, partly accounting for the differences in its mRNA level [48]. *Chalk5* is considered to be the major QTL affecting the chalkiness variation in most Indica rice varieties.

The *WCR1* (*White-Core Rate 1*) gene, encoding an F-box protein, is a recently discovered gene that negatively regulates grain chalkiness [49]. It reduces chalkiness by inhibiting the accumulation of reactive oxygen species (ROS) and delaying programmed cell death (PCD) in the endosperm. Additionally, a functional SNP (A/G) located at −1696 bp in the *WCR1* promoter region has been associated with both WCR phenotype and nucleotide variation in Asian-cultivated rice accessions. This functional SNP can affect transcription factor binding sites, and molecular markers developed based on this SNP can be used to efficiently identify individuals carrying desirable alleles during selection, potentially speeding up the breeding process by allowing more efficient screening of large populations [49].

Through the process of long-term breeding practices, it has been discovered that slender-grain rice varieties generally possess better appearance qualities such as low chalkiness, while wide-grain varieties tend to have poorer quality. Studies have also shown that grain shape and chalkiness are often co-related, with cloned grain shape QTLs frequently exhibiting chalkiness as a co-effect [61]. The correlation between grain shape and chalkiness has been confirmed, with certain QTLs such as *gw2* [27] and *GS2* [62] showing a significant increase in chalkiness while enhancing grain width/weight, and others such as *GL7/GW7* [36,37], *gw8* [35], and *gs9* [39] reducing chalkiness percentage by increasing grain length and decreasing grain width. A recent study identified that the grain width QTL *GW5/GSE5* has pleiotropic effects on both chalkiness and grain shape [47]. Introducing the *GSE5^{ZJB}* allele to the Indica rice varieties Zhenshan 97B showed lower chalkiness and longer but smaller grains. Therefore, increasing grain length provides an effective strategy for reducing chalkiness.

Unlike the diverse requirements for grain shape, low chalkiness is a common demand in rice breeding. The development of molecular markers for fine-mapped QTLs related to chalkiness is important in breeding practices. Introducing QTLs or genes with significant genetic effects into the Indica and Japonica cultivars can be an effective strategy for producing superior quality rice with a high yield (Table 1). However, high yield and superior quality are often contradictory and difficult to reconcile. The mining of *Chalk5*, which is tightly linked to *GW5* and *GS5*, has provided genetic and molecular evidence that has helped to overcome this contradiction [48]. On the one hand, in Indica rice, the haplotype block comprising *chalk5*, *gs5*, and *GW5*, which are tightly linked, leads to the formation of slender, low-chalkiness grains. On the other hand, in Japonica rice, the combination of *Chalk5*, *GS5*, and *gw5* genes results in the development of wide grains that exhibit high chalkiness. Therefore, breeders generally select slender grain seeds to achieve the coordination of high yield and superior quality in Indica rice. Alternatively, the use of *WCR1^A* from the Japonica variety represents a novel way to overcome the negative association between quality and yield for variety improvement, especially for Indica varieties. Introducing *WCR1^A* from the Japonica variety into the widely planted Indica accession 9311 has resulted in lower chalkiness, higher grain length, and increased grain yield [49].

To enhance appearance quality, the primary objectives should be centered on reducing grain chalkiness, enhancing transparency, and generating grains with the appropriate size and shape that fulfill the demands of diverse consumers. The screening and pyramiding of superior alleles of grain shape and chalkiness will be beneficial for producing excellent grain appearance lines [63].

4. Genetics of Eating and Cooking Quality

The eating and cooking quality (ECQ) of the rice grain is a fundamental component that reflects the flavor of rice. Good ECQ is represented by good palatability, an aromatic scent, a white and shiny appearance, a pliable and supple texture, non-stickiness, and no hardening or rebounding after cooling [10]. Artificial taste testing is not accurate for evaluating ECQ; therefore, three classic physicochemical indicators are commonly used as references: amylose content (AC), gel consistency (GC), and gelatinization temperature (GT) [64]. Starch, the primary component of rice, is divided into amylose and amylopectin based on distinct molecular structures, and amylose is considered the most significant determinant of rice quality [65]. Rice with moderate AC (14~20%) has a soft and fluffy texture which is preferred by most consumers, while glutinous rice is best with a low AC (<2%). GC is a crucial parameter for assessing the textural properties of cooked rice and the stickiness of cooled paste for cooked rice flour. A softer GC is often preferred by consumers. In addition, the GT of rice starch is a key determinant of the rice cooking time. Rice with lower gelatinization temperature (<70 °C) requires less water and a shorter time for rice cooking, making it more popular. These characteristics are interrelated and jointly determine the ECQ of rice; AC is negatively correlated with GC and GT value, while GC is positively correlated with the GT value [66–68].

The endosperm, which mainly comprises starch and protein, constitutes the primary edible part of the rice seed. An evaluation of rice ECQ is primarily related to the physicochemical characteristics of starch. At the molecular level, genes involved in endosperm starch synthesis and regulation potentially play a significant role in the formation of rice ECQ [69,70]. For detailed information related to starch biosynthesis, please refer to the excellent review and book chapters published recently [71,72]. Here, we shall concentrate solely on several crucial genes that dictate rice quality. The biosynthesis of starch in the endosperm of cereals is a highly conserved pathway, involving several key enzymes such as adenosine diphosphate (ADP)-glucose pyrophosphorylase (AGPase), granule-bound starch synthase (GBSS), soluble starch synthase (SS), starch-branching enzyme (SBE), and starch debranching enzyme (DBE) [72,73]. AGPase is responsible for the synthesis of ADP glucose (ADPG), which serves as the primary substrate for starch synthesis [70]. ADPG is then transported to the amyloplast to synthesize amylose and amylopectin. In brief, amylose is synthesized mainly by GBSS I, whereas amylopectin is synthesized by a combination of multiple isoforms of SS, SBE, and DBE [70]. The complexity of the amylose and amylopectin composition, which is controlled by multiple isozymes of enzymes, makes it difficult to genetically dissect ECQ [64].

4.1. Wx Determines the Amylose Content

In rice, GBSSI is encoded by the *Wx* (*Waxy*) gene, which is mainly responsible for the synthesis of amylose (Figure 1). Variations in allelic forms of *Wx* determine the amylose content in rice grain and serve as a decisive factor in regulating ECQ [64,74,75]. *Wx* is located on chromosome 6 and consists of 14 exons and 13 introns [76]. It is specifically expressed in the endosperm of developing rice seeds, with the transcription start site located in the first exon and the translation start codon located in the second exon [77,78]. A 3.3 kb pre-mRNA is first transcribed, which is then spliced to form a mature mRNA of 2.3 kb that is further translated into GBSSI. Glutinous rice and high-AC varieties produce mRNA of 3.3 kb and 2.3 kb, respectively, while the intermediate-AC varieties exhibit a combination of the two [77]. Any changes in the structure of the *Wx* exons and introns can affect both the expression level and protein function. There are multiple alleles of the *Wx* gene in rice, each with distinct single nucleotide polymorphisms (SNPs) and varying amylose content (AC) levels [54,79–86]. The Wx^a and Wx^{lv} alleles have AC levels greater than 25%, while Wx^{in} has AC levels between 18% and 22%, and Wx^b has AC levels between 15% and 18%. Other alleles such as Wx^{mw}, Wx^{la}, Wx^{mq}, Wx^{mp}, Wx^{op}, and Wx^{hp} have AC levels ranging from approximately 10% to 15%. The *wx* allele has the lowest AC level of around 2%. Two major alleles, Wx^a and Wx^b, are predominantly distributed in Indica and Japonica rice varieties, respectively, and evolved from the ancestral Wx^{lv} allele that originated from wild rice. The original Wx^{lv} allele underwent mutations to become Wx^a in the Indica subspecies and Wx^b in the Japonica subspecies. Later, Wx^{lv} mutated to Wx^{in} in Indica, and Wx^b mutated to Wx^{mq} in Japonica as a result of artificial mutagenesis [85]. Based on the distinct characteristics of these various Wx alleles, molecular markers can be used to assist in the targeted improvement of rice AC and ECQ. For example, a functional marker for Wx^{in} has been developed to screen lines with intermediate AC, which is identified by a SNP (A to C) in exon 6 that distinguishes high and intermediate amylose varieties [82,87].

In addition to its impact on rice AC, Wx also serves as the major gene for controlling GC [88,89]. The dominant QTL for GC, $qGC6$, has been confirmed to be located at the Wx locus [90]. GC and AC have been shown to have a significant negative correlation. Therefore, Wx exhibits pleiotropy and acts as a major gene affecting the ECQ of rice.

4.2. ALK and Wx Influence Gelatinization Temperature

The gelatinization temperature (GT) is another important indicator of rice ECQ and is determined by the *alkali degeneration* (*ALK*) gene located near the Wx locus on chromosome 6. *ALK*, the dominant QTL for GT, encodes the SSIIa enzyme, which plays a critical role in the formation of amylopectin medium-length branching chains [91] (Figure 1). A comparison of the *ALK* gene sequences in different rice varieties has revealed that two amino acid substitutions within the coding region of the gene are responsible for changes in SSIIa enzyme activity. These changes affect the synthesis of amylopectin medium-length branching chains, alter the crystal structure of starch granules, and ultimately result in changes in the GT [91–93]. Previous studies have identified at least three SNPs in exon 8 of the *ALK* gene that are tightly associated with GT variation, including positions Ex8-733 bp (A/G), Ex8-864 bp (G/T), and Ex8-865 bp (C/T) [81,94,95]. These SNPs generate three haplotypes, including ALK^a (A-GC) and ALK^b (G-TT), which control low GT, and ALK^c (G-GC), which controls high GT [68,96].

The Wx gene plays a major role in regulating both AC and GC and also has a minor effect on GT. Allelic variations in the Wx gene could significantly affect the activity of SSIIa during grain filling. By pyramiding different haplotypes of Wx and *ALK*, a better ECQ has been achieved for lower AC, lower GT, and soft GC [97]. *ALK* also has an impact on AC, GC, and pasting properties. The loss-of-function mutant *ssIIa* exhibits higher AC with a reduced GT compared to the wild type [95]. Taken together, the combination of allelic variants of Wx and *ALK* genes are responsible for most of the phenotypic variations in GT and AC in rice varieties, and these genes are key factors in determining the ECQ of rice.

4.3. Aroma

The aroma of rice is a key attribute of its eating and sensory quality, and it can have a significant impact on market price. Currently, rice varieties that exhibit a distinctive aroma are highly sought after by producers and consumers. Extensive research has revealed the presence of hundreds of volatile compounds in rice, among which 2-acetyl-1-pyrroline (2-AP) is considered to be the principal compound responsible for the fragrance of rice grains. Extensive studies have focused on its genetic regulation. Now, it is well known that fragrance in rice results from the loss of function of the *betaine aldehyde dehydrogenase 2* (*badh2*) gene on chromosome 8 [98,99] (Figure 1), which leads to the enhancement of 2-AP biosynthesis in fragrant rice [100]. The *badh2* gene is composed of 15 exons and 14 introns, and its non-functional recessive alleles display a range of variations throughout the sequence. Currently, more than 10 alleles of *badh2* have been identified, such as *badh2-E7*, *badh2-E2*, *badh2-E4-5*, *badh2-p-50UTR*, and *badh2-p* [98,100–105]. An 8 bp deletion in exon 7 (*badh2-E7*) is the predominant allele in most aromatic varieties including the famous Jasmine and Basmati fragrant rice [98,100], and a haplotype analysis has revealed that the *badh2* allele in Indica varieties originated from Japonica rice [102,106]. The nucleotide diversity of *badh2* allows the development of specific molecular markers to discriminate aromatic and non-aromatic rice cultivars by PCR amplification [101,103,107], which greatly facilitated the selection and breeding of aromatic rice cultivars.

4.4. Genetic Improvement of Rice Eating and Cooking Quality

As the core of improving rice quality, the goal of ECQ improvement is to obtain rice varieties that possess desirable characteristics such as good palatability, aromatic scent, white and shiny appearance, pliable and supple texture, no stickiness, and no hardening or rebounding after cooling [10]. Fortunately, the relevant genes that determine these traits have been cloned, and their abundant allelic variations have been utilized by breeders for ECQ improvements through marker-assisted selection (MAS) (Table 1).

Most consumers prefer medium-grain rice with a relatively soft texture; therefore, improving the cooking and eating quality of rice usually focuses on reducing amylose content (AC). MAS of the *Wx* alleles has been widely used to improve grain quality for a long time. Decades ago, a DNA marker for Wx^{mq} was applied to varietal identification in Japan [79]. Later, several novel *Wx* alleles were utilized to breed rice varieties with better eating quality in Japan [108–110]. For example, the Japonica rice variety Yumepirika, which has a low amylose content of 16.1% due to the presence of the *Wx1-1* allele (a 37 bp deletion in intron 10), is an example of a rice variety which known for its sticky texture. It has been developed and assessed as a superior quality rice variety for the Japanese market [109,110]. In China, the Wx^b allele is the most widely used gene in the ECQ improvement process, especially in hybrid rice. The poor quality of the widely used female parent of hybrids, Zhenshan 97, was improved by updating the *Wx* locus, which led to reduced AC and increased GC and GT [51]. Using MAS, the Wx^b allele was introgressed into the maintainer lines of Longtefu B and Zhenshan 97B, resulting in the development of low-amylose content maintainers and restorers with improved cooking and eating quality. The allelic combination significantly reduced the AC and improved cooking and eating quality [52]. Similarly, the hybrid variety Xieyou 57 was also improved by the incorporation of the Wx^b allele, leading to better characteristics in terms of ECQ parameters [53]. Recently, the *Wx* genotyping for thirty-six main parents of hybrid rice in China revealed that only Wx^a, Wx^{lv}, and Wx^b alleles existed in these main parents, and the allelic combination gradually changed to Wx^b/Wx^b as the quality improved [111]. In addition to Wx^b, the Wx^{mq} allele that controls low AC has been successfully used for superior quality rice breeding in China over the past decade. The low-AC rice is also called semi-glutinous rice or soft rice [50]. Consumers prefer this type of rice due to its favorable eating quality, which combines the softness of glutinous rice and the elasticity of non-glutinous rice. When cooked, this rice has a soft texture and excellent ability to puff up. A series of Japonica varieties, such as Nangeng 46, Nangeng 5055, and Nangeng 9108, with low AC (10–15%), have been produced through the combination of high-yield Japonica varieties with the good quality japonica rice "Kantou 194," which harbors the Wx^{mq} allele [50,112]. It should be noted that the low AC in soft rice does not necessarily mean better quality, as grains with excessively low AC often exhibit dark or dull endosperm appearances and reduced transparency, leading to a decline in overall appearance quality. However, the recently discovered Wx^{mw} allele shows promise as a solution to this issue, as it not only contributes to a relatively low AC of approximately 14% but also improves endosperm transparency, potentially enhancing the appearance quality of soft rice [54].

For aroma improvement, *badh2* currently stands as the only target gene in rice breeding programs. To this end, several gene-specific molecular markers, highly correlated with 2-AP levels, have been developed for aroma enrichment by MAS [113,114]. The utilization of CRISPR/Cas9 gene editing has led to the generation of a succession of *badh2* null alleles, resulting in the successful cultivation of numerous high-aroma rice cultivars [115–118]. Nevertheless, it should be noted that the *badh2* gene cannot fully account for all the phenotypic variations observed in aromatic rice cultivars. Some of these cultivars, despite not having any functional badh2 gene variant, possess relatively elevated levels 2-AP levels and exhibit distinct aromatic characteristics [10,119]. Furthermore, the 2-AP content varies significantly among different aromatic rice types and cultivars, with some cultivars containing low 2-AP levels [119]. Hence, it is reasonable to speculate that there may exist other genes or chemical constituents that could influence or regulate the aromatic properties of rice and that further analysis of its genetic mechanisms is required for the effective utilization of this trait.

Typically, enhancements in ECQ traits have been achieved through the manipulation of a single gene. However, ECQ is a complex trait, and pyramiding multiple beneficial alleles is necessary for ECQ improvements. Additionally, *ALK* has also been proven to be a good target for rice ECQ improvement by altering rice amylopectin [96]. In a previous study, Wx^a, ALK^b, and *badh2-E7* were introgressed into a key maintainer line II-32B with poor quality using MAS. As a result, it exhibited low AC and GT with aroma as expected [55]. Similarly, it has been proposed that the coordinated expression of *ALK* and *Wx* is a feasible approach for rice ECQ improvement [120]. Through rational design, new rice varieties have been developed by introducing favorable alleles from Nipponbare (*Wx* and *ALK*) that control good ECQ and alleles from 9311 (*GS3* and *GW5*) that control good appearance traits into the Indica rice variety Teqing. As a result, the appearance, ECQ, and yield were promoted simultaneously [121].

5. Genetics of Nutritional Quality

In addition to starch, brown rice contains other nutrients such as proteins, storage lipids, as well as trace amounts of amino acids, minerals, vitamins, and phytochemicals [122]. Milled rice leaves only endosperm with the removal of embryo and bran layers through polish processing. Therefore, the nutritional quality of rice grain is determined by the major components of the polished grain, which are proteins and lipids [10] (Figure 1).

5.1. Protein

Rice grains have protein as their second main component. Protein content (PC) varies across different rice varieties and is typically found to be between 5% and 16%, with Indica rice generally having a slightly higher content (from 2% to 3%) compared to Japonica rice [123–125]. Grain protein in rice is mainly composed of glutelins, prolamins, globulins, and albumins. Among them, glutelin, as the most abundant storage protein, has the highest nutritional value because of its high digestibility and lysine content [126]. Any significant changes in glutelin content will certainly affect rice quality. Protein content is a key factor not only for evaluating the nutritional quality of rice but also for its eating and cooking quality [127]. Previous research indicates that higher protein content generally leads to reduced ECQ, while rice grain with lower protein content generally exhibits better ECQ [128–130].

There have been significant efforts aimed at understanding the genetic basis of rice PC. Many QTLs for this trait have been detected [56,131–139]. However, PC is very susceptible to environmental conditions, and almost all studies have used phenotyping data from only a single environment. Thus, the detected QTLs for rice PC are frequently inconsistent because of the different environments or populations researchers used. To date, only two major QTLs, *qPC1* and *qGPC-10*, which underlay natural variation and control PC in rice, were map-based cloned and functionally characterized [125,140]. The *qPC1*, encoding a putative amino acid transporter OsAAP6, controls PC by regulating the synthesis and accumulation of storage proteins and starch [140]. It functions as a positive regulator of PC in rice. By increasing the expression of this gene in rice varieties with low PC, the total amount of amino acids and PC can be increased, which can ultimately improve the nutritional quality of the grain. A genetic variation analysis revealed that two nucleotide changes in the *OsAAP6* 5′-UTR seem to be associated with PC diversity mainly in Indica cultivars [140]. The *qGPC-10/OsGluA2* is another positive PC regulator which encodes a glutelin type-A2 precursor [125]. *OsGluA2* enhances PC by increasing the total amount of glutelin content. Similar to *qPC1*, no variation has been found in the coding regions of this gene, but one SNP was beendetected upstream of the coding regions. This functional SNP could account for the expression level differences between the *OsGluA2* alleles and it clarifies two haplotypes, $OsGluA2^{LET}$ and $OsGluA2^{HET}$. $OsGluA2^{LET}$ is the low PC type which mainly residues in Japonica cultivars, while the high PC type $OsGluA2^{HET}$ is in Indica cultivars [125].

Generally, a high grain protein content is thought to be favorable for nutritional value, while high PC usually leads to densely structured rice grains, and thus, poor palatability [141,142]. Hence, balancing nutrition quality and eating and cooking quality in rice grain is a feasible strategy in breeding practice. Developing rice varieties with desirable ECQ has led to a focus on reducing the protein content in rice grains. For example, many famous commercial varieties with good ECQ usually contain PC less than 7%, such as Koshihikari in Japan and Kongyu131 in northern China [125]. Little progress has been achieved in the breeding improvement of rice PC due to limited gene resources. The identified QTL *qPC-1* has been applied to rice quality improvement [56]. When the Habataki allele of *qPC-1* was introduced into a Japonica background, it resulted in decreased protein content but improved eating and cooking quality (Table 1). The same allele was also utilized in the genetic improvement of Nangeng 46 to increase its palatability [57]. *OsAPP6* and *OsGluA2* are the potential candidates known so far for the manipulation of rice PC known. The low expression alleles, *OsAPP6* and the $OsGluA2^{LET}$, are promising target genes for low PC rice breeding through MAS. Still, more valuable gene resources are needed for rice nutritional quality and eating and cooking quality improvement.

5.2. Lipids

Lipids are mainly stored in rice embryo and aleurone, which are the bran layers. The lipid content in rice endosperm is very low. Rice lipids contain a high proportion of triacylglycerols (TAGs) along with a smaller amount of phospholipids (PLs) and free fatty acids [143]. The main fatty acids include palmitic acid, oleic acid (OA), and linoleic acid (LA). Among them, the unsaturated fatty acids, OA and LA, are good for human health because they are essential nutrients that cannot be synthesized by the human body [8,17]. Lipids not only affect rice nutritional quality but also influence the ECQ. Phospholipids and glycolipids can form complexes with amylose and amylopectin, reducing the expansibility of starch and increasing the GT, thereby affecting the texture of cooked rice [144,145]. In addition, the unsaturated fatty acids of milled grain have been reported to contribute to the rice aroma [146]. Fragrant rice varieties generally have a higher concentration of unsaturated fatty acids in their grains. Therefore, increasing the content of unsaturated fatty acids can be a potential target for improving both the rice ECQ and the NQ. Recently, new progress has been achieved in deciphering the genetic basis of oil biosynthesis in rice grains. A GWAS research on oil composition and oil concentration identified natural variants in four genes (*PAL6*, *LIN6*, *MYR2*, and *ARA6*) involved in oil metabolism and they showed variation among rice subspecies [147]. A rice grain nutrition QTL, *NET* (*Nutrition, Eating, Taste*), regulates lipid content and accumulation of nutrient metabolites, such as vitamins, amino acids, and polyphenols, thus, influencing the taste of rice grain [148]. Another GWAS study identified a key glycerolipid-related gene, *OsLP1* (diacylglycerol choline phosphotransferase) that contributes to variations in saturated TAG [149]. The allelic variation in the *OsLP1* sequence between Indica and Japonica results in different saturated TAG levels. Although there is no evidence for their use in MAS, these genes/QTLs described above showed clear diversities among Indica and Japonica subpopulations and could be used as biomarkers to facilitate breeding for enhanced oil and grain quality in the future.

6. External Factors That Affect Rice Grain Quality

In addition to the main genetic factors related to the variety, the improvement of rice quality is largely influenced by the environment and cultivation practices. For instance, the northeast region of China is a famous production area for superior quality Japonica rice, characterized by both excellent appearance and taste quality. This is largely due to the long growth period, low occurrence of high-temperature weather, and high organic matter content in the soil. Conversely, various factors in the southern region of China, such as short growth periods, extreme weather, and heavy use of fertilizers, are not conducive to improving the quality of rice. Conversely, various factors in the southern region of China, such as short growth periods, the occurrence of extreme weather, and heavy use of fertilizers, are not conducive to improving the quality of rice. As a result, the rice produced in this region often has poor appearance quality and high protein content. Additionally, the expression of fragrance genes is easily influenced by environmental factors such as soil, water, temperature, and light [102,119]. Therefore, many fragrant rice varieties lose their fragrance when grown outside their original production area. Most of the quality traits of rice are quantitative traits that are highly susceptible to environmental influences, which makes it challenging to improve rice quality through conventional methods. Adverse environmental factors, such as drought, low temperature, high temperature, and salinity stress, not only affect rice yield, but also have a significant impact on its appearance quality, eating and cooking quality, milling quality, and nutritional quality [150]. For instance, drought stress at the flowering stage has been shown to induce no significant effect on the appearance and nutritional quality except for increased grain chalkiness [151]. Low temperatures during the flowering period have been reported to improve milling and nutritional qualities but reduce the cooking and eating quality of late-season rice in southern China [152]. Salinity adversely affects the grain quality by decreasing the head rice recovery, amylose content and by increasing the chalky rice rate [153].

6.1. High Temperature Predominantly Affects Rice Quality

Traditionally, drought, salinity, and related stresses could be managed manually through appropriate field management, but temperature cannot be controlled artificially. Temperature stress, especially high temperature (HT), is the primary ecological factor that affects rice quality. During the entire growth period of rice, the grain filling stage has the greatest impact on rice quality, and this stage often coincides with high temperatures. In years with an extreme climate, high temperature can result in poor grain filling, decreased milling quality, and increased chalkiness. Therefore, in recent years, exploring the genetic mechanisms underlying the decline in rice quality caused by high temperatures has been a hot research topic.

During grain filling, HT stress can alter the chemical composition of rice grains, including starch and storage proteins, which, in turn, affects the quality of the rice [154] (Figure 2). Since starch is the main component of the rice grain, most of the research has focused on the effects of HT on starch. HT can reduce the content and change the structure of grain starch, leading to a reduction in amylose content and short-chain amylopectin, an increase in long-chain amylopectin, and an increase in gelatinization temperature [155–157]. Genetically, HT-induced impairment of starch accumulation is primarily due to decreased expression of genes involved in starch biosynthesis, particularly *GBSSI* and *BEIIb* [158], and a subsequent decline in their corresponding enzyme activities [159]. The decrease in amylose content in rice endosperm caused by HT has been observed for a long time [160]. The main reason for this symptom is the decrease in Wx/GBSSI activity [158,161,162]. The transcription of the *Wx* gene is inhibited under HT, but the splicing efficiency of the first intron of the *Wx* gene is higher in the HT-tolerant variety. This leads to an increase in the percentage of the large isoform of *Wx* pre-mRNA, which shows higher enzyme activity, and likely stabilizes amylose content at high temperatures [162]. In addition, suppression of the HT-induced *OsMADS7* gene in endosperm increases *Wx* expression and stabilizes amylose content under HT conditions [163] (Figure 2). Apart from GBSSI, the expressions and activities of amylopectin-related enzymes are also influenced by HT, resulting in an altered chain length distribution in amylopectin [158,164,165] (Figure 2).

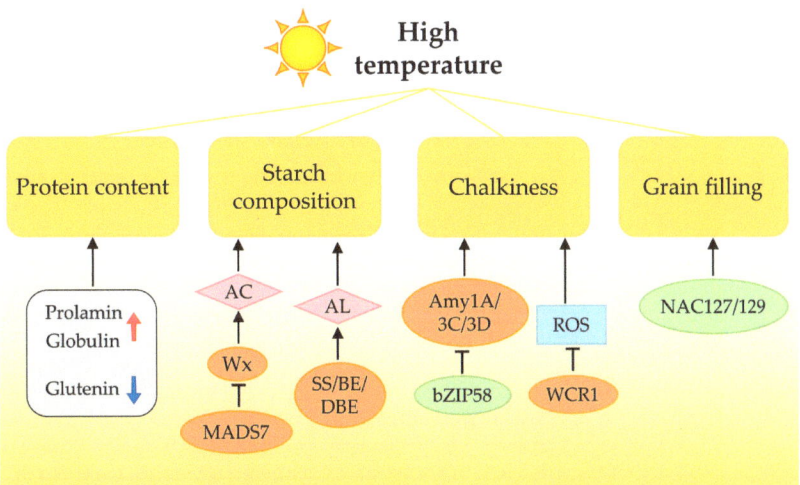

Figure 2. High temperature (HT) negatively affects rice grain quality. During grain filling, HT can alter the chemical composition of starch and storage proteins in rice grains through various aspects. Moreover, HT-induced chalkiness is the most common and severe symptom during the reproductive stage. Additionally, several transcription factors regulate HT-induced chalkiness. AC, amylose content; AL, amylopectin length; ROS, reactive oxygen species. The up arrow in red and the down arrow in blue represent increase and decrease of protein contents, respectively.

HT may also change the composition of seed storage proteins (Figure 2). At the early filling stage, HT stress can increase the accumulation of all classes of storage proteins, but decrease the accumulation of prolamins during maturation [154]. However, during the later grain-filling stage, the expression of grain protein-related genes decreases under HT stress, which disrupts the normal folding process of storage proteins. This leads to a decrease in the accumulation of 13 kD prolamin and globulin, and an increase in glutenin content in mature rice, which negatively affects the eating quality of rice [126,154,158].

Chalkiness is the most severe symptom caused by HT stress during the grain-filling stage, and HT is considered to be the major climatic factor conferring the chalky trait of rice [166,167]. HT at the milky stage of grain filling has the greatest influence on rice grain chalkiness [166]. The formation of chalkiness under HT is mainly related to changes in grain components, which have been discussed above. Additionally, some other components have been reported to affect the formation of chalkiness in recent years (Figure 2). HT upregulates the expression of α-amylase genes and its enzyme activity during seed maturation [158,168]. Suppressing α-amylase genes expression has been shown to result in a reduction in chalky grain formation under HT [168]. By contrast, the overexpressing of α-amylase genes, such as *Amy1A/3C/3D*, has been found to result in varying degrees of grain chalkiness [169]. In addition, ROS (reactive oxygen species) is also involved in inducing chalkiness under HT. An increased content of ROS and the expression of *NADPH oxidase* genes are associated with chalkiness occurrence in grains exposed to HT, accompanied by the increased activity of α-amylase [170]. Furthermore, the F-box protein WCR1 was recently found to reduce grain chalkiness through the MT2b-dependent ROS scavenging pathways in rice endosperm [49] (Figure 2). Additionally, several transcription factors have been reported to regulate HT-induced chalkiness, including *OsbZIP58*, *ONAC127*, and *ONAC129* [171,172] (Figure 2).

Although there have been some advances in understanding the genetic and molecular mechanisms underlying the decrease in rice quality caused by HT, few natural gene resources have been reported for the genetic improvement of rice quality under HT stress. To date, a few QTLs associated with HT tolerance during the reproductive and grain-filling stages have been identified in rice [162,173–176], which could be potential genetic resources to facilitate the development of high-temperature-tolerant rice varieties through molecular breeding. Strengthening the breeding of HT-tolerant rice varieties by screening for varietal differences in grain quality when ripened under high temperature is of crucial importance for coping with climate change.

6.2. Field Management Effects on Grain Quality

Field management has relatively minor effects on rice yield and quality compared to genetic and environmental factors. The application of nitrogen fertilizers increases grain yield dramatically but causes a series of problems, including declined rice quality. Numerous studies have examined the impact of nitrogen application on rice quality, and the consensus is that nitrogen application is generally positively correlated with rice milling quality and nutritional quality [177,178], but has negative effects on rice appearance quality [179] and eating and cooking quality [178,180]. However, the appropriate application of nitrogen can maintain and improve rice quality [179,181]. For example, the application of a moderate nitrogen rate (210–260 kg/ha) remarkably increased the milling quality and nutritional quality and resulted in moderate rice eating and cooking quality in a Japonica cultivar [181]. Nitrogen application has the greatest impact on eating and cooking quality, mainly through its influence on protein content and starch properties. High nitrogen levels can increase amylose content, reduce amylopectin branching, and increase protein content, leading to a hard texture of cooked rice and reduced palatability [182–185]. Therefore, to balance rice yield and quality, the use of chemical fertilizers should be controlled within a reasonable range, and biochar application may provide an alternative solution. The application of biochar has been widely recognized as a beneficial practice for improving rice yield while reducing nitrogen fertilizer usage [184]. More recently, several studies have investigated the potential impact of biochar application on rice grain quality. It has been suggested that biochar application can positively influence various grain quality traits, including milled rice rate, starch viscosity attributes, eating quality, and grain appearance, with the specific effects varying depending on the dosage of biochar used [184,186,187]. Furthermore, the combined application of biochar and a low nitrogen rate (135 kg/ha) has emerged as a promising approach for enhancing both yield and grain quality in a sustainable manner [184].

Environmental conditions and cultural managements belong to the preharvest factors which have significant effects on rice grain quality, while postharvest operations, such as seed drying, storage conditions, and milling processes, have been used to maintain desirable rice grain quality [188]. These postharvest factors play a critical role in determining the commercial quality and value of rice and should receive more attention.

7. Perspectives

7.1. Deepening Basic Research on Rice Quality Traits

In the past decade, significant progress has been made in the functional genomics research of rice quality, including the cloning and molecular regulation of important genes related to rice grain quality traits. However, rice grain quality is composed of multiple traits, and there are universal interactions among different quality traits and between quality and environmental factors. Our understanding of these scientific issues is still insufficient, which greatly limits the genetic improvement of rice quality. In the future, there are still many key scientific questions to be answered in the basic research of rice grain quality.

The fundamental research on rice grain quality aims to address the trade-off between rice yield and quality. Grain shape is a crucial trait that is deeply studied in rice quality research, as it is a key factor in achieving high yields and superior quality simultaneously. Numerous essential genes that control the shape and quality of rice grains, as well as significant biosynthetic and regulatory pathways, have been discovered in the past few decades. Moreover, some varieties have been successfully bred with excellent rice quality and high yield, but most of the favorable genes come from common varieties used in production. Rare genes such as *GL3.1*, *GL7/GW7*, and *GS2* have been cloned and provide favorable resources for rice quality improvement, but they often come with adverse effects such as increased chalkiness and reduced grain weight. Grain shape is the result of the interaction of multiple genes, but the mechanisms underlying their interactions are still poorly understood. There is still a major challenge in understanding the regulatory mechanisms that control these traits. Specifically, identifying the upstream and downstream components of these pathways and constructing a comprehensive genetic and molecular regulatory network for rice is necessary for a more thorough understanding of grain development and improvement.

Appearance quality is an important trait of rice grain quality. Any rice grain with a transparent endosperm and low chalkiness is considered to be of high quality and will be highly favored by consumers, regardless of its grain shape. Therefore, chalkiness is a major trait and important research direction in rice grain quality improvement and basic research. Chalkiness is related to the genetic factors of the variety itself and is also highly susceptible to high temperature stress. Because *Chalk5* and the major-effect gene *GS5* of grain width are closely linked in the Indica varieties, appropriately increasing grain length and reducing grain width to maintain grain weight is an effective way to break the unfavorable linkage and reduce chalkiness. Moreover, the exploration of the genetic basis of rice grain quality under high temperature stress is relatively limited, as it is more challenging to measure grain quality than grain yield. It is necessary to screen abundant natural variation germplasm or artificially induced mutants and to explore materials and their excellent genes that are tolerant to high temperature, and thus, solve the problem of HT-induced chalkiness formation.

Aroma is a symbol of premium rice, but there are issues with unclear genetic mechanisms and unstable fragrance in the breeding process of fragrant rice. Some fragrant rice varieties without functional loss of *badh2* alleles also have high 2-AP content, indicating that rice may have new functional genes and aromatic substances that control fragrance [119,189]. In the future, it will be important to focus on the collection and identification of fragrant rice resources. These resources can be used to discover new varieties that are not controlled by the *badh2* gene. Additionally, it is important to identify fragrant rice that exhibits stable or significant changes in aroma under different environmental conditions. By exploring these resources, it may be possible to discover new aromatic substances and genes, as well as their regulatory mechanisms.

The milling quality of rice, which is mainly measured by brown rice recovery, milled rice recovery, and head rice recovery, is closely related to grain shape and chalkiness. Generally, the milling quality of short and round grains with low chalkiness is higher than that of long and slender grains with high chalkiness [186]. Rice milling quality directly affects the yield of polished rice, but it is easily influenced by factors such as milling methods and storage conditions. Related studies on functional genomic research of milling quality have not received sufficient attention, and most studies are only limited to the initial mapping of related QTLs, with no critical genes being cloned yet. However, a recent GWAS study presented a possible molecular basis of milling quality [190]. Specifically, the study highlighted the pivotal role of the *Wx* gene in regulating head rice recovery by modulating the amorphous and semi-crystalline layered structure of starch granules. This finding suggested that the amylose content, which is controlled by the *Wx* gene, is closely linked to milling quality, along with the traditional determinants of grain shape and chalkiness. In the future, more attention should be paid to the identification and functional analysis of key genes related to rice milling quality. By doing so, we can improve efficiency and ensure the stable production of superior quality rice.

7.2. Application of New Technology to Grain Quality Improvement

Map-based gene cloning is a classic and effective method for analyzing the genetic basis of rice quality, but it is time-consuming and cumbersome, and it may be difficult to mine natural gene resources that control quality traits. Valuable variation is present in natural accessions, which is why GWAS approaches have been employed in basic research and breeding practices for superior quality rice [191,192]. Furthermore, the resequencing of 3010 rice accessions has made it possible to use GWAS approaches to identify minor alleles for grain quality from diverse rice collections [193].

Although transgenic technology has played a pivotal role in crop improvement, such as the successful creation of vitamin A-enriched "Golden Rice" and iron-enriched rice [194,195], policy limitations have restricted the application of this technology in rice breeding. However, recent innovations in CRISPR/Cas technology, such as prime editing and base editing, provide important technical support for crop improvement. Improving rice grain quality through gene-editing technology is a fast and efficient method that can accelerate the process of rice quality improvement. Gene-editing technology is now widely applied to directly generate new rice varieties with improved grain quality by manipulating elite genes, including *GS3, GW2, GW5, TGW6, GS9, Wx, OsAAP, badh2*, and *FAD2-1*, indicating that genome editing has significant potential to improve the quality of rice grains [7,112,196].

7.3. Directions of Breeding Rice Variety with Superior Grain Quality

As a staple food, the eating and cooking quality, which determines whether it is delicious or not, is the core of rice quality, while the appearance quality is directly linked to the commercial valueof rice. Thus, breeding rice varieties with both superior appearance and eating quality represents the primary focus of superior quality rice breeding. Long-grain rice, characterized by lower chalkiness, low amylose content, and low protein content, usually presents excellent appearance quality and good taste. As such, recently, breeders in China have devoted considerable attention to long-grain Japonica and super-long-grain Indica rice. The varying preferences for rice taste exist in different populations among various regions and countries, which can be achieved by fine tuning the amylose content through introducing new *Wx* alleles, starch synthesis-related genes such as *SSIIa, SSIIIa*, or gene loci that moderately increase AC. The influence of Jasmine rice from Thailand has led to aroma becoming a premium quality trait, with significant economic bonuses for rice farmers. Hence, developing fragrant rice varieties has become an essential aspect that cannot be disregarded.

The quality of the rice grain in conventional Japonica rice is often superior to that of Indica rice. Although hybrid rice has yield advantages, its quality is often inferior to that of conventional rice. Therefore, strengthening the breeding of superior quality conventional and hybrid Indica rice varieties is an important direction for future rice breeding. It is worth noting that hybrid Japonica rice, which has received little attention, has potential advantages in yield and disease resistance over conventional Japonica rice, despite its small planting area. Currently, compared with conventional Japonica rice, hybrid Japonica rice has higher chalkiness and lower head rice rate, and the research and promotion of superior quality hybrid Japonica rice is an important breakthrough for increasing grain production in Japonica rice areas.

Furthermore, with the improvement of living standards and the pursuit of nutrition and health by consumers, the development of functional rice with special nutritional value or special needs for specific populations, such as rice with low glutelin content, high resistant starch, and high γ-aminobutyric acid (GABA), should be promoted [197–199].

In the pursuit of improving rice quality, the objectives may vary depending on the demands of different regional markets and consumers, but the fundamental strategy for achieving this goal is relatively uniform (Figure 3), that is, in summary, fully utilizing cloned rice quality-related QTLs/genes and employing various breeding methods, pyramiding multiple elite alleles to enhance head rice recovery and ECQ while maintaining high yield. On this foundation, enhancing the appearance quality and aroma of rice grains becomes feasible, resulting in an output characterized by uniform shape, high translucency, and an appealing fragrance.

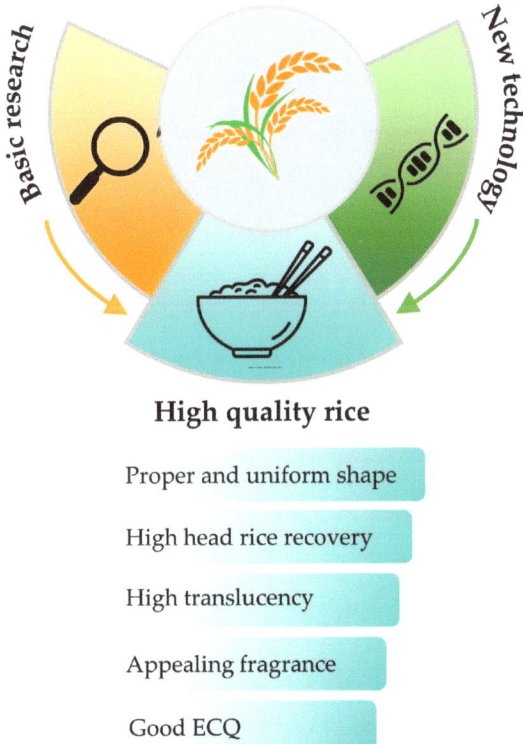

Figure 3. Breeding rice variety with superior grain quality.

Author Contributions: Writing—original draft preparation, D.G. and F.H.; Writing—review, editing, and financial support, G.Y.; Collection of information and data, X.Z., Y.C., R.L., J.Y., M.Z. and W.Z. All authors have read and agreed to the published version of the manuscript.

Funding: This work was supported by earmarked funding for the project from the President's Fund Project of Liaoning Academy of Agricultural Sciences (2023QN2408), the Liaoning Provincial Doctoral Research Start-up Fund in 2022 (2022-BS-050), the Liaoning Academy of Agricultural Sciences Collaborative Innovation Plan 2022 "Revelation and Commanding" project (2022XTCX0501003), the China Agriculture Research System (CARS-01-55), and the Applied Basic Research Project of Liaoning Province (2022JH2/101300283).

Conflicts of Interest: The authors declare no conflict of interest.

References

1. Rao, Y.; Li, Y.; Qian, Q. Recent progress on molecular breeding of rice in China. *Plant Cell Rep.* **2014**, *33*, 551–564. [CrossRef] [PubMed]
2. Yu, T.-Q.; Jiang, W.; Ham, T.-H.; Chu, S.-H.; Lestari, P.; Lee, J.-H.; Kim, M.-K.; Xu, F.-R.; Han, L.; Dai, L.-Y. Comparison of grain quality traits between japonica rice cultivars from Korea and Yunnan Province of China. *J. Crop Sci. Biotechnol.* **2008**, *11*, 135–140.
3. Chen, Y.; Wang, M.; Ouwerkerk, P.B. Molecular and environmental factors determining grain quality in rice. *Food Energy Secur.* **2012**, *1*, 111–132. [CrossRef]
4. Bao, J. Genes and QTLs for rice grain quality improvement. In *Rice*; InTech: London, UK, 2014; pp. 239–278.
5. Custodio, M.C.; Cuevas, R.P.; Ynion, J.; Laborte, A.G.; Velasco, M.L.; Demont, M. Rice quality: How is it defined by consumers, industry, food scientists, and geneticists? *Trends Food Sci. Technol.* **2019**, *92*, 122–137. [CrossRef] [PubMed]
6. Zhou, H.; Xia, D.; He, Y. Rice grain quality—Traditional traits for high quality rice and health-plus substances. *Mol. Breed.* **2020**, *40*, 1. [CrossRef]
7. Cheng, J.; Lin, X.; Long, Y.; Zeng, Q.; Zhao, K.; Hu, P.; Peng, J. Rice grain quality: Where we are and where to go? *Adv. Agron.* **2022**, *172*, 211–252.

8. Li, P.; Chen, Y.-H.; Lu, J.; Zhang, C.-Q.; Liu, Q.-Q.; Li, Q.-F. Genes and their molecular functions determining seed structure, components, and quality of rice. *Rice* **2022**, *15*, 18. [CrossRef]
9. Zhou, H.; Yun, P.; He, Y. Rice appearance quality. In *Rice*; Elsevier: Amsterdam, The Netherlands, 2019; pp. 371–383.
10. Fitzgerald, M.A.; McCouch, S.R.; Hall, R.D. Not just a grain of rice: The quest for quality. *Trends Plant Sci.* **2009**, *14*, 133–139. [CrossRef]
11. Lu, S.; Luh, B.S. Properties of the rice caryopsis. In *Rice: Volume I. Production/Volume II. Utilization*; Springer: Boston, MA, USA, 1991; pp. 389–419. [CrossRef]
12. Huang, R.; Jiang, L.; Zheng, J.; Wang, T.; Wang, H.; Huang, Y.; Hong, Z. Genetic bases of rice grain shape: So many genes, so little known. *Trends Plant Sci.* **2013**, *18*, 218–226. [CrossRef]
13. Zuo, J.; Li, J. Molecular genetic dissection of quantitative trait loci regulating rice grain size. *Annu. Rev. Genet.* **2014**, *48*, 99–118. [CrossRef]
14. Li, N.; Li, Y. Signaling pathways of seed size control in plants. *Curr. Opin. Plant Biol.* **2016**, *33*, 23–32. [CrossRef] [PubMed]
15. Li, N.; Xu, R.; Li, Y. Molecular Networks of Seed Size Control in Plants. *Annu. Rev. Plant Biol.* **2019**, *70*, 435–463. [CrossRef] [PubMed]
16. Zhao, D.; Zhang, C.; Li, Q.; Liu, Q. Genetic control of grain appearance quality in rice. *Biotechnol. Adv.* **2022**, *60*, 108014. [CrossRef] [PubMed]
17. Ren, D.; Ding, C.; Qian, Q. Molecular bases of rice grain size and quality for optimized productivity. *Sci. Bull.* **2023**, *68*, 314–350.
18. Fan, C.; Xing, Y.; Mao, H.; Lu, T.; Han, B.; Xu, C.; Li, X.; Zhang, Q. GS3, a major QTL for grain length and weight and minor QTL for grain width and thickness in rice, encodes a putative transmembrane protein. *Theor. Appl. Genet.* **2006**, *112*, 1164–1171. [CrossRef]
19. Takano-Kai, N.; Jiang, H.; Kubo, T.; Sweeney, M.; Matsumoto, T.; Kanamori, H.; Padhukasahasram, B.; Bustamante, C.; Yoshimura, A.; Doi, K.; et al. Evolutionary history of GS3, a gene conferring grain length in rice. *Genetics* **2009**, *182*, 1323–1334. [CrossRef] [PubMed]
20. Mao, H.; Sun, S.; Yao, J.; Wang, C.; Yu, S.; Xu, C.; Li, X.; Zhang, Q. Linking differential domain functions of the GS3 protein to natural variation of grain size in rice. *Proc. Natl. Acad. Sci. USA* **2010**, *107*, 19579–19584. [CrossRef]
21. Trusov, Y.; Chakravorty, D.; Botella, J.R. Diversity of heterotrimeric G-protein gamma subunits in plants. *BMC Res. Notes* **2012**, *5*, 608. [CrossRef]
22. Sun, S.; Wang, L.; Mao, H.; Shao, L.; Li, X.; Xiao, J.; Ouyang, Y.; Zhang, Q. A G-protein pathway determines grain size in rice. *Nat. Commun.* **2018**, *9*, 851. [CrossRef]
23. Qi, P.; Lin, Y.S.; Song, X.J.; Shen, J.B.; Huang, W.; Shan, J.X.; Zhu, M.Z.; Jiang, L.; Gao, J.P.; Lin, H.X. The novel quantitative trait locus GL3.1 controls rice grain size and yield by regulating Cyclin-T1;3. *Cell Res.* **2012**, *22*, 1666–1680. [CrossRef]
24. Zhang, X.; Wang, J.; Huang, J.; Lan, H.; Wang, C.; Yin, C.; Wu, Y.; Tang, H.; Qian, Q.; Li, J.; et al. Rare allele of OsPPKL1 associated with grain length causes extra-large grain and a significant yield increase in rice. *Proc. Natl. Acad. Sci. USA* **2012**, *109*, 21534–21539. [CrossRef] [PubMed]
25. Gao, X.; Zhang, X.; Lan, H.; Huang, J.; Wang, J.; Zhang, H. The additive effects of GS3 and qGL3 on rice grain length regulation revealed by genetic and transcriptome comparisons. *BMC Plant Biol.* **2015**, *15*, 156. [CrossRef] [PubMed]
26. Xia, D.; Zhou, H.; Liu, R.; Dan, W.; Li, P.; Wu, B.; Chen, J.; Wang, L.; Gao, G.; Zhang, Q.; et al. GL3.3, a Novel QTL Encoding a GSK3/SHAGGY-like Kinase, Epistatically Interacts with GS3 to Produce Extra-long Grains in Rice. *Mol. Plant* **2018**, *11*, 754–756. [CrossRef] [PubMed]
27. Song, X.J.; Huang, W.; Shi, M.; Zhu, M.Z.; Lin, H.X. A QTL for rice grain width and weight encodes a previously unknown RING-type E3 ubiquitin ligase. *Nat. Genet.* **2007**, *39*, 623–630. [CrossRef]
28. Shomura, A.; Izawa, T.; Ebana, K.; Ebitani, T.; Kanegae, H.; Konishi, S.; Yano, M. Deletion in a gene associated with grain size increased yields during rice domestication. *Nat. Genet.* **2008**, *40*, 1023–1028. [CrossRef]
29. Weng, J.; Gu, S.; Wan, X.; Gao, H.; Guo, T.; Su, N.; Lei, C.; Zhang, X.; Cheng, Z.; Guo, X.; et al. Isolation and initial characterization of GW5, a major QTL associated with rice grain width and weight. *Cell Res.* **2008**, *18*, 1199–1209. [CrossRef]
30. Duan, P.; Xu, J.; Zeng, D.; Zhang, B.; Geng, M.; Zhang, G.; Huang, K.; Huang, L.; Xu, R.; Ge, S.; et al. Natural Variation in the Promoter of GSE5 Contributes to Grain Size Diversity in Rice. *Mol. Plant* **2017**, *10*, 685–694. [CrossRef]
31. Liu, J.; Chen, J.; Zheng, X.; Wu, F.; Lin, Q.; Heng, Y.; Tian, P.; Cheng, Z.; Yu, X.; Zhou, K.; et al. GW5 acts in the brassinosteroid signalling pathway to regulate grain width and weight in rice. *Nat. Plants* **2017**, *3*, 17043. [CrossRef]
32. Zhang, X.F.; Yang, C.Y.; Lin, H.X.; Wang, J.W.; Xue, H.W. Rice SPL12 coevolved with GW5 to determine grain shape. *Sci. Bull.* **2021**, *66*, 2353–2357. [CrossRef]
33. Li, Y.; Fan, C.; Xing, Y.; Jiang, Y.; Luo, L.; Sun, L.; Shao, D.; Xu, C.; Li, X.; Xiao, J.; et al. Natural variation in GS5 plays an important role in regulating grain size and yield in rice. *Nat. Genet.* **2011**, *43*, 1266–1269. [CrossRef]
34. Xu, C.; Liu, Y.; Li, Y.; Xu, X.; Xu, C.; Li, X.; Xiao, J.; Zhang, Q. Differential expression of GS5 regulates grain size in rice. *J. Exp. Bot.* **2015**, *66*, 2611–2623. [CrossRef]
35. Wang, S.; Wu, K.; Yuan, Q.; Liu, X.; Liu, Z.; Lin, X.; Zeng, R.; Zhu, H.; Dong, G.; Qian, Q.; et al. Control of grain size, shape and quality by OsSPL16 in rice. *Nat. Genet.* **2012**, *44*, 950–954. [CrossRef] [PubMed]

36. Wang, S.; Li, S.; Liu, Q.; Wu, K.; Zhang, J.; Wang, S.; Wang, Y.; Chen, X.; Zhang, Y.; Gao, C.; et al. The OsSPL16-GW7 regulatory module determines grain shape and simultaneously improves rice yield and grain quality. *Nat. Genet.* **2015**, *47*, 949–954. [CrossRef] [PubMed]
37. Wang, Y.; Xiong, G.; Hu, J.; Jiang, L.; Yu, H.; Xu, J.; Fang, Y.; Zeng, L.; Xu, E.; Xu, J.; et al. Copy number variation at the GL7 locus contributes to grain size diversity in rice. *Nat. Genet.* **2015**, *47*, 944–948. [CrossRef] [PubMed]
38. Zhou, Y.; Miao, J.; Gu, H.; Peng, X.; Leburu, M.; Yuan, F.; Gu, H.; Gao, Y.; Tao, Y.; Zhu, J.; et al. Natural Variations in SLG7 Regulate Grain Shape in Rice. *Genetics* **2015**, *201*, 1591–1599. [CrossRef] [PubMed]
39. Zhao, D.S.; Li, Q.F.; Zhang, C.Q.; Zhang, C.; Yang, Q.Q.; Pan, L.X.; Ren, X.Y.; Lu, J.; Gu, M.H.; Liu, Q.Q. GS9 acts as a transcriptional activator to regulate rice grain shape and appearance quality. *Nat. Commun.* **2018**, *9*, 1240. [CrossRef]
40. Zhao, D.-s.; Liu, J.-y.; Ding, A.-q.; Zhang, T.; Ren, X.-y.; Zhang, L.; Li, Q.-f.; Fan, X.-l.; Zhang, C.-q.; Liu, Q.-q. Improving grain appearance of erect-panicle japonica rice cultivars by introgression of the null gs9 allele. *J. Integr. Agric.* **2021**, *20*, 2032–2042. [CrossRef]
41. Lu, L.; Shao, D.; Qiu, X.; Sun, L.; Yan, W.; Zhou, X.; Yang, L.; He, Y.; Yu, S.; Xing, Y. Natural variation and artificial selection in four genes determine grain shape in rice. *New. Phytol.* **2013**, *200*, 1269–1280. [CrossRef]
42. Fan, C.; Yu, S.; Wang, C.; Xing, Y. A causal C-A mutation in the second exon of GS3 highly associated with rice grain length and validated as a functional marker. *Theor. Appl. Genet.* **2009**, *118*, 465–472. [CrossRef]
43. Wang, C.; Chen, S.; Yu, S. Functional markers developed from multiple loci in GS3 for fine marker-assisted selection of grain length in rice. *Theor. Appl. Genet.* **2011**, *122*, 905–913. [CrossRef]
44. Nan, J.; Feng, X.; Wang, C.; Zhang, X.; Wang, R.; Liu, J.; Yuan, Q.; Jiang, G.; Lin, S. Improving rice grain length through updating the GS3 locus of an elite variety Kongyu 131. *Rice* **2018**, *11*, 21. [CrossRef] [PubMed]
45. Liu, Q.; Han, R.; Wu, K.; Zhang, J.; Ye, Y.; Wang, S.; Chen, J.; Pan, Y.; Li, Q.; Xu, X.; et al. G-protein betagamma subunits determine grain size through interaction with MADS-domain transcription factors in rice. *Nat. Commun.* **2018**, *9*, 852. [CrossRef]
46. Huang, X.; Qian, Q.; Liu, Z.; Sun, H.; He, S.; Luo, D.; Xia, G.; Chu, C.; Li, J.; Fu, X. Natural variation at the DEP1 locus enhances grain yield in rice. *Nat. Genet.* **2009**, *41*, 494–497. [CrossRef] [PubMed]
47. Jiang, L.; Zhong, H.; Jiang, X.; Zhang, J.; Huang, R.; Liao, F.; Deng, Y.; Liu, Q.; Huang, Y.; Wang, H.; et al. Identification and Pleiotropic Effect Analysis of GSE5 on Rice Chalkiness and Grain Shape. *Front. Plant Sci.* **2021**, *12*, 814928. [CrossRef] [PubMed]
48. Li, Y.; Fan, C.; Xing, Y.; Yun, P.; Luo, L.; Yan, B.; Peng, B.; Xie, W.; Wang, G.; Li, X.; et al. Chalk5 encodes a vacuolar H(+)-translocating pyrophosphatase influencing grain chalkiness in rice. *Nat. Genet.* **2014**, *46*, 398–404. [CrossRef] [PubMed]
49. Wu, B.; Yun, P.; Zhou, H.; Xia, D.; Gu, Y.; Li, P.; Yao, J.; Zhou, Z.; Chen, J.; Liu, R.; et al. Natural variation in WHITE-CORE RATE 1 regulates redox homeostasis in rice endosperm to affect grain quality. *Plant Cell* **2022**, *34*, 1912–1932. [CrossRef]
50. Wang, C.; Zhang, Y.; Zhu, Z.; Chen, T.; Zhao, L.; Lin, J.; Zhou, L. Development of a new japonica rice variety Nan-jing 46 with good eating quality by marker assisted selection. *Rice Genom. Genet.* **2010**, *1*. [CrossRef]
51. Zhou, P.H.; Tan, Y.F.; He, Y.Q.; Xu, C.G.; Zhang, Q. Simultaneous improvement for four quality traits of Zhenshan 97, an elite parent of hybrid rice, by molecular marker-assisted selection. *Theor. Appl. Genet.* **2003**, *106*, 326–331. [CrossRef]
52. Liu, Q.Q.; Li, Q.F.; Cai, X.L.; Wang, H.M.; Tang, S.Z.; Yu, H.X.; Wang, Z.Y.; Gu, M.H. Molecular marker-assisted selection for improved cooking and eating quality of two elite parents of hybrid rice. *Crop Sci.* **2006**, *46*, 2354–2360. [CrossRef]
53. Ni, D.; Zhang, S.; Chen, S.; Xu, Y.; Li, L.; Li, H.; Wang, Z.; Cai, X.; Li, Z.; Yang, J. Improving cooking and eating quality of Xieyou57, an elite indica hybrid rice, by marker-assisted selection of the Wx locus. *Euphytica* **2011**, *179*, 355–362. [CrossRef]
54. Zhang, C.; Yang, Y.; Chen, S.; Liu, X.; Zhu, J.; Zhou, L.; Lu, Y.; Li, Q.; Fan, X.; Tang, S.; et al. A rare Waxy allele coordinately improves rice eating and cooking quality and grain transparency. *J. Integr. Plant Biol.* **2021**, *63*, 889–901. [CrossRef] [PubMed]
55. Jin, L.; Lu, Y.; Shao, Y.; Zhang, G.; Xiao, P.; Shen, S.; Corke, H.; Bao, J. Molecular marker assisted selection for improvement of the eating, cooking and sensory quality of rice (*Oryza sativa* L.). *J. Cereal Sci.* **2010**, *51*, 159–164. [CrossRef]
56. Yang, Y.; Guo, M.; Li, R.; Shen, L.; Wang, W.; Liu, M.; Zhu, Q.; Hu, Z.; He, Q.; Xue, Y. Identification of quantitative trait loci responsible for rice grain protein content using chromosome segment substitution lines and fine mapping of qPC-1 in rice (*Oryza sativa* L.). *Mol. Breed.* **2015**, *35*, 1–9. [CrossRef]
57. Yang, Y.; Shen, Z.; Xu, C.; Guo, M.; Li, Y.; Zhang, Y.; Zhong, C.; Sun, Y.; Yan, C. Genetic improvement of panicle-erectness japonica rice toward both yield and eating and cooking quality. *Mol. Breed.* **2020**, *40*, 1–12. [CrossRef]
58. Wu, K.; Xu, X.; Zhong, N.; Huang, H.; Yu, J.; Ye, Y.; Wu, Y.; Fu, X. The rational design of multiple molecular module-based assemblies for simultaneously improving rice yield and grain quality. *J. Genet. Genom.* **2018**, *45*, 337–341. [CrossRef]
59. Yang, T.; Gu, H.; Yang, W.; Liu, B.; Liang, S.; Zhao, J. Artificially Selected Grain Shape Gene Combinations in Guangdong Simiao Varieties of Rice (*Oryza sativa* L.). *Rice* **2023**, *16*, 3. [CrossRef]
60. Lisle, A.; Martin, M.; Fitzgerald, M. Chalky and translucent rice grains differ in starch composition and structure and cooking properties. *Cereal Chem.* **2000**, *77*, 627–632. [CrossRef]
61. Gong, J.; Miao, J.; Zhao, Y.; Zhao, Q.; Feng, Q.; Zhan, Q.; Cheng, B.; Xia, J.; Huang, X.; Yang, S.; et al. Dissecting the Genetic Basis of Grain Shape and Chalkiness Traits in Hybrid Rice Using Multiple Collaborative Populations. *Mol. Plant* **2017**, *10*, 1353–1356. [CrossRef]
62. Hu, J.; Wang, Y.; Fang, Y.; Zeng, L.; Xu, J.; Yu, H.; Shi, Z.; Pan, J.; Zhang, D.; Kang, S.; et al. A Rare Allele of GS2 Enhances Grain Size and Grain Yield in Rice. *Mol. Plant* **2015**, *8*, 1455–1465. [CrossRef]

63. Ayaad, M.; Han, Z.; Zheng, K.; Hu, G.; Abo-Yousef, M.; Sobeih, S.E.S.; Xing, Y. Bin-based genome-wide association studies reveal superior alleles for improvement of appearance quality using a 4-way MAGIC population in rice. *J. Adv. Res.* **2021**, *28*, 183–194. [CrossRef]
64. Tian, Z.; Qian, Q.; Liu, Q.; Yan, M.; Liu, X.; Yan, C.; Liu, G.; Gao, Z.; Tang, S.; Zeng, D.; et al. Allelic diversities in rice starch biosynthesis lead to a diverse array of rice eating and cooking qualities. *Proc. Natl. Acad. Sci. USA* **2009**, *106*, 21760–21765. [CrossRef] [PubMed]
65. Duan, M.; Sun, S.S. Profiling the expression of genes controlling rice grain quality. *Plant Mol. Biol.* **2005**, *59*, 165–178. [CrossRef] [PubMed]
66. Li, H.; Gilbert, R.G. Starch molecular structure: The basis for an improved understanding of cooked rice texture. *Carbohydr. Polym.* **2018**, *195*, 9–17. [CrossRef] [PubMed]
67. Wang, L.Q.; Liu, W.J.; Xu, Y.; He, Y.Q.; Luo, L.J.; Xing, Y.Z.; Xu, C.G.; Zhang, Q. Genetic basis of 17 traits and viscosity parameters characterizing the eating and cooking quality of rice grain. *Theor. Appl. Genet.* **2007**, *115*, 463–476. [CrossRef] [PubMed]
68. Zhang, C.; Yang, Y.; Chen, Z.; Chen, F.; Pan, L.; Lu, Y.; Li, Q.; Fan, X.; Sun, Z.; Liu, Q. Characteristics of Grain Physicochemical Properties and the Starch Structure in Rice Carrying a Mutated ALK/SSIIa Gene. *J. Agric. Food Chem.* **2020**, *68*, 13950–13959. [CrossRef]
69. Jeon, J.S.; Ryoo, N.; Hahn, T.R.; Walia, H.; Nakamura, Y. Starch biosynthesis in cereal endosperm. *Plant Physiol. Biochem.* **2010**, *48*, 383–392. [CrossRef]
70. Pfister, B.; Zeeman, S.C. Formation of starch in plant cells. *Cell. Mol. Life Sci.* **2016**, *73*, 2781–2807. [CrossRef]
71. Bao, J. Rice starch. In *Rice*; Elsevier: Amsterdam, The Netherlands, 2019; pp. 55–108.
72. Huang, L.; Tan, H.; Zhang, C.; Li, Q.; Liu, Q. Starch biosynthesis in cereal endosperms: An updated review over the last decade. *Plant Commun.* **2021**, *2*, 100237. [CrossRef]
73. Zeeman, S.C.; Kossmann, J.; Smith, A.M. Starch: Its metabolism, evolution, and biotechnological modification in plants. *Annu. Rev. Plant Biol.* **2010**, *61*, 209–234. [CrossRef]
74. Ball, S.G.; van de Wal, M.H.; Visser, R.G. Progress in understanding the biosynthesis of amylose. *Trends Plant Sci.* **1998**, *3*, 462–467. [CrossRef]
75. Gu, M.-H.; Liu, Q.-Q.; Yan, C.-J.; Tang, S.-Z. Grain quality of hybrid rice: Genetic variation and molecular improvement. *Accel. Hybrid. Rice Dev. Los. Banos (Philipp.) Int. Rice Res. Inst.* **2010**, 345–356.
76. Wang, Z.Y.; Wu, Z.L.; Xing, Y.Y.; Zheng, F.G.; Guo, X.L.; Zhang, W.G.; Hong, M.M. Nucleotide sequence of rice waxy gene. *Nucleic Acids Res.* **1990**, *18*, 5898. [CrossRef] [PubMed]
77. Wang, Z.Y.; Zheng, F.Q.; Shen, G.Z.; Gao, J.P.; Snustad, D.P.; Li, M.G.; Zhang, J.L.; Hong, M.M. The amylose content in rice endosperm is related to the post-transcriptional regulation of the waxy gene. *Plant J.* **1995**, *7*, 613–622. [CrossRef]
78. Cai, X.L.; Wang, Z.Y.; Xing, Y.Y.; Zhang, J.L.; Hong, M.M. Aberrant splicing of intron 1 leads to the heterogeneous 5' UTR and decreased expression of waxy gene in rice cultivars of intermediate amylose content. *Plant J.* **1998**, *14*, 459–465. [CrossRef] [PubMed]
79. Sato, H.; Suzuki, Y.; Sakai, M.; Imbe, T. Molecular characterization of Wx-mq, a novel mutant gene for low-amylose content in endosperm of rice (*Oryza sativa* L.). *Breed. Sci.* **2002**, *52*, 131–135. [CrossRef]
80. Wanchana, S.; Toojinda, T.; Tragoonrung, S.; Vanavichit, A. Duplicated coding sequence in the waxy allele of tropical glutinous rice (*Oryza sativa* L.). *Plant Sci.* **2003**, *165*, 1193–1199. [CrossRef]
81. Bao, J.S.; Corke, H.; Sun, M. Nucleotide diversity in starch synthase IIa and validation of single nucleotide polymorphisms in relation to starch gelatinization temperature and other physicochemical properties in rice (*Oryza sativa* L.). *Theor. Appl. Genet.* **2006**, *113*, 1171–1183. [CrossRef]
82. Mikami, I.; Uwatoko, N.; Ikeda, Y.; Yamaguchi, J.; Hirano, H.Y.; Suzuki, Y.; Sano, Y. Allelic diversification at the wx locus in landraces of Asian rice. *Theor. Appl. Genet.* **2008**, *116*, 979–989. [CrossRef]
83. Liu, L.; Ma, X.; Liu, S.; Zhu, C.; Jiang, L.; Wang, Y.; Shen, Y.; Ren, Y.; Dong, H.; Chen, L.; et al. Identification and characterization of a novel Waxy allele from a Yunnan rice landrace. *Plant Mol. Biol.* **2009**, *71*, 609–626. [CrossRef]
84. Yang, J.; Wang, J.; Fan, F.J.; Zhu, J.Y.; Chen, T.; Wang, C.L.; Zheng, T.Q.; Zhang, J.; Zhong, W.G.; Xu, J.L. Development of AS-PCR marker based on a key mutation confirmed by resequencing of Wx-mp in Milky P rincess and its application in japonica soft rice (*Oryza sativa* L.) breeding. *Plant Breed.* **2013**, *132*, 595–603. [CrossRef]
85. Zhang, C.; Zhu, J.; Chen, S.; Fan, X.; Li, Q.; Lu, Y.; Wang, M.; Yu, H.; Yi, C.; Tang, S.; et al. Wx(lv), the Ancestral Allele of Rice Waxy Gene. *Mol. Plant* **2019**, *12*, 1157–1166. [CrossRef] [PubMed]
86. Zhou, H.; Xia, D.; Zhao, D.; Li, Y.; Li, P.; Wu, B.; Gao, G.; Zhang, Q.; Wang, G.; Xiao, J.; et al. The origin of Wx(la) provides new insights into the improvement of grain quality in rice. *J. Integr. Plant Biol.* **2021**, *63*, 878–888. [CrossRef] [PubMed]
87. Zhou, L.; Chen, S.; Yang, G.; Zha, W.; Cai, H.; Li, S.; Chen, Z.; Liu, K.; Xu, H.; You, A. A perfect functional marker for the gene of intermediate amylose content Wx-in in rice (*Oryza sativa* L.). *Crop Breed. Appl. Biotechnol.* **2018**, *18*, 103–109. [CrossRef]
88. Tan, Y.F.; Li, J.X.; Yu, S.B.; Xing, Y.Z.; Xu, C.G.; Zhang, Q. The three important traits for cooking and eating quality of rice grains are controlled by a single locus in an elite rice hybrid, Shanyou 63. *Theor. Appl. Genet.* **1999**, *99*, 642–648. [CrossRef]
89. Tian, R.; Jiang, G.-H.; Shen, L.-H.; Wang, L.-Q.; He, Y.-Q. Mapping quantitative trait loci underlying the cooking and eating quality of rice using a DH population. *Mol. Breed.* **2005**, *15*, 117–124. [CrossRef]

90. Su, Y.; Rao, Y.; Hu, S.; Yang, Y.; Gao, Z.; Zhang, G.; Liu, J.; Hu, J.; Yan, M.; Dong, G.; et al. Map-based cloning proves qGC-6, a major QTL for gel consistency of japonica/indica cross, responds by Waxy in rice (*Oryza sativa* L.). *Theor. Appl. Genet.* **2011**, *123*, 859–867. [CrossRef]
91. Gao, Z.; Zeng, D.; Cui, X.; Zhou, Y.; Yan, M.; Huang, D.; Li, J.; Qian, Q. Map-based cloning of the ALK gene, which controls the gelatinization temperature of rice. *Sci. China C. Life Sci.* **2003**, *46*, 661–668. [CrossRef]
92. Nakamura, Y.; Francisco, P.B., Jr.; Hosaka, Y.; Sato, A.; Sawada, T.; Kubo, A.; Fujita, N. Essential amino acids of starch synthase IIa differentiate amylopectin structure and starch quality between japonica and indica rice varieties. *Plant Mol. Biol.* **2005**, *58*, 213–227. [CrossRef]
93. Gao, Z.; Zeng, D.; Cheng, F.; Tian, Z.; Guo, L.; Su, Y.; Yan, M.; Jiang, H.; Dong, G.; Huang, Y.; et al. ALK, the key gene for gelatinization temperature, is a modifier gene for gel consistency in rice. *J. Integr. Plant Biol.* **2011**, *53*, 756–765. [CrossRef]
94. Waters, D.L.; Henry, R.J.; Reinke, R.F.; Fitzgerald, M.A. Gelatinization temperature of rice explained by polymorphisms in starch synthase. *Plant Biotechnol. J.* **2006**, *4*, 115–122. [CrossRef]
95. Miura, S.; Crofts, N.; Saito, Y.; Hosaka, Y.; Oitome, N.F.; Watanabe, T.; Kumamaru, T.; Fujita, N. Starch Synthase IIa-Deficient Mutant Rice Line Produces Endosperm Starch With Lower Gelatinization Temperature Than Japonica Rice Cultivars. *Front. Plant Sci.* **2018**, *9*, 645. [CrossRef]
96. Chen, Z.; Lu, Y.; Feng, L.; Hao, W.; Li, C.; Yang, Y.; Fan, X.; Li, Q.; Zhang, C.; Liu, Q. Genetic Dissection and Functional Differentiation of ALK(a) and ALK(b), Two Natural Alleles of the ALK/SSIIa Gene, Responding to Low Gelatinization Temperature in Rice. *Rice* **2020**, *13*, 39. [CrossRef]
97. Xiang, X.; Kang, C.; Xu, S.; Yang, B. Combined effects of Wx and SSIIa haplotypes on rice starch physicochemical properties. *J. Sci. Food Agric.* **2017**, *97*, 1229–1234. [CrossRef]
98. Bradbury, L.M.; Fitzgerald, T.L.; Henry, R.J.; Jin, Q.; Waters, D.L. The gene for fragrance in rice. *Plant Biotechnol. J.* **2005**, *3*, 363–370. [CrossRef]
99. Chen, S.; Wu, J.; Yang, Y.; Shi, W.; Xu, M. The fgr gene responsible for rice fragrance was restricted within 69kb. *Plant Sci.* **2006**, *171*, 505–514. [CrossRef]
100. Chen, S.; Yang, Y.; Shi, W.; Ji, Q.; He, F.; Zhang, Z.; Cheng, Z.; Liu, X.; Xu, M. Badh2, encoding betaine aldehyde dehydrogenase, inhibits the biosynthesis of 2-acetyl-1-pyrroline, a major component in rice fragrance. *Plant Cell* **2008**, *20*, 1850–1861. [CrossRef]
101. Shi, W.; Yang, Y.; Chen, S.; Xu, M. Discovery of a new fragrance allele and the development of functional markers for the breeding of fragrant rice varieties. *Mol. Breed.* **2008**, *22*, 185–192. [CrossRef]
102. Kovach, M.J.; Calingacion, M.N.; Fitzgerald, M.A.; McCouch, S.R. The origin and evolution of fragrance in rice (*Oryza sativa* L.). *Proc. Natl. Acad. Sci. USA* **2009**, *106*, 14444–14449. [CrossRef]
103. Shao, G.; Tang, A.; Tang, S.; Luo, J.; Jiao, G.; Wu, J.; Hu, P. A new deletion mutation of fragrant gene and the development of three molecular markers for fragrance in rice. *Plant Breed.* **2011**, *130*, 172–176. [CrossRef]
104. Shi, Y.; Zhao, G.; Xu, X.; Li, J. Discovery of a new fragrance allele and development of functional markers for identifying diverse fragrant genotypes in rice. *Mol. Breed.* **2014**, *33*, 701–708. [CrossRef]
105. Bindusree, G.; Natarajan, P.; Kalva, S.; Madasamy, P. Whole genome sequencing of *Oryza sativa* L. cv. Seeragasamba identifies a new fragrance allele in rice. *PLoS ONE* **2017**, *12*, e0188920. [CrossRef]
106. Shao, G.; Tang, S.; Chen, M.; Wei, X.; He, J.; Luo, J.; Jiao, G.; Hu, Y.; Xie, L.; Hu, P. Haplotype variation at Badh2, the gene determining fragrance in rice. *Genomics* **2013**, *101*, 157–162. [CrossRef]
107. Myint, K.M.; Arikit, S.; Wanchana, S.; Yoshihashi, T.; Choowongkomon, K.; Vanavichit, A. A PCR-based marker for a locus conferring the aroma in Myanmar rice (*Oryza sativa* L.). *Theor. Appl. Genet.* **2012**, *125*, 887–896. [CrossRef]
108. Chuba, M.; Sakurada, H.; Yuki, K.; Sano, T.; Chuba, R.; Sato, K.; Yokoo, N.; Honma, T.; Sato, S.; Miyano, H. Breeding of a new rice cultivar with low amylose content "Yukinomai" (Yamagata84). *Bull. Agric. Res. Yamagata Prefect.* **2006**, *38*, 1–23.
109. Ando, I.; Sato, H.; Aoki, N.; Suzuki, Y.; Hirabayashi, H.; Kuroki, M.; Shimizu, H.; Ando, T.; Takeuchi, Y. Genetic analysis of the low-amylose characteristics of rice cultivars Oborozuki and Hokkai-PL9. *Breed. Sci.* **2010**, *60*, 187–194. [CrossRef]
110. Fujino, K.; Hirayama, Y.; Kaji, R. Marker-assisted selection in rice breeding programs in Hokkaido. *Breed. Sci.* **2019**, *69*, 383–392. [CrossRef]
111. Shao, Y.; Peng, Y.; Mao, B.; Lv, Q.; Yuan, D.; Liu, X.; Zhao, B. Allelic variations of the Wx locus in cultivated rice and their use in the development of hybrid rice in China. *PLoS ONE* **2020**, *15*, e0232279. [CrossRef]
112. Sreenivasulu, N.; Zhang, C.; Tiozon, R.N., Jr.; Liu, Q. Post-genomics revolution in the design of premium quality rice in a high-yielding background to meet consumer demands in the 21st century. *Plant Commun.* **2022**, *3*, 100271. [CrossRef]
113. Golestan Hashemi, F.S.; Rafii, M.Y.; Razi Ismail, M.; Mohamed, M.T.; Rahim, H.A.; Latif, M.A.; Aslani, F. Opportunities of marker-assisted selection for rice fragrance through marker-trait association analysis of microsatellites and gene-based markers. *Plant Biol.* **2015**, *17*, 953–961. [CrossRef]
114. Addison, C.K.; Angira, B.; Kongchum, M.; Harrell, D.L.; Baisakh, N.; Linscombe, S.D.; Famoso, A.N. Characterization of Haplotype Diversity in the BADH2 Aroma Gene and Development of a KASP SNP Assay for Predicting Aroma in U.S. Rice. *Rice* **2020**, *13*, 47. [CrossRef]
115. Ashokkumar, S.; Jaganathan, D.; Ramanathan, V.; Rahman, H.; Palaniswamy, R.; Kambale, R.; Muthurajan, R. Creation of novel alleles of fragrance gene OsBADH2 in rice through CRISPR/Cas9 mediated gene editing. *PLoS ONE* **2020**, *15*, e0237018. [CrossRef] [PubMed]

116. Tang, Y.; Abdelrahman, M.; Li, J.; Wang, F.; Ji, Z.; Qi, H.; Wang, C.; Zhao, K. CRISPR/Cas9 induces exon skipping that facilitates development of fragrant rice. *Plant Biotechnol. J.* **2021**, *19*, 642–644. [CrossRef] [PubMed]
117. Hui, S.; Li, H.; Mawia, A.M.; Zhou, L.; Cai, J.; Ahmad, S.; Lai, C.; Wang, J.; Jiao, G.; Xie, L.; et al. Production of aromatic three-line hybrid rice using novel alleles of BADH2. *Plant Biotechnol. J.* **2022**, *20*, 59–74. [CrossRef] [PubMed]
118. Imran, M.; Shafiq, S.; Tang, X. CRISPR-Cas9-mediated editing of BADH2 gene triggered fragrance revolution in rice. *Physiol. Plant* **2023**, *175*, e13871. [CrossRef] [PubMed]
119. Sakthivel, K.; Sundaram, R.M.; Shobha Rani, N.; Balachandran, S.M.; Neeraja, C.N. Genetic and molecular basis of fragrance in rice. *Biotechnol. Adv.* **2009**, *27*, 468–473. [CrossRef]
120. Huang, L.; Gu, Z.; Chen, Z.; Yu, J.; Chu, R.; Tan, H.; Zhao, D.; Fan, X.; Zhang, C.; Li, Q.; et al. Improving rice eating and cooking quality by coordinated expression of the major starch synthesis-related genes, SSII and Wx, in endosperm. *Plant Mol. Biol.* **2021**, *106*, 419–432. [CrossRef]
121. Zeng, D.; Tian, Z.; Rao, Y.; Dong, G.; Yang, Y.; Huang, L.; Leng, Y.; Xu, J.; Sun, C.; Zhang, G.; et al. Rational design of high-yield and superior-quality rice. *Nat. Plants* **2017**, *3*, 17031. [CrossRef]
122. Zhao, M.; Lin, Y.; Chen, H. Improving nutritional quality of rice for human health. *Theor. Appl. Genet.* **2020**, *133*, 1397–1413. [CrossRef]
123. Lin, R.; Luo, Y.; Liu, D.; Huang, C. Determination and analysis on principal qualitative characters of rice germplasm. In *Rice germplasm resources in China*; Agricultural Science and Technology Publisher of China: Beijing, China, 1993; pp. 83–93.
124. Zhou, L.; Liu, Q.; Zhang, C.; Xu, Y.; Tang, S.; Gu, M. Variation and distribution of seed storage protein content and composition among different rice varieties. *Acta Agron. Sin.* **2009**, *35*, 884–891. [CrossRef]
125. Yang, Y.; Guo, M.; Sun, S.; Zou, Y.; Yin, S.; Liu, Y.; Tang, S.; Gu, M.; Yang, Z.; Yan, C. Natural variation of OsGluA2 is involved in grain protein content regulation in rice. *Nat. Commun.* **2019**, *10*, 1949. [CrossRef]
126. He, W.; Wang, L.; Lin, Q.; Yu, F. Rice seed storage proteins: Biosynthetic pathways and the effects of environmental factors. *J. Integr. Plant Biol.* **2021**, *63*, 1999–2019. [CrossRef] [PubMed]
127. Xinkang, L.; Chunmin, G.; Lin, W.; Liting, J.; Xiangjin, F.; Qinlu, L.; Zhengyu, H.; Chun, L. Rice Storage Proteins: Focus on Composition, Distribution, Genetic Improvement and Effects on Rice Quality. *Rice Sci.* **2023**, *30*, 207–221. [CrossRef]
128. Li, H.; Yang, J.; Yan, S.; Lei, N.; Wang, J.; Sun, B. Molecular causes for the increased stickiness of cooked non-glutinous rice by enzymatic hydrolysis of the grain surface protein. *Carbohydr. Polym.* **2019**, *216*, 197–203. [CrossRef] [PubMed]
129. Zhang, H.; Jang, S.G.; Lar, S.M.; Lee, A.R.; Cao, F.Y.; Seo, J.; Kwon, S.W. Genome-Wide Identification and Genetic Variations of the Starch Synthase Gene Family in Rice. *Plants* **2021**, *10*, 1154. [CrossRef] [PubMed]
130. Xiong, Q.; Sun, C.; Wang, R.; Wang, R.; Wang, X.; Zhang, Y.; Zhu, J. The Key Metabolites in Rice Quality Formation of Conventional japonica Varieties. *Curr. Issues Mol. Biol.* **2023**, *45*, 990–1001. [CrossRef]
131. Tan, Y.; Sun, M.; Xing, Y.; Hua, J.; Sun, X.; Zhang, Q.; Corke, H. Mapping quantitative trait loci for milling quality, protein content and color characteristics of rice using a recombinant inbred line population derived from an elite rice hybrid. *Theor. Appl. Genet.* **2001**, *103*, 1037–1045. [CrossRef]
132. Wang, L.; Zhong, M.; Li, X.; Yuan, D.; Xu, Y.; Liu, H.; He, Y.; Luo, L.; Zhang, Q. The QTL controlling amino acid content in grains of rice (*Oryza sativa*) are co-localized with the regions involved in the amino acid metabolism pathway. *Mol. Breed.* **2008**, *21*, 127–137. [CrossRef]
133. Lou, J.; Chen, L.; Yue, G.; Lou, Q.; Mei, H.; Xiong, L.; Luo, L. QTL mapping of grain quality traits in rice. *J. Cereal Sci.* **2009**, *50*, 145–151. [CrossRef]
134. Ye, G.; Liang, S.; Wan, J. QTL mapping of protein content in rice using single chromosome segment substitution lines. *Theor. Appl. Genet.* **2010**, *121*, 741–750. [CrossRef]
135. Zheng, L.; Zhang, W.; Chen, X.; Ma, J.; Chen, W.; Zhao, Z.; Zhai, H.; Wan, J. Dynamic QTL analysis of rice protein content and protein index using recombinant inbred lines. *J. Plant Biol.* **2011**, *54*, 321–328. [CrossRef]
136. Zheng, L.; Zhang, W.; Liu, S.; Chen, L.; Liu, X.; Chen, X.; Ma, J.; Chen, W.; Zhao, Z.; Jiang, L. Genetic relationship between grain chalkiness, protein content, and paste viscosity properties in a backcross inbred population of rice. *J. Cereal Sci.* **2012**, *56*, 153–160. [CrossRef]
137. Cheng, L.; Xu, Q.; Zheng, T.; Ye, G.; Luo, C.; Xu, J.; Li, Z. Identification of stably expressed quantitative trait loci for grain yield and protein content using recombinant inbred line and reciprocal introgression line populations in rice. *Crop Sci.* **2013**, *53*, 1437–1446. [CrossRef]
138. Kashiwagi, T.; Munakata, J. Identification and characteristics of quantitative trait locus for grain protein content, TGP12, in rice (*Oryza sativa* L.). *Euphytica* **2018**, *214*, 165. [CrossRef]
139. Chattopadhyay, K.; Behera, L.; Bagchi, T.B.; Sardar, S.S.; Moharana, N.; Patra, N.R.; Chakraborti, M.; Das, A.; Marndi, B.C.; Sarkar, A.; et al. Detection of stable QTLs for grain protein content in rice (*Oryza sativa* L.) employing high throughput phenotyping and genotyping platforms. *Sci. Rep.* **2019**, *9*, 3196. [CrossRef]
140. Peng, B.; Kong, H.; Li, Y.; Wang, L.; Zhong, M.; Sun, L.; Gao, G.; Zhang, Q.; Luo, L.; Wang, G.; et al. OsAAP6 functions as an important regulator of grain protein content and nutritional quality in rice. *Nat. Commun.* **2014**, *5*, 4847. [CrossRef]
141. Hamaker, B.R.; Griffin, V.K. Effect of disulfide bond-containing protein on rice starch gelatinization and pasting. *Cereal Chem.* **1993**, *70*, 377–380.
142. Martin, M.; Fitzgerald, M. Proteins in rice grains influence cooking properties! *J. Cereal Sci.* **2002**, *36*, 285–294. [CrossRef]

143. Yoshida, H.; Tanigawa, T.; Yoshida, N.; Kuriyama, I.; Tomiyama, Y.; Mizushina, Y. Lipid components, fatty acid distributions of triacylglycerols and phospholipids in rice brans. *Food Chem.* **2011**, *129*, 479–484. [CrossRef]
144. Tong, C.; Liu, L.; Waters, D.L.; Huang, Y.; Bao, J. The contribution of lysophospholipids to pasting and thermal properties of nonwaxy rice starch. *Carbohydr. Polym.* **2015**, *133*, 187–193. [CrossRef]
145. Concepcion, J.C.T.; Calingacion, M.; Garson, M.J.; Fitzgerald, M.A. Lipidomics reveals associations between rice quality traits. *Metabolomics* **2020**, *16*, 54. [CrossRef]
146. Concepcion, J.C.T.; Ouk, S.; Riedel, A.; Calingacion, M.; Zhao, D.; Ouk, M.; Garson, M.J.; Fitzgerald, M.A. Quality evaluation, fatty acid analysis and untargeted profiling of volatiles in Cambodian rice. *Food Chem.* **2018**, *240*, 1014–1021. [CrossRef] [PubMed]
147. Zhou, H.; Xia, D.; Li, P.; Ao, Y.; Xu, X.; Wan, S.; Li, Y.; Wu, B.; Shi, H.; Wang, K.; et al. Genetic architecture and key genes controlling the diversity of oil composition in rice grains. *Mol. Plant* **2021**, *14*, 456–469. [CrossRef] [PubMed]
148. Li, Y.; Yang, Z.; Yang, C.; Liu, Z.; Shen, S.; Zhan, C.; Lyu, Y.; Zhang, F.; Li, K.; Shi, Y.; et al. The NET locus determines the food taste, cooking and nutrition quality of rice. *Sci. Bull.* **2022**, *67*, 2045–2049. [CrossRef] [PubMed]
149. Hong, J.; Rosental, L.; Xu, Y.; Xu, D.; Orf, I.; Wang, W.; Hu, Z.; Su, S.; Bai, S.; Ashraf, M.; et al. Genetic architecture of seed glycerolipids in Asian cultivated rice. *Plant Cell. Environ.* **2023**, *46*, 1278–1294. [CrossRef] [PubMed]
150. Sreenivasulu, N.; Butardo, V.M., Jr.; Misra, G.; Cuevas, R.P.; Anacleto, R.; Kavi Kishor, P.B. Designing climate-resilient rice with ideal grain quality suited for high-temperature stress. *J. Exp. Bot.* **2015**, *66*, 1737–1748. [CrossRef] [PubMed]
151. Yang, X.; Wang, B.; Chen, L.; Li, P.; Cao, C. The different influences of drought stress at the flowering stage on rice physiological traits, grain yield, and quality. *Sci. Rep.* **2019**, *9*, 3742. [CrossRef]
152. Huang, M.; Cao, J.; Liu, Y.; Zhang, M.; Hu, L.; Xiao, Z.; Chen, J.; Cao, F. Low-temperature stress during the flowering period alters the source–sink relationship and grain quality in field-grown late-season rice. *J. Agron. Crop Sci.* **2021**, *207*, 833–839. [CrossRef]
153. Zheng, C.; Liu, C.; Liu, L.; Tan, Y.; Sheng, X.; Yu, D.; Sun, Z.; Sun, X.; Chen, J.; Yuan, D. Effect of salinity stress on rice yield and grain quality: A meta-analysis. *Eur. J. Agron.* **2023**, *144*, 126765. [CrossRef]
154. Lin, C.J.; Li, C.Y.; Lin, S.K.; Yang, F.H.; Huang, J.J.; Liu, Y.H.; Lur, H.S. Influence of high temperature during grain filling on the accumulation of storage proteins and grain quality in rice (*Oryza sativa* L.). *J. Agric. Food Chem.* **2010**, *58*, 10545–10552. [CrossRef]
155. Counce, P.; Bryant, R.; Bergman, C.; Bautista, R.; Wang, Y.J.; Siebenmorgen, T.; Moldenhauer, K.; Meullenet, J.F. Rice milling quality, grain dimensions, and starch branching as affected by high night temperatures. *Cereal Chem.* **2005**, *82*, 645–648. [CrossRef]
156. Aboubacar, A.; Moldenhauer, K.A.; McClung, A.M.; Beighley, D.H.; Hamaker, B.R. Effect of growth location in the United States on amylose content, amylopectin fine structure, and thermal properties of starches of long grain rice cultivars. *Cereal Chem.* **2006**, *83*, 93–98. [CrossRef]
157. Zhang, C.; Zhou, L.; Zhu, Z.; Lu, H.; Zhou, X.; Qian, Y.; Li, Q.; Lu, Y.; Gu, M.; Liu, Q. Characterization of Grain Quality and Starch Fine Structure of Two Japonica Rice (*Oryza Sativa*) Cultivars with Good Sensory Properties at Different Temperatures during the Filling Stage. *J. Agric. Food Chem.* **2016**, *64*, 4048–4057. [CrossRef] [PubMed]
158. Yamakawa, H.; Hirose, T.; Kuroda, M.; Yamaguchi, T. Comprehensive expression profiling of rice grain filling-related genes under high temperature using DNA microarray. *Plant Physiol.* **2007**, *144*, 258–277. [CrossRef] [PubMed]
159. Jiang, H.; Dian, W.; Wu, P. Effect of high temperature on fine structure of amylopectin in rice endosperm by reducing the activity of the starch branching enzyme. *Phytochemistry* **2003**, *63*, 53–59. [CrossRef]
160. Umemoto, T.; Terashima, K. Research note: Activity of granule-bound starch synthase is an important determinant of amylose content in rice endosperm. *Funct. Plant Biol.* **2002**, *29*, 1121–1124. [CrossRef] [PubMed]
161. Lin, S.K.; Chang, M.C.; Tsai, Y.G.; Lur, H.S. Proteomic analysis of the expression of proteins related to rice quality during caryopsis development and the effect of high temperature on expression. *Proteomics* **2005**, *5*, 2140–2156. [CrossRef] [PubMed]
162. Zhang, H.; Duan, L.; Dai, J.S.; Zhang, C.Q.; Li, J.; Gu, M.H.; Liu, Q.Q.; Zhu, Y. Major QTLs reduce the deleterious effects of high temperature on rice amylose content by increasing splicing efficiency of Wx pre-mRNA. *Theor. Appl. Genet.* **2014**, *127*, 273–282. [CrossRef]
163. Zhang, H.; Xu, H.; Feng, M.; Zhu, Y. Suppression of OsMADS7 in rice endosperm stabilizes amylose content under high temperature stress. *Plant Biotechnol. J.* **2018**, *16*, 18–26. [CrossRef]
164. Fan, X.; Li, Y.; Lu, Y.; Zhang, C.; Li, E.; Li, Q.; Tao, K.; Yu, W.; Wang, J.; Chen, Z.; et al. The interaction between amylose and amylopectin synthesis in rice endosperm grown at high temperature. *Food Chem.* **2019**, *301*, 125258. [CrossRef]
165. Zhao, Q.; Ye, Y.; Han, Z.; Zhou, L.; Guan, X.; Pan, G.; Asad, M.A.; Cheng, F. SSIIIa-RNAi suppression associated changes in rice grain quality and starch biosynthesis metabolism in response to high temperature. *Plant Sci.* **2020**, *294*, 110443. [CrossRef]
166. Tashiro, T.; Wardlaw, I. The effect of high temperature on kernel dimensions and the type and occurrence of kernel damage in rice. *Aust. J. Agric. Res.* **1991**, *42*, 485–496. [CrossRef]
167. Tsukaguchi, T.; Iida, Y. Effects of assimilate supply and high temperature during grain-filling period on the occurrence of various types of chalky kernels in rice plants (*Oryza sativa* L.). *Plant Prod. Sci.* **2008**, *11*, 203–210. [CrossRef]
168. Hakata, M.; Kuroda, M.; Miyashita, T.; Yamaguchi, T.; Kojima, M.; Sakakibara, H.; Mitsui, T.; Yamakawa, H. Suppression of alpha-amylase genes improves quality of rice grain ripened under high temperature. *Plant Biotechnol. J.* **2012**, *10*, 1110–1117. [CrossRef] [PubMed]
169. Nakata, M.; Fukamatsu, Y.; Miyashita, T.; Hakata, M.; Kimura, R.; Nakata, Y.; Kuroda, M.; Yamaguchi, T.; Yamakawa, H. High Temperature-Induced Expression of Rice alpha-Amylases in Developing Endosperm Produces Chalky Grains. *Front. Plant Sci.* **2017**, *8*, 2089. [CrossRef] [PubMed]

170. Suriyasak, C.; Harano, K.; Tanamachi, K.; Matsuo, K.; Tamada, A.; Iwaya-Inoue, M.; Ishibashi, Y. Reactive oxygen species induced by heat stress during grain filling of rice (*Oryza sativa* L.) are involved in occurrence of grain chalkiness. *J. Plant Physiol.* **2017**, *216*, 52–57. [CrossRef]
171. Xu, H.; Li, X.; Zhang, H.; Wang, L.; Zhu, Z.; Gao, J.; Li, C.; Zhu, Y. High temperature inhibits the accumulation of storage materials by inducing alternative splicing of OsbZIP58 during filling stage in rice. *Plant Cell Env.* **2020**, *43*, 1879–1896. [CrossRef]
172. Ren, Y.; Huang, Z.; Jiang, H.; Wang, Z.; Wu, F.; Xiong, Y.; Yao, J. A heat stress responsive NAC transcription factor heterodimer plays key roles in rice grain filling. *J. Exp. Bot.* **2021**, *72*, 2947–2964. [CrossRef]
173. Nevame, A.Y.M.; Emon, R.M.; Malek, M.A.; Hasan, M.M.; Alam, M.A.; Muharam, F.M.; Aslani, F.; Rafii, M.Y.; Ismail, M.R. Relationship between High Temperature and Formation of Chalkiness and Their Effects on Quality of Rice. *Biomed. Res. Int.* **2018**, *2018*, 1653721. [CrossRef]
174. Yang, W.; Liang, J.; Hao, Q.; Luan, X.; Tan, Q.; Lin, S.; Zhu, H.; Liu, G.; Liu, Z.; Bu, S.; et al. Fine mapping of two grain chalkiness QTLs sensitive to high temperature in rice. *Rice* **2021**, *14*, 33. [CrossRef]
175. Murata, K.; Iyama, Y.; Yamaguchi, T.; Ozaki, H.; Kidani, Y.; Ebitani, T. Identification of a novel gene (Apq1) from the indica rice cultivar 'Habataki' that improves the quality of grains produced under high temperature stress. *Breed. Sci.* **2014**, *64*, 273–281. [CrossRef]
176. Park, J.R.; Kim, E.G.; Jang, Y.H.; Kim, K.M. Screening and identification of genes affecting grain quality and spikelet fertility during high-temperature treatment in grain filling stage of rice. *BMC Plant Biol.* **2021**, *21*, 263. [CrossRef] [PubMed]
177. Leesawatwong, M.; Jamjod, S.; Kuo, J.; Dell, B.; Rerkasem, B. Nitrogen fertilizer increases seed protein and milling quality of rice. *Cereal Chem.* **2005**, *82*, 588–593. [CrossRef]
178. Zhou, C.; Huang, Y.; Jia, B.; Wang, Y.; Wang, Y.; Xu, Q.; Li, R.; Wang, S.; Dou, F. Effects of cultivar, nitrogen rate, and planting density on rice-grain quality. *Agronomy* **2018**, *8*, 246. [CrossRef]
179. Zhu, D.-w.; Zhang, H.-c.; Guo, B.-w.; Ke, X.; Dai, Q.-g.; Wei, H.-y.; Hui, G.; Hu, Y.-j.; Cui, P.-y.; Huo, Z.-y. Effects of nitrogen level on yield and quality of japonica soft super rice. *J. Integr. Agric.* **2017**, *16*, 1018–1027. [CrossRef]
180. Gu, J.; Chen, J.; Chen, L.; Wang, Z.; Zhang, H.; Yang, J. Grain quality changes and responses to nitrogen fertilizer of japonica rice cultivars released in the Yangtze River Basin from the 1950s to 2000s. *Crop J.* **2015**, *3*, 285–297. [CrossRef]
181. Liang, H.; Gao, S.; Ma, J.; Zhang, T.; Wang, T.; Zhang, S.; Wu, Z. Effect of nitrogen application rates on the nitrogen utilization, yield and quality of rice. *Food Nutr. Sci.* **2021**, *12*, 13–27. [CrossRef]
182. Cao, X.; Sun, H.; Wang, C.; Ren, X.; Liu, H.; Zhang, Z. Effects of late-stage nitrogen fertilizer application on the starch structure and cooking quality of rice. *J. Sci. Food Agric.* **2018**, *98*, 2332–2340. [CrossRef]
183. Liang, H.; Tao, D.; Zhang, Q.; Zhang, S.; Wang, J.; Liu, L.; Wu, Z.; Sun, W. Nitrogen fertilizer application rate impacts eating and cooking quality of rice after storage. *PLoS ONE* **2021**, *16*, e0253189. [CrossRef]
184. Ali, I.; Iqbal, A.; Ullah, S.; Muhammad, I.; Yuan, P.; Zhao, Q.; Yang, M.; Zhang, H.; Huang, M.; Liang, H.; et al. Effects of Biochar Amendment and Nitrogen Fertilizer on RVA Profile and Rice Grain Quality Attributes. *Foods* **2022**, *11*, 625. [CrossRef]
185. Xia, D.; Wang, Y.; Shi, Q.; Wu, B.; Yu, X.; Zhang, C.; Li, Y.; Fu, P.; Li, M.; Zhang, Q.; et al. Effects of Wx Genotype, Nitrogen Fertilization, and Temperature on Rice Grain Quality. *Front. Plant Sci.* **2022**, *13*, 901541. [CrossRef]
186. Gong, D.; Xu, X.; Wu, L.; Dai, G.; Zheng, W.; Xu, Z. Effect of biochar on rice starch properties and starch-related gene expression and enzyme activities. *Sci. Rep.* **2020**, *10*, 16917. [CrossRef]
187. Chen, L.; Guo, L.; Deng, X.; Pan, X.; Liao, P.; Xiong, Q.; Gao, H.; Wei, H.; Dai, Q.; Zeng, Y. Effects of biochar on rice yield, grain quality and starch viscosity attributes. *J. Sci. Food Agric.* **2023**. [CrossRef] [PubMed]
188. Koornneef, M.; Reuling, G.; Karssen, C.M. The isolation and characterization of abscisic acid-insensitive mutants of Arabidopsis thaliana. *Physiol. Plant* **1984**, *61*, 377–383. [CrossRef]
189. Fitzgerald, M.A.; Sackville Hamilton, N.R.; Calingacion, M.N.; Verhoeven, H.A.; Butardo, V.M. Is there a second fragrance gene in rice? *Plant Biotechnol. J.* **2008**, *6*, 416–423. [CrossRef] [PubMed]
190. Deng, Z.; Liu, Y.; Gong, C.; Chen, B.; Wang, T. Waxy is an important factor for grain fissure resistance and head rice yield as revealed by a genome-wide association study. *J. Exp. Bot.* **2022**, *73*, 6942–6954. [CrossRef] [PubMed]
191. Si, L.; Chen, J.; Huang, X.; Gong, H.; Luo, J.; Hou, Q.; Lu, T.; Zhu, J.; Shangguan, Y.; et al. OsSPL13 controls grain size in cultivated rice. *Nat. Genet.* **2016**, *48*, 447–456. [CrossRef] [PubMed]
192. Xiao, N.; Pan, C.; Li, Y.; Wu, Y.; Cai, Y.; Lu, Y.; Wang, R.; Yu, L.; Shi, W.; Kang, H.; et al. Genomic insight into balancing high yield, good quality, and blast resistance of japonica rice. *Genome Biol.* **2021**, *22*, 283. [CrossRef]
193. Wang, W.; Mauleon, R.; Hu, Z.; Chebotarov, D.; Tai, S.; Wu, Z.; Li, M.; Zheng, T.; Fuentes, R.R.; Zhang, F.; et al. Genomic variation in 3,010 diverse accessions of Asian cultivated rice. *Nature* **2018**, *557*, 43–49. [CrossRef]
194. Ye, X.; Al-Babili, S.; Kloti, A.; Zhang, J.; Lucca, P.; Beyer, P.; Potrykus, I. Engineering the provitamin A (beta-carotene) biosynthetic pathway into (carotenoid-free) rice endosperm. *Science* **2000**, *287*, 303–305. [CrossRef]
195. Wirth, J.; Poletti, S.; Aeschlimann, B.; Yakandawala, N.; Drosse, B.; Osorio, S.; Tohge, T.; Fernie, A.R.; Gunther, D.; Gruissem, W.; et al. Rice endosperm iron biofortification by targeted and synergistic action of nicotianamine synthase and ferritin. *Plant Biotechnol. J.* **2009**, *7*, 631–644. [CrossRef]
196. Kumar, K.; Gambhir, G.; Dass, A.; Tripathi, A.K.; Singh, A.; Jha, A.K.; Yadava, P.; Choudhary, M.; Rakshit, S. Genetically modified crops: Current status and future prospects. *Planta* **2020**, *251*, 91. [CrossRef] [PubMed]

197. Kusaba, M.; Miyahara, K.; Iida, S.; Fukuoka, H.; Takano, T.; Sassa, H.; Nishimura, M.; Nishio, T. Low glutelin content1: A dominant mutation that suppresses the glutelin multigene family via RNA silencing in rice. *Plant Cell* **2003**, *15*, 1455–1467. [CrossRef] [PubMed]
198. Zhou, H.; Wang, L.; Liu, G.; Meng, X.; Jing, Y.; Shu, X.; Kong, X.; Sun, J.; Yu, H.; Smith, S.M.; et al. Critical roles of soluble starch synthase SSIIIa and granule-bound starch synthase Waxy in synthesizing resistant starch in rice. *Proc. Natl. Acad. Sci. USA* **2016**, *113*, 12844–12849. [CrossRef] [PubMed]
199. Zhao, G.C.; Xie, M.X.; Wang, Y.C.; Li, J.Y. Molecular Mechanisms Underlying gamma-Aminobutyric Acid (GABA) Accumulation in Giant Embryo Rice Seeds. *J. Agric. Food Chem.* **2017**, *65*, 4883–4889. [CrossRef]

Disclaimer/Publisher's Note: The statements, opinions and data contained in all publications are solely those of the individual author(s) and contributor(s) and not of MDPI and/or the editor(s). MDPI and/or the editor(s) disclaim responsibility for any injury to people or property resulting from any ideas, methods, instructions or products referred to in the content.

Article

Genome-Wide Analysis of the HD-Zip Gene Family in Chinese Cabbage (*Brassica rapa* subsp. *pekinensis*) and the Expression Pattern at High Temperatures and in Carotenoids Regulation

Lian Yin [1], Yudong Sun [1], Xuehao Chen [2], Jiexia Liu [2], Kai Feng [2], Dexu Luo [1], Manyi Sun [3], Linchuang Wang [1], Wenzhao Xu [1], Lu Liu [1] and Jianfeng Zhao [1,*]

1. Vegetable Research and Development Center, Huaiyin Institute of Agricultural Science in Xuhuai Area of Jiangsu Province, Huaian 223001, China; yinlian1996@163.com (L.Y.)
2. College of Horticulture and Landscape Architecture, Yangzhou University, Yangzhou 225009, China
3. State Key Laboratory of Crop Genetics and Germplasm Enhancement, College of Horticulture, Nanjing Agricultural University, Nanjing 210095, China
* Correspondence: zhaold1977@163.com

Citation: Yin, L.; Sun, Y.; Chen, X.; Liu, J.; Feng, K.; Luo, D.; Sun, M.; Wang, L.; Xu, W.; Liu, L.; et al. Genome-Wide Analysis of the HD-Zip Gene Family in Chinese Cabbage (*Brassica rapa* subsp. *pekinensis*) and the Expression Pattern at High Temperatures and in Carotenoids Regulation. *Agronomy* 2023, 13, 1324. https://doi.org/10.3390/agronomy13051324

Academic Editor: Zhiyong Li

Received: 29 March 2023
Revised: 27 April 2023
Accepted: 5 May 2023
Published: 9 May 2023

Copyright: © 2023 by the authors. Licensee MDPI, Basel, Switzerland. This article is an open access article distributed under the terms and conditions of the Creative Commons Attribution (CC BY) license (https://creativecommons.org/licenses/by/4.0/).

Abstract: HD-Zip, a special class of transcription factors in high plants, has a role in plant development and responding to external environmental stress. Heat stress has always been an important factor affecting plant growth, quality, and yield. Carotenoid content is also an important factor affecting the color of the inner leaf blades of Chinese cabbage. In this study, the genomes of three Brassicaceae plants were selected: Chinese cabbage (*Brassica rapa* subsp. *pekinensis*), *Brassica oleracea*, and *Brassica napus*. We identified 93, 96, and 184 *HD-Zip* genes in the *B. rapa*, *B. oleracea*, and *B. napus*, respectively. The HD-Zip gene family was classified into four subfamilies based on phylogeny: I, II, III, and IV. The results of cis-acting element analysis suggested that HD-Zip family genes may participate in various biological processes, such as pigment synthesis, cell cycle regulation, defense stress response, etc. Conserved motifs prediction revealed that three motifs exist among the four HD-Zip gene families and that different motifs exhibit significant effects on the structural differences in HD-Zips. Synteny, Ks, and 4DTv results displayed that genome-wide triplication events act in HD-Zip gene family expansion. Transcriptome data showed that 18 genes responded (>1.5-fold change) to heat stress in Chinese cabbage, and 14 of 18 genes were from the HD-Zip I subfamily. Three genes had up-regulation, and eight genes had down-regulation in high-carotenoid-content Chinese cabbage. The *BraA09g011460.3C* expression level was up-regulated after heat stress treatment and significantly reduced in varieties with high carotenoid content, indicating its potential for heat stress tolerance and carotenoid content regulation. This study provided important gene resources for the subsequent breeding of Chinese cabbage.

Keywords: *HD-Zip* genes; *Brassica rapa* subsp. *Pekinensis*; high temperature; carotenoids; comparative genomics

1. Introduction

The HD-Zip family, a transcription factor gene family in plants, consists of the HD (homeodomain) and LZ (leucine zipper) structural domains. In addition to higher plants, the HD-Zip protein was investigated in Pteridophyta [1] and Bryophyta [2]. Until now, the HD-Zip genes have been widely and systematically studied in *Arabidopsis thaliana* (*Arabidopsis*), cassava (*Manihot esculenta*), and maize (*Zea mays*) [3–5]. Numerous reports have shown that the HD-Zip gene family is closely associated with plant growth and tolerance to environmental stress [6]. In view of the existing reports, HD-Zips are usually divided into four groups (I–IV) [7]. HD-Zip I members only consist of HD and LZ domains without other motifs. The expression of HD-Zip I gene members is particularly regulated by abiotic stress, such as light and temperature [8]. Group I genes can also promote the fruit coloration process by regulating the carotenoid content [9]. Group II members have been widely proven in their light quality change response and shade aversion response, and group III has been extensively studied in embryonic plant development [8,10]. Furthermore, Turchi et al. [11] pointed out the interaction of the transcription factors of groups II and III in the auxin regulation of plant development. Group IV proteins were involved in plant morphogenesis, mainly displayed in the plant differentiation of the epidermis and epidermal cells and stomatal development [12,13].

Chinese cabbage (*Brassica rapa* subsp. *Pekinensis*) is an herbaceous plant in Brassicaceae with a leafy head. It is a specialty and important vegetable in China with a large cultivation area [14]. Heat stress is one factor affecting production, distribution, and quality. Chinese cabbage prefers cold and cool, and high temperatures affect its growth and development [14]. Therefore, it is of great theoretical and practical importance to research mechanisms and adaptation under high temperatures. Zhang et al. [15] selected heat-tolerant and heat-sensitive Chinese cabbage and measured physiological indicators using transcriptome. Several genes (*Prx50*, *Prx52*, *Prx54*, *SOD1*, and *SOD2*) related to reactive oxygen species (ROS) scavenging were identified, which were significantly up-regulated in heat-tolerant varieties. In addition, Quan et al. [16] found that glycine betaine (GB) and β-aminobutyric acid (BABA) could improve photosynthetic performance and antioxidant enzyme activity under high temperatures to alleviate heat stress on Chinese cabbage. Based on the sequenced Chinese cabbage genome, Huang et al. [17] identified 30 heat shock factors (Hsfs) that function in several organs of Chinese cabbage and also hypothesized that Hsfs might be essential in the developmental regulation of the underground parts of Chinese cabbage.

Transcription factors control the plant's abiotic stress response and regulate the expression of the many downstream target genes of various metabolic processes [18]. Up to the present, the function of HD-Zips under heat stress in other species has been studied. In lily (*Lilium longiflorum*), the *LlHB16* gene can positively regulate heat resistance by linking the heat response pathway and ABA signaling [19]. Li et al. [20] identified 43 *HD-Zip* genes on 12 chromosomes of potato (*Solanum tuberosum* L.) and found that the *StHOX2* gene was significantly up-regulated in the root. To investigate how HD-Zips regulate the heat stress tolerance of the radish (*Raphanus sativus* L.), *RsHDZ17* was isolated from HD-Zip group I and overexpressed in *Arabidopsis*. It was found that the gene enhanced the heat stress tolerance of radishes by improving photosynthesis and enhancing the scavenging activity of reactive oxygen [21]. Wang et al. [22] detected the transcription levels of group I genes in heat-tolerant and heat-sensitive perennial ryegrass lines and demonstrated that *LpHOX21* was positively associated with heat tolerance. Taken together, *HD-Zip* genes are essential in the plant lifecycle, including growth, development, reproduction, differentiation, and morphogenesis [23]. However, the effects of the HD-Zip gene family in *B. rapa* are still being unraveled.

In this study, we identified the HD-Zip gene family in *B. rapa*, and phylogenetic tree construction, motif and cis-element prediction, and collinearity and Ka/Ks analysis were carried out. Simultaneously, the transcription patterns of *HD-Zip* genes under heat stress and Chinese cabbage varieties with different carotenoid content were analyzed. The results identified several candidate genes related to heat stress and carotenoid accumulation. We provided a foundation for research on *HD-Zip genes* in the mechanism of carotenoid accumulation and heat stress response in Chinese cabbage leaf blades.

2. Materials and Methods

2.1. Plant Materials, Heat Stress Treatment, and Carotenoid Content Measurement

The Chinese cabbage seeds of the hybrid one-generation 'Gailiang Qingza 3' were purchased from Qingdao International Seedling Co. (Qingdao, China), and the high-generation inbred line '54' were conserved in the Huaian Key Laboratory for Facility Vegetables (33°53′ N, 119°04′ E). Under heat stress treatment, the seeds were sowed in 96-hole cell trays, transplanted into 32 cm diameter pots at 30 days, and the 40-old-day seedlings were cultivated in a light incubator (40 °C for 16 h in the daytime, 30 °C for 8 h at night, with a relative humidity of 60%). The leaf blades of the Chinese cabbages were taken at 0, 4, 8, and 10 day, separately, and the samples were immediately frozen in liquid nitrogen. Under normal conditions, the seeds were sowed in 96-hole cell trays on 8 August 2022 and transplanted into the field on 2 September. The head of the outer (the fifth part) and inner leaves of the 90-day-old line of the '54' cabbage and 'Gailiang Qingza 3' were collected to determine the carotenoid content. Carotenoid content was measured according to previous methods [24]. Thus, 50 mg of the sample was ground into a powder by ball mill was extracted with 0.5 mL of a hexane/acetone/ethanol (1:1:1, $v/v/v$) mixture containing 0.1% BHT (butylated hydroxytoluene), vortexed for 20 min at room temperature, centrifuged at 12,000 r/min for 5 min, and the supernatant was extracted and then repeated twice. The combined supernatant was redissolved with 100 uL of a methanol/methyl tert-butyl ether mixture, and the content of carotenoids was determined by LC-MS/MS after filtration.

2.2. GenomeWide Identification of HD-Zip Genes

All protein sequences of *B. rapa*, *B. oleracea*, and *B. napus* were downloaded from the Brassicaceae Database (http://www.brassicadb.cn/, accessed on 8 October 2022) for *HD-Zip* gene identification. Candidate HD-Zip genes were searched against known HD-Zip protein sequences using the BLASTP program. Hidden Markov Model (HMM) profiles of the homeodomain (PF00046) and the leucine zipper domain (PF02183) were downloaded from the PFAM database [25] for search by HMMER3.0 [26]. All candidate sequences were further examined using the NCBI-CDD database (https://www.ncbi.nlm.nih.gov/Structure/cdd/wrpsb.cgi, accessed on 8 October 2022) to confirm the presence of the HD and LZ domains.

2.3. Multiple Sequence Alignment and Phylogenetic Analysis

All HD-Zip protein sequences in *B. rapa*, *Arabidopsis*, *B. oleracea*, and *B. napus* were aligned with ClustalW2 [27]. The tree was constructed by using IQTREE software [28] (version 1.6) with the max likelihood (ML) method, and 1000 ultrafast bootstraps were estimated. The model was selected using the 'MF' function.

2.4. Gene Structure, Motif, and Cis-Regulatory Elements Analysis

Gene structural information for *HD-Zip* genes was extracted from whole genome data and displayed using Tbtools software [29]. The MEME website (http://meme-suite.org/tools/meme, accessed on 10 October 2022) was used to identify the motif sequence. The upstream 2000 bp regions of the *HD-Zip* genes were extracted, and Plant CARE software (http://bioinformatics.psb.ugent.be/webtools/plantcare/html/, accessed on 9 October 2022) was used to identify the cis-regulatory elements of the promoter region of the *HD-Zip* genes.

2.5. Synteny Analysis of HD-Zip Genes

For analyzing *HD-Zip* gene duplication events, BLASTP was used to make an all-vs-all BLAST search (top five matches and e-value of 1×10^{-5}) with all protein sequences as input data. The BLAST output and the whole genome annotation file were imported to MCScanX [30] for homologous pairs and syntenic regions identification. Via duplicate gene classifier [30], *HD-Zip genes* were classified into five duplication types [30].

2.6. Calculating the Ka, Ks, and 4DTv of HD-Zip Paralogs

The Ka, Ks, and Ka/Ks ratios of *HD-Zip* paralog gene pairs were calculated using ParaAT (v2.0) [31] and Kaks_calculator (v2.0) [32]. The ParaAT2.0 software was used to compare the coding and nucleotide sequences of the *HD-Zip* genes in Brassicaceae, and Ka, Ks, and Ka/Ks values were calculated via KaKs_calculator2.0 software [32] with the model set to the MYN model [33]. Additionally, 4DTv (4-fold synonymous third-codon transversion) was used to estimate the genetic distances of synteny gene pairs. We calculated 4DTv values of *HD-Zip* paralog pairs using an in-house Python script.

2.7. Expression Pattern Analysis of HD-Zip Genes

Transcriptome data of the highly inbred line '268' with heat stress treatment were obtained from Yue et al. [34]. The transcriptome data of 'QZ' and '54' were sequenced on the Illumina NovSeq6000 platform with 150 bp pair-end sequences (Illumina, California, USA). The raw data were trimmed using Trimmomatic (v0.39) [35]. Then, the high-quality reads were mapped to the reference genome (Chiifu_V3.0) [36] using HISAT2 (v2.2.1) [37]. StringTie (v2.1.7) [38] was used to quantify the read count and calculate the Fragments Per Kilobase of the exon model per million mapped fragments (FPKM). The expression heat map of the HD-Zip family was analyzed with TBtool software.

Extraction of total RNA and cDNA was performed using an RNA simple total RNA Kit (Tiangen, Beijing, China) and the Prime Script RT Reagent Kit (TaKaRa, Dalian, China). The expression analysis of *HD-Zip* genes was detected by quantitative real-time PCR analysis (qRT-PCR) using the SYBR GREEN method with *BrActin1* as the internal reference gene [34]. The primers were designed by Primer Premier 6.0 (Supplemental Table S1). The formulation of the qRT-PCR reaction system (20 μL) included 10 μL SYBR Premix Ex Taq, 1 μL cDNA, 1 μL forward/reverse primers, and 7 μL ddH$_2$O. Two-step qRT-PCR amplification conditions were set: 95 °C for 5 min, 55 cycles at 95 °C for 3 s, 60 °C for 10 s, and 72 °C for 30 s, followed by 72 °C for 3 min. Three replicates were set in each reaction. The relative transcription levels of genes were calculated with $2^{-\Delta\Delta Ct}$ methods.

3. Results

3.1. Whole-Genome Identification of HD-Zip Genes in Brassicaceae Plants

HD-Zip genes in the *B. rapa*, *B. oleracea*, and *B. napus* genomes were identified by BLAST search and hmmsearch functions based on Hmmer 3.0 software. The HD-Zip protein contains a homeodomain (HD) and a leucine zipper (LZ) domain. A total of 93 *HD-Zip* genes were identified from the whole genome of *B. rapa* (AA, 2n = 20) (Table 1). Meanwhile, 96 and 184 genes were identified from *B. oleracea* (CC, 2n = 18) and *B. napus* (AACC, 2n = 38), respectively. The number of HD-Zip genes in *B. napus* was 1.98- and 1.92-fold higher than that in *B. rapa* and *B. oleracea*, respectively. The relative molecular weights of the HD-Zip family genes of Chinese cabbage ranged from a minimum of 17.93 kD (*BraA02g014240*) to a maximum of 99.25 kD (*BraA09g015080*), with theoretical pI of 4.54 to 10.24 (Supplemental Table S2).

Table 1. Summary of the HD-zip gene family in three Brassica species.

Species	Genome Size	Chromosome Number (2n)	Whole Gene Number	HD-Zip Gene Number			
				I	II	III	IV
Brassica rapa	351.06	20	46,250	39	18	10	26
Brassica oleracea	561.16	18	59,064	41	20	10	25
Brassica napus	924	38	108,190	74	38	19	53

3.2. Phylogenetic Analysis of the HD-Zip Genes

The phylogenetic tree was constructed using the full-length HD-Zip proteins of *B. rapa* and *Arabidopsis* (Figure 1a). The result showed that a total of 141 HD-Zip proteins from the two species were phylogenetically categorized into four subgroups and further named I, II, III, and IV based on the classification of *HD-Zip* genes in *Arabidopsis* (Figure 1b). From the four subfamilies, 39 (I), 18 (II), 10 (III), and 26 (IV) *HD-Zip* genes were obtained in the *B. rapa* genome, respectively, and subfamily I has a higher percentage of *HD-Zip* genes than classes II, III, and IV. Meanwhile, 41 (I), 20 (II), 10 (III), and 25 (IV) *HD-Zip* genes were identified from the *B. oleracea* genome (Figure S1), and 74 (I), 38 (II), 19 (III), and 53 (IV) *HD-Zip* genes identified from the *B. napus* genome (Figure S2). These results suggested that subfamily I contained the greatest number of *HD-Zip* genes in the three *Brassicaceae* species. Many *AtHD-Zip* genes contained at least two homology *HD-Zip* genes in the three *Brassicaceae* species, suggesting that genome-wide duplication has led to HD-Zip family expansion [39].

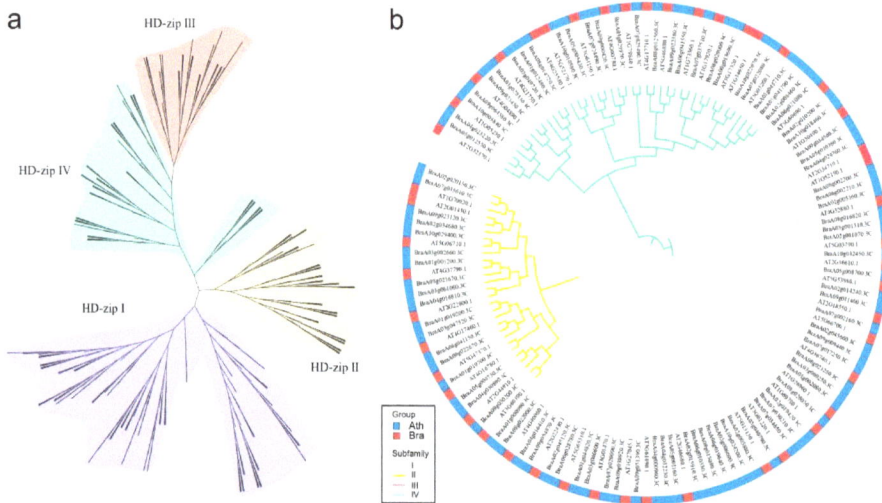

Figure 1. Phylogenetic tree and the classification of *B. rapa, B. oleracea,* and *B. napus*. (**a**) A phylogenetic tree of the HD-Zip proteins of *B. rapa, B. oleracea, B. napus,* and *Arabidopsis*. The tree was constructed by using IQTREE software (version 1.6) with the max likelihood (ML) method and 1000 ultrafast bootstraps. Purple represents the HD-Zip subfamily I, yellow represents II, red represents III, and cyan represents IV. (**b**) A phylogenetic tree of the HD-Zip proteins of *B. rapa* and *Arabidopsis*. Purple branches indicate the HD-Zip subfamily I, yellow branches indicate II, red branches indicate III, and cyan branches indicate IV. Red rectangles indicate genes coming from *Arabidopsis*, and blue squares indicate genes coming from *B. rapa*.

3.3. Conserved Motif Analysis and Gene Structural Analysis of HD-Zip Genes

Gene structures and conserved motifs of the 93 HD-Zips in *B. rapa* were predicted, and a total of 20 motifs were discovered using the MEME tool. The genes in the same subfamily possessed the same motif structure, such as class II and III (Figure 2a). It is noteworthy that three motifs (motifs 1, 2, and 3) appeared in all *HD-Zip* genes. Most subfamily I genes contained only three motifs, except for four genes (*BraA02g005880.3C, BraA01g024200.3C, BraA03g006690.3C,* and *BraA09g015080.3C*). Compared with HD-Zip I, motif 19 appeared at the C terminal of the HD-Zip II genes. HD-Zip III and IV genes had many more types of motifs than HD-Zip I and II. All HD-Zip III genes contained fourteen motifs, and fifteen types of motifs appeared in nearly all of HD-Zip IV. In addition, motifs 1, 2, and 3 also appeared in all genes of *B. napus* (Figure S3a). However, only motif 1 and motif 2 appeared in most *HD-Zip* genes of *B. oleracea*, and motif 3 was not in HD-Zip IV (Figure S4a). Consistently, the number of introns and exons of the HD-Zip III and IV genes was much more than in HD-Zip I and II in the three Brassicaceae species (Figure 2b, Figures S3b and S4b). These results suggested that the number of motifs and exons may result in the divergence of four subfamilies.

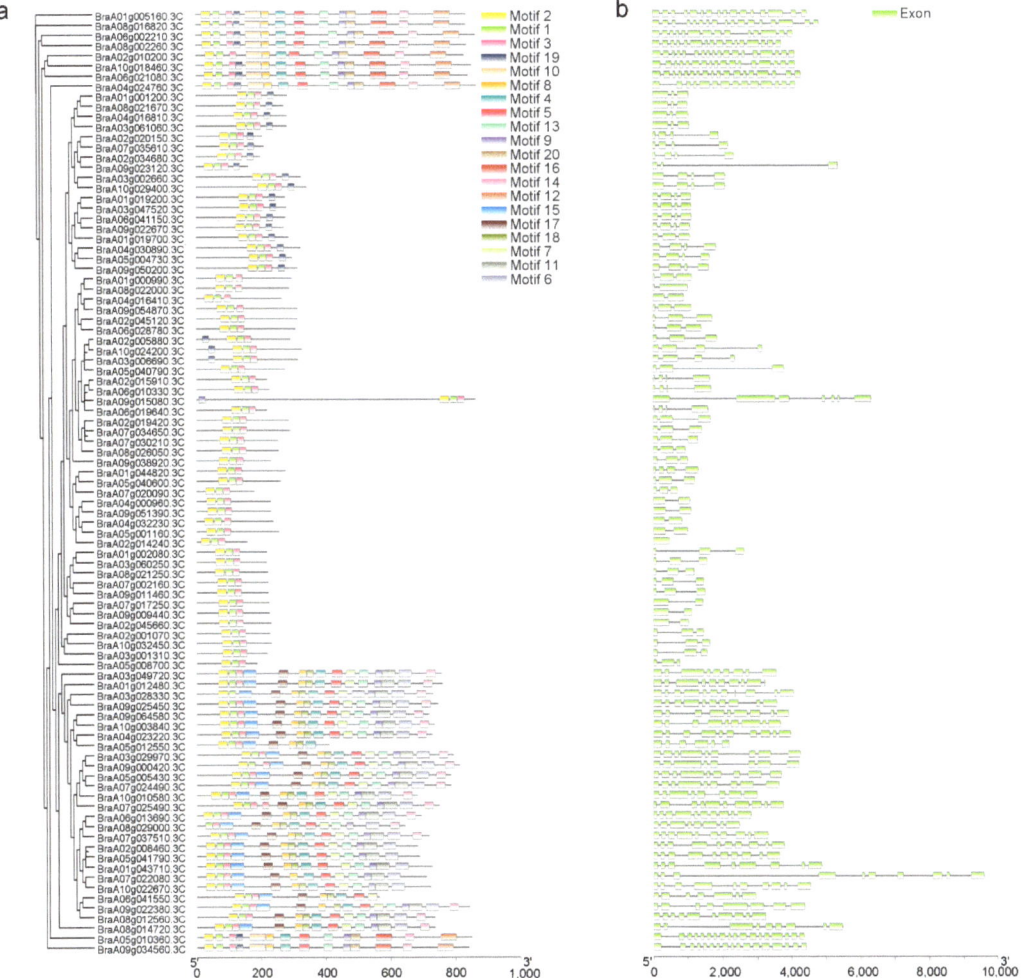

Figure 2. The conserved motifs and gene structure of HD-Zip proteins in *B. rapa*. (**a**) A phylogenetic tree of HD-Zip proteins was constructed using the Max likelihood (ML) method. The distribution of conserved motifs across *B. rapa* HD-Zip proteins. (**b**) The gene structure of *HD-Zip* genes in *B. rapa*, including intron and exon genes. The black lines indicate introns, and the green squares indicate a coding sequence (CDS).

3.4. Cis-Acting Elements Analysis in the Putative Promoter of HD-Zip Genes

In general, cis-acting elements in the promoter region can influence gene function and response for several environment adaptations [40]. In this study, the upstream 2000 bp sequence of the *HD-Zip* genes in *B. rapa* was extracted for cis-acting element analysis, and 18 types of cis-acting elements appeared in three species (Figure 3, Figures S5 and S6). It is obvious that most of the elements in *HD-Zips* were related to the light-responsive element, anaerobic induction, and MeJA-responsiveness, suggesting that the expression of *HD-Zip* genes may be regulated by various factors. Previous studies reported that *HD-Zip* genes could regulate auxin perception or auxin response and further regulate plant development [11,41]. For instance, 45.16% (42/93) of genes contained an auxin-responsive element, suggesting that these genes may be important in the auxin-response pathway and plant growth (Figure 3). In addition, 43.01% (40/93) genes contained "defense and stress responsiveness" elements, 48.39% (45/93) genes contained "low-temperature responsiveness" elements, and 48.39% (45/93) genes contained "MYB binding site involved in drought inducibility" elements, suggesting that *HD-Zip* genes may respond to various environmental factors.

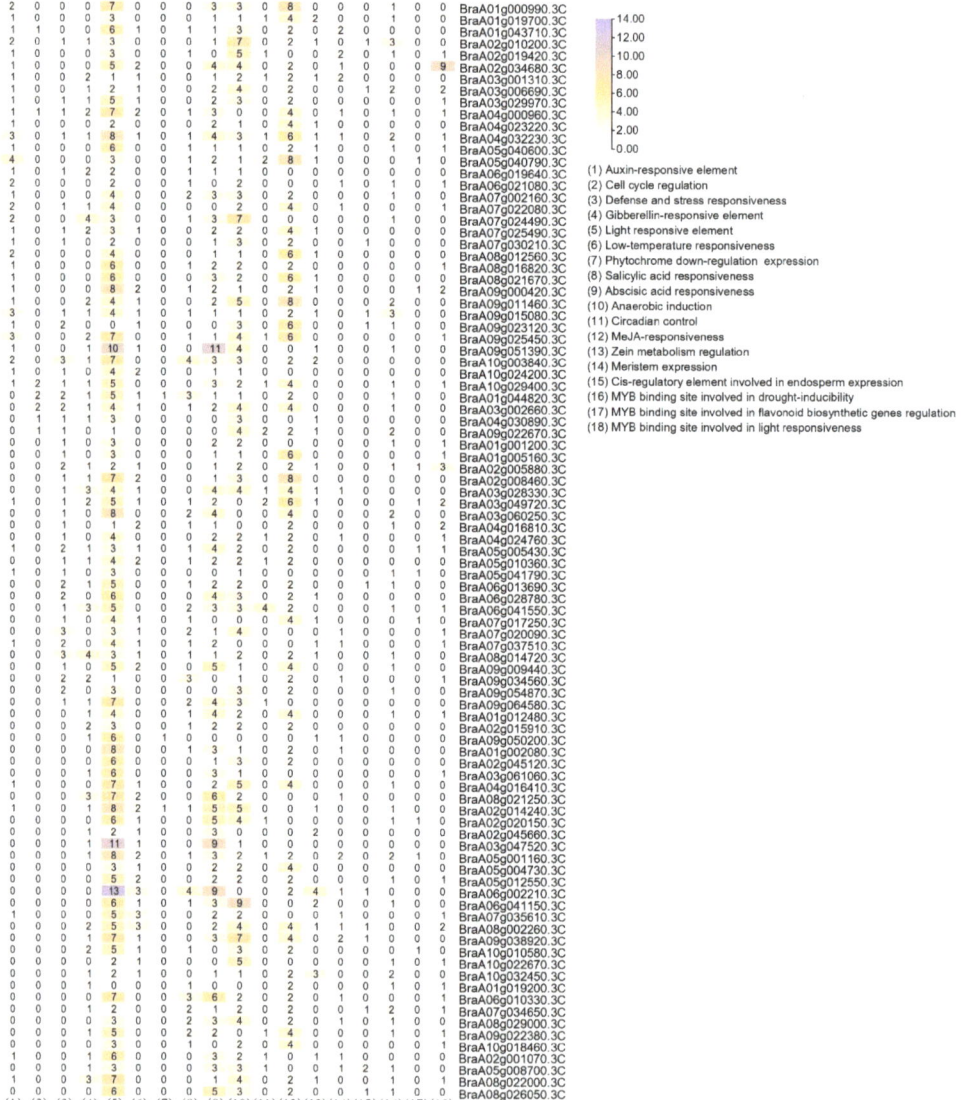

Figure 3. The cis-acting elements prediction on putative promoters of *HD-Zip* genes. The number of cis-acting elements on putative promoters of *HD-Zip* genes. A total of eighteen cis-acting elements were investigated in our study.

3.5. Chromosome Location and Gene Family Expansion Analysis of HD-Zip Genes in Brassicaceae Plants

Based on the *HD-Zip* chromosome location information of Chinese cabbage, 93 genes were unevenly distributed on 10 chromosomes (Figure 4a, Supplemental Table S2). Seven to fourteen *HD-Zip* genes were distributed on each chromosome, and chromosome A09 contained the most *HD-Zip* genes. HD-Zip I, II, and IV genes were distributed on all 10 chromosomes, and group 3 subfamily members were not discovered on chromosome 3 or 7. In *B. oleracea*, seven to fifteen genes were unevenly distributed on each of nine chromosomes (Supplemental Table S3), and chromosome 3 contained the most *HD-Zip* genes (Figure 4b). In *B. napus*, 184 genes were discovered on 19 chromosomes (Figure 4c), with a gene number ranging from six to sixteen on each chromosome (Supplemental Table S4).

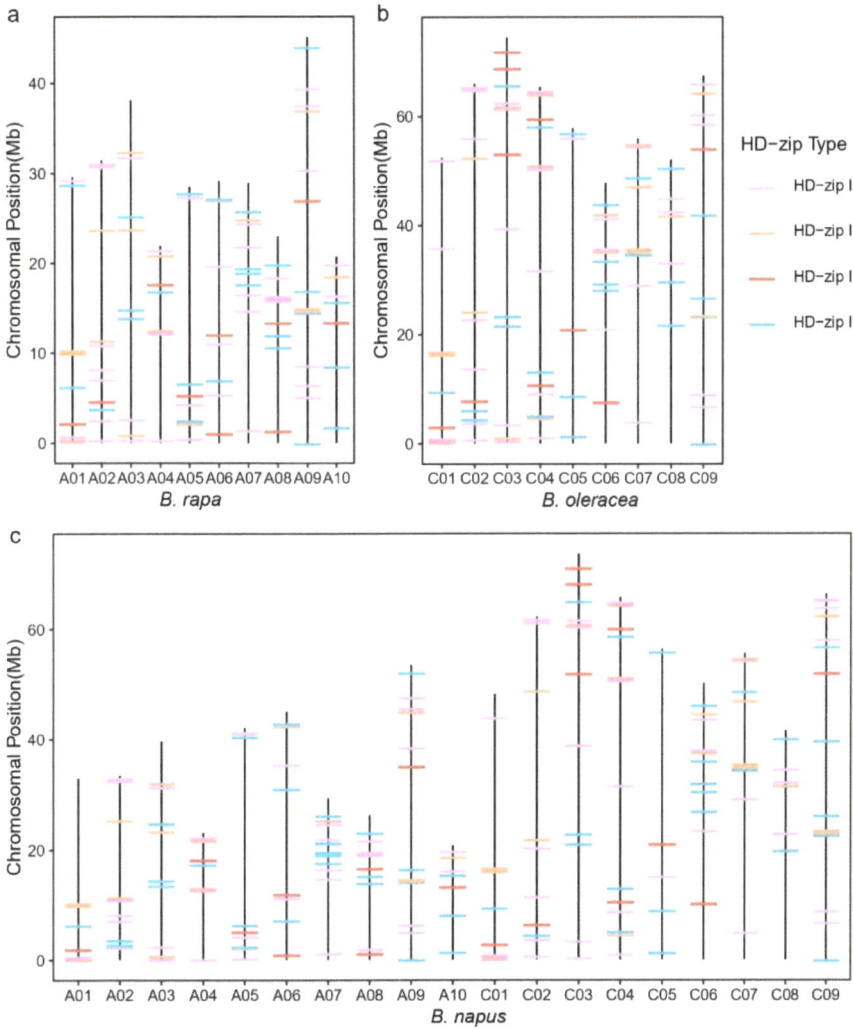

Figure 4. Chromosome location of *HD-Zip* genes in *B. rapa* (**a**), *B. oleracea* (**b**), and *B. napus* (**c**). Different colors represent different types of *HD-Zip* genes. Red (subfamily III), cyan (subfamily IV),

yellow (subfamily II), and purple (subfamily I). The x-axis represents the chromosome number, and the y-axis represents chromosome length.

The results of collinearity analysis within the genome showed that there were 76.34% (71/96) of *HD-Zip* genes in *B. rapa* and 75.53% (71/94) in *B. oleracea* that have paralogs (Figure 5a,b, Supplemental Table S5). In *B. napus*, 75.53% of genes in the A genome and 81.11% of genes in the C genome have paralogs (Figure 5c). Duplication gene classification analysis showed that 77.42% (72/93) of genes were classified into segmental/whole genome duplication (WGD), and 11.83% (11/93) of genes were dispersed and duplicated (Figure 5d). In *B. oleracea* and *B. napus*, more than 75% of genes were classified into segmental/whole genome duplication (WGD). These results suggest that HD-Zip gene family expansion was mainly caused by the WGD in the three *Brassica* species.

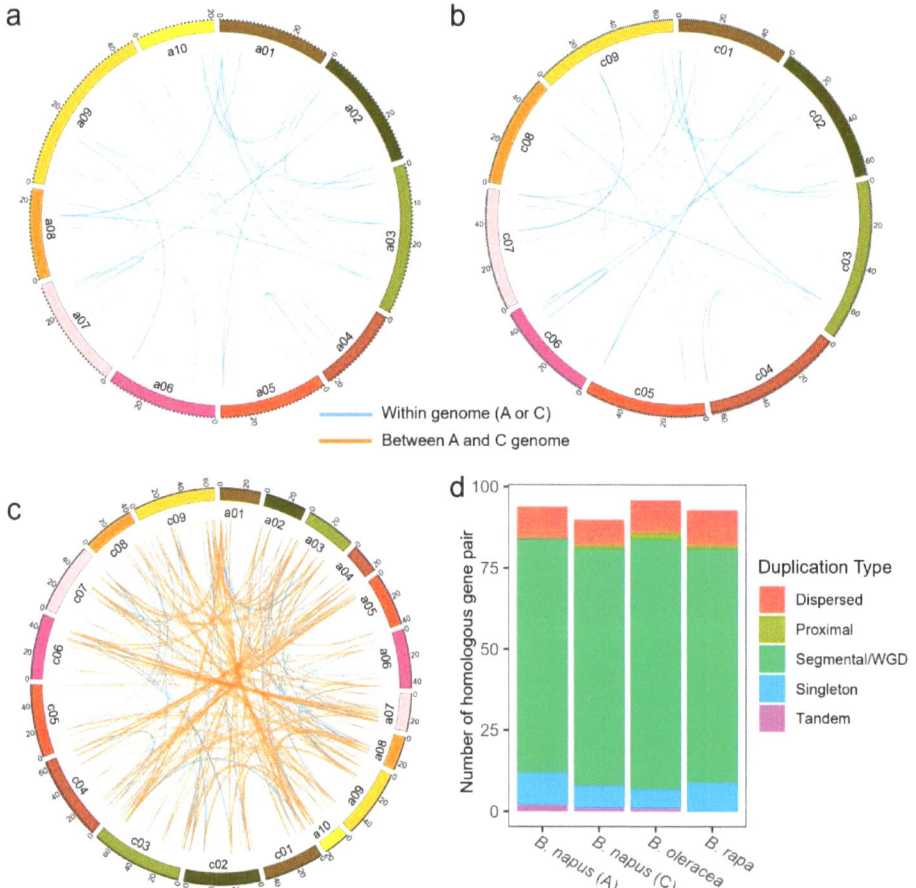

Figure 5. Synteny analysis of the *HD-Zip* genes in *B. rapa* (**a**), *B. napus* (**b**), and *B. oleracea* (**c**). The letter 'a' represents the chromosome of the 'A' genome in *B. rapa* and *B. napus*, and the letter 'c' represents the chromosome of the 'C' genome in *B. oleracea* and *B. napus*. Blue lines mean collinearity relationships among genes within a genome (within A genome or within C genome), and orange lines mean collinearity relationships among genes between A and C genomes. (**d**) Classification of duplication gene pairs in three *Brassica* species. The details of the classification are displayed in the method section.

3.6. Estimating Dates and Driving Forces for the Evolution of the HD-Zip Gene Family

Many *HD-Zip* genes were involved in collinear regions, indicating that many genes underwent duplication events. In order to reveal which duplication event drove the evolution of the HD-Zip gene family, we calculated 4DTv and Ks values to evaluate the date of duplication events and further calculated Ka/Ks ratios to determine which selective pressures drove the evolution of *HD-Zip* genes in the three *Brassica* species. We found obvious 4DTv peaks within the genome at 0.10 to 0.13 (Figure 6a), and Ks peaked at 0.27 to 0.35 (Figure 6b), which is consistent with the ancient whole-genome triplication (WGT) event (Ks = 0.34) that occurred approximately 16 million years ago [42]. Two peaks were observed between the A and C genomes in *B. napus*; one was consistent with the peak of a WGT event, and another ranged from 0.07 to 0.12, consistent with the divergence of the A and C genomes [42]. These results suggested that HD-Zip gene family expansion was mainly driven by WGT events. The results of the Ka/Ks ratios of HD-Zip paralogs showed that most HD-Zip gene pairs were concentrated between 0.1 and 0.3 (Figure 6c), suggesting that *HD-Zip* genes were mainly under purifying selection in the three species. It was notable that the ratio of five orthologs between the A and C genomes in *B. napus* was more than one, indicating that these gene pairs were under positive selection.

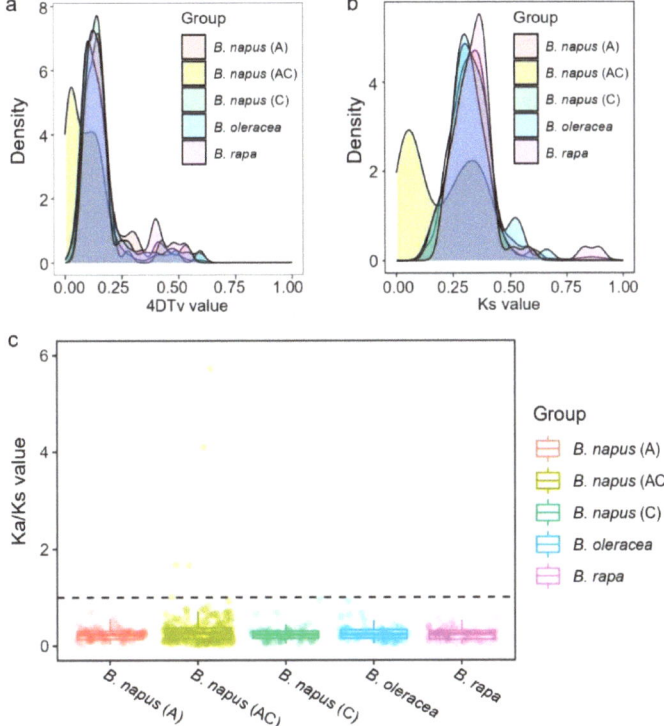

Figure 6. The evolution analysis of the HD-Zip gene family. The distribution of 4DTv (**a**) and Ks (**b**) values of homologous gene pairs in three species. (**c**) The distribution Ka/Ks values of homology gene pairs. 'A' means homology gene pairs in the A genome, 'C' means homology gene pairs in the C genome, and 'AC' means homology gene pairs between the A and C genomes.

3.7. Expression Patterns of HD-Zip Genes in Different Chinese Cabbage Varieties

Comparative transcriptome analysis was widely used to identify the potential role of gene function and provide a foundation for further functional analysis [43]. Previous studies found that *HD-Zip* genes participate in various abiotic stresses, such as heat, drought, and salt stress [44–46]. To reveal the *HD-Zip* genes related to the heat stress response in Chinese cabbage, transcriptome sequencing analysis was performed using a heat-resistant inbred line '268', and the FPKM values were counted to represent the expression level (Figure 7). A total of 59 *HD-Zip* genes were expressed (FPKM > 1) in at least one sample, and 18 genes had more than a 1.5-fold mean FPKM difference between the three heat treatments (HT-4 day, 8 day, 10 day) and the CK (Control treatment) group. Thirteen of the 18 genes were highly expressed under heat treatment, 11 of which were from subfamily I, indicating that HD-Zip I genes have a positive function under heat stress. *BraA07g002160.3C* showed a 7.57-fold up-regulation under heat treatment and had the highest expression at 10 days of heat treatment (HT-10). *BraA05g001160.3C* had higher expression levels at three periods under heat treatments than the CK. In addition, *BraA09g022670.3C* was up-regulated under heat stress treatment (HT-8 day, 10 day). The expression level of *BraA07g022080.3C*, belonging to the HD-Zip IV subfamily, was 12.66-fold up-regulated upon heat stress treatment. We further selected eight different expression genes for qRT-PCR analysis in two *B. rapa* varieties ('QZ' and '54'). Six genes were up-regulated when 'QZ' was under heat stress treatment, and five genes showed up-regulation in '54'. *BraA05g040600.3C* showed significant up-regulation for heat stress treatment (from 2 day to 10 day) in 'QZ', but no up-regulation was observed in '54', suggesting that *BraA05g040600.3C* has different expression patterns between the two varieties. The findings suggest that these genes may be important for the heat tolerance of Chinese cabbage.

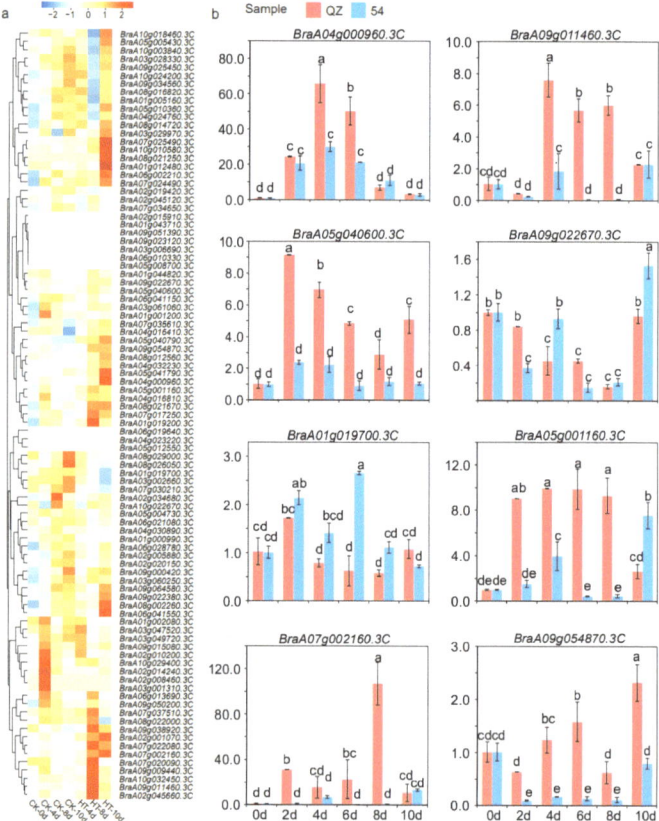

Figure 7. Expression pattern of *HD-zip* genes in *B. rapa* ('268', a heat-resistant inbred line) under heat stress treatment. (**a**) The heat maps were plotted by the R/Pheatmap package. A red color indicates

high expression, blue indicates low expression, and gray indicates no data. 'CK' means normal treatment, and 'HT' means heat treatment (40 °C). The FPKM value was normalized by using the z-score method. (**b**) qRT-PCR analysis of the eight different expression genes. Two *B. rapa* varieties ('QZ' and '54') under heat stress treatment (0 d, 2 d, 4 d, 6 d, 8 d, and 10 d) were used for qRT-PCR analysis. SPSS software was used to analyze the difference in expression level at the 0.05 level. The yellow inner-leaf variety is represented by '54', and 'QZ' represents 'Gailiang Qiuza 3' (white inner-leaf variety). Different lower case letters indicate significant difference at 0.05 level.

A previous study found that the *HD-Zip* gene can regulate the carotenoid content of fruits and generate different fruit colors. To reveal the potential role of *HD-Zip* genes in the regulation of leaf carotenoid content, the leaf blades of two Chinese cabbage varieties (yellow leaf: '54'; white leaf: 'Gailiang Qiuza 3', QZ) were selected (Figure 8a). These have significantly (p value = 0.0026) different carotenoid contents (Figure 8b), and transcriptome data analysis was performed. The results showed that fifteen *HD-Zip* genes have more than a 2-fold change between 'QZ' and '54' (FPKM value > 1 in at least one sample) (Figure 8c). Three genes in '54' showed higher expression levels than in QZ, and 12 genes showed lower expression levels. Interestingly, all three up-regulated genes in '54' belong to the HD-Zip I subfamily. *BraA09g015080.3C* has a high expression level (FPKM > 20) in '54', but it was not expressed in QZ. *BraA02g019420.3C* and *BraA04g000960.3C* showed 5.57-fold and 2.46-fold up-regulation in '54', respectively. The qRT-PCR analysis was performed, and a high correlation between the transcriptome and qRT-PCR results suggests the accuracy of the transcriptome. These results suggest three *HD-Zip* genes may have positive potential in the carotenoid content regulation of Chinese cabbage.

Interestingly, *BraA09g011460.3C*, belonging to the HD-zip I subfamily, had up-regulated expression after eight days of heat treatment, and it also had high expression after heat treatment in 'QZ' and '54'. These results suggest *BraA09g011460.3C* has potential involvement in the heat stress tolerance of *B. rapa*. In addition, it had lower expression levels in '54' (mean FPKM = 0.57) than in 'QZ' (mean FPKM = 3.31) and showed a significant (p-value = 9.82E-4) negative correlation with the total carotenoid content. The phylogenetic and BLAST search results showed that *BraA09g011460.3C* was orthologous with *AT2G18550.1* (*ATHB21*), which can promote the expression of *AtNCED3* [47]. NCED encode 9-CIS-EPOXICAROTENOIDDIOXIGENASE has been proven to lead to carotenoid cleavage [48] and the accumulation of abscisic acid (ABA), which is involved in acquired thermotolerance [49]. Therefore, we speculate that *BraA09g011460.3C* may improve heat stress tolerance and decrease the carotenoid content of *B. rapa*.

Figure 8. *Cont.*

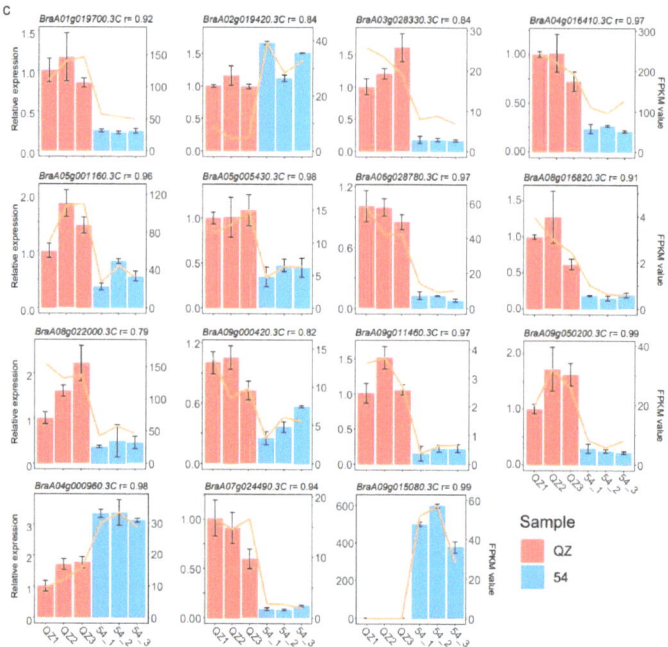

Figure 8. Leaf color (**a**) and carotenoid content (**b**) in two Chinese cabbage varieties. The yellow inner-leaf variety is represented by '54', and 'QZ' represents 'Gailiang Qiuza 3' (white inner-leaf variety). The three replicated are represented by '1, 2, and 3'. (**c**) Expression profiles of 15 genes in two Chinese cabbages by qRT-PCR analysis and RNA-seq data. Lines represent the value of FPKM, and bars represent the value of the relative expression level.

4. Discussion

4.1. Whole-Genome Identification and Phylogenetic Analysis of HD-Zip Genes in Chinese Cabbage

The *HD-Zip* genes encoding the HD and LZ domains are important in plant development and abiotic stress tolerance [50]. Thus far, HD-Zip transcription factors have been studied in plants such as *Arabidopsis* [3], Physic nut [51], peach [52], and cucumber [53]. Chinese cabbage, a member of *Brassica*, is an important vegetable with rich nutrients and high yield. In this study, we performed relatively rigorous criteria to identify candidate members of the *HD-Zip* genes in *B. rapa* and two other *Brassica* species. According to the criteria, 93 *HD-Zip* genes were identified in *B. rapa*, 96 in *B. oleracea*, and 184 in *B. napus*, and the expansion of *HD-Zip* genes in *B. napus* can be explained by *B. napus* (AACC, $2n = 38$) originating from the natural hybridization of *B. rapa* (AA, $2n = 20$) and *B. oleracea* (CC, $2n = 18$) [54].

Based on the phylogenetic results, a total of 93 *HD-Zip* genes of *B. rapa* were classified into four subfamilies (HD-Zip I, II, III, and IV). The percentage of *HD-Zip* I genes reached 41.9%, 42.7%, and 40.2% in *B. rapa*, *B. oleracea*, and *B. napus*, which was higher than the three other subfamilies. This percentage was slightly higher than *Arabidopsis* (35.41%), suggesting that more *HD-Zip* I genes were retained after the WGT event. *HD-Zip* I genes can participate in various abiotic stress response tolerances, such as salt tolerance in apples [55] and physic nuts [51], thermotolerance in lilies [19], and drought tolerance in rice [56]. Therefore, we speculate that a high percentage of *HD-Zip* genes may improve the abiotic stress tolerance and environmental adaptation of the three *Brassica* species.

4.2. The Evolution History of the HD-Zip Gene Family

The expansion of gene families was mainly driven by gene duplication, like and dispersed, and tandem or whole genome duplication [56]. In the evolution process of the Chinese cabbage genome, a WGT event occurred approximately 16 million years ago, which resulted in the explosion of genes in *Brassica* species. The gene duplication type classification and Ka/Ks value peaks suggest that the HD-Zip gene family was primarily expanded by a WGT event, and many duplication genes were retained. No gene was classified into tandem duplication, which was different from the NBS gene family, wherein 43.3% of genes were formed by tandem duplication, suggesting a different expansion pattern between the HD-Zip and NBS gene families [57]. The ratios of Ka/Ks were mostly less than 1, indicating that HD-Zip genes within this species are mainly subject to purifying selection. Interestingly, five orthologous gene pairs between the A and C genomes in *B. napus* showed positive selection, which can facilitate the prevalence of advantageous traits for a particular species' evolution [58,59].

4.3. The Potential Roles of Chinese Cabbage HD-Zip Transcription Factors

Growing evidence suggests that *HD-Zip* genes are important in a plant's abiotic stress tolerance. Typically, gene expression differences in different samples can be explored, and functional genes of plants can be mined based on transcriptome sequencing analysis [60,61]. In a previous study, *MdHB7* (belonging to the HD-Zip I subfamily) could improve tolerance to salinity in apples [55]. Overexpression of *Zmhdz10* in maize could enhance a plant's tolerance to drought and salt stress and increase susceptivity to ABA [62]. In addition, overexpression of *RsHDZ17* could improve heat tolerance in radishes [63]. *LlHB16*, the HD-Zip I gene in lilies, could regulate the basal heat-response pathway and ABA signal to promote thermotolerance [19]. In *Eucalyptus grandis*, *EgHD-Zip37* also showed a response to temperature changes [45]. The 14 *HD-Zip* genes in *B. rapa* exhibited different expression levels during heat stress treatment. Of these, 11 belonged to the HD-Zip I gene family, suggesting their importance in Chinese cabbage thermotolerance. Interestingly, the large number of *HD-Zip* genes containing stress-response elements in promoter regions also indicates their importance in other stresses.

The yellow inner leaf of Chinese cabbage is an important agronomic trait, which is primarily caused by a high carotenoid content, such as lutein, β-carotene, or lycopene [63]. A lot of carotene synthesis genes were up-regulated in the yellow leafy head cultivar [64]. In watermelon, differences in carotenoid content resulted in differences in the flesh color among watermelon varieties, and a large number of *HD-Zip* genes showed different expression levels, suggesting their potential role in carotenoid regulation [65]. In addition, different transcript levels of carotenoid biosynthesis genes and *HD-Zip* genes were identified between citruses, with fruits having color variations [66]. In our study, 11 *HD-Zip* genes showed different expression patterns between yellow- and white-inner leaf varieties, and all of the three up-regulated genes belonged to the HD-Zip I subfamily. *CaATHB-12* belongs to the HD-Zip I subfamily, and its overexpression could increase the carotenoid content under normal conditions in *Capsicum annuum* [7]. These results suggest that *HD-Zip* genes may also participate in carotenoid regulation, and the three up-regulated HD-Zip I genes may play positive roles in yellow-inner-leaf Chinese cabbage. A previous study found that the *AtHB21* gene could regulate the *NCED* gene, which could decrease the carotenoid content [46,47,67] and promote ABA accumulation to improve heat stress tolerance [19]. *BraA09g011460.3C* was homologous to *AtHB21* and was highly expressed after heat treatment and lowly expressed in high-carotenoid varieties, suggesting a regulatory effect on heat tolerance and the carotenoid content in *B. rapa*.

5. Conclusions

In our study, a total of 93, 96, and 184 *HD-Zip* genes were identified in the *B. rapa*, *B. oleracea*, and *B. napus* genomes, and they were further divided into four subfamilies (I to IV) based on gene structure. The results of the gene structure and evolutionary trajectories predicted that members in the same subclade possessed the same motif structure, and motifs 1, 2, and 3 were commonly present in all HD-Zip family genes. Notably, stress-response-related elements and hormone-response elements were discovered in the promoters of all three of these genes. Collinearity and Ks results showed that HD-Zip gene family expansion was driven by WGT events, and most *HD-Zip* genes were under purifying selection. Transcriptome analysis identified 14 genes showing different expressions between the CK and heat treatment groups, and 11 of the 14 genes were from the HD-Zip I subfamily. In addition, three up-regulated genes were identified in high-carotenoid-content Chinese cabbages. *BraA09g011460.3C* showed up-regulation after heat treatment and low expression in high-carotenoid-content varieties, suggesting its regulation in heat stress tolerance and the carotenoid content in *B. rapa*.

Supplementary Materials: The following supporting information can be downloaded at: https://www.mdpi.com/article/10.3390/agronomy13051324/s1, Figure S1: A phylogenetic tree of HD-Zip proteins of *B. oleracea* and *Arabidopsis*. Purple branches indicate HD-Zip subfamily I, yellow branches indicate II, red branches indicate III and cyan branches indicate IV. Red squares indicate genes coming from *Arabidopsis* and blue squares indicate genes coming from *B. napus*. Figure S2: A phylogenetic tree of HD-Zip proteins of *B. napus* and *Arabidopsis*. Purple branches indicate HD-Zip subfamily I, yellow branches indicate II, red branches indicate III and cyan branches indicate IV. Red squares indicate genes coming from *Arabidopsis* and blue squares indicate genes coming from *B. oleracea*. Figure S3: The phylogenetic relationship, conserved motifs and gene structure of HD-Zip proteins in *B. oleracea*. (a) A phylogenetic tree of HD-Zip proteins was constructed using the Max likelihood (ML) method and 1000 ultrafast bootstraps. (b) The distribution of conserved motifs across *B. oleracea* HD-Zip proteins. A total of 20 motifs were predicted using MEME tool. (c) The gene structure of *HD-Zip* genes in *B. oleracea*, including intron and exon. The black lines indicate intron and green squares indicates coding sequence (CDS). Figure S4: The phylogenetic relationship, conserved motifs and gene structure of HD-Zip proteins in *B. napus*. (a) A phylogenetic tree of HD-Zip proteins was constructed using the Max likelihood (ML) method and 1000 ultrafast bootstraps. (b) The distribution of conserved motifs across *B. napus* HD-Zip proteins. A total of 20 motifs were predicted using MEME tool. (c) The gene structure of *HD-Zip* genes in *B. napus*, including intron and exon. The black lines indicate intron and yellow squares indicates coding sequence (CDS). Figure S5: The cis-acting elements predication on putative promoters of *HD-Zip* genes in *B. oleracea*. The number of cis-acting elements on putative promoters of *HD-Zip* genes. A total of eighteen cis-acting elements were investigated in our study, including: (1) Auxin-responsive element; (2) Cell cycle regulation; (3) Defense and stress responsiveness; (4) Gibberellin-responsive element; (5) Light responsive element; (6) Low-temperature responsiveness; (7) Phytochrome down-regulation expression; (8) Salicylic acid responsiveness; (9) Abscisic acid responsiveness; (10) Anaerobic induction; (11) Circadian control; (12) MeJA-responsiveness; (13) Zein metabolism regulation; (14) Meristem expression; (15) Cis-regulatory element involved in endosperm expression; (16) MYB binding site involved in drought-inducibility; (17) MYB binding site involved in flavonoid biosynthetic genes regulation; (18) MYB binding site involved in light responsiveness. Figure S6: The cis-acting elements predication on putative promoters of *HD-Zip* genes in *B. napus*. The number of cis-acting elements on putative promoters of *HD-Zip* genes. A total of eighteen cis-acting elements were investigated in our study, including: (1) Auxin-responsive element; (2) Cell cycle regulation; (3) Defense and stress responsiveness; (4) Gibberellin-responsive element; (5) Light responsive element; (6) Low-temperature responsiveness; (7) Phytochrome down-regulation expression; (8) Salicylic acid responsiveness; (9) Abscisic acid responsiveness; (10) Anaerobic induction; (11) Circadian control; (12) MeJA-responsiveness; (13) Zein metabolism regulation; (14) Meristem expression; (15) Cis-regulatory element involved in endosperm expression; (16) MYB binding site involved in drought-inducibility; (17) MYB binding site involved in flavonoid biosynthetic genes regulation; (18) MYB binding site involved in light responsiveness. Supplemental Table S1: Primers of *HD-Zip* genes in *B. rapa*. Supplemental Table S2: The information of *HD-Zip genes* in *B. rapa*. Supplemental Table S3: The information of *HD-Zip* genes in *B. oleracea*. Supplemental Table S4: The information of *HD-Zip* genes in *B. napus*. Supplemental Table S5: The duplication information of *HD-Zip* genes in three species.

Author Contributions: J.Z. and L.Y. initiated and designed the research; L.Y., L.W., W.X. and L.L. performed the experiments; L.Y., Y.S., X.C., D.L. and M.S. analyzed the data; L.Y. and J.Z. wrote the paper; J.Z. contributed to and edited the paper; J.Z., Y.S., X.C., J.L. and K.F. revised the paper; J.L. and K.F. polished the language. All authors have read and agreed to the published version of the manuscript.

Funding: This research was supported by the Independent Innovation Fund Project of Agricultural Science and Technology in Jiangsu Province (CX(21)2020), the Seed Industry Revitalization Project of Jiangsu Province (JBGS(2021)073), and the Research and Development Fund of Huai'an Academy of Agricultural Sciences (HNY202131).

Conflicts of Interest: The authors declare no conflict of interest.

References

1. Aso, K.; Kato, M.; Banks, J.A.; Hasebe, M. Characterization of homeodomain-leucine zipper genes in the fern Ceratopteris richardii and the evolution of the homeodomain-leucine zipper gene family in vascular plants. *Mol. Biol. Evol.* **1999**, *16*, 544–552. [CrossRef]
2. Sakakibara, K.; Nishiyama, T.; Sumikawa, N.; Kofuji, R.; Murata, T.; Hasebe, M. Involvement of auxin and a homeodomain-leucine zipper I gene in rhizoid development of the moss *Physcomitrella patens*. *Development* **2003**, *130*, 4835–4846. [CrossRef] [PubMed]
3. Kamata, N.; Okada, H.; Komeda, Y.; Takahashi, T. Mutations in epidermis-specific HD-ZIP IV genes affect floral organ identity in *Arabidopsis thaliana*. *Plant J.* **2013**, *75*, 430–440. [CrossRef] [PubMed]
4. Ding, Z.; Fu, L.; Yan, Y.; Tie, W.; Xia, Z.; Wang, W.; Peng, M.; Hu, W.; Zhang, J.; Zhang, J. Genome-wide characterization and expression profiling of HD-Zip gene family related to abiotic stress in cassava. *PLoS ONE* **2017**, *12*, e0173043. [CrossRef]
5. Vernoud, V.; Laigle, G.; Rozier, F.; Meeley, R.B.; Perez, P.; Rogowsky, P.M. The HD-ZIP IV transcription factor OCL4 is necessary for trichome patterning and anther development in maize. *Plant J.* **2009**, *59*, 883–894. [CrossRef] [PubMed]
6. Sharif, R.; Raza, A.; Chen, P.; Li, Y.; El-Ballat, E.M.; Rauf, A.; Hano, C.; El-Esawi, M.A. HD-ZIP Gene Family: Potential roles in improving plant growth and regulating stress-responsive mechanisms in plants. *Genes* **2021**, *12*, 1256. [CrossRef]
7. Elhiti, M.; Stasolla, C. Structure and function of homodomain-leucine zipper (HD-Zip) proteins. *Plant Signal Behav.* **2009**, *4*, 86–88. [CrossRef]
8. Harris, J.C.; Hrmova, M.; Lopato, S.; Langridge, P. Modulation of plant growth by HD-Zip class I and II transcription factors in response to environmental stimuli. *New Phytol.* **2011**, *190*, 823–837. [CrossRef]
9. Zhang, R.X.; Zhu, W.C.; Cheng, G.X.; Yu, Y.N.; Li, Q.H.; ul Haq, S.; Said, F.; Gong, Z.H. A novel gene, *CaATHB-12*, negatively regulates fruit carotenoid content under cold stress in *Capsicum annuum*. *Food Nutr. Res.* **2020**, *64*, 3729. [CrossRef]
10. Roodbarkelari, F.; Groot, E.P. Regulatory function of homeodomain-leucine zipper (HD-ZIP) family proteins during embryogenesis. *New Phytol.* **2017**, *213*, 95–104. [CrossRef]
11. Turchi, L.; Baima, S.; Morelli, G.; Ruberti, I. Interplay of HD-Zip II and III transcription factors in auxin-regulated plant development. *J. Exp. Bot.* **2015**, *66*, 5043–5053. [CrossRef]
12. Takada, S.; Takada, N.; Yoshida, A. ATML1 promotes epidermal cell differentiation in *Arabidopsis* shoots. *Development* **2013**, *140*, 1919–1923. [CrossRef]
13. Ursache, R.; Miyashima, S.; Chen, Q.; Vatén, A.; Nakajima, K.; Carlsbecker, A.; Zhao, Y.; Helariutta, Y.; Dettmer, J. Tryptophan-dependent auxin biosynthesis is required for HD-ZIP III-mediated xylem patterning. *Development* **2014**, *141*, 1250–1260. [CrossRef]
14. Sun, X.X.; Feng, D.; Liu, M.; Qin, R.; Li, Y.; Lu, Y.; Zhang, X.; Wang, Y.; Shen, S.; Ma, W.; et al. Single-cell transcriptome reveals dominant subgenome expression and transcriptional response to heat stress in Chinese cabbage. *Genome. Biol.* **2022**, *23*, 1–19. [CrossRef]
15. Zhang, L.; Dai, Y.; Yue, L.; Chen, G.; Yuan, L.; Zhang, S.; Li, F.; Zhang, H.; Li, G.; Zhu, S.; et al. Heat stress response in Chinese cabbage (*Brassica rapa* L.) revealed by transcriptome and physiological analysis. *Peerj* **2022**, *10*, e13427. [CrossRef]
16. Quan, J.; Zheng, W.; Wu, M.; Shen, Z.; Tan, J.; Li, Z.; Zhu, B.; Hong, S.B.; Zhao, Y.; Zhu, Z.; et al. Glycine betaine and beta-aminobutyric acid mitigate the detrimental effects of heat stress on Chinese cabbage (*Brassica rapa* L. ssp. *pekinensis*) Seedlings with Improved Photosynthetic Performance and Antioxidant System. *Plants* **2022**, *11*, 1213. [CrossRef]
17. Song, X.; Liu, G.; Duan, W.; Liu, T.; Huang, Z.; Ren, J.; Hou, X. Genome-wide identification, classification and expression analysis of the heat shock transcription factor family in Chinese cabbage. *Mol. Genet. Genomics.* **2014**, *289*, 541–551. [CrossRef]
18. Nakashima, K.; Ito, Y.; Yamaguchi-Shinozaki, K. Transcriptional regulatory networks in response to abiotic stresses in *Arabidopsis* and grasses. *Plant Physiol.* **2009**, *149*, 88–95. [CrossRef]
19. Wu, Z.; Li, T.; Zhang, D.; Teng, N. Lily HD-Zip I transcription factor LlHB16 promotes thermotolerance by activating LlHSFA2 and LlMBF1c. *Plant Cell Physiol.* **2022**, *63*, 1729–1744. [CrossRef]
20. Li, W.; Dong, J.; Cao, M.; Gao, X.; Wang, D.; Liu, B.; Chen, Q. Genome-wide identification and characterization of HD-ZIP genes in potato. *Gene* **2019**, *697*, 103–117. [CrossRef]
21. Wang, K.; Xu, L.; Wang, Y.; Ying, J.; Li, J.; Dong, J.; Li, C.; Zhang, X.; Liu, L. Genome-wide characterization of homeodomain-leucine zipper genes reveals *RsHDZ17* enhances the heat tolerance in radish (*Raphanus sativus* L.). *Physiol. Plantarum.* **2022**, *174*, e13789. [CrossRef] [PubMed]

22. Wang, J.; Zhuang, L.; Zhang, J.; Yu, J.; Yang, Z.; Huang, B. Identification and characterization of novel homeodomain leucine zipper (HD-Zip) transcription factors associated with heat tolerance in perennial ryegrass. *Environ. Exp. Bot.* **2019**, *160*, 1–11. [CrossRef]
23. Zhao, J.G.; Lu, Z.G.; Wang, L.; Jin, B. Plant responses to heat stress: Physiology, transcription, noncoding RNAs, and epigenetics. *Int. J. Mol. Sci.* **2020**, *22*, 117. [CrossRef]
24. Inbaraj, B.S.; Lu, H.; Hung, C.F.; Wu, W.B.; Lin, C.L.; Chen, B.H. Determination of carotenoids and their esters in fruits of *Lycium barbarum* Linnaeus by HPLC-DAD-APCI-MS. *J. Pharm. Biomed.* **2008**, *47*, 812–818. [CrossRef] [PubMed]
25. Mistry, J.; Chuguransky, S.; Williams, L.; Qureshi, M.; Salazar, G.A.; Sonnhammer, E.L.L.; Tosatto, S.C.E.; Paladin, L.; Raj, S.; Richardson, L.J.; et al. Pfam: The protein families database in 2021. *Nucleic. Acids Res.* **2021**, *49*, D412–D419. [CrossRef]
26. Johnson, L.S.; Eddy, S.R.; Portugaly, E. Hidden Markov model speed heuristic and iterative HMM search procedure. *BMC Bioinform.* **2010**, *11*, 431. [CrossRef]
27. Larkin, M.A.; Blackshields, G.; Brown, N.P.; Chenna, R.; McGettigan, P.A.; McWilliam, H.; Valentin, F.; Wallace, I.M.; Wilm, A.; Lopez, R.; et al. Clustal W and Clustal X version 2.0. *Bioinformatics* **2007**, *23*, 2947–2948. [CrossRef]
28. Nguyen, L.T.; Schmidt, H.A.; von Haeseler, A.; Minh, B.Q. IQ-TREE: A fast and effective stochastic algorithm for estimating maximum-likelihood phylogenies. *Mol. Biol. Evol.* **2015**, *32*, 268–274. [CrossRef]
29. Chen, C.; Chen, H.; Zhang, Y.; Thomas, H.R.; Frank, M.H.; He, Y.; Xia, R. TBtools: An integrative toolkit developed for interactive analyses of big biological data. *Mol. Plant.* **2020**, *13*, 1194–1202. [CrossRef]
30. Wang, Y.; Tang, H.; Debarry, J.D.; Tan, X.; Li, J.; Wang, X.; Lee, T.H.; Jin, H.; Marler, B.; Guo, H.; et al. MCScanX: A toolkit for detection and evolutionary analysis of gene synteny and collinearity. *Nucleic. Acids Res.* **2012**, *40*, e49. [CrossRef]
31. Zhang, Z.; Xiao, J.; Wu, J.; Zhang, H.; Liu, G.; Wang, X.; Dai, L. ParaAT: A parallel tool for constructing multiple protein-coding DNA alignments. *Biochem. Biophys. Res. Commun.* **2012**, *419*, 779–781. [CrossRef]
32. Wang, D.; Zhang, Y.; Zhang, Z.; Zhu, J.; Yu, J. KaKs_Calculator 2.0: A toolkit incorporating gamma-series methods and sliding window strategies. *Genom. Proteom. Bioinform.* **2010**, *8*, 77–80. [CrossRef]
33. Zhang, Z.; Li, J.; Yu, J. Computing Ka and Ks with a consideration of unequal transitional substitutions. *BMC Evol. Biol.* **2006**, *6*, 44. [CrossRef]
34. Yue, L.X.; Li, G.; Dai, Y.; Sun, X.; Li, F.; Zhang, S.; Zhang, H.; Sun, R.; Zhang, S. Gene co-expression network analysis of the heat-responsive core transcriptome identifies hub genes in *Brassica rapa*. *Planta* **2021**, *253*, 1–23. [CrossRef]
35. Bolger, A.M.; Lohse, M.; Usadel, B. Trimmomatic: A flexible trimmer for Illumina sequence data. *Bioinformatics* **2014**, *30*, 2114–2120. [CrossRef]
36. Zhang, Z.; Guo, J.; Cai, X.; Li, Y.; Xi, X.; Lin, R.; Liang, J.; Wang, X.; Wu, J. Improved reference genome annotation of *Brassica rapa* by pacific biosciences RNA sequencing. *Front. Plant Sci.* **2022**, *13*, 841618. [CrossRef]
37. Kim, D.; Paggi, J.M.; Park, C.; Bennett, C.; Salzberg, S.L. Graph-based genome alignment and genotyping with HISAT2 and HISAT-genotype. *Nat. Biotechnol.* **2019**, *37*, 907–915. [CrossRef]
38. Pertea, M.; Pertea, G.M.; Antonescu, C.M.; Chang, T.C.; Mendell, J.T.; Salzberg, S.L. StringTie enables improved reconstruction of a transcriptome from RNA-seq reads. *Nat. Biotechnol.* **2015**, *33*, 290–295. [CrossRef]
39. Cai, X.; Chang, L.; Zhang, T.; Chen, H.; Zhang, L.; Lin, R.; Liang, J.; Wu, J.; Freeling, M.; Wang, X. Impacts of allopolyploidization and structural variation on intraspecific diversification in *Brassica rapa*. *Genome. Biol.* **2021**, *22*, 166. [CrossRef]
40. Zhao, J.; Zhai, Z.; Li, Y.; Geng, S.; Song, G.; Guan, J.; Jia, M.; Wang, F.; Sun, G.; Feng, N. Genome-wide identification and expression profiling of the TCP family genes in spike and grain development of wheat (*Triticum aestivum* L.). *Front. Plant Sci.* **2018**, *9*, 1282. [CrossRef]
41. He, G.H.; Liu, P.; Zhao, H.X.; Sun, J.Q. The HD-ZIP II transcription factors regulate plant architecture through the auxin pathway. *Int. J. Mol. Sci.* **2020**, *21*, 3250. [CrossRef] [PubMed]
42. Cheng, F.; Sun, R.; Hou, X.; Zheng, H.; Zhang, F.; Zhang, Y.; Liu, B.; Liang, J.; Zhuang, M.; Liu, Y.; et al. Subgenome parallel selection is associated with morphotype diversification and convergent crop domestication in *Brassica rapa* and *Brassica oleracea*. *Nat. Genet.* **2016**, *48*, 1218–1224. [CrossRef] [PubMed]
43. Wang, C.; Song, B.; Dai, Y.; Zhang, S.; Huang, X. Genome-wide identification and functional analysis of U-box E3 ubiquitin ligases gene family related to drought stress response in Chinese white pear (*Pyrus bretschneideri*). *Bmc Plant Biol.* **2021**, *21*, 235. [CrossRef] [PubMed]
44. Zhang, S.X.; Haider, I.; Kohlen, W.; Jiang, L.; Bouwmeester, H.; Meijer, A.H.; Schluepmann, H.; Liu, C.M.; Ouwerkerk, P.B. Function of the HD-Zip I gene *Oshox22* in ABA-mediated drought and salt tolerances in rice. *Plant Mol. Biol.* **2012**, *80*, 571–585. [CrossRef]
45. Zhang, J.S.; Wu, J.; Guo, M.; Aslam, M.; Wang, Q.; Ma, H.; Li, S.; Zhang, X.; Cao, S. Genome-wide characterization and expression profiling of *Eucalyptus grandis* HD-Zip gene family in response to salt and temperature stress. *Bmc Plant Biol.* **2020**, *20*, 1–15. [CrossRef]
46. Wang, D.; Gong, Y.; Li, Y.; Nie, S.M. Genome-wide analysis of the homeodomain-leucine zipper family in *Lotus japonicus* and the overexpression of *LjHDZ7* in *Arabidopsis* for salt tolerance. *Front. Plant Sci.* **2022**, *13*, 955199. [CrossRef]
47. Gonzalez-Grandio, E.; Pajoro, A.; Franco-Zorrilla, J.M.; Tarancón, C.; Immink, R.G.; Cubas, P. Abscisic acid signaling is controlled by a BRANCHED1/HD-ZIP I cascade in *Arabidopsis* axillary buds. *Proc. Natl. Acad. Sci. USA* **2017**, *114*, E245–E254. [CrossRef]

48. Sun, Y.; Bai, P.P.; Gu, K.J.; Yang, S.Z.; Lin, H.Y.; Shi, C.G.; Zhao, Y.P. Dynamic transcriptome and network-based analysis of yellow leaf mutant *Ginkgo biloba*. *Bmc Plant Biol.* **2022**, *22*, 465. [CrossRef]
49. Zhou, H.; Wang, Y.; Zhang, Y.; Xiao, Y.; Liu, X.; Deng, H.; Lu, X.; Tang, W.; Zhang, G. Comparative analysis of heat-tolerant and heat-susceptible rice highlights the role of *OsNCED1* gene in heat stress tolerance. *Plants* **2022**, *11*, 1062. [CrossRef]
50. Li, Y.; Yang, Z.; Zhang, Y.; Guo, J.; Liu, L.; Wang, C.; Wang, B.; Han, G. The roles of HD-ZIP proteins in plant abiotic stress tolerance. *Front. Plant Sci.* **2022**, *13*, 1027071. [CrossRef]
51. Tang, Y.; Wang, J.; Bao, X.; Liang, M.; Lou, H.; Zhao, J.; Sun, M.; Liang, J.; Jin, L.; Li, G.; et al. Genome-wide identification and expression profile of *HD-ZIP* genes in physic nut and functional analysis of the *JcHDZ16* gene in transgenic rice. *Bmc Plant Biol.* **2019**, *19*, 298. [CrossRef]
52. Wang, Z.; Wu, X.; Zhang, B.; Xiao, Y.; Guo, J.; Liu, J.; Chen, Q.; Peng, F. Genome-wide identification, bioinformatics and expression analysis of HD-Zip gene family in peach. *Bmc Plant Biol.* **2023**, *23*, 122. [CrossRef]
53. Sharif, R.; Xie, C.; Wang, J.; Cao, Z.; Zhang, H.; Chen, P.; Li, Y.H. Genome wide identification, characterization and expression analysis of HD-ZIP gene family in *Cucumis sativus* L. under biotic and various abiotic stresses. *Int. J. Biol. Macromol.* **2020**, *158*, 502–520. [CrossRef]
54. Song, J.M.; Guan, Z.; Hu, J.; Guo, C.; Yang, Z.; Wang, S.; Liu, D.; Wang, B.; Lu, S.; Zhou, R.; et al. Eight high-quality genomes reveal pan-genome architecture and ecotype differentiation of *Brassica napus*. *Nat. Plants* **2020**, *6*, 34–45. [CrossRef]
55. Zhao, S.; Wang, H.; Jia, X.; Gao, H.; Mao, K.; Ma, F. The HD-Zip I transcription factor MdHB7-like confers tolerance to salinity in transgenic apple (*Malus domestica*). *Physiol. Plant* **2021**, *172*, 1452–1464. [CrossRef]
56. Qiao, X.; Li, Q.; Yin, H.; Qi, K.; Li, L.; Wang, R.; Zhang, S.; Paterson, A.H. Gene duplication and evolution in recurring polyploidization-diploidization cycles in plants. *Genome. Biol.* **2019**, *20*, 38. [CrossRef]
57. Yu, J.Y.; Tehrim, S.; Zhang, F.; Tong, C.; Huang, J.; Cheng, X.; Dong, C.; Zhou, Y.; Qin, R.; Hua, W.; et al. Genome-wide comparative analysis of NBS-encoding genes between *Brassica* species and *Arabidopsis thaliana*. *Bmc Genom.* **2014**, *15*, 3. [CrossRef]
58. Yang, Z.H.; Nielsen, R. Codon-substitution models for detecting molecular adaptation at individual sites along specific lineages. *Mol. Biol. Evol.* **2002**, *19*, 908–917. [CrossRef]
59. Sabeti, P.C.; Schaffner, S.F.; Fry, B.; Lohmueller, J.; Varilly, P.; Shamovsky, O.; Palma, A.; Mikkelsen, T.S.; Altshuler, D.; Lander, E.S. Positive natural selection in the human lineage. *Science* **2006**, *312*, 1614–1620. [CrossRef]
60. Umezawa, T.; Fujita, M.; Fujita, Y.; Yamaguchi-Shinozaki, K.; Shinozaki, K. Engineering drought tolerance in plants: Discovering and tailoring genes to unlock the future. *Curr. Opin. Biotechnol.* **2006**, *17*, 113–122. [CrossRef]
61. Valliyodan, B.; Nguyen, H.T. Understanding regulatory networks and engineering for enhanced drought tolerance in plants. *Curr. Opin. Plant Biol.* **2006**, *9*, 189–195. [CrossRef] [PubMed]
62. Zhao, Y.; Ma, Q.; Jin, X.; Peng, X.; Liu, J.; Deng, L.; Yan, H.; Sheng, L.; Jiang, H.; Cheng, B. A novel maize homeodomain-leucine zipper (HD-Zip) I gene, *Zmhdz10*, positively regulates drought and salt tolerance in both rice and *Arabidopsis*. *Plant Cell Physiol.* **2014**, *55*, 1142–1156. [CrossRef] [PubMed]
63. Li, Y.; Fan, Y.; Jiao, Y.; Wu, J.; Zhang, Z.; Yu, X.; Ma, Y. Transcriptome profiling of yellow leafy head development during the heading stage in Chinese cabbage (*Brassica rapa* subsp. *pekinensis*). *Physiol. Plant* **2019**, *165*, 800–813. [CrossRef] [PubMed]
64. Jung, H.J.; Manoharan, R.K.; Park, J.I.; Chung, M.Y.; Lee, J.; Lim, Y.P.; Hur, Y.; Nou, I.S. Identification of yellow pigmentation genes in *Brassica rapa* ssp. *pekinensis* using Br300 microarray. *Int. J. Genom.* **2014**, *2014*, 204969. [CrossRef]
65. Yuan, P.; Umer, M.J.; He, N.; Zhao, S.; Lu, X.; Zhu, H.; Gong, C.; Diao, W.; Gebremeskel, H.; Kuang, H.; et al. Transcriptome regulation of carotenoids in five flesh-colored watermelons (*Citrullus lanatus*). *Bmc Plant Biol.* **2021**, *21*, 203. [CrossRef]
66. Huang, Q.; Liu, J.; Hu, C.; Wang, N.; Zhang, L.; Mo, X.; Li, G.; Liao, H.; Huang, H.; Ji, S.; et al. Integrative analyses of transcriptome and carotenoids profiling revealed molecular insight into variations in fruits color of *Citrus Reticulata* Blanco induced by transplantation. *Genomics* **2022**, *114*, 110291. [CrossRef]
67. Jiang, F.; Zhang, J.; Wang, S.; Yang, L.; Luo, Y.; Gao, S.; Zhang, M.; Wu, S.; Hu, S.; Sun, H.; et al. The apricot (*Prunus armeniaca* L.) genome elucidates Rosaceae evolution and beta-carotenoid synthesis. *Hortic. Res.* **2019**, *6*, 128. [CrossRef]

Disclaimer/Publisher's Note: The statements, opinions and data contained in all publications are solely those of the individual author(s) and contributor(s) and not of MDPI and/or the editor(s). MDPI and/or the editor(s) disclaim responsibility for any injury to people or property resulting from any ideas, methods, instructions or products referred to in the content.

Article

Phylogenetic Analyses and Transcriptional Survey Reveal the Characteristics, Evolution, and Expression Profile of NBS-Type Resistance Genes in Papaya

Qian Jiang [1,2], Yu Wang [1], Aisheng Xiong [3], Hui Zhao [1], Ruizong Jia [1], Mengyao Li [4], Huaming An [2], Changmian Ji [1,*] and Anping Guo [1,*]

[1] Hainan Key Laboratory for Biosafety Monitoring and Molecular Breeding in Off-Season Reproduction Regions, Institute of Tropical Bioscience and Biotechnology & Sanya Research Institute, Chinese Academy of Tropical Agricultural Sciences, Haikou 571101, China
[2] Agricultural College, Guizhou University, Guiyang 550025, China
[3] College of Horticulture, Nanjing Agricultural University, Nanjing 210095, China
[4] College of Horticulture, Sichuan Agricultural University, Chengdu 611130, China
* Correspondence: jichangmian@itbb.org.cn (C.J.); guoanping@itbb.org.cn (A.G.)

Citation: Jiang, Q.; Wang, Y.; Xiong, A.; Zhao, H.; Jia, R.; Li, M.; An, H.; Ji, C.; Guo, A. Phylogenetic Analyses, and Transcriptional Survey Reveal the Characteristics, Evolution, and Expression Profile of NBS-Type Resistance Genes in Papaya. *Agronomy* **2023**, *13*, 970. https://doi.org/10.3390/agronomy13040970

Academic Editors: Zhiyong Li, Chaolei Liu and Jiezheng Ying

Received: 3 February 2023
Revised: 22 March 2023
Accepted: 23 March 2023
Published: 25 March 2023

Copyright: © 2023 by the authors. Licensee MDPI, Basel, Switzerland. This article is an open access article distributed under the terms and conditions of the Creative Commons Attribution (CC BY) license (https://creativecommons.org/licenses/by/4.0/).

Abstract: *Carica papaya* maintains an abnormally small but complete NLR family while showing weak disease resistance. To better understand their origin, evolution, and biological function, we identified 59 NLR genes via a customized RGAugury and investigated their characteristics, evolutionary history, and expression profiles based on the improved papaya genome and large-scale RNA-seq data. The results indicated that duplication is a major evolutionary force driving the formation of the papaya NLR family. Synteny analyses of papaya and other angiosperms showed that both insertion and inheritance-derived NLRs are present in papaya. Transcriptome-based expression and network analyses revealed that NLRs are actively involved in biotic stress responses. For example, a papaya-specific inserted TNL was up-regulated strongly by the fungal infection. Both transcriptome and qRT-PCR analyses confirmed the expression divergence of an RNL and an RCNL, a pair of tandem duplication genes involved in different co-expression modules. Furthermore, we observed an inserted gene cluster composed of five duplicated CNLs, showing dosage effects and functional differentiation of disease-resistance genes during evolution. This research will enhance our knowledge of the special NLR family in papaya, which may serve as a model plant for disease-resistance genetic studies.

Keywords: *Carica papaya*; NLRs; insertion-derived; biotic stress; expression divergence; transcriptomic network

1. Introduction

Papaya (*Carica papaya* L.) is a tropical fruit that originated in Central America and benefits tropical and subtropical regions for its high commercial values [1,2]. Papaya fruits have a high nutritional value with abundant vitamin C, vitamin A, and carotenoids [3,4]. Papaya latex contains two important proteolytic enzymes: papain and chymopapain, which are widely used in food, chemicals, and pharmaceutical industries [5–7]. The draft genome of papaya was first reported in 2008, which revealed that the papaya genome contains fewer genes than most angiosperms. Specifically, significantly fewer disease-resistance genes were identified from the papaya genome [1]. In 2022, two updated genomes of the transgenic SunUp and its progenitor Sunset papaya were released [2]. The genome quality was substantially improved, and the lack of disease-resistance genes was reconfirmed in the updated genome release [2]. Among major diseases in papaya-growing regions, PRSV (papaya ringspot virus) is the most destructive and widespread virus to papaya, but no PRSV resistance has been found in papaya germplasm so far [2]. Although genetic engineering made papaya resistant to PRSV infestation [8], the rapid evolution of PRSV and the diversity in lineages led to the loss of transgenic papaya resistance in some regions [9].

The disease resistance (R) gene is a crucial breakthrough in plant disease resistance breeding [10,11]. Intracellular R proteins are often members of the NLR (nucleotide-binding site leucine-rich repeat receptor) family, which played important roles in effector-triggered immunity (ETI) [12–15]. Full-length NLR proteins usually contain a specific N-terminal domain, a conserved nucleotide-binding site (NBS) domain, and an extremely variable C-terminal leucine-rich repeat (LRR) domain, and could be divided into three subclasses based on the N-terminal domain: Toll/Interleukin-1 receptor (TIR)-NBS-LRR (TNL), resistance to powdery mildew8 (RPW8)-NBS-LRR (RNL), and coiled-coil (CC)-NBS-LRR (CNL) [11,13,15,16]. NLR gene copy number and subclass composition often varied greatly in angiosperms for the rapid gene loss and gain [17]. Liu et al. [17] surveyed 305 angiosperms and found that 93 species have no TNL, 23 species have no RNL, and only 1 species has no CNL. Moreover, TNL gene loss events occur frequently throughout dicots and magnoliid diversification, and the TNL is even absent in all monocots [17–19]. It is certain that papaya contains all three subclasses of NLRs and the number is comparatively lower than major angiosperms, although previous studies have reported slightly different numbers of NLR genes (50, 54, and 55) [1,17,20,21]. In terms of disease resistance of papaya, there is resistance against *Colletotrichum brevisporum* [22,23], but no resistance against PRSV in papaya germplasm [2]. The unique NLR gene family composition of papaya attracts us to study their characteristics, evolution, and functional roles.

Here, we identified 59 NLR genes from the improved papaya genome using a customized RGAugury pipeline and performed comprehensive analysis on them, including structural composition, sequence diversity, chromosomal distribution, and phylogenetic analysis. Gene duplication inspection and the syntenic analysis helped us to trace the evolutionary origin of papaya NLRs. Additionally, the expression profile and divergence of NLRs among transcriptome experiments were carefully investigated to elucidate their potential biological roles and functional differentiation under different biotic and abiotic stresses.

2. Materials and Methods

2.1. Identification and Similarity Comparison of NLRs

For the identification of NLRs, we customized the RGAugury pipeline [24] with two major changes: (1) The identification of powdery mildew8 (RPW8) domain was added. (2) NLRs in the original "others" group were assigned to classes, which are named by abbreviated terms of domains. To sum up, the customized pipeline mainly included the following steps: First, potential RGAs (Resistance gene analogs) were obtained by blast to the RGAdb [24]. Second, NBS, TIR, and RPW8 domains were identified by pfam_scan [25]. Third, LRR and CC domains were identified by InterProScan [26] and nCoils [27]. Finally, all identified results were summarized. In the study, TIR, RPW8, CC, NBS, and LRR domains were represented by their initials T, R, C, N, and L, respectively. The customized RGAugury is available in a git repository at https://github.com/reductase4/papaya_NLRs (accessed on 31 January 2023).

In this study, we collected genomes of three basal angiosperms and six eudicots to compare the NLR family of papaya with other angiosperms (for genome data source, see Table S1). The three basal angiosperms (*Nymphaea colorata*, *Nymphaea thermarum*, and *Euryale ferox*) and six eudicots (*Vitis vinifera*, *Prunus persica*, *Citrus sinensis*, *Theobroma cacao*, *Tarenaya hassleriana*, and *Arabidopsis thaliana*) keep syntenic associations with papaya, and they are gradually close to papaya in taxon phylogeny (Figure 1). Then we performed the customized RGAugury pipeline on each genome with the command line: `RGAugury_modified.pl -p longest_transcript.pep.fa -d SUPERFAMILY, SMART, gene3d, Pfam`. Gene IDs and classes of 59 NLRs from papaya are listed in Table S2. Regarding similarity comparison, we applied EMBOSS Needle with the Needleman–Wunsch algorithm [28] to generate the optimal global alignment of two sequences and calculate the similarity of pairwise sequences.

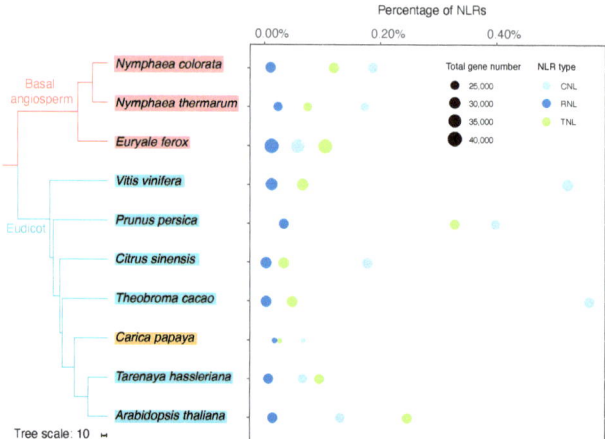

Figure 1. Phylogeny of papaya and selected species with the genome-wide proportion of NLRs (CNL%, RNL%, TNL%). Phylogenomic relationships of species refer to http://www.timetree.org (accessed on 15 November 2022) [29] and research of water lily genome [30].

2.2. Duplication Modes Identification and Chromosomal Localization of Papaya NLRs

DupGen_finder was used to identify genome-wide duplication modes of papaya, including whole-genome, tandem, proximal, transposed, and dispersed duplication (WGD, TD, PD, TRD, DSD) [31]. According to DupGen_finder's description [31], three basal angiosperms (*N. colorata*, *N. thermarum*, and *E. ferox*) were used as the outgroup for the transposed duplication detection. We wrote python scripts to extract gene pairs involving at least one NLR gene from genome-wide duplication identification, in which genes are allowed to be reused in duplication gene pairs. Gene density of chromosomes was calculated by a window size of 100 kb. Finally, papaya chromosomes with NLR loci and duplication associations were visualized by TBtools [32].

2.3. Phylogenetic Analysis of NLRs in Papaya

Referring to the phylogenetic analysis of NLR proteins performed by Gao et al. [33] and Shao et al. [16], we used the NBS domain to construct the phylogenetic tree and NBS genes from Rhodophyta were used as an outgroup in this study. All NLR proteins shared the conserved domain NBS, and Rhodophyta is outside the green plants. A tree of all the full-length NLRs found in papaya and nine other selected species was built in order to illustrate the phylogeny of NLRs in angiosperms. To fully investigate the phylogeny of NLRs in papaya, a tree of full-length NLRs was also constructed.

The NBS domain sequences of NLRs were aligned using the UPP algorithm [34], and the aligned sequences were trimmed using trimAl with the "-gappyout" option [35]. FastTree [36,37] was used to construct the approximately-maximum-likelihood phylogenetic tree with the "-pseudo" and "-gamma" options. The reliability of the tree was estimated by the Shimodaira–Hasegawa test [38] with 1000 resamples.

2.4. Analysis of Conserved Motifs, Domains, and Gene Structures of NLRs in Papaya

Conserved motifs were discovered using MEME Suite 5.5.0 [39]. Conserved domains were re-confirmed in NCBI Conserved Domain Database (CDD) by Batch CD-Search interface with an expected value threshold of 0.05 [40]. We visualized the phylogenetic tree of NLRs with their conserved motifs, domains, and gene structures using TBtools [32].

2.5. Estimation of Divergent Times

We constructed multiple alignments for protein-coding DNA sequences of duplication gene pairs by ParaAT [41] and formatted them into AXT formats. Ks values were calculated by KaKs_Calculator 2.0 [42] with the YN model [43]. According to previous research [44,45], the formula $T = K/2r$ was used to calculate the divergence times, where T represents the divergence time, K means the divergence distance (using Ks values here), r indicates the mutation rate of substitution per site per year. We used a mutation rate of 12×10^{-9}, which we collected from *A. thaliana* under high temperature (29 °C) [46]. Considering that papaya is long-lived in high-temperature environments and R genes evolve rapidly, the mutation rate should be higher, and actual divergent times should be over-estimated.

2.6. Micro-Synteny Analysis of NLRs among Species

Pairwise synteny of papaya and other selected genomes was obtained from the following command line: `python -m jcvi.compara.catalog ortholog --full`. We observed and compared synteny near the loci of all NLRs, and visualized the local synteny around typical NLRs by command line: `python -m jcvi.graphics.synteny`. The analysis was performed following instructions in https://github.com/tanghaibao/jcvi/wiki/MCscan-(Python-version) (accessed on 23 November 2022).

2.7. Expression Pattern and Co-Expression Analysis of NLRs in Papaya

Transcriptome sequencing data of five projects were downloaded from NCBI (PRJNA560275, PRJNA591254, PRJNA692338, PRJNA352643, and PRJNA470602), including 52 papaya samples of different tissues under biotic and abiotic stresses, such as three viruses, fungi, ethylene, and drought stresses (Table S6) [3,22,23,47,48]. We performed the quality control of the downloaded RNA-seq data by fastp [49], then aligned clean reads to the sunset genome by HISAT2 and assemble transcripts in StringTie [50]. Principal Component Analysis (PCA) was used to evaluate the biological reproducibility of treatments. The gene expression levels were calculated with Fragments Per kb per Million reads (FPKM) values. Differentially expressed genes (DEGs) were detected by DESeq2 with cutoffs of FDR < 0.05 and abs(fold change) > 2 [51,52]. Heatmaps of gene expression level and fold change values of NLRs were generated by R scripts. Weighted gene co-expression network analysis (WGCNA) was applied to cluster highly correlated genes into modules [53–55]. The co-expression network was illustrated by Cytoscape [56].

2.8. Real-Time Quantitative PCR Analysis of NLRs in PRSV Infected Papaya

Three-month seedlings of 'Tainong No. 2' papaya were prepared for inoculation. A 10% w/v inoculum was made by grinding the PRSV-infected leaves with 0.05 M phosphate buffer (pH 7.0). PBS buffer was used as a control inoculum. Inoculum was applied to the rubbed areas and rinsed away with distilled water after 10 min. After 7 days, fresh leaves of the same part were sampled and stored at −80 °C.

Total RNA was isolated from the control and PRSV-infected leaves using Spin Column Plant Total RNA Purification Kit (Sangon Biotech, Shanghai, China). Extracted RNA was reverse transcribed into cDNA using PrimeScriptTM II 1st Strand cDNA Synthesis Kit (TaKaRa, Beijing, China) according to the manufacturer's instructions. qRT-PCR was performed on the Applied Biosystems 7500 Real-Time PCR System (Thermo Fisher Scientific Inc.) using 2 × Q3 SYBR qPCR Master Mix (Universal). The qRT-PCR system was 20 µL, including 0.8 µL of forward and reverse primers (10 µM), 1 µL of cDNA (50 ng/µL), 10 µL of 2 × Q3 SYBR qPCR Master Mix, and 9.2 µL of ddH$_2$O. The relative expression was analyzed using the $2^{-\Delta\Delta C_T}$ method [57] with papain as the housekeeping gene. Specific primers of genes were listed in Table S7. Three biological replicates and three technical replicates were used in this study.

3. Results

3.1. A Set of Complete and Simplified NLRs in Papaya

To make the identification results appropriate to the analysis in this study, we modified the RGAugury pipeline [24]. A total of 59 NLRs were identified from the improved papaya genome [2], including 15 CNLs, 8 T(C)NLs, 5 R(C)NLs, and 31 partial (single or two domains) NLRs (Tables 1 and S2). The results showed papaya NLRs in this study than in previous studies [1,17,20,21], which should attribute to the recent optimization of identification methods and the substantial improvement in genome quality.

To further compare NLR genes in papaya with those in other angiosperms, we surveyed the NLR genes of nine angiosperms (*N. colorata*, *N. thermarum*, *E. ferox*, *V. vinifera*, *P. persica*, *C. sinensis*, *T. cacao*, *T. hassleriana*, and *A. thaliana*), which are gradually close to papaya in taxon phylogeny (Figure 1 and Table S1). Consistent with the previous report, papaya has fewer NLR genes than other angiosperms [1]. Furthermore, we compared the genome-wide proportion of three subclasses of NLR genes (TNL%, CNL%, and RNL%) in papaya with those in other species, and found that papaya has the lowest TNL% (papaya: 0.03%, other species: 0.03~0.33%), a relative low CNL% (papaya: 0.07%, other species: 0.06~0.56%) and a medium RNL% (papaya: 0.02%, other species: 0.00~0.03%) (Figure 1 and Table 1). The proportion of RNL was the lowest among the three subclasses of NLRs in all surveyed species (Figure 1). The proportion of CNL was usually higher than that of TNL in most species, except for one basal angiosperm (*E. ferox*) and two brassicales (*A. thaliana* and *T. hassleriana*) (Figure 1). These results are consistent with the previous study of 305 angiosperms, in which CNL is the predominant NLR subclass in 268 species, and TNL predominates 36 species [17]. Moreover, Liu et al. [17] also observed that RNL is a minority subclass, which accounts for no more than 10% of all NLRs.

Similar to most angiosperms, papaya retains members of all three NLR subclasses, and CNL predominates in three NLR subclasses (Figure 1). However, each NLR subclass of papaya is simplified, containing only very few genes but maintaining an acceptable genome-wide proportion (Figure 1 and Table 1). In summary, papaya contains a set of complete and simplified NLRs, which is unique among angiosperms and makes papaya a suitable model plant for studying basic disease-resistance genes.

Table 1. The number of NLRs identified in papaya and selected species.

Species	Total	NLR	NBS	CN	TN	RN	NL	TCN	RCN	CNL	TNL	RNL	TCNL	RCNL
Nymphaea colorata	28,438	255	63	17	12	2	57	0	1	53	34	3	9	4
Nymphaea thermarum	25,461	157	21	15	10	2	31	1	1	44	19	6	3	4
Euryale ferox	40,049	134	8	9	8	0	13	1	0	23	42	5	16	9
Vitis vinifera	31,845	439	52	41	8	2	126	0	1	166	21	4	7	11
Prunus persica	26,873	374	24	7	17	0	105	3	1	107	88	9	11	2
Citrus sinensis	24,456	112	7	12	2	0	25	0	1	53	10	1	0	1
Theobroma cacao	29,181	275	15	13	3	0	63	0	0	163	14	1	0	3
Carica papaya	**22,416**	59	10	3	1	0	16	1	0	15	6	4	2	1
Tarenaya hassleriana	27,396	76	7	1	4	0	15	0	0	18	26	2	3	0
Arabidopsis thaliana	27,628	166	4	1	14	1	25	3	1	36	68	4	9	0

3.2. Different Similarity Distribution of Three NLR Subclasses

To explore the difference in genetic diversity of NLRs in angiosperms, we conducted a meticulous evaluation of the pairwise similarity of amino acid sequences for three NLR subclasses (Figure 2). As aforementioned, we observed that papaya has significantly fewer CNLs and TNLs than other species. However, what is interesting is that the similarity distribution of CNLs in papaya is similar to other species, while the similarity distribution of TNLs in papaya is different to other species, which lacks a similarity distribution higher than 78% and lower than 38%. Furthermore, the median of similarity among CNLs of papaya (26.4%) is lower than that of other species (31.3~40%), while the median of similarity among TNLs of papaya (48.1%) is higher than that of other species (27.2~44.6%) (Figure 2 and Table S3). RNLs are rare in all species, and even absent in *C. sinensis* and *T. cacao*.

However, there are four RNLs in papaya, and their median value of similarity in papaya (26.15%) is much lower than that in other eudicots (50.3~70.9%) (Figure 2 and Table S3).

The lack of high similarity distribution of NLRs is consistent with the previous study of the papaya genome [1]. This was proposed to be a consequence of the lack of recent genome duplication in the papaya genome, which is atypical of other angiosperms; see the study of Ming et al. [1]. The absence of TNL gene pairs with low similarity in papaya, which is not common in angiosperms, suggests that the TNL-lost event had occurred in papaya. Additionally, the small size of RNLs and irregular distribution of RNL similarity indicated that the loss of RNL is a continuous process in angiosperm genomes.

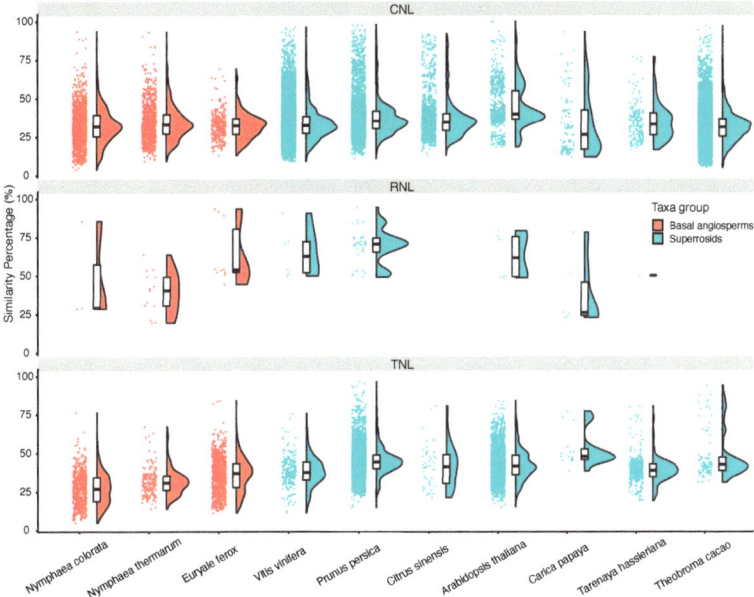

Figure 2. Similarity comparison of NLRs (CNL, RNL, TNL) in papaya and selected species.

3.3. Chromosome Distribution and Duplication Modes of NLRs in Papaya

Figure 3 exhibits locations of NLRs on nine chromosomes of papaya, with the least numbers of NLRs on Chr01 (one RNL) and Chr09 (one NL and one TNL). Chr02 contains five NLRs, Chr03 contains seven NLRs, Chr08 contains eight NLRs, while Chr04, Chr05, Chr06, and Chr07 contains nine NLRs, respectively (Figure 3). Most NLR genes are clustered at loci with relatively high gene density on chromosomes (Figure 3). Duplication gene pairs involving at least one NLR gene were extracted from the identification of genome-wide duplication. We identified 79 duplication gene pairs involving at least 1 NLR gene (51 DSD, 18 TD, 8 PD, 1 WGD, and 1 TRD) and observed that gene duplication events frequently occur at NLR-clustered loci (Figure 3 and Table S4). For example, five CNLs clustered on Chr06 (*sunset06G0024840*, *sunset06G0024850*, *sunset06G0024860*, *sunset06G0024870*, and *sunset06G0024880*) were identified as a segment of consecutive tandem duplication genes (Table S4). One TNL and two TCNLs clustered on Chr02 (*sunset02G0020440*, *sunset02G0020450*, and *sunset02G0020470*) were identified as proximal duplication genes (Table S4). The potential functional differentiation and expression divergence of these duplicated genes still remains unclear.

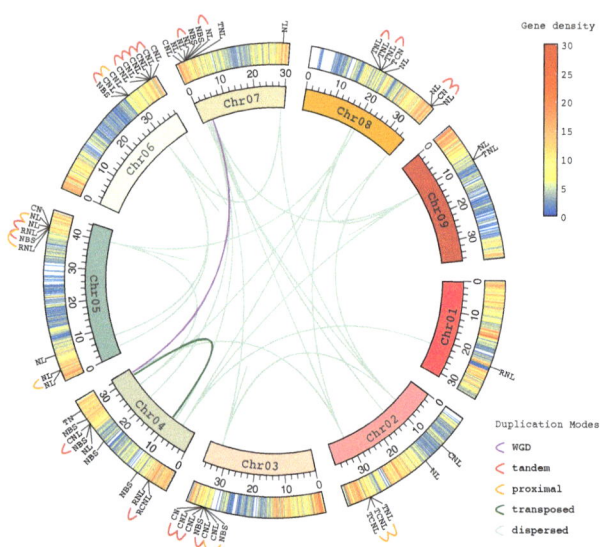

Figure 3. Physical locations and duplication modes of NLRs in papaya genome.

3.4. Conserved Sequences, Gene Structures, and Phylogeny of NLRs in Papaya

Three subclasses of full-length NLRs were found clustered in distinct clades of the tree after we studied the phylogeny of these genes in papaya and other selected species (Figure A1). To clearly show the phylogeny and structures of papaya NLRs, we constructed a maximum likelihood phylogenetic tree using 28 full-length papaya NLRs and 2 NBS genes of Rhodophyta as the outgroup (Figure 4A). The two NBS genes, *BWQ96_09803* from *Gracilariopsis chorda* and *CHC_T00003089001* from *Chondrus crispus*, were chosen as the outgroup of the papaya NLR tree because they are the closest to NLR genes among the outgroup genes in Figure A1. Meanwhile, we analyzed the conserved motifs, conserved domains, and gene structures for NLRs on the tree (Figure 4B–D). The phylogeny, conserved sequences, and gene structures showed that the RCNL and TCNL are highly related to RNL and TNL, respectively (Figure 4A–D). Additionally, we noticed that two RNL genes (*sunset05G0021430* and *sunset05G0021370*) fall inside the CNL clade (Figures 4A and A1). The classification of these two genes deserves more discussion because a segment of their N-terminal sequence could match both the CC and RPW8 domains simultaneously in the results of pfam_scan.

Overall, members of the same NLR subclass generally shared similar compositions of protein and gene structures, suggesting similar functions among them (Figure 4A–D). For conserved motifs, motif 1 was present in all NBS domains while motif 9 was present in all TIR domains on the tree (Figure 4B,C). The CC and LRR domains were more variable, with motif 6 typically present in CC domains and motif 5 regularly repeated in LRR domains (Figure 4B,C). The majority of the domains identified by RGAugury, which combines several databases and domain-specific identification tools [24], could be confirmed in the CDD database, with the exception of a small number of CC and PRW8 domains (Figure 4C). Conserved domain analysis showed that all TNLs in papaya contain the PPP1R42 (protein phosphatase 1 regulatory subunit 42) domain architecture, which influences centrosome separation, interacts with PP1 (protein phosphatase-1), and promotes its activity [58,59] (Figure 4C). Moreover, most papaya TNLs had the C-JID domain (C-terminal jelly roll/Ig-like), which has been demonstrated to be identified by pathogen effectors specifically [60] (Figure 4C). Gene structure analysis indicated that most NLRs clustered in the same clade displayed similar gene structures, except for a few CNLs with abnormally large intron insertions (Figure 4D).

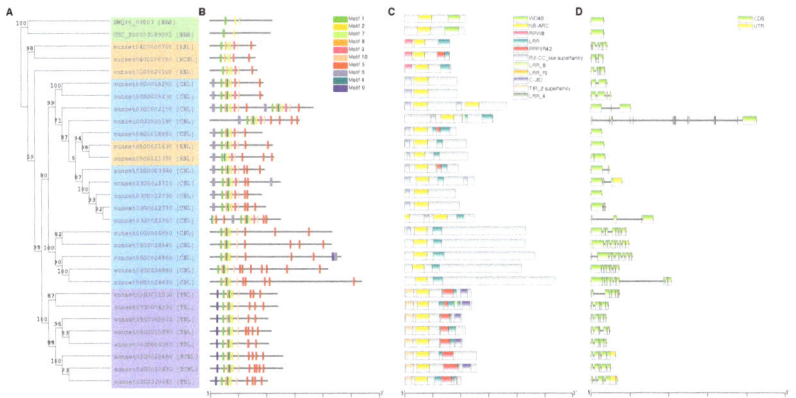

Figure 4. Phylogeny, conserved motifs, domains, and gene structures of papaya NLRs (CNL, RNL, TNL). (**A**) Phylogeny of papaya NLRs. NBS genes from Rhodophyta were used as the outgroup. The numbers at the nodes mean branch support values, which are evaluated by the Shimodaira–Hasegawa test. Green indicates the NBS subclass. Orange indicates the RNL subclass. Blue indicates the CNL subclass. Purple indicates the TNL subclass. (**B**) Conserved motifs of papaya NLRs. (**C**) Conserved domains of papaya NLRs. (**D**) Gene structures of papaya NLRs.

3.5. Inheritance- and Insertion-Origin of NLRs in Papaya Genome

To explore the evolutionary origin of papaya NLRs, local syntenic relationships of full-length NLR genes were investigated among selected angiosperms, covering basal angiosperms to eudicots (Table S5). By analysis, syntenic orthologs of RNL (*sunset01G0020100*) were discovered in genomes of *V. vinifera*, *P. persica*, *C. sinensis*, and *T. cacao*, demonstrating that this gene is conservative in dicots (Figure 5A). Additionally, papaya and basal angiosperms (*N. colorata* and *N. thermarum*) retained the syntenic relationship of RNL (*sunset04G0006760*) and RCNL (*sunset04G0006750*), suggesting that these two genes are derived from basal angiosperms (Figure 5B). Table S5 showed that papaya is the only eudicot that inherited both genes from basal angiosperms in this study. Moreover, we observed an inversion in the conserved syntenic region of Chr04: 5.01–5.14 Mb in papaya, which comprises the RNL and RCNL genes (Figure 5B). These findings revealed that papaya RNL genes are usually acquired via inheritance, which is the most typical origin of functional genes in plants.

In contrast to the dominance of inheritance in RNL genes, the local synteny analysis revealed that all TNL and most CNL genes are insertion-derived in papaya (Table S5). Compared to adjacent orthologous genes among selected species, a papaya TNL (*sunset07G0004990*) was found to be an insertion-derived gene located at Chr07: 3.78-3.99 Mb of papaya (Figure 6A). Additionally, a cluster of duplicated CNL genes (*sunset06G0024840*, *sunset06G0024850*, *sunset06G0024860*, *sunset06G0024870*, and *sunset06G0024880*) were found to be inserted into Chr06: 37.15–36.91 Mb of the papaya genome (Figure 6C). The upstream of these five CNLs on the papaya genome matched best with Chr06: 1.54–1.71 Mb region of *V. vinifera*, while the downstream matched best with Chr14: 4.72–5.08 Mb region of *V. vinifera* (Figure 6C). This syntenic relationship around the inserted loci was also kept in other dicots, demonstrating the confident insertion-origin of papaya-specific CNLs. Notably, the inserted loci were reshaped by a chromosome rearrangement during the chromosome evolution of papaya (Figure 6C). We calculated synonymous substitution rates (Ks) values of the five duplicated CNLs and estimated their divergent times based on a mutation rate of 12×10^{-9} base substitutions per site per generation [46] (Table 2). Since previous research estimated that Caricaceae diverge from its sister clade around 35.5–39.4 Mya and papaya diverge from its sister clade around 24.2–25.1 Mya [61,62], we deduced that *sunset06G0024860* diverge around the divergent time of Caricaceae, while clade I and clade II diverge around the divergent time of papaya (Figure 6B). Moreover, *sunset06G0024870* and *sunset06G0024880* were generated by the most recent duplication event (Figure 6B).

Figure 5. Inheritance-derived NLRs in papaya. (**A**) Conservation of a single RNL (*sunset01G0020100*). (**B**) Conservation of RNL (*sunset04G0006760*) and RCNL (*sunset04G0006750*).

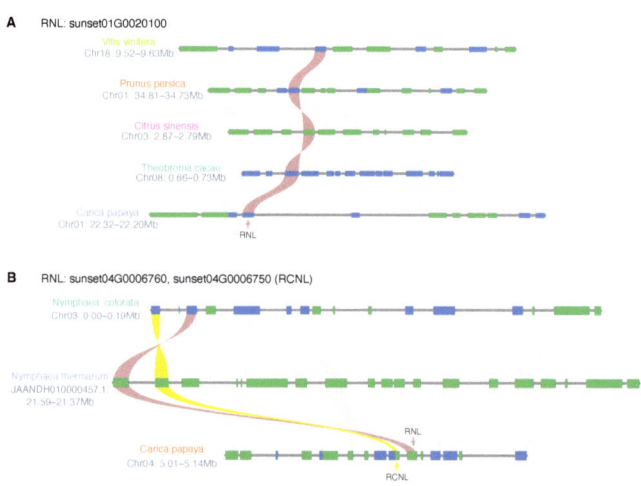
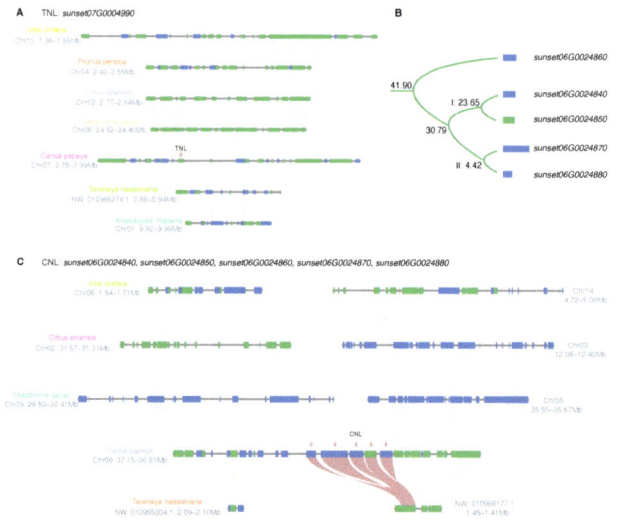

Figure 6. Insertion of NLRs in papaya. (**A**) Insertion of a single TNL (*sunset07G0004990*). (**B**) Evolution model of five duplicated CNLs. The numbers at the nodes indicated divergent times (Mya) estimated in Table 2. (**C**) Insertion of five CNLs in the papaya genome (*sunset06G0024840*, *sunset06G0024850*, *sunset06G0024860*, *sunset06G0024870*, *sunset06G0024880*).

Table 2. Duplication modes and divergent times of a five-CNL-cluster in papaya.

Duplicate 1	Type 1	Duplicate 2	Type 2	E-Value	Duplication Mode	Ks	Divergent Time (Mya)
sunset06G0024860	CNL	sunset06G0024870	CNL	8.56×10^{-82}	tandem	1.01	41.90
sunset06G0024840	CNL	sunset06G0024860	CNL	0	dispersed	0.93	38.81
sunset06G0024860	CNL	sunset06G0024880	CNL	2.62×10^{-82}	dispersed	0.87	36.19
sunset06G0024850	CNL	sunset06G0024860	CNL	3.64×10^{-84}	tandem	0.87	36.06
sunset06G0024850	CNL	sunset06G0024870	CNL	0	dispersed	0.74	30.79
sunset06G0024840	CNL	sunset06G0024850	CNL	2.30×10^{-83}	tandem	0.57	23.65
sunset06G0024870	CNL	sunset06G0024880	CNL	0	tandem	0.11	4.42

3.6. Expression of NLR Genes in Papaya under Biotic and Abiotic Stresses

We investigated the expression levels of NLR genes under biotic and abiotic stresses to mine their functions. A total of 52 papaya samples' RNA-seq data were collected from 5 NCBI projects, including treatments of PRSV, papaya mosaic virus (PAPMV), papaya leaf-distortion mosaic virus (PLDMV), fungi (*C. brevisporum*), ethylene, drought, tissues of leaf, root, and sap (Figure A2 and Table S6). We found that more NLR genes are up-regulated by biotic stresses and more NLR genes are down-regulated by abiotic stresses (Figure 7). This was consistent with the recognition that NLRs are involved in the process of pathogen resistance [63,64]. Additionally, more NLRs responded to fungal infection than to PRSV, PAPMV, and PLDMV infection (Figure 7). For fungal infection, more NLRs responded to stress in resistant papaya than that in susceptible papaya (Figure 7). The papaya-specific inserted TNL (*sunset07G0004990*) was strongly up-regulated by the infection of fungi and viruses, and especially highly expressed in fungi-infected papaya (Figures 7 and A2).

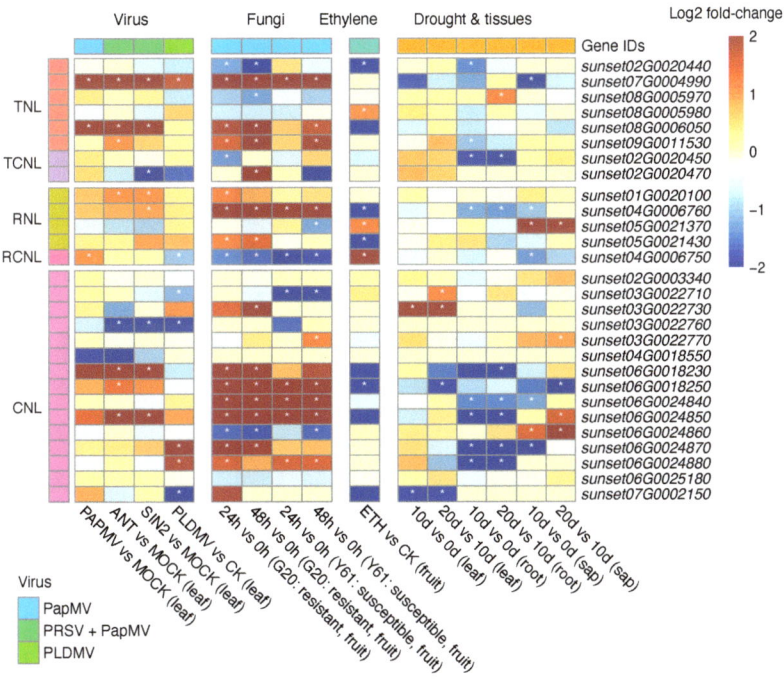

Figure 7. Log2 fold-change expression of NLRs in papaya under biotic and abiotic stresses. DEGs are marked by white starts with Log2 fold-change > 1 and P_{FDR} < 0.05. ANT indicates 'Antagonism' samples, which are inoculated with simultaneous PapMV and PRSV [47]. SIN2 indicates 'Sinergism2' samples, which are inoculated with stepwise PapMV and PRSV [47].

As shown in Figure 5B, the tandem duplication genes RNL(*sunset04G0006760*) and RCNL (*sunset04G0006750*) were both conserved from basal angiosperms. However, the RNL was up-regulated, whereas the RCNL was down-regulated by fungal infection (Figure 7). Meanwhile, the RNL is highly expressed in virus-infected papaya samples, and the RCNL is highly expressed in papaya sap (Figure A2). Among five CNL genes of the papaya-specific CNL cluster, which derived from tandem duplication and insertion, four (*sunset06G0024840, sunset06G0024850, sunset06G0024870,* and *sunset06G0024880*) were up-regulated by fungal infection to varying degrees, whereas one (*sunset06G0024860*) was down-regulated (Figure 7). These findings showed the dosage effect and expression divergence of duplicated genes, both of which contribute to the functional diversity of disease-resistance genes in papaya.

The expression levels of duplication genes discussed above were analyzed in PRSV-infected papaya by qRT-PCR technology. The results showed that the relative expression level of RNL (*sunset04G0006760*) is strongly up-regulated by PRSV infection, whereas the relative expression levels of RCNL (*sunset04G0006750*) are not significantly different between control and PRSV infected papaya (Figure 8A). Among five CNL genes of the papaya-specific CNL cluster, the relative expression levels of three (*sunset06G0024840, sunset06G0024870,* and *sunset06G0024880*) in PRSV infected papaya were significantly higher than in control (Figure 8B). The relative expression level of *sunset06G0024850* in PRSV infected papaya is higher than in the control, but not significantly (Figure 8B). Moreover, the relative expression levels of *sunset06G0024860* were not significantly different between control and treatment (Figure 8B). Although these results are slightly inconsistent with those of transcriptome analysis of virus-infected papaya, they also showed the dosage effect and expression divergence of duplicated genes as above. The 'Antagonism' and 'Sinergism2' samples of PRJNA560275 were papaya inoculated with simultaneous or step-wise PapMV and PRSV, and samples were collected from systemic leaves at 60 dpi (day post infection) [47]. In this study, we inoculated papaya seedlings with single PRSV and collected samples at 7 dpi. Therefore, both the virus inoculation and sampling time point contributed to the inconsistency.

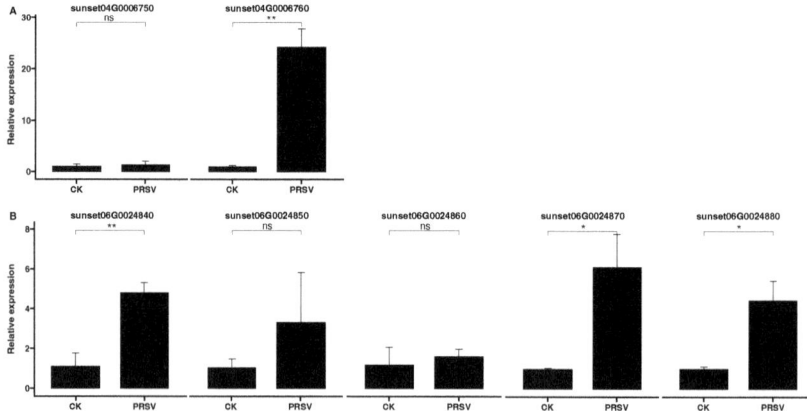

Figure 8. Real-time quantitative PCR analysis of NLR genes in PRSV infected papaya. (**A**) Relative expression levels of RNL duplication gene pairs. (**B**) Relative expression levels of CNL duplication gene pairs. Error bars indicate the standard deviation. Statistical significance is evaluated by T-test and represented by symbols (ns: $p > 0.05$, *: $p <= 0.05$, and **: $p <= 0.01$).

3.7. Co-Expression Network Analysis of NLRs in Papaya Infected by Fungi

To investigate the potential interactions involving NLR genes, WGCNA was performed based on transcription data from papaya samples infected by fungi. As demonstrated in Figure 9A, the gene set in the lightcyan module was significantly correlated with the papaya anthracnose resistant/susceptible traits ($r = 0.88$). Gene set in the midnightblue module showed a strong positive correlation ($r = 0.8$) with fungal infection or not, while the brown module showed a strong positive correlation ($r = 0.9$) with infection time (0/24/48 h). In addition, it is also worth noting that the turquoise module is strongly negatively correlated with fungal infection ($r = -0.89$) and infection time ($r = -0.88$). We found 11 NLRs involved in these 8 modules, including 3 (NBS: *sunset03G0022700*, CNL: *sunset03G0022710*, and RCNL: *sunset04G0006750*) in black, 3 (RNL: *sunset04G0006760*, CNL: *sunset06G0018250*, and CNL: *sunset06G0024840*) in midnightblue, 4 (CNL: *sunset06G0024850*, CNL: *sunset06G0024860*, CNL: *sunset06G0025180*, and TNL: *sunset08G0006050*) in turquoise, and 1 (TNL: *sunset07G0004990*) in brown. The gene and eigengene expressions of turquoise, brown, and midnightblue modules are presented in Figure 9B–D. Eigengenes of the brown and midnightblue module were up-regulated by fungal infection (Figure 9C,D), while that of the turquoise module was down-regulated (Figure 9B). The hub-network of the midnightblue module was constructed, and two CNLs (*sunset06G0018250* and *sunset06G0024840*) and one RNL (*sunset04G0006760*) were found to be core hub-genes for fungal infection (Figure 9E and Table S8). The hub-network and hub-genes of the co-expression genes would be beneficial to improve the fungal resistance for papaya breeding.

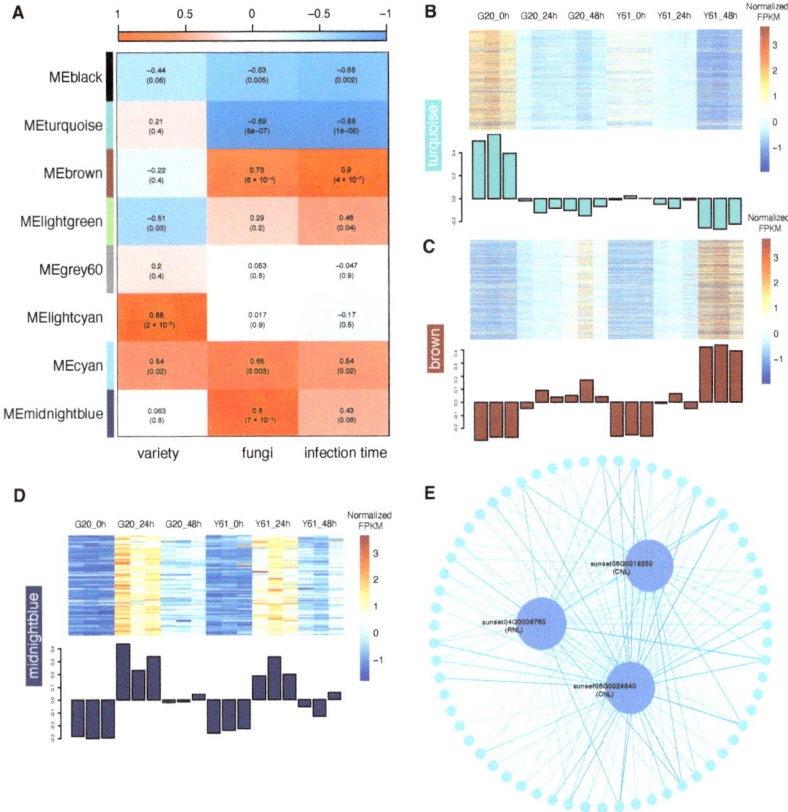

Figure 9. Gene co-expression network analysis of NLRs in papaya. (**A**) Correlation of gene modules and traits. (**B–D**) Eigengene and total gene expression of turquoise, brown, and midnightblue modules. (**E**) Gene co-expression network of three NLRs in the midnightblue module.

4. Discussion

In this study, we have identified and classified NLR family members more accurately than the previous research on papaya NLRs [20] due to the increasing awareness of crucial domain architectures in NLR and the development of NLR identification tools. In addition, excellent tools (e.g. NLR-Annotator [65], NLGenomeSweeper [66], and Homology-based R-gene Prediction (HRP) [11]) have also been recently developed to facilitate the NLR identification. For example, NLGenomeSweeper complements R genes that have not been annotated in the genome [66], and HRP has a high performance in full-length R-gene discovery [11]. We believe that these tools will help us further improve the accuracy of gene recognition and the quality of subsequent bioinformatic mining.

Based on the distribution of sequence similarity, we propose that both the loss of NLRs and the lack of recent genome duplication lead to the smaller NLR family of papaya compared to typical angiosperms. The similar subclasses composition and genome-wide proportion of papaya and typical eudicots revealed that the papaya NLR family is a simplified set of NLRs in typical eudicots. The study of papaya NLRs will help us better understand the complex and diverse disease-resistance genes in eudicots, which comprise the largest species diversity of living angiosperms [67,68].

Gene duplication is an important mechanism driving gene diversification and genetic innovation, and its roles in species evolution and adaptation have been well recognized [69,70]. Genes involved in signaling, transport, and metabolism tend to retain copies after duplication more than those involved in the maintenance of genome stability and organelle function [69]. As we know, NLR is a large and diverse family that activates ETI in response to pathogen effectors and mediates immune signaling subsequently [14,64]. In this study, we found that DSD, TD, and PD are the main duplication modes in papaya that contribute to the expansion of the NLR family. DSD is prevalent among plant genomes, which generate gene copies with unclear mechanisms, neither colinear nor neighboring [31]. For the remaining duplication modes with known mechanisms, TD and PD dominated papaya NLRs, while WGD and TRD contributed little. Among 50 large gene families investigated in *A. thaliana*, the NLR family was classified into the family group containing moderate TD and low WGD [71]. In most angiosperms, WGD accompanied by massive TRD usually contributed more than TD and PD [31,72]. Despite the absence of recent genome duplication in papaya, gene pairs produced from WGD (1698 pairs) and TRD (4374 pairs) were significantly more numerous than those from TD (1192 pairs) and PD (696 pairs). This implies that various duplication modes have contributed differently to the papaya NLR gene set than the papaya genome. In addition, the research on plant gene duplication found that more than 63% of gene pairs derived from duplication events showed expression divergence [31]. The expression divergence of duplication pairs was well confirmed in three TD-derived NLR pairs of RNL (*sunset04G0006760*) and RCNL (*sunset04G0006750*), CNL (*sunset06G0024860*) and CNL (*sunset06G0024850*), and CNL (*sunset06G0024860*) and CNL (*sunset06G0024870*).

Natural evolution involved the fission and fusion of chromosomes, which resulted in genome rearrangement and assembly [73]. Gene deletion and insertion occurred frequently in recombination regions [74]. The papaya genome suffered chromosome fission and fusion at the region of Chr06: 37.15–36.91 Mb, because this region matched with two collinear segments on *V. vinifera*, *C. sinensis*, *T. cacao*, and *T. hassleriana* genomes. A cluster of tandem duplicated CNLs (*sunset06G0024840*, *sunset06G0024850*, *sunset06G0024860*, *sunset06G0024870*, and *sunset06G0024880*) were discovered to be inserted into the two segments of this recombination region. Three genes in this cluster were involved in two co-expression modules, which correlated strongly to the fungal infection. Dosage effect and expression divergence were both observed in this tandem duplication CNL-cluster.

Plant NLRs are typically classified into two functional roles in pathogen-induced immune responses: sensor NLRs that recognize pathogen effectors and helper NLRs that assist other NLRs in triggering the immune response [64,75]. To catch up with the evolution of pathogens, it is more urgent for sensor NLRs than for helper NLRs to accelerate gene diversity and genetic novelty. RNL is an ancient and conserved subclass of plant NLRs and is preferred as a helper [16,75]. We speculate that the simplified NLR family maintains the essential ETI in papaya by preserving a few conserved RNLs and CNLs as 'helpers' and inserting relatively abundant TNLs and CNLs for recognizing variable pathogen effectors. In general, papaya lacks germplasm, which is disease resistant [2]. As a result, it has evolved into other resistant pathways, such as the excretion of latex-containing proteins and enzymes that have demonstrated toxicity [76].

5. Conclusions

In this study, 59 NLR genes were identified from the improved papaya genome, including 15 CNLs, 8 T(C)NLs, 5 R(C)NLs, and 31 partial NLRs. Compared to typical angiosperms, papaya retains complete and simplified subclasses of NLR genes, making papaya a suitable plant model for studying basic disease-resistance genes. The absence of NLR gene pairs with low similarity indicates the loss of ancestral NLR genes. Moreover, members of the same subclass are close on the phylogenetic tree, and they have similar conserved sequences, gene structures, and functional recognition sites, suggesting their similar biological functions. Pairwise synteny analysis revealed the inheritance- and insertion-derived NLR genes. One tandem duplication gene pair (RNL *sunset04G0006760* and RCNL *sunset04G0006750*) is found to be inheritance-derived. Meanwhile, we observed an inserted gene cluster composed of five duplicated CNL genes. The insertion loci of the CNL cluster are found to have been reshaped by a chromosome rearrangement during the chromosome evolution of papaya. We further estimated their divergent time and established an evolutionary model. Expression and transcriptomic network analysis revealed that papaya NLR genes actively respond to biotic stress. Both the public RNA-seq data and qRT-PCR experiment confirmed the dosage effect and functional differentiation of the papaya NLR genes. This study gives new perspectives on the evolution of NLR genes and provides a basis for the disease-resistant breeding of papaya.

Supplementary Materials: The following supporting information can be downloaded at: https://www.mdpi.com/article/10.3390/agronomy13040970/s1, Table S1: Data source of plant genomes; Table S2: NLRs identified in papaya genome; Table S3: Summary of pairwise similarity (%) of NLRs; Table S4: Duplication modes of NLRs in papaya; Table S5: Syntenic paralogs of NLRs in papaya; Table S6: Data sources of transcriptome; Table S7: Gene-specific primers for qRT-PCR; Table S8: Network edges of midnightblue module.

Author Contributions: Conceptualization, A.G., C.J. and Q.J.; methodology, A.X.; software, Q.J. and Y.W.; formal analysis, Q.J., Y.W. and H.Z.; investigation, A.X., M.L. and H.A.; resources, R.J. and M.L.; data curation, Q.J., Y.W. and C.J.; writing—original draft preparation, Q.J.; writing—review and editing, A.G., C.J. and Q.J.; supervision, A.G. and H.A.; funding acquisition, Q.J. and C.J. All authors have read and agreed to the published version of the manuscript.

Funding: This work was supported by Hainan Provincial Natural Science Foundation of China (No. 321QN289), Hainan Major Science and Technology Project (No. ZDKJ202002), Central Public-interest Scientific Institution Basal Research Fund for Chinese Academy of Tropical Agricultural Sciences (No. 1630052020007 and No. 1630052021017) the innovation platform for Academiniculs of Hainan Province and Hainan Postdoctoral Research Funding Project.

Institutional Review Board Statement: Not applicable.

Informed Consent Statement: Not applicable.

Data Availability Statement: Transcriptional data were downloaded from NCBI. Sunset genome and annotation files were downloaded from Genome Warehouse of China National Center for Bioinformation. Other genome and annotation files were downloaded from public databases listed in Table S1. Scripts were available in a git repository at: https://github.com/reductase4/papaya_NLRs (accessed on 31 January 2023).

Conflicts of Interest: The authors declare no conflict of interest. The funding body had no role in the design of the study, collection, analysis, interpretation of data, and writing the manuscript.

Abbreviations

The following abbreviations are used in this manuscript:

NLR	Nucleotide-binding site (NBS) leucine-rich repeat (LRR) receptor
ETI	Effector-triggered immunity
TIR	Toll/Interleukin-1 receptor
RPW8	Resistance to Powdery Mildew Locus 8
CC	Coiled-coil
Mya	Millions of years ago
Ks	Synonymous substitution rates
PRSV	Papaya ringspot virus
PAPMV	Papaya mosaic virus
PLDMV	Papaya leaf-distortion mosaic virus
WGCNA	Weighted gene co-expression network analysis
WGD	Whole-genome duplication
TD	Tandem duplication
PD	Proximal duplication
TRD	Transposed duplication
DSD	Dispersed duplication
PCA	Principal Component Analysis
FPKM	Fragments Per kb per Million reads
DEG	Differentially expressed genes

Appendix A

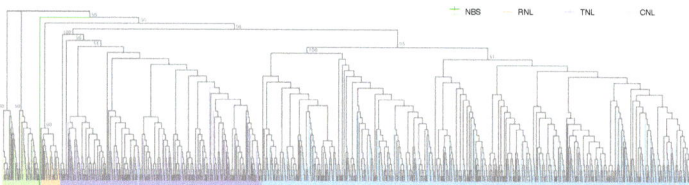

Figure A1. Phylogeny of NBS domains of NLRs. NLRs of three subclasses (RNL, TNL, CNL) of papaya and nine selected species were used to construct the NBS domain tree, in which NBS genes from Rhodophyta were used as the outgroup. The dark green indicates the two NBS genes used as the outgroup in Figure 4A. The numbers at the nodes mean branch support values, which are evaluated by the Shimodaira–Hasegawa test.

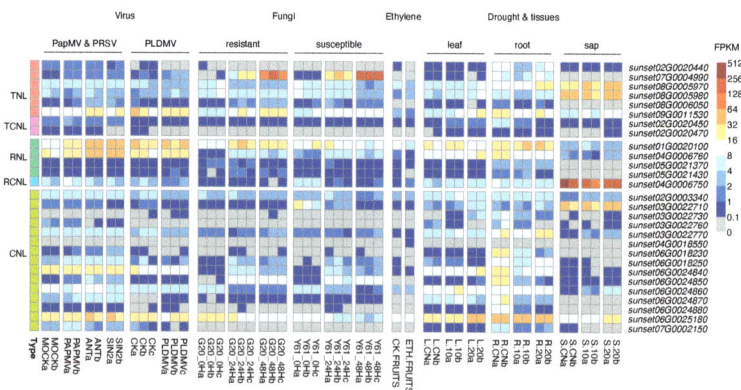

Figure A2. FPKM values of NLRs in papaya under biotic and abiotic stresses. ANT indicates 'Antagonism' samples, which are inoculated with simultaneous PapMV and PRSV [47]. SIN2 indicates 'Sinergism2' samples, which are inoculated with stepwise PapMV and PRSV [47].

References

1. Ming, R.; Hou, S.; Feng, Y.; Yu, Q.; Dionne-Laporte, A.; Saw, J.H.; Senin, P.; Wang, W.; Ly, B.V.; Lewis, K.L.T.; et al. The draft genome of the transgenic tropical fruit tree papaya (*Carica papaya* Linnaeus). *Nature* **2008**, *452*, 991–996. [CrossRef]
2. Yue, J.; VanBuren, R.; Liu, J.; Fang, J.; Zhang, X.; Liao, Z.; Wai, C.M.; Xu, X.; Chen, S.; Zhang, S.; et al. SunUp and Sunset genomes revealed impact of particle bombardment mediated transformation and domestication history in papaya. *Nat. Genet.* **2022**, *54*, 715–724. [CrossRef]
3. Shen, Y.H.; Lu, B.G.; Feng, L.; Yang, F.Y.; Geng, J.J.; Ming, R.; Chen, X.J. Isolation of ripening-related genes from ethylene/1-MCP treated papaya through RNA-seq. *BMC Genom.* **2017**, *18*, 671. [CrossRef]
4. Chan-León, A.C.; Estrella-Maldonado, H.; Dubé, P.; Ortiz, G.F.; Espadas-Gil, F.; May, C.T.; Prado, J.R.; Desjardins, Y.; Santamaría, J.M. The high content of β-carotene present in orange-pulp fruits of *Carica papaya* L. is not correlated with a high expression of the CpLCY-β2 gene. *Food Res. Int.* **2017**, *100*, 45–56. [CrossRef]
5. Liu, J.; Sharma, A.; Niewiara, M.J.; Singh, R.; Ming, R.; Yu, Q. Papain-like cysteine proteases in *Carica papaya*: Lineage-specific gene duplication and expansion. *BMC Genom.* **2018**, *19*, 26. [CrossRef] [PubMed]
6. Liu, Y.; Huang, C.C.; Wang, Y.; Xu, J.; Wang, G.; Bai, X. Biological evaluations of decellularized extracellular matrix collagen microparticles prepared based on plant enzymes and aqueous two-phase method. *Regen. Biomater.* **2021**, *8*, rbab002. [CrossRef]
7. Hafid, K.; John, J.; Sayah, T.M.; Domínguez, R.; Becila, S.; Lamri, M.; Dib, A.L.; Lorenzo, J.M.; Gagaoua, M. One-step recovery of latex papain from *Carica Papaya* using three phase partitioning and its use as milk-clotting and meat-tenderizing agent. *Int. J. Biol. Macromol.* **2020**, *146*, 798–810. [CrossRef]
8. Lobato-Gómez, M.; Hewitt, S.; Capell, T.; Christou, P.; Dhingra, A.; Girón-Calva, P.S. Transgenic and genome-edited fruits: Background, constraints, benefits, and commercial opportunities. *Hortic. Res.* **2021**, *8*, 166. [CrossRef] [PubMed]
9. Wu, Z.; Mo, C.; Zhang, S.; Li, H. Characterization of papaya ringspot virus isolates infecting transgenic papaya 'Huanong No. 1' in South China. *Sci. Rep.* **2018**, *8*, 8206. [CrossRef] [PubMed]
10. Wang, W.; Feng, B.; Zhou, J.M.; Tang, D. Plant immune signaling: Advancing on two frontiers. *J. Integr. Plant Biol.* **2020**, *62*, 2–24. [CrossRef]
11. Andolfo, G.; Dohm, J.C.; Himmelbauer, H. Prediction of NB-LRR resistance genes based on full-length sequence homology. *Plant J. Cell Mol. Biol.* **2022**, *110*, 1592–1602. [CrossRef]
12. Jones, J.D.G.; Dangl, J.L. The plant immune system. *Nature* **2006**, *444*, 323–329. [CrossRef]
13. Dangl, J.L.; Horvath, D.M.; Staskawicz, B.J. Pivoting the plant immune system from dissection to deployment. *Science* **2013**, *341*, 746–751. [CrossRef] [PubMed]
14. Lolle, S.; Stevens, D.; Coaker, G. Plant NLR-triggered immunity: From receptor activation to downstream signaling. *Curr. Opin. Immunol.* **2020**, *62*, 99–105. [CrossRef] [PubMed]
15. Delaux, P.M.; Schornack, S. Plant evolution driven by interactions with symbiotic and pathogenic microbes. *Science* **2021**, *371*, eaba6605. [CrossRef]
16. Shao, Z.Q.; Xue, J.Y.; Wang, Q.; Wang, B.; Chen, J.Q. Revisiting the Origin of Plant NBS-LRR Genes. *Trends Plant Sci.* **2019**, *24*, 9–12. [CrossRef]
17. Liu, Y.; Zeng, Z.; Zhang, Y.M.; Li, Q.; Jiang, X.M.; Jiang, Z.; Tang, J.H.; Chen, D.; Wang, Q.; Chen, J.Q.; et al. An angiosperm NLR Atlas reveals that NLR gene reduction is associated with ecological specialization and signal transduction component deletion. *Mol. Plant* **2021**, *14*, 2015–2031. [CrossRef] [PubMed]
18. Shao, Z.Q.; Xue, J.Y.; Wu, P.; Zhang, Y.M.; Wu, Y.; Hang, Y.Y.; Wang, B.; Chen, J.Q. Large-Scale Analyses of Angiosperm Nucleotide-Binding Site-Leucine-Rich Repeat Genes Reveal Three Anciently Diverged Classes with Distinct Evolutionary Patterns. *Plant Physiol.* **2016**, *170*, 2095–2109. [CrossRef]
19. Li, Q.; Jiang, X.M.; Shao, Z.Q. Genome-Wide Analysis of NLR Disease Resistance Genes in an Updated Reference Genome of Barley. *Front. Genet.* **2021**, *12*, 694682. [CrossRef]
20. Porter, B.W.; Paidi, M.; Ming, R.; Alam, M.; Nishijima, W.T.; Zhu, Y.J. Genome-wide analysis of *Carica papaya* reveals a small NBS resistance gene family. *Mol. Genet. Genom. MGG* **2009**, *281*, 609–626. [CrossRef]
21. Ngou, B.P.M.; Heal, R.; Wyler, M.; Schmid, M.W.; Jones, J.D.G. Concerted expansion and contraction of immune receptor gene repertoires in plant genomes. *Nat. Plants* **2022**, *8*, 1146–1152. [CrossRef]
22. Yang, M.; Zhou, C.; Yang, H.; Kuang, R.; Huang, B.; Wei, Y. Identification of WRKY transcription factor genes in papaya and response of their expresion to *Colletotrichum brevisporum* infection. *J. Northwest A F Univ. (Nat. Sci. Ed.)* **2022**, *50*, 127–138+154. [CrossRef]
23. Yang, M.; Zhou, C.; Yang, H.; Kuang, R.; Liu, K.; Huang, B.; Wei, Y. Comparative transcriptomics and genomic analyses reveal differential gene expression related to *Colletotrichum brevisporum* resistance in papaya (*Carica papaya* L.). *Front. Plant Sci.* **2022**, *13*, 1038598. [CrossRef] [PubMed]
24. Li, P.; Quan, X.; Jia, G.; Xiao, J.; Cloutier, S.; You, F.M. RGAugury: A pipeline for genome-wide prediction of resistance gene analogs (RGAs) in plants. *BMC Genom.* **2016**, *17*, 852. [CrossRef]
25. Mistry, J.; Chuguransky, S.; Williams, L.; Qureshi, M.; Salazar, G.A.; Sonnhammer, E.L.L.; Tosatto, S.C.E.; Paladin, L.; Raj, S.; Richardson, L.J.; et al. Pfam: The protein families database in 2021. *Nucleic Acids Res.* **2021**, *49*, D412–D419. [CrossRef]
26. Jones, P.; Binns, D.; Chang, H.Y.; Fraser, M.; Li, W.; McAnulla, C.; McWilliam, H.; Maslen, J.; Mitchell, A.; Nuka, G.; et al. InterProScan 5: Genome-scale protein function classification. *Bioinformatics* **2014**, *30*, 1236–1240. [CrossRef] [PubMed]

27. Lupas, A.; Van Dyke, M.; Stock, J. Predicting coiled coils from protein sequences. *Science* **1991**, *252*, 1162–1164. [CrossRef] [PubMed]
28. Madeira, F.; Park, Y.M.; Lee, J.; Buso, N.; Gur, T.; Madhusoodanan, N.; Basutkar, P.; Tivey, A.R.N.; Potter, S.C.; Finn, R.D.; et al. The EMBL-EBI search and sequence analysis tools APIs in 2019. *Nucleic Acids Res.* **2019**, *47*, W636–W641. [CrossRef]
29. Kumar, S.; Suleski, M.; Craig, J.M.; Kasprowicz, A.E.; Sanderford, M.; Li, M.; Stecher, G.; Hedges, S.B. TimeTree 5: An Expanded Resource for Species Divergence Times. *Mol. Biol. Evol.* **2022**, *39*, msac174. [CrossRef]
30. Zhang, L.; Chen, F.; Zhang, X.; Li, Z.; Zhao, Y.; Lohaus, R.; Chang, X.; Dong, W.; Ho, S.Y.W.; Liu, X.; et al. The water lily genome and the early evolution of flowering plants. *Nature* **2020**, *577*, 79–84. [CrossRef]
31. Qiao, X.; Li, Q.; Yin, H.; Qi, K.; Li, L.; Wang, R.; Zhang, S.; Paterson, A.H. Gene duplication and evolution in recurring polyploidization-diploidization cycles in plants. *Genome Biol.* **2019**, *20*, 38. [CrossRef]
32. Chen, C.; Chen, H.; Zhang, Y.; Thomas, H.R.; Frank, M.H.; He, Y.; Xia, R. TBtools: An Integrative Toolkit Developed for Interactive Analyses of Big Biological Data. *Mol. Plant* **2020**, *13*, 1194–1202. [CrossRef] [PubMed]
33. Gao, Y.; Wang, W.; Zhang, T.; Gong, Z.; Zhao, H.; Han, G.Z. Out of Water: The Origin and Early Diversification of Plant R-Genes. *Plant Physiol.* **2018**, *177*, 82–89. [CrossRef]
34. Nguyen, N.P.D.; Mirarab, S.; Kumar, K.; Warnow, T. Ultra-large alignments using phylogeny-aware profiles. *Genome Biol.* **2015**, *16*, 124. [CrossRef]
35. Capella-Gutiérrez, S.; Silla-Martínez, J.M.; Gabaldón, T. trimAl: A tool for automated alignment trimming in large-scale phylogenetic analyses. *Bioinformatics* **2009**, *25*, 1972–1973. [CrossRef]
36. Price, M.N.; Dehal, P.S.; Arkin, A.P. FastTree: Computing large minimum evolution trees with profiles instead of a distance matrix. *Mol. Biol. Evol.* **2009**, *26*, 1641–1650. [CrossRef] [PubMed]
37. Price, M.N.; Dehal, P.S.; Arkin, A.P. FastTree 2–approximately maximum-likelihood trees for large alignments. *PloS ONE* **2010**, *5*, e9490. [CrossRef] [PubMed]
38. Shimodaira, H.; Hasegawa, M. Multiple Comparisons of Log-Likelihoods with Applications to Phylogenetic Inference. *Mol. Biol. Evol.* **1999**, *16*, 1114. [CrossRef]
39. Bailey, T.L.; Johnson, J.; Grant, C.E.; Noble, W.S. The MEME Suite. *Nucleic Acids Res.* **2015**, *43*, W39–W49. [CrossRef]
40. Lu, S.; Wang, J.; Chitsaz, F.; Derbyshire, M.K.; Geer, R.C.; Gonzales, N.R.; Gwadz, M.; Hurwitz, D.I.; Marchler, G.H.; Song, J.S.; et al. CDD/SPARCLE: The conserved domain database in 2020. *Nucleic Acids Res.* **2020**, *48*, D265–D268. [CrossRef]
41. Zhang, Z.; Xiao, J.; Wu, J.; Zhang, H.; Liu, G.; Wang, X.; Dai, L. ParaAT: A parallel tool for constructing multiple protein-coding DNA alignments. *Biochem. Biophys. Res. Commun.* **2012**, *419*, 779–781. [CrossRef] [PubMed]
42. Wang, D.; Zhang, Y.; Zhang, Z.; Zhu, J.; Yu, J. KaKs_Calculator 2.0: A toolkit incorporating gamma-series methods and sliding window strategies. *Genom. Proteom. Bioinform.* **2010**, *8*, 77–80. [CrossRef]
43. Yang, Z.; Nielsen, R. Estimating synonymous and nonsynonymous substitution rates under realistic evolutionary models. *Mol. Biol. Evol.* **2000**, *17*, 32–43. [CrossRef]
44. Nystedt, B.; Street, N.R.; Wetterbom, A.; Zuccolo, A.; Lin, Y.C.; Scofield, D.G.; Vezzi, F.; Delhomme, N.; Giacomello, S.; Alexeyenko, A.; et al. The Norway spruce genome sequence and conifer genome evolution. *Nature* **2013**, *497*, 579–584. [CrossRef] [PubMed]
45. Liu, Y.; Wang, S.; Li, L.; Yang, T.; Dong, S.; Wei, T.; Wu, S.; Liu, Y.; Gong, Y.; Feng, X.; et al. The Cycas genome and the early evolution of seed plants. *Nat. Plants* **2022**, *8*, 389–401. [CrossRef]
46. Belfield, E.J.; Brown, C.; Ding, Z.J.; Chapman, L.; Luo, M.; Hinde, E.; Van Es, S.W.; Johnson, S.; Ning, Y.; Zheng, S.J.; et al. Thermal stress accelerates *Arabidopsis thaliana* mutation rate. *Genome Res.* **2021**, *31*, 40–50. [CrossRef]
47. Vargas-Mejía, P.; Vega-Arreguín, J.; Chávez-Calvillo, G.; Ibarra-Laclette, E.; Silva-Rosales, L. Differential Accumulation of Innate- and Adaptive-Immune-Response-Derived Transcripts during Antagonism between Papaya Ringspot Virus and Papaya Mosaic Virus. *Viruses* **2020**, *12*, 230. [CrossRef]
48. Gamboa-Tuz, S.D.; Pereira-Santana, A.; Zamora-Briseño, J.A.; Castano, E.; Espadas-Gil, F.; Ayala-Sumuano, J.T.; Keb-Llanes, M.Á.; Sanchez-Teyer, F.; Rodríguez-Zapata, L.C. Transcriptomics and co-expression networks reveal tissue-specific responses and regulatory hubs under mild and severe drought in papaya (*Carica papaya* L.). *Sci. Rep.* **2018**, *8*, 14539. [CrossRef] [PubMed]
49. Chen, S.; Zhou, Y.; Chen, Y.; Gu, J. fastp: An ultra-fast all-in-one FASTQ preprocessor. *Bioinformatics* **2018**, *34*, i884–i890. [CrossRef] [PubMed]
50. Pertea, M.; Kim, D.; Pertea, G.M.; Leek, J.T.; Salzberg, S.L. Transcript-level expression analysis of RNA-seq experiments with HISAT, StringTie and Ballgown. *Nat. Protoc.* **2016**, *11*, 1650–1667. [CrossRef] [PubMed]
51. Anders, S.; McCarthy, D.J.; Chen, Y.; Okoniewski, M.; Smyth, G.K.; Huber, W.; Robinson, M.D. Count-based differential expression analysis of RNA sequencing data using R and Bioconductor. *Nat. Protoc.* **2013**, *8*, 1765–1786. [CrossRef]
52. Liu, S.; Wang, Z.; Zhu, R.; Wang, F.; Cheng, Y.; Liu, Y. Three Differential Expression Analysis Methods for RNA Sequencing: Limma, EdgeR, DESeq2. *J. Vis. Exp.* **2021**, *175*, e62528. [CrossRef]
53. Zhang, B.; Horvath, S. A general framework for weighted gene co-expression network analysis. *Stat. Appl. Genet. Mol. Biol.* **2005**, *4*, 17. [CrossRef] [PubMed]
54. Langfelder, P.; Horvath, S. WGCNA: An R package for weighted correlation network analysis. *BMC Bioinform.* **2008**, *9*, 559. [CrossRef] [PubMed]
55. Langfelder, P.; Luo, R.; Oldham, M.C.; Horvath, S. Is my network module preserved and reproducible? *PLoS Comput. Biol.* **2011**, *7*, e1001057. [CrossRef] [PubMed]

56. Shannon, P.; Markiel, A.; Ozier, O.; Baliga, N.S.; Wang, J.T.; Ramage, D.; Amin, N.; Schwikowski, B.; Ideker, T. Cytoscape: A software environment for integrated models of biomolecular interaction networks. *Genome Res.* **2003**, *13*, 2498–2504. [CrossRef]
57. Livak, K.J.; Schmittgen, T.D. Analysis of relative gene expression data using real-time quantitative PCR and the $2^{-\Delta\Delta C_T}$ Method. *Methods* **2001**, *25*, 402–408. [CrossRef]
58. Wang, R.; Sperry, A.O. Identification of a novel Leucine-rich repeat protein and candidate PP1 regulatory subunit expressed in developing spermatids. *BMC Cell Biol.* **2008**, *9*, 9. [CrossRef]
59. DeVaul, N.; Wang, R.; Sperry, A.O. PPP1R42, a PP1 binding protein, regulates centrosome dynamics in ARPE-19 cells. *Biol. Cell* **2013**, *105*, 359–371. [CrossRef]
60. Ma, S.; Lapin, D.; Liu, L.; Sun, Y.; Song, W.; Zhang, X.; Logemann, E.; Yu, D.; Wang, J.; Jirschitzka, J.; et al. Direct pathogen-induced assembly of an NLR immune receptor complex to form a holoenzyme. *Science* **2020**, *370*, eabe3069. [CrossRef]
61. Antunes Carvalho, F.; Renner, S.S. A dated phylogeny of the papaya family (Caricaceae) reveals the crop's closest relatives and the family's biogeographic history. *Mol. Phylogenetics Evol.* **2012**, *65*, 46–53. [CrossRef] [PubMed]
62. Rockinger, A.; Sousa, A.; Carvalho, F.A.; Renner, S.S. Chromosome number reduction in the sister clade of *Carica Papaya* with concomitant genome size doubling. *Am. J. Bot.* **2016**, *103*, 1082–1088. [CrossRef] [PubMed]
63. Cui, H.; Tsuda, K.; Parker, J.E. Effector-triggered immunity: From pathogen perception to robust defense. *Annu. Rev. Plant Biol.* **2015**, *66*, 487–511. [CrossRef] [PubMed]
64. Wang, J.; Song, W.; Chai, J. Structure, biochemical function, and signaling mechanism of plant NLRs. *Mol. Plant* **2023**, *16*, 75–95. [CrossRef] [PubMed]
65. Steuernagel, B.; Witek, K.; Krattinger, S.G.; Ramirez-Gonzalez, R.H.; Schoonbeek, H.J.; Yu, G.; Baggs, E.; Witek, A.I.; Yadav, I.; Krasileva, K.V.; et al. The NLR-Annotator Tool Enables Annotation of the Intracellular Immune Receptor Repertoire. *Plant Physiol.* **2020**, *183*, 468–482. [CrossRef]
66. Toda, N.; Rustenholz, C.; Baud, A.; Le Paslier, M.C.; Amselem, J.; Merdinoglu, D.; Faivre-Rampant, P. NLGenomeSweeper: A Tool for Genome-Wide NBS-LRR Resistance Gene Identification. *Genes* **2020**, *11*, 333. [CrossRef]
67. Friis, E.M.; Pedersen, K.R.; Crane, P.R. The emergence of core eudicots: New floral evidence from the earliest Late Cretaceous. *Proc. Biol. Sci.* **2016**, *283*, 20161325. [CrossRef]
68. Guo, X.; Fang, D.; Sahu, S.K.; Yang, S.; Guang, X.; Folk, R.; Smith, S.A.; Chanderbali, A.S.; Chen, S.; Liu, M.; et al. Chloranthus genome provides insights into the early diversification of angiosperms. *Nat. Commun.* **2021**, *12*, 6930. [CrossRef]
69. Li, Z.; Defoort, J.; Tasdighian, S.; Maere, S.; Van de Peer, Y.; De Smet, R. Gene Duplicability of Core Genes Is Highly Consistent across All Angiosperms. *Plant Cell* **2016**, *28*, 326–344. [CrossRef]
70. Glenfield, C.; Innan, H. Gene Duplication and Gene Fusion Are Important Drivers of Tumourigenesis during Cancer Evolution. *Genes* **2021**, *12*, 1376. [CrossRef]
71. Cannon, S.B.; Mitra, A.; Baumgarten, A.; Young, N.D.; May, G. The roles of segmental and tandem gene duplication in the evolution of large gene families in *Arabidopsis thaliana*. *BMC Plant Biol.* **2004**, *4*, 10. [CrossRef]
72. Yang, F.S.; Nie, S.; Liu, H.; Shi, T.L.; Tian, X.C.; Zhou, S.S.; Bao, Y.T.; Jia, K.H.; Guo, J.F.; Zhao, W.; et al. Chromosome-level genome assembly of a parent species of widely cultivated azaleas. *Nat. Commun.* **2020**, *11*, 5269. . [CrossRef] [PubMed]
73. Wang, K.; de la Torre, D.; Robertson, W.E.; Chin, J.W. Programmed chromosome fission and fusion enable precise large-scale genome rearrangement and assembly. *Science* **2019**, *365*, 922–926. [CrossRef] [PubMed]
74. Ling, H.Q.; Ma, B.; Shi, X.; Liu, H.; Dong, L.; Sun, H.; Cao, Y.; Gao, Q.; Zheng, S.; Li, Y.; et al. Genome sequence of the progenitor of wheat A subgenome *Triticum urartu*. *Nature* **2018**, *557*, 424–428. [CrossRef] [PubMed]
75. Tamborski, J.; Krasileva, K.V. Evolution of Plant NLRs: From Natural History to Precise Modifications. *Annu. Rev. Plant Biol.* **2020**, *71*, 355–378. [CrossRef]
76. Konno, K.; Hirayama, C.; Nakamura, M.; Tateishi, K.; Tamura, Y.; Hattori, M.; Kohno, K. Papain protects papaya trees from herbivorous insects: Role of cysteine proteases in latex. *Plant J. Cell Mol. Biol.* **2004**, *37*, 370–378. [CrossRef] [PubMed]

Disclaimer/Publisher's Note: The statements, opinions and data contained in all publications are solely those of the individual author(s) and contributor(s) and not of MDPI and/or the editor(s). MDPI and/or the editor(s) disclaim responsibility for any injury to people or property resulting from any ideas, methods, instructions or products referred to in the content.

Article

Identifying QTLs Related to Grain Filling Using a Doubled Haploid Rice (*Oryza sativa* L.) Population

So-Myeong Lee [1,†], Nkulu Rolly Kabange [1,†], Ju-Won Kang [1], Youngho Kwon [1], Jin-Kyung Cha [1], Hyeonjin Park [1], Ki-Won Oh [1], Jeonghwan Seo [2], Hee-Jong Koh [3] and Jong-Hee Lee [1,*]

[1] Department of Southern Area Crop Science, National Institute of Crop Science, RDA, Miryang 50424, Republic of Korea
[2] Crop Breeding Division, National Institute of Crop Science, RDA, Jeonju 55365, Republic of Korea
[3] Department of Plant Science, Seoul National University, Seoul 08826, Republic of Korea
* Correspondence: ccriljh@korea.kr; Tel.: +82-53-350-1168; Fax: +82-55-352-3059
† These authors contributed equally to this work.

Citation: Lee, S.-M.; Kabange, N.R.; Kang, J.-W.; Kwon, Y.; Cha, J.-K.; Park, H.; Oh, K.-W.; Seo, J.; Koh, H.-J.; Lee, J.-H. Identifying QTLs Related to Grain Filling Using a Doubled Haploid Rice (*Oryza sativa* L.) Population. *Agronomy* 2023, 13, 912. https://doi.org/10.3390/agronomy13030912

Academic Editors: Zhiyong Li, Chaolei Liu and Jiezheng Ying

Received: 22 February 2023
Revised: 13 March 2023
Accepted: 17 March 2023
Published: 19 March 2023

Copyright: © 2023 by the authors. Licensee MDPI, Basel, Switzerland. This article is an open access article distributed under the terms and conditions of the Creative Commons Attribution (CC BY) license (https://creativecommons.org/licenses/by/4.0/).

Abstract: Grain filling is an important trait of rice that affects the yield of grain-oriented crop species with sink capacity-related traits. Here, we used a doubled haploid (DH) population derived from a cross between 93-11 (P1, *indica*) and Milyang352 (P2, *japonica*) to investigate quantitative traits loci (QTLs) controlling grain filling in rice employing the Kompetitive allele-specific PCR (KASP) markers. The mapping population was grown under early-, normal-, and late-cultivation periods. The phenotypic evaluation revealed that spikelet number per panicle positively correlated with the grain-filling ratio under early cultivation conditions. Notably, three significant QTLs associated with the control of grain filling, *qFG3*, *qFG5-1*, and *qFG5-2*, were identified. Genes harbored by these QTLs are linked with diverse biological processes and molecular functions. Likewise, genes associated with abiotic stress response and transcription factor activity and redox homeostasis were detected. Genes such as *MYB*, *WRKY60*, and *OsSh1* encoding transcription factor, β-catenin, and the tubulin FtsZ, as well as those encoding cytochrome P450, would play a forefront role in controlling grain filling under early cultivation conditions. Our results suggest that *qFG3*-related genes could mediate the transition between grain filling and abiotic stress response mechanisms. Fine-mapping these QTLs would help identify putative candidate genes for downstream functional characterization.

Keywords: grain filling; yield; quantitative trait locus; rice

1. Introduction

Rice is a staple food crop feeding nearly 50% of the global population. The improved field management and genetics of rice have brought breakthroughs in rice yield over the past decades [1–3]. Nevertheless, developing rice varieties with a high-yielding potential is still at the center of interest in many breeding programs [4], which could be partly explained by the increasing global population and emerging environmental challenges, employing various breeding methods [3,5–7]. Of the various traits associated with rice yield, grain filling is regarded as a limiting factor for rice productivity and quality. Currently, many rice breeders are trying to improve rice yield by developing 'super rice' cultivars that have high-yield potential due to the large spikelet number per panicle; however, their poor grain filling of inferior spikelets renders the goal unattainable. The filling ratio of the rice grain is the final factor determining rice yield, which is closely associated with the sink and source ability. Moreover, further studies are still needed to address the problem of breeding rice cultivars with high-yield rice cultivars and high grain-filling ratios in molecular approaches [8,9].

Several reports support that grain filling and starch metabolism in plants are tightly linked [10–12]. Similarly, an efficient grain-filling rate favors a high grain weight [8,13]. The starch metabolism in rice has been comprehensively investigated [11,14–20]. As per some evidence, rice sink strength and grain filling have an antagonistic relationship. According to Chen et al. [21], rice plants with large panicles have a large sink capacity, while exhibiting a slow and low spikelet filling, especially inferior spikelets in the panicle. This is ultimately due to the asynchronous grain filling, in which sucrose synthase (SUS), among other enzymes, and abscisic acid (ABA) are said to play an important role [10]. Efforts toward increasing rice yield through the enlargement of sink size or strength often result in sink-source imbalance, leading to a decrease in the source productivity or grain filling of the rice grain [22].

Nagata et al. [23] reported that the *indica* allele at a QTL associated with sink size and ripening in rice, mapped on chromosomes 1 and 6, improved rice yield by increasing the number of primary and secondary rachis without causing any apparent decrease in grain filling. Under these conditions, maintaining sink productivity even when the number of spikelets per panicle increased would be more favorable. According to San et al. [24], a near-isogenic line (NIL derived from a cross between the *indica* cultivar Takanari and *japonica* cultivar Koshihikari) carrying the *indica* allele at *qLIA3* showed more tilted leaves than Koshihikari, with the latter showing a greater photosynthesis efficiency at the ripening stage. In addition to the factors mentioned above, nitrogen use efficiency (NUE) has proven to be a crucial factor in sink productivity. In this regard, Yamamoto et al. [25] reported that *qCHR1* associated with nitrogen transport mapped on Chr 1 improved nitrogen accumulation and distribution to leaves during ripening, but no change in the yield was observed. Meanwhile, an RIL carrying the *japonica* allele at *LSCHL4*, a gene reported in the narrowed-leaf mutant of rice, recorded a nearly 18.7% increase in yield higher than the *indica* counterpart. The *japonica* allele at *LSCHL4* conferred a high leaf chlorophyll content, improved panicle type, and enlarged flag leaf size [26].

This study aimed to investigate genetic regions controlling high-grain filling rice under early cultivation conditions. To achieve that, a doubled haploid (DH) population was used along with Kompetitive allele-specific PCR (KASP) markers and Fluidigm markers. Several QTLs associated with grain filling were detected, of which three QTLs (*qFG3*, *qFG5-1*, and *qFG5-2*) with high LOD scores were identified as significant in the early cultivation environment.

2. Materials and Methods

2.1. Plant Materials and Growth Conditions

A total of 117 DH lines developed through anther culture, derived from an F1 plant crossed between the typical *indica* rice cultivar 93-11 (a high-yielding cultivar originated from China) and the *japonica* cultivar Milyang352 (a Korean cultivar with moderate yield and early maturity), was used to conduct the experiments. The population was developed in summer 2017 in Miryang, Korea. Prior to transplanting, seeds were surface sterilized with a diluted chemical solution (copper hydroxide 0.012%, Ipconazole 0.016%, 32 °C) and soaked for 48 h to induce germination. Germinated seeds were sown and grown for about 3 weeks in 50-well trays, followed by transplanting in an open field in Miryang, Korea. Each DH line was transplanted in order of entry no. of each line, once for each cultivation season in 4 rows. The row length was 4 m, the hills within the row were spaced by 15 cm (26 plants per each line), and the space between rows was 30 cm.

The experiments were conducted during three consecutive years, with two cultivation seasons in 2018 and one cultivation season in 2019 and 2021. During the first year (2018), the mapping population was cultivated in early transplanting (first half of May) and late transplanting (first half of July) under a normal nitrogen cultivation regime. For the last two years, 2019 and 2021, plants were grown during the regular rice cultivation season in Korea (first half of June).

2.2. Phenotypic Measurements

To assess the grain-filling ratio, rice panicles from each plant were harvested 45 days after heading in triplicate, and threshed manually in a plastic box to prevent loss of spikelets.

Then, the filled grains in this study were defined as grains sinking into the NaCl solution (2.7 M) [27]. The filled grain ratio (FG) was calculated using the following formula, FG = [(number of filled grains/total number of grains per panicle) × 100], and was expressed in percentage. Other agronomic traits were also investigated: thousands grain weight (TGW = [(average filled grain weight/number of samples) × 100]), spikelet number per panicle (SN), and panicle number (PN) per plant. Flowering time was determined by the date when 40–50% of panicles emerged in a line. All samples were collected from the inside rows, excluding the border rows, to avoid the border effects on the traits studied or competition between lines. A correlation analysis between the grain-filling ratio and other traits was performed by analysis protocol in GraphPad Prism 7.00. A GxE analysis of FG was conducted by R (version 4.2.2) and "which-won-where" view of the GGE biplot were made by R Studio package "ggplot2".

2.3. DNA Extraction and Molecular Markers Analysis

The genomic DNA was extracted from leaf samples using the CTAB method [28], with slight modifications. In essence, snap-frozen samples (with liquid nitrogen for a few seconds) were ground to a fine powder in 2 mL Eppendorf (Eppendorf, T2795) tubes with 3 mm stainless steel beads (Masuda, Cat. 12-410-032, Tokyo, Japan). Next, 700 µL of DNA Extraction buffer (D2026, Lot D2622Z12H, Biosesang, Seongnam-si, Korea) was added to the ground leaf sample in 2 mL tube, and the mixture was vortexed, followed by incubation at 65 °C for 30 min in a dry oven. Then, 500 µL PCI solution (phenol:chloroform:isoamylalcohol = 25:24:1, Sigma-Aldrich, St. Louis, MO, USA) was added, followed by gentle mixing by inversion and centrifugation for 15 min at 13,000 rpm. The supernatant (500 µL) was transferred to fresh 1.5 mL tubes, isopropanol (500 µL) was added, and the mixture was incubated for 1 h at −20 °C. Soon after, tubes were centrifuged at 13,000 rpm to pellet down the DNA. The pellets were washed by adding 70% ethanol (EtOH, 1 mL), the tubes were centrifuged at 13,000 rpm for 10 min, and EtOH was discarded. DNA pellets were dried at room temperature and later resuspended in 100 µL 1× TE buffer (10 mM Tris-HCL, pH 8.0; 2.5 mM EDTA). DNA samples were quantified using nano-drop (ND1000 spectrophotometer, Mettler Toledo, Greifensee, Switzerland).

For the construction of the molecular map, KASP marker amplification and allelic discrimination were performed using the Nexar system (LGC Douglas Scientific, Alexandria, VA, USA) at the Seed Industry Promotion Center (Gimje, Republic of Korea) of the Foundation of Agri. Tech. Commercialization & Transfer in Korea. An aliquot (0.8 µL) of 2X master mix (LGC Genomics, London, UK), 0.02 µL of 106 KASP assay mix (LGC Genomics), containing one KASP SNP marker in each mix, and 5 ng genomic DNA template were mixed in a 1.6 µL KASP reaction mixture in a 384-well array tape. KASP amplification was performed as described by Cheon et al. [29].

A total of 240 SNP markers, including 106 KASP SNP markers and 134 Fluidigm SNP markers, were used for genotyping the DH population, and genotype data of 230 polymorphic SNP markers were used for final mapping (Figure S1, Table S1–S3) [29,30]. Fluidigm markers for SNP genotypes were determined using the BioMark™ HD system (Fluidigm, San Francisco, CA, USA) and 96.96 Dynamic Array IFC (96.96 IFC) chip according to the manufacturer's instructions at the National Instrumentation Center for Environmental Management (NICEM), Seoul National University (Pyeongchang, Republic of Korea). The genotyping results were obtained using the Fluidigm SNP Genotyping Analysis software. All genotype calls were manually confirmed, and any errors in homozygous or heterozygous clusters were curated.

2.4. Linkage Mapping and QTL Analysis

The genotype and phenotype data, comprising 230 polymorphic SNP markers (96 KASP markers and 134 Fluidigm markers) and 117 DH lines, were used to perform linkage mapping and QTL analysis with IciMapping software v.4.1 for a bi-parental population (position mapping and Kosambi mapping functions were selected) in order to detect genetic loci associated with grain-filling traits in rice. The logarithm of the odds (LOD) threshold 3.0 was selected for detecting significant QTLs based on Akond et al. [31]. The proportion of observed phenotypic variance explained by each QTL and the corresponding additive effects were also estimated.

2.5. Gene Ontology Search of qFG3, qFG5-1, and qFG5-2-Related Genes

The genome browser in the rice genome annotation project database (http://rice.uga.edu/cgi-bin/gbrowse/rice/, accessed on 29 August 2022) and the PlantPAN 3.0 database (http://plantpan.itps.ncku.edu.tw/search.php#species, accessed on 29 August 2022) were used to identify the locus ID and the basic annotation of the genes within the QTL region. The Plant Transcription Factor Database (http://planttfdb.gao-lab.org/, accessed on 29 August 2022) was used to verify whether a particular gene encodes a transcription factor or not, while the funRiceGenes database (https://funricegenes.github.io/geneKeyword.table.txt, accessed on 29 August 2022) was used to search for published reports on target genes.

3. Results

3.1. Differential Phenotypic Response of Parental Lines and Doubled Haploid Lines

The parental lines (93-11, a typical *indica* ssp. and Milyang352, a typical *japonica* ssp.) and their derived population were evaluated for their phenotypic responses in different cultivation conditions. The traits studied were PN per plant, SN, grain-filling ratio, and TGW. As shown in Figure 1, 93-11 (P1) and Milyang352 (P2) had differential phenotypic responses for all traits studied. In essence, 93-11 had a higher number of panicles per plant than Milyang352 under early and late cultivation conditions in 2018 and normal cultivation in 2019. However, under normal cultivation conditions in 2021, 93-11 and Milyang352 exhibited an opposite phenotypic response (Figure 1A, Table 1). SN recorded opposite phenotypic patterns in the two different cultivation seasons of 2018 and the normal cultivation seasons in 2019 and 2021 (Figure 1B). The grain-filling ratio of 93-11 was consistently higher in all cultivation conditions compared to Milyang352 (Figure 1C). In the same way, the TGW of 93-11 was much lower under early (2018), normal (2021), and late (2018) cultivation conditions, whereas parental lines grown under normal cultivation conditions in 2019 showed an opposite pattern (Figure 1D).

Table 1. Grain properties of parental lines (93-11 and Milyang352) in different cropping seasons (*: significant ($p < 0.05$), **: highly significant ($p < 0.01$), ns: not significant).

Traits	Cultivation Seasons							
	Early 2018		Normal 2019		Normal 2021		Late 2018	
	93-11	Milyang 352	93-11	Milyang 352	93-11	Milyang 352	93-11	Milyang 352
Panicle number per plant	14 ± 1.5 *	10 ± 2.6	11 ± 0.3 ns	10 ± 1.1	8 ± 1.8 ns	9 ± 0.7	12 ± 0.2 ns	11 ± 0.9
Spikelet number per panicle	173 ± 6.2 **	120 ± 3.5	119 ± 2.6 **	100 ± 4.4	157 ± 4.0 **	220 ± 5.3	136 ± 3.6 **	173 ± 3.5
Grain filling (%)	86.1 ± 0.4 **	57.5 ± 1.7	93.4 ± 2.6 **	78.0 ± 8.9	91.7 ± 8.2 **	66.1 ± 8.7	74.4 ± 7.2 **	61.1 ± 4.6
Thousand-grain weight (g)	19.6 ± 0.9 **	23.1 ± 0.9	25.2 ± 1.0 **	23.2 ± 1.2	22.7 ± 1.1 **	24.2 ± 1.6	20.8 ± 0.8 ns	23.3 ± 1.7

When analyzing the phenotype of the mapping population relative to their parental lines (Tables 1 and 2), we observed that the majority of the DH lines had a Milyang352-like PN per plant under normal cultivation conditions in 2019. In contrast, a higher percentage of DH lines grown under normal cultivation conditions in 2021 showed a 93-11-like PN. Regarding the SN, the DH lines grown under early cultivation season in 2018 recorded a 93-11-like phenotype. As for the grain-filling ratio, a majority of the DH lines recorded a relatively low filling ratio (Milyang352-like) in all cultivation conditions, except for the late season of 2018 (Figure 1C, Table 2). From Table 2, we could see that DH lines had a relatively high TGW under early cultivation in 2018.

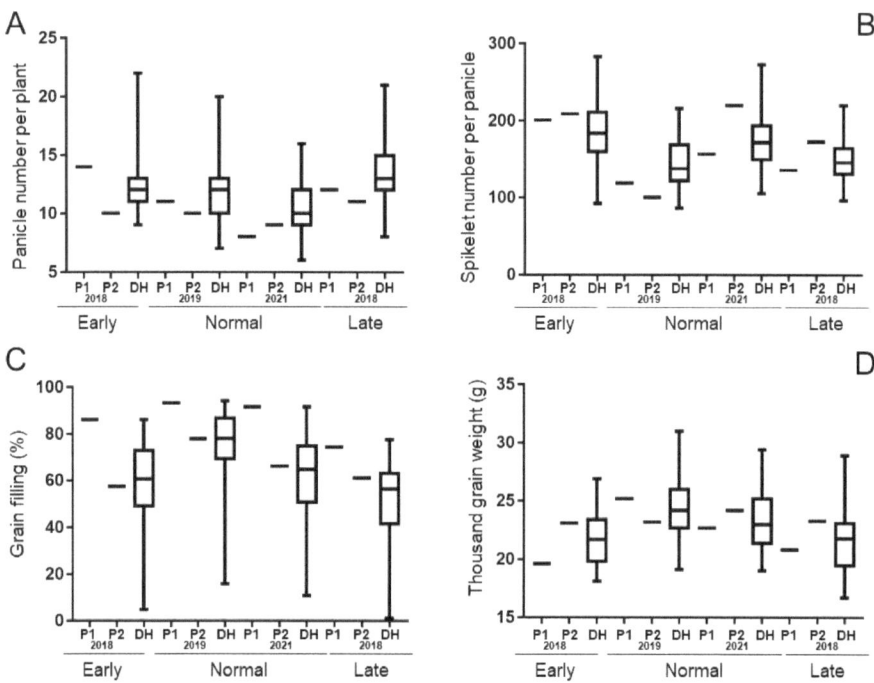

Figure 1. Phenotypic distribution of DH population of traits in different cropping seasons (P1: 93-11, P2: Milyang352). Phenotypes of normal season of 2021 had different distribution with normal season of 2019 except grain filling ratio. (**A**) Panicle number per plant; (**B**) Spikelet number per panicle; (**C**) Grain filling; (**D**) Thousand grain weight.

Table 2. Grain properties of the mapping population in different cultivation seasons. a, b, ab, c: Similarity based on *T*-test result (a: similar with 93-11, b: similar with Milyang352, ab: no difference with both parental lines, c: different with both parental lines).

Traits		Cultivation Seasons			
		Early 2018	Normal 2019	Normal 2021	Late 2018
Panicle number per plant	Average ± SD	13 ± 2.3 [ab]	12 ± 2.3 [b]	10 ± 1.8 [a]	13 ± 2.3 [c]
	Range	9–22	7–20	6–16	8–21
Spikelet number per panicle	Average ± SD	183 ± 40.2 [a]	144 ± 28.9 [c]	173 ± 31.4 [c]	149 ± 26.4 [c]
	Range	92–283	22–216	105–273	96–220
Grain filling (%)	Average ± SD	58.7 ± 16.7 [b]	75.3 ± 13.9 [b]	62.4 ± 16.4 [b]	50.8 ± 26.5 [c]
	Range	4.7–86.1	15.7–94.1	10.8–91.7	1.0–77.6
Thousand-grain weight (g)	Average ± SD	21.7 ± 2.2 [b]	24.4 ± 2.5 [ab]	23.3 ± 2.4 [ab]	21.7 ± 2.7 [ab]
	Range	18.1–26.9	19.1–31.0	19.0–29.4	16.7–28.9

3.2. Detected QTLs Associated with Grain Filling and Other Traits

To perform linkage mapping and QTL analysis, the genotype and phenotype data were employed. Among 240 SNP markers, genotype data of a total of 230 SNP markers were used for mapping, and 69 markers were deviated (Table S3). As indicated in Table 3 and Figure 2, a total of 56 QTLs were detected (all years and cropping seasons considered). Of this number, thirteen QTLs (seven under early cropping, four under normal cropping, and two under late cropping seasons of (Korean conditions) associated with grain filling (hereinafter referred to as FG) in rice were identified and mapped on five chromosomes in the present populations (Chr 3, 5, 8, 10, and 12). The *qFGL5-2*, detected in the DH population grown under late cropping season, mapped on Chr 5, had the highest logarithm of the odds (LOD) value of 15.16 and a phenotypic variation explained (PVE) of 21.66%. Other significant QTLs, *qFGE3*, *qFGE8-1* and *qFGE10*, recorded LOD and PVE values of 13.24 and 15.95%, 10.45 and 11.82%, and 8.72 and 10.42%, respectively. The findings of the GxE analysis on FG indicate that the early season of 2018 and normal season of 2019 exhibited congruent characteristics. Consequently, it is possible to discern *qFG3*, which encompasses *qFGE3* and *qFGN3*, within a particular environmental setting (Figure 3 and Table S4).

Table 3. Detected quantitative traits loci in different environments.

Cultivation Season	Trait	QTL Name	Position (cM)	Left Marker	Right Marker	LOD	PVE (%)	Add	Left CI	Right CI
Early Season	Spikelet Number Per Panicle	qSNE2	264	KJ02_047	KJ02_053	4.50	14.08	−13.75	262.5	265.5
		qSNE4	159	id4009823	cmb0432.2	4.23	14.43	−14.86	152.5	170.5
		qSNE9	60	ae09005437	KJ09_071	3.52	12.73	−13.56	51.5	66.5
	Panicle Number Per Plant	qPNE2	268	KJ02_053	KJ02_057	5.05	12.25	0.76	265.5	272.5
		qPNE4	160	id4009823	cmb0432.2	7.87	20.67	1.06	154.5	167.5
		qPNE7	98	ad07001853	KJ07_021	3.95	9.39	−0.66	93.5	105.5
		qPNE9	42	id9004072	ae09005437	3.72	9.12	0.72	36.5	47.5
	Percentage of Filled Grain	qFGE3	133	ad03013905	ad03014175	13.24	15.95	8.14	129.5	135.5
		qFGE5	43	KJ05_017	KJ05_019	10.88	12.53	7.23	39.5	45.5
		qFGE8-1	56	id8003584	KJ08_053	10.45	11.82	7.02	52.5	58.5
		qFGE8-2	86	KJ08_085	ae08007378	7.44	7.86	−5.79	82.5	88.5
		qFGE8-3	109	GW8-AG	id8007764	4.17	4.18	−4.25	107.5	109
		qFGE10	70	cmb1016.4	wd10003790	8.72	10.42	7.00	58.5	75.5
		qFGE12	108	KJ12_059	cmb1224.0	3.65	3.57	−3.81	103.5	112.5
	Thousand Grain Weight	qTGWE1-1	230	KJ01_121	KJ01_125	4.90	3.17	−0.63	221.5	238.5
		qTGWE1-2	285	P1193	KJ01_129	4.32	2.77	−0.44	275.5	285.5
		qTGWE2	252	KJ02_039	KJ02_043	8.37	8.12	−0.75	238.5	262.5
		qTGWE3-1	30	ad03000001	KJ03_007	10.60	7.41	0.71	22.5	32.5
		qTGWE3-2	183	KJ03_069	Hd6-AT	10.58	7.64	−0.73	172.5	187.5
		qTGWE5	38	id5002497	KJ05_013	19.80	16.94	−1.08	32.5	39.5
		qTGWE7	173	KJ07_067	cmb0723.0	9.24	6.25	0.68	171.5	175.5
		qTGWE8	109	GW8-AG	id8007764	10.88	7.58	0.74	107.5	109
		qTGWE10	130	KJ10_049	id10007384	12.66	9.33	−0.79	127.5	130
		qTGWE12-1	14	cmb1202.4	KJ12_007	9.75	6.69	0.67	9.5	23.5
		qTGWE12-2	127	cmb1226.0	id12010130	6.26	3.95	−0.52	122.5	131
Normal Season (2019)	Spikelet Number Per Panicle	qSNN4	192	KJ04_093	id4012434	4.51	10.72	−12.05	186.5	192
		qSNN8	88	ae08007378	id8006751	5.77	14.29	−13.20	82.5	91.5
		qSNN12	56	id12003700	KJ12_041	6.09	14.87	−13.73	51.5	60.5
	Panicle Number Per Plant	qPNN2	267	KJ02_053	KJ02_057	4.22	11.83	0.73	265.5	272.5
		qPNN4	162	cmb0432.2	cmb0434.1	4.25	12.26	0.79	154.5	169.5
		qPNN5	102	ad05008445	id5010886	3.52	11.47	−0.77	82.5	112.5
		qPNN7	106	KJ07_021	id7001155	4.99	14.68	−0.81	98.5	114.5

Table 3. Cont.

Cultivation Season	Trait	QTL Name	Position (cM)	Left Marker	Right Marker	LOD	PVE (%)	Add	Left CI	Right CI
	Percentage of Filled Grain	qFGN3	133	ad03013905	ad03014175	5.17	5.97	5.11	128.5	138.5
		qFGN5-1	38	id5002497	KJ05_013	11.05	14.40	7.86	31.5	39.5
		qFGN5-2	130	KJ05_063	cmb0526.3	6.33	7.41	−6.95	127.5	130.5
		qFGN5-3	139	KJ05_071	id5014265	11.48	15.05	9.90	138.5	140.5
	Thousand Grain Weight	qTGWN2	290	ae02004877	cmb0232.7	7.53	15.18	−1.22	284.5	296.5
		qTGWN8	109	GW8-AG	id8007764	6.55	12.48	1.10	107.5	109
		qTGWN12	9	id12000076	cmb1202.4	6.33	12.92	1.08	2.5	13.5
Late Season	Spikelet Number Per Panicle	qSNL3	82	id3005168	ah03001094	4.67	8.07	−8.16	74.5	91.5
		qSNL4	150	ah04001252	id4009823	12.51	23.17	−14.62	146.5	154.5
		qSNL5	150	id5014265	cmb0529.7	3.29	5.07	−7.17	149.5	150
		qSNL7	158	id7003072	KJ07_067	11.93	22.70	−14.07	154.5	163.5
	Panicle Number Per Plant	qPNL2	254	KJ02_039	KJ02_043	4.26	13.21	0.83	227.5	262.5
		qPNL3	196	Hd6-AT	id3015453	5.42	12.65	0.83	193.5	209.5
		qPNL4	147	ah04001252	id4009823	7.81	19.81	1.09	140.5	149.5
		qPNL7	105	KJ07_021	id7001155	3.07	7.19	−0.61	98.5	108.5
	Percentage of Filled Grain	qFGL5-1	130	KJ05_063	cmb0526.3	6.79	8.08	−9.99	127.5	130.5
		qFGL5-2	139	KJ05_071	id5014265	15.16	21.66	16.38	138.5	141.5
	Thousand Grain Weight	qTGWL2	245	KJ02_039	KJ02_043	4.83	12.30	−0.98	221.5	262.5
		qTGWL3-1	53	id3003462	id3005168	6.19	8.76	0.83	47.5	63.5
		qTGWL3-2	230	ah03002520	cmb0336.5	6.59	8.57	−0.82	221.5	233
		qTGWL5	39	KJ05_013	KJ05_017	10.89	15.44	−1.10	38.5	41.5
		qTGWL7-1	0	cmb0700.1	KJ07_011	4.72	5.91	−0.68	0.0	21.5
		qTGWL7-2	160	id7003072	KJ07_067	5.80	8.36	0.84	152.5	166.5
		qTGWL10	130	KJ10_049	id10007384	7.84	10.66	−0.91	123.5	130

(a) Cultivation seasons of the mapping population: early cultivation, standard rice cultivation, late cultivation; (b) rice traits used for the QTL analysis; (c) chromosome number; (d) QTL associated with traits studied; (e) absolute position of detected QTLs; (f) left flanking markers; (g) right flanking markers; (h) logarithm of the odds (LOD) scores; (i) phenotypic variation explained (PVE) by the QTLs expressed in percentage; (j) additive effects: the negative value indicates that the allele from Milyang352 increased the corresponding trait value, while the position value denotes that the allele from 93-11 increased the trait value.

It is also found that qFGE8-3 and QTLs related to TGW (qTGWE8 and qTGWN8) are flanked by the same markers (GW8-1AG and id8007764) but are mapped at different positions of the same chromosome (Chr 8). Similarly, another set of two adjacent QTLs, qFGE5 and qFGN5-1, qFGN5-2 and qFGL5-1, and qFGN5-3 and qFGL5-2, were mapped on Chr 5 and here proposed to govern grain filling in rice.

Data in Table 3 also indicate that QTLs qSNL5 is adjacent to qFGN5-3 and qFGL5-2 on chromosome 5 (27.80 Mbp) and flanked by KJ05_063 and cmb0529.7 marker. When comparing QTLs detected in different environments (Figure 2 and Figure S3), we could see that qFGE5 is adjacent to qTGWE5 and qTGWL5 controlling TGW under early and late cultivation conditions (flanking markers: id5002497 and KJ05_019).

Furthermore, ten QTLs controlling the SN were mapped on eight chromosomes of rice (Chr 2-5, 7-9, 12), of which number qSNL4 and qSNL7 recorded the highest LOD scores and PVE percentage.

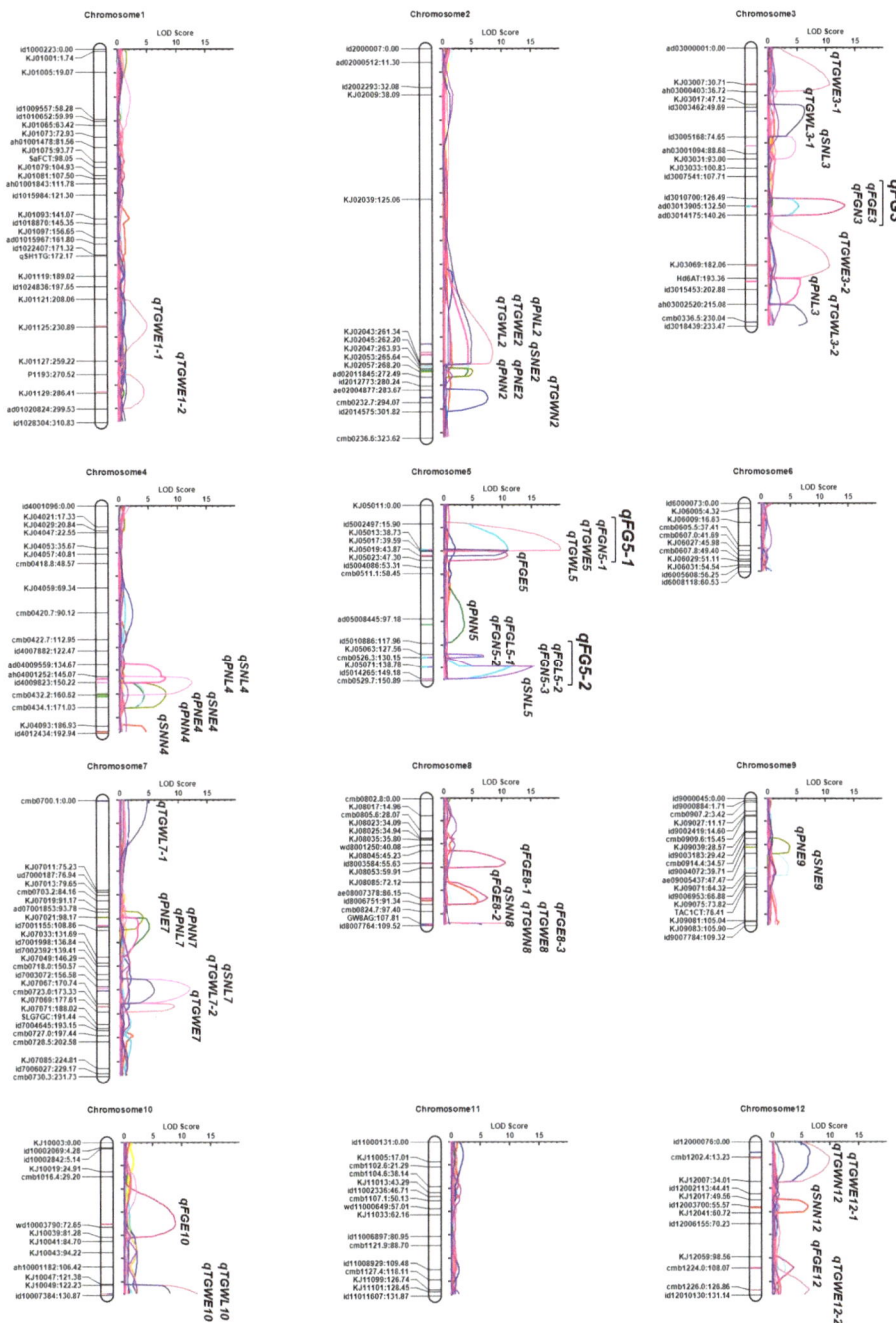

Figure 2. Linkage map of grain traits-related QTLs. *qFGE3* and *qFGN3* are repeatedly detected on close region of Chromosome 3 in different cultivation seasons.

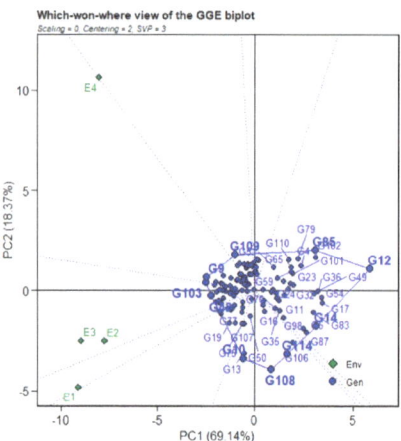

Figure 3. GxE analysis result of FG (E1: Early season, E2: Normal season of 2019, E3: Normal season of 2021, E4: Late season of 2018). The result indicates *qFG3* is detected under certain Environment.

Moreover, PN and TGW are widely recognized as important rice traits that reflect rice productivity. Here, data in Table 3 show that a total of twelve QTLs associated with the control of PN were detected (all cultivation conditions). Among them, two set of QTLs, *qPNE2* and *qPNN2*, and *qPNN7* and *qPNN7*, mapped on Chr 2 and Chr 7 of the mapping population (Table 3), were flanked by the same markers (KJ02_053 and KJ02_057, KJ07_021 and id7001155), respectively. Similarly, twenty-one QTLs associated with TGW were mapped on eight chromosomes (Chr 1–3, 5, 7, 8, 10, and 12). Interestingly, *qTGWE5* and *qTGWL5* flanked by KJ05_013 and KJ05_017 KASP markers coincided with a major QTL (*qGW-5-1/qGT-51/qGLWR-5-1*, 24 cM) recently proposed to govern grain shape (width, thickness, and grain length–width ratio) in rice [32]. Based on this observation, we could say that TGW and grain shape traits would be controlled by the same genetic locus, implying that both traits could be tightly related.

3.3. Genes Harbored by qFG3, qFG5-1 and qFG5-2 QTLs Associated with Grain Filling in Rice

To calculate recombination, 210 Markers from 230 genotyping data of polymorphic SNP markers were mapped and grouped into 12 linkage groups covering 2116.6 cM (Table S1). Twelve QTLs associated with grain filling (FG) were detected in different cultivation periods. We were particularly interested in unveiling the identity of genes located in QTLs for grain filling under early cultivation conditions. In this regard, among the seven QTLs for FG in early cultivation, we identified two detected by our study for deep mining. The other remaining five QTLs have already been reported by previous studies [32]. As Kang et al. (2020) reported, those QTLs are found to be associated with QTLs related to grain shape or weight. Thus, a deep search of gene ontology fetching from several genome annotation databases was freely available online. As shown in Table S6, the result of this search reveals the biological processes and the molecular functions characterizing genes harbored by *qFG3*. In essence, genes encoding a tubulin/FtsZ domain-containing protein, DNA-directed RNA polymerase II subunit RPB2, β-subunit, Chaperone protein DnaJ [33,34], are associated with cell division, reproduction and embryo development, and flower development, respectively. The *qFG3* also harbors genes related to cell differentiation and cellular component organization, including *OsSh1* [35–37], *SWI1* [38–44], and Os03g45260, similar to the Arabidopsis Membrin 11 (*AtMEMB11*).

In the same genetic locus, genes annotated as being involved in ion transport or ion binding are found, including Os03g44484, Os03g44630, Os03g45370, Os03g44636, and Os03g44660. Another set of genes located in the *qFG3* region, said to be involved in the transcriptional regulation of specific genes or translation, are Os03g44540, Os03g44780, Os03g44820 [45], Os03g44900 [46], Os03g45400, and Os03g45450.

Likewise, a category of genes involved in the metabolic process (Os03g44500 and Os03g45390, involved in carbohydrate metabolism), catalytic process (Os03g45194) [47–49], and biosynthetic process (Os03g45320 and Os03g45420) were identified [50]. We could, additionally, find two genes encoding a cytochrome P450 protein (Os03g44740 and Os03g45619). The QTL *qFG3* equally harbors genes belonging to the family of proteins of unknown function (DUF266, Os03g44580) [51,52], encoding an IQ calmodulin-binding motif domain (Os03g44610) [53–55], a peptidase S54 rhomboid domain-containing protein (*OsRhmbd9*, Os03g44830, having a serine-type endopeptidase activity) [56], a C2 domain-containing protein (Os03g44890, having a transferase activity) [57], a protease inhibitor/seed storage/LTP protein precursor (*LTPL91*, Os03g44950) [58,59], or encoding an ankyrin repeat domain-containing protein 50 (Os03g45920) [60,61].

From the above phenotype of the mapping population, we were interested in assessing the eventual effect of *qFG3* on different traits evaluated. From data in Figure S4A,C and Table 3, it is confirmed that alleles from 93-11 had significantly enhanced the observed phenotypic variation of SN and grain-filling ratio in the early season. However, for SN, the alleles from Milyang352 contributed to the observed phenotypic variation (Figure S4B, Table 3). Unlike other traits, a non-significant difference was found between the effects of alleles from 93-11 and Milyang352 for TGW (Figure S4D).

Intriguingly, the majority of genes included in QTLs *qFG3*, *qFG5-1* and *qFG5-2*, in addition to their involvement in developmental and cell differentiation processes, are proposed to be involved in the adaptive response mechanism toward abiotic stress or the oxidation-reductase process (Table S5 and S6).

4. Discussion

4.1. Differential Grain-Filling Ratio between Parents and DH Lines

Rice genotypes exhibiting a high grain-filling ratio are generally preferred to those with an opposite phenotype, mainly because a high grain filling ratio is expected to result in high yield. Sink strength and grain filling are two of the major determinants of rice yield, with carbohydrates from photosynthetic assimilates and non-structural carbohydrates (NSC) present in culms and leaf sheaths playing important roles [62]. However, Zhang et al. [37] suggested that, although grain filling in rice largely consists of starch accumulation (about 90% of the dry weight of grains), carbohydrate supply may not be the only limiting factor for poor grain filling, especially for inferior spikelets as compared with superior ones.

Here, we recorded a differential grain-filling ratio of the parental lines and their derived mapping population cultivated in different seasons. The majority of the DH lines exhibited a low grain-filling ratio similar to that observed in Milyang352 (P2, *japonica*) in all cultivation seasons, while only a few DH lines showed a P1-like grain-filling pattern. The highest grain-filling ratio (75.3%) was obtained in the normal rice cultivation season in Korea in 2019.

In their study, Sekhar et al. [63] observed that grain filling in rice was significantly influenced by factors such as spikelet fertility, coupled with panicle compactness and ethylene production. However, our data did not detect any strong correlation between grain-filling ratio and other traits in all cultivation seasons (Figure S2).

4.2. A Novel QTL Associated with Grain Filling Is Identified

Several QTLs reported to control grain filling have been found and mapped to several chromosomes of rice. Some of these QTLs are also associated with the control of other grain traits, including grain shape and size, TGW (*GW2* [64], *qSW5/GW5*, *GW7/GL7*, and *GW8*, *GS3*, 5, and 6 [65–76]), spikelet fertility (*qSFP1.1*, *qSFP3.1*, *qSFP6.1*, *qSFP8.1*), panicle compactness (*qIGS3.2*, *qIGS4.1*), and ethylene production (*qETH1.2*, *qETH3.1*, *qETH4.1*, *4.2*, *qETH6.1*, *6.2*). Our study identified seven QTLs associated with rice grain filling, of which number *qFG3* is novel, and was detected in both early and normal cultivation seasons. *qFG3* is therefore regarded as a stable QTL, and could serve as a useful source for downstream breeding.

We could see that *qFG3* harbors genes associated with cell division, elongation, and differentiation, including a member of the tubulin superfamily (tubulin/FtsZ, major components of the cytoskeleton of eukaryotes). In plants, tubulin contributes to the formation of microtubules [77] that, in turn, control cell division, growth polarity, intracellular trafficking and communications, cell-wall deposition, and cell morphogenesis [78–80]. As primary components of microtubules [77], plant tubulins (α, β, and γ) provide structural support for overall cell shape as well as the transport and positioning of organelles [78]. Other genes encoding α-tubulin, *Srs5* (*small and round seed 5*) [81] and β-tubulin (*OsTUB8*) [82] are associated with seed formation, elongation or size [83].

The gene ontology search of *qFG3*-related genes showed that Os03g44420 (a tubulin/FtsZ domain-containing protein) or glutamyl-tRNA synthetase amidotransferase C subunit [44] are involved in cell division and differentiation, or ion transport. Another set of two genes encoding β-catenin repeat protein (Os03g45420) or an MYB-like DNA-binding domain-containing protein (Os03g45197) is proposed to be involved in the regulation of chlorophyll and ABA biosynthesis processes or photosynthesis. A relationship between grain filling, yield and chlorophyll content has been established. Reports suggest that rice grain yield is highly dependent on the photosynthetic assimilation of leaves during the grain-filling stage, where about 60–80% of the nutrients required for grain filling are contributed by the photosynthesis of source leaves during the same period. In contrast, Chen et al. [84] investigated the genetic diversity of six rice cultivars for their grain-filling rate and revealed that the contribution of photosynthesis to rice grain filling would be genotype-dependent.

4.3. QTL qFG3 Harbors Genes with Transcription Factor Activity

In all biological systems, gene regulation is governed through a combinational action of multiple regulatory proteins, including transcription factors (TFs) [85]. As shown in Table S6, four genes encoding TFs were found within the genetic region covered by *qFG3*. These genes encode Nf-Y-A subunit (Os03g44540), CCR4-NOT (Os03g44900), antitermination NusB domain-containing protein (Os03g45400), WRKY60 (Os03g45450), and YABBY domain-containing protein (*OsSh1*, shattering1) [35]. Several studies have shown that genes containing the YABBY domain functionally interact with other genes or biological compounds to regulate plant growth and development [86]. However, Dai et al. [36] demonstrated that a *WUSCHEL-LIKE HOMEOBOX* gene acts as a negative regulator of a *YABBY* gene required for rice leaf development.

4.4. qFG3-Related Genes Are Associated with Abiotic Stress Response Mechanisms

Plant growth and development are often favored under normal conditions, while slow and impaired growth is commonly observed when plants experience stress conditions. In this study, a set of genes located in the *qFG3* QTL region is proposed to be involved in the adaptive response mechanisms (Table S6). It is well established that reactive oxygen (ROS) and nitrogen (RNS) species play important roles in maintaining regular plant growth under normal conditions, such as seed development and germination, shoot and flower development, etc. [87]. ROS and RNS are by-products of aerobic metabolism generated in various cellular compartments, including chloroplasts [88,89], mitochondria [90,91], and peroxysomes [92,93]. Within this perspective, *OsIPMPDH* (Os03g45320), having an oxidoreductase activity and magnesium ion binding, and those encoding cytochrome P450 protein, coupled with Os03g44620 and Os03g44630, can be targets to investigate the relationship between grain filling and stress response in rice.

5. Conclusions

Grain filling is a determinant trait for the yield of cereal crop species, such as rice. This study identified QTLs associated with grain filling in rice under different cultivation seasons. Among them, *qFG3*, *qFG5-1*, and *qFG5-2* were significant, of which *qFG3* recorded a high LOD score. Genes harbored by *qFG3* are associated with cell division, embryo and post-embryo development, photosynthesis, and starch synthesis, among others. Genes such as MYB, WRKY60, and *OsSh1* encoding TFs, β-catenin, and the tubulin FtsZ have interesting predicted molecular functions. Another set of genes are proposed to be involved in abiotic stress signaling or adaptive response mechanisms. *qFG3* is a QTL that can be detected in a specific environment. It can serve as a target QTL region for downstream breeding, including the fine mapping and functional characterization of putative candidate genes for their roles during grain filling in rice.

Supplementary Materials: The following supporting information can be downloaded at: https://www.mdpi.com/article/10.3390/agronomy13030912/s1, Figure S1: Example of allele discrimination plot of polymorphic SNP marker (KJ11_13, Table S2).; Figure S2: Correlation between grain properties and yield component with normal fertilization level (y: FG, x: SN or PN or TGW, SN: spikelet number per panicle, PN: panicle number per plant, TGW: thousand-grain weight (g), FG: percentage of filled grain (%), E: early season, N: normal season, L: late season, R2: coefficient of determination).; Figure S3: Weather conditions prevailing during 2018, 2019, and 2021 cultivation seasons.; Figure S4: The distribution of grain properties of the two groups differed by the genotype of *qFG3* in different fertilization levels and cropping seasons.; Table S1: cM of each chromosome.; Table S2: List of SNP markers.; Table S3: Information on the physical distance between SNP markers.; Table S4: Distribution of flowering time of DH population according to the genotype of *qFG3*.; Table S5: List of candidate genes; Table S6: Putative candidate genes located within *qFG3*.

Author Contributions: Conceptualization, J.-H.L.; methodology, J.-W.K., J.-H.L. and N.R.K.; software, S.-M.L. and N.R.K.; validation, J.-H.L.; formal analysis, S.-M.L., N.R.K., Y.K., J.-K.C. and H.P.; investigation, S.-M.L., J.-W.K. and J.S.; resources, J.-W.K., J.S. and J.-H.L.; data curation, S.-M.L. and N.R.K.; writing—original draft preparation, S.-M.L. and N.R.K.; writing—review and editing, J.-H.L.; visualization, J.-H.L.; supervision, J.-H.L., H.-J.K. and K.-W.O.; project administration and funding acquisition, J.-H.L. All authors have read and agreed to the published version of the manuscript.

Funding: This research was funded by the Rural Development Administration, Republic of Korea, grant number PJ01428201.

Data Availability Statement: Not applicable.

Acknowledgments: We are grateful for the support from the Rural Development Administration (RDA) and Crop Molecular Breeding lab in Seoul National University (SNU), Republic of Korea.

Conflicts of Interest: The authors declare no conflict of interest.

References

1. Khan, M.H.; Dar, Z.A.; Dar, S.A. Breeding strategies for improving rice yield—A review. *Agric. Sci.* **2015**, *6*, 467. [CrossRef]
2. Peng, S.; Khush, G.S.; Virk, P.; Tang, Q.; Zou, Y. Progress in ideotype breeding to increase rice yield potential. *Field Crops Res.* **2008**, *108*, 32–38. [CrossRef]
3. Xing, Y.; Zhang, Q. Genetic and molecular bases of rice yield. *Annu. Rev. Plant Biol.* **2010**, *61*, 421–442. [CrossRef]
4. Okamura, M.; Arai-Sanoh, Y.; Yoshida, H.; Mukouyama, T.; Adachi, S.; Yabe, S.; Nakagawa, H.; Tsutsumi, K.; Taniguchi, Y.; Kobayashi, N. Characterization of high-yielding rice cultivars with different grain-filling properties to clarify limiting factors for improving grain yield. *Field Crops Res.* **2018**, *219*, 139–147. [CrossRef]
5. Zha, X.; Luo, X.; Qian, X.; He, G.; Yang, M.; Li, Y.; Yang, J. Over-expression of the rice LRK1 gene improves quantitative yield components. *Plant Biotechnol. J.* **2009**, *7*, 611–620. [CrossRef]
6. Rao, Y.; Li, Y.; Qian, Q. Recent progress on molecular breeding of rice in China. *Plant Cell Rep.* **2014**, *33*, 551–564. [CrossRef]
7. Li, D.; Wang, L.; Wang, M.; Xu, Y.Y.; Luo, W.; Liu, Y.J.; Xu, Z.H.; Li, J.; Chong, K. Engineering OsBAK1 gene as a molecular tool to improve rice architecture for high yield. *Plant Biotechnol. J.* **2009**, *7*, 791–806. [CrossRef] [PubMed]
8. Zhang, W.; Cao, Z.; Zhou, Q.; Chen, J.; Xu, G.; Gu, J.; Liu, L.; Wang, Z.; Yang, J.; Zhang, H. Grain filling characteristics and their relations with endogenous hormones in large-and small-grain mutants of rice. *PLoS ONE* **2016**, *11*, e0165321. [CrossRef] [PubMed]
9. Yang, J.; Zhang, J. Grain-filling problem in 'super' rice. *J. Exp. Bot.* **2010**, *61*, 1–5. [CrossRef] [PubMed]
10. Tang, T.; Xie, H.; Wang, Y.; Lü, B.; Liang, J. The effect of sucrose and abscisic acid interaction on sucrose synthase and its relationship to grain filling of rice (*Oryza sativa* L.). *J. Exp. Bot.* **2009**, *60*, 2641–2652. [CrossRef]

11. Su, J.-C. Starch synthesis and grain filling in rice. In *Developments in Crop Science*; Elsevier: Amsterdam, The Netherlands, 2000; Volume 26, pp. 107–124.
12. Jiang, Z.; Chen, Q.; Chen, L.; Yang, H.; Zhu, M.; Ding, Y.; Li, W.; Liu, Z.; Jiang, Y.; Li, G. Efficiency of sucrose to starch metabolism is related to the initiation of inferior grain filling in large panicle rice. *Front. Plant Sci.* **2021**, *12*, 732867. [CrossRef] [PubMed]
13. Yabe, S.; Yoshida, H.; Kajiya-Kanegae, H.; Yamasaki, M.; Iwata, H.; Ebana, K.; Hayashi, T.; Nakagawa, H. Description of grain weight distribution leading to genomic selection for grain-filling characteristics in rice. *PLoS ONE* **2018**, *13*, e0207629. [CrossRef]
14. Isshiki, M.; Nakajima, M.; Satoh, H.; Shimamoto, K. *dull*: Rice mutants with tissue-specific effects on the splicing of the waxy pre-mRNA. *Plant J.* **2000**, *23*, 451–460. [CrossRef]
15. Ise, K.; Akama, Y.; Horisue, N.; Nakane, A.; Yokoo, M.; Ando, I.; Hata, T.; Suto, M.; Numaguchi, K.; Nemoto, H. Milky Queen, a new high-quality rice cultivar with low amylose content in endosperm. *Bull. Natl. Inst. Crop Sci.* **2001**, *2*, 39–61.
16. He, P.; Li, S.; Qian, Q.; Ma, Y.; Li, J.; Wang, W.; Chen, Y.; Zhu, L. Genetic analysis of rice grain quality. *Theor. Appl. Genet.* **1999**, *98*, 502–508. [CrossRef]
17. Han, Y.; Xu, M.; Liu, X.; Yan, C.; Korban, S.S.; Chen, X.; Gu, M. Genes coding for starch branching enzymes are major contributors to starch viscosity characteristics in waxy rice (*Oryza sativa* L.). *Plant Sci.* **2004**, *166*, 357–364. [CrossRef]
18. Butardo, V.M., Jr.; Daygon, V.D.; Colgrave, M.L.; Campbell, P.M.; Resurreccion, A.; Cuevas, R.P.; Jobling, S.A.; Tetlow, I.; Rahman, S.; Morell, M. Biomolecular analyses of starch and starch granule proteins in the high-amylose rice mutant Goami 2. *J. Agric. Food Chem.* **2012**, *60*, 11576–11585. [CrossRef]
19. Bao, J.; Corke, H.; Sun, M. Microsatellites, single nucleotide polymorphisms and a sequence tagged site in starch-synthesizing genes in relation to starch physicochemical properties in nonwaxy rice (*Oryza sativa* L.). *Theor. Appl. Genet.* **2006**, *113*, 1185–1196. [CrossRef] [PubMed]
20. Aluko, G.; Martinez, C.; Tohme, J.; Castano, C.; Bergman, C.; Oard, J. QTL mapping of grain quality traits from the interspecific cross *Oryza sativa* × *O. glaberrima*. *Theor. Appl. Genet.* **2004**, *109*, 630–639. [CrossRef]
21. Chen, L.; Deng, Y.; Zhu, H.; Hu, Y.; Jiang, Z.; Tang, S.; Wang, S.; Ding, Y. The initiation of inferior grain filling is affected by sugar translocation efficiency in large panicle rice. *Rice* **2019**, *12*, 75. [CrossRef]
22. Akita, S. Improving yield potential in tropical rice. In *Progress in Irrigated Rice Research*; International Rice Research Institute: Manila, Philippines, 1989; pp. 41–73.
23. Nagata, K.; Fukuta, Y.; Shimizu, H.; Yagi, T.; Terao, T. Quantitative trait loci for sink size and ripening traits in rice (*Oryza sativa* L.). *Breed. Sci.* **2002**, *52*, 259–273. [CrossRef]
24. San, N.S.; Ootsuki, Y.; Adachi, S.; Yamamoto, T.; Ueda, T.; Tanabata, T.; Motobayashi, T.; Ookawa, T.; Hirasawa, T. A near-isogenic rice line carrying a QTL for larger leaf inclination angle yields heavier biomass and grain. *Field Crops Res.* **2018**, *219*, 131–138. [CrossRef]
25. Yamamoto, T.; Suzuki, T.; Suzuki, K.; Adachi, S.; Sun, J.; Yano, M.; Ookawa, T.; Hirasawa, T. Characterization of a genomic region that maintains chlorophyll and nitrogen contents during ripening in a high-yielding stay-green rice cultivar. *Field Crops Res.* **2017**, *206*, 54–64. [CrossRef]
26. Zhang, G.-H.; Li, S.-Y.; Wang, L.; Ye, W.-J.; Zeng, D.-L.; Rao, Y.-C.; Peng, Y.-L.; Hu, J.; Yang, Y.-L.; Xu, J. LSCHL4 from japonica cultivar, which is allelic to NAL1, increases yield of indica super rice 93-11. *Mol. Plant* **2014**, *7*, 1350–1364. [CrossRef]
27. Ha, W.; Kim, H.; Choi, H.; Lim, S.; Seo, H.S.; Lim, M. Heritability Estimates of Sink and Source Characters by Fin-fill$_5$ Correlation in Rice. *Korean J. Crop Sci.* **2002**, *47*, 151–156.
28. Murray, M.; Thompson, W. Rapid isolation of high molecular weight plant DNA. *Nucleic Acids Res.* **1980**, *8*, 4321–4326. [CrossRef] [PubMed]
29. Cheon, K.-S.; Baek, J.; Cho, Y.-i.; Jeong, Y.-M.; Lee, Y.-Y.; Oh, J.; Won, Y.J.; Kang, D.-Y.; Oh, H.; Kim, S.L. Single nucleotide polymorphism (SNP) discovery and kompetitive allele-specific PCR (KASP) marker development with Korean japonica rice varieties. *Plant Breed. Biotechnol.* **2018**, *6*, 391–403. [CrossRef]
30. Seo, J.; Lee, G.; Jin, Z.; Kim, B.; Chin, J.H.; Koh, H.-J. Development and application of indica–japonica SNP assays using the Fluidigm platform for rice genetic analysis and molecular breeding. *Mol. Breed.* **2020**, *40*, 39. [CrossRef]
31. Akond, Z.; Alam, M.J.; Hasan, M.N.; Uddin, M.S.; Alam, M.; Mollah, M.N.H. A comparison on some interval mapping approaches for QTL detection. *Bioinformation* **2019**, *15*, 90. [CrossRef] [PubMed]
32. Kang, J.-W.; Kabange, N.R.; Phyo, Z.; Park, S.-Y.; Lee, S.-M.; Lee, J.-Y.; Shin, D.; Cho, J.H.; Park, D.-S.; Ko, J.-M. Combined linkage mapping and genome-wide association study identified QTLs associated with grain shape and weight in rice (*Oryza sativa* L.). *Agronomy* **2020**, *10*, 1532. [CrossRef]
33. Zhu, X.; Liang, S.; Yin, J.; Yuan, C.; Wang, J.; Li, W.; He, M.; Wang, J.; Chen, W.; Ma, B. The DnaJ OsDjA7/8 is essential for chloroplast development in rice (*Oryza sativa*). *Gene* **2015**, *574*, 11–19. [CrossRef]
34. Zhong, X.; Yang, J.; Shi, Y.; Wang, X.; Wang, G.L. The DnaJ protein OsDjA6 negatively regulates rice innate immunity to the blast fungus *Magnaporthe oryzae*. *Mol. Plant Pathol.* **2018**, *19*, 607–614. [CrossRef] [PubMed]
35. Malik, N.; Ranjan, R.; Parida, S.K.; Agarwal, P.; Tyagi, A.K. Mediator subunit OsMED14_1 plays an important role in rice development. *Plant J.* **2020**, *101*, 1411–1429. [CrossRef]
36. Lin, Z.; Li, X.; Shannon, L.M.; Yeh, C.-T.; Wang, M.L.; Bai, G.; Peng, Z.; Li, J.; Trick, H.N.; Clemente, T.E. Parallel domestication of the Shattering1 genes in cereals. *Nat. Genet.* **2012**, *44*, 720–724. [CrossRef] [PubMed]

37. Dai, M.; Hu, Y.; Zhao, Y.; Liu, H.; Zhou, D.-X. A *WUSCHEL-LIKE HOMEOBOX* gene represses a YABBY gene expression required for rice leaf development. *Plant Physiol.* **2007**, *144*, 380–390. [CrossRef]
38. Zhang, Z.; Zhao, H.; Tang, J.; Li, Z.; Li, Z.; Chen, D.; Lin, W. A proteomic study on molecular mechanism of poor grain-filling of rice (*Oryza sativa* L.) inferior spikelets. *PLoS ONE* **2014**, *9*, e89140. [CrossRef] [PubMed]
39. Yuan, X.; Wang, H.; Bi, Y.; Yan, Y.; Gao, Y.; Xiong, X.; Wang, J.; Li, D.; Song, F. ONAC066, a stress-responsive NAC transcription activator, positively contributes to rice immunity against *Magnaprothe oryzae* through modulating expression of OsWRKY62 and three cytochrome P450 genes. *Front. Plant Sci.* **2021**, *12*, 749186. [CrossRef]
40. Yang, W.; Gao, M.; Yin, X.; Liu, J.; Xu, Y.; Zeng, L.; Li, Q.; Zhang, S.; Wang, J.; Zhang, X. Control of rice embryo development, shoot apical meristem maintenance, and grain yield by a novel cytochrome p450. *Mol. Plant* **2013**, *6*, 1945–1960. [CrossRef] [PubMed]
41. Magome, H.; Nomura, T.; Hanada, A.; Takeda-Kamiya, N.; Ohnishi, T.; Shinma, Y.; Katsumata, T.; Kawaide, H.; Kamiya, Y.; Yamaguchi, S. CYP714B1 and CYP714B2 encode gibberellin 13-oxidases that reduce gibberellin activity in rice. *Proc. Natl. Acad. Sci. USA* **2013**, *110*, 1947–1952. [CrossRef] [PubMed]
42. Luo, A.; Qian, Q.; Yin, H.; Liu, X.; Yin, C.; Lan, Y.; Tang, J.; Tang, Z.; Cao, S.; Wang, X. EUI1, encoding a putative cytochrome P450 monooxygenase, regulates internode elongation by modulating gibberellin responses in rice. *Plant Cell Physiol.* **2006**, *47*, 181–191. [CrossRef]
43. Guo, F.; Endo, M.; Yamaguchi, T.; Uchino, A.; Sunohara, Y.; Matsumoto, H.; Iwakami, S. Investigation of clomazone-tolerance mechanism in a long-grain cultivar of rice. *Pest Manag. Sci.* **2021**, *77*, 2454–2461. [CrossRef] [PubMed]
44. Fang, N.; Xu, R.; Huang, L.; Zhang, B.; Duan, P.; Li, N.; Luo, Y.; Li, Y. SMALL GRAIN 11 controls grain size, grain number and grain yield in rice. *Rice* **2016**, *9*, 64. [CrossRef]
45. Yang, X.; Li, G.; Tian, Y.; Song, Y.; Liang, W.; Zhang, D. A rice glutamyl-tRNA synthetase modulates early anther cell division and patterning. *Plant Physiol.* **2018**, *177*, 728–744. [CrossRef] [PubMed]
46. Chou, W.-L.; Chung, Y.-L.; Fang, J.-C.; Lu, C.-A. Novel interaction between CCR4 and CAF1 in rice CCR4–NOT deadenylase complex. *Plant Mol. Biol.* **2017**, *93*, 79–96. [CrossRef]
47. Ren, D.; Rao, Y.; Yu, H.; Xu, Q.; Cui, Y.; Xia, S.; Yu, X.; Liu, H.; Hu, H.; Xue, D. MORE FLORET1 encodes a MYB transcription factor that regulates spikelet development in rice. *Plant Physiol.* **2020**, *184*, 251–265. [CrossRef] [PubMed]
48. Cao, W.-L.; Chu, R.-Z.; Zhang, Y.; Luo, J.; Su, Y.-Y.; Xie, L.-J.; Zhang, H.-S.; Wang, J.-F.; Bao, Y.-M. OsJAMyb, a R2R3-type MYB transcription factor, enhanced blast resistance in transgenic rice. *Physiol. Mol. Plant Pathol.* **2015**, *92*, 154–160. [CrossRef]
49. Bao, Y.-M.; Sun, S.-J.; Li, M.; Li, L.; Cao, W.-L.; Luo, J.; Tang, H.-J.; Huang, J.; Wang, Z.-F.; Wang, J.-F. Overexpression of the Qc-SNARE gene OsSYP71 enhances tolerance to oxidative stress and resistance to rice blast in rice (*Oryza sativa* L.). *Gene* **2012**, *504*, 238–244. [CrossRef]
50. Sikdar, M.; Kim, J. Isolation of a gene encoding 3-isopropylmalate dehydrogenase from rice. *Russ. J. Plant Physiol.* **2011**, *58*, 190–196. [CrossRef]
51. Zhou, Y.; Li, S.; Qian, Q.; Zeng, D.; Zhang, M.; Guo, L.; Liu, X.; Zhang, B.; Deng, L.; Liu, X. BC10, a DUF266-containing and Golgi-located type II membrane protein, is required for cell-wall biosynthesis in rice (*Oryza sativa* L.). *Plant J.* **2009**, *57*, 446–462. [CrossRef]
52. Li, K.; Chen, Y.; Luo, Y.; Huang, F.; Zhao, C.; Cheng, F.; Xiang, X.; Pan, G. A 22-bp deletion in OsPLS3 gene encoding a DUF266-containing protein is implicated in rice leaf senescence. *Plant Mol. Biol.* **2018**, *98*, 19–32. [CrossRef]
53. Zhang, L.; Liu, B.-F.; Liang, S.; Jones, R.L.; Lu, Y.-T. Molecular and biochemical characterization of a calcium/calmodulin-binding protein kinase from rice. *Biochem. J.* **2002**, *368*, 145–157. [CrossRef]
54. Koo, S.C.; Choi, M.S.; Chun, H.J.; Shin, D.B.; Park, B.S.; Kim, Y.H.; Park, H.-M.; Seo, H.S.; Song, J.T.; Kang, K.Y. The calmodulin-binding transcription factor OsCBT suppresses defense responses to pathogens in rice. *Mol. Cells* **2009**, *27*, 563–570. [CrossRef] [PubMed]
55. Kim, M.C.; Lee, S.H.; Kim, J.K.; Chun, H.J.; Choi, M.S.; Chung, W.S.; Moon, B.C.; Kang, C.H.; Park, C.Y.; Yoo, J.H. Mlo, a modulator of plant defense and cell death, is a novel calmodulin-binding protein: Isolation and characterization of a rice Mlo homologue. *J. Biol. Chem.* **2002**, *277*, 19304–19314. [CrossRef]
56. Ricachenevsky, F.K.; Sperotto, R.A.; Menguer, P.K.; Fett, J.P. Identification of Fe-excess-induced genes in rice shoots reveals a WRKY transcription factor responsive to Fe, drought and senescence. *Mol. Biol. Rep.* **2010**, *37*, 3735–3745. [CrossRef]
57. Fu, S.; Fu, L.; Zhang, X.; Huang, J.; Yang, G.; Wang, Z.; Liu, Y.-G.; Zhang, G.; Wu, D.; Xia, J. OsC2DP, a novel C2 domain-containing protein is required for salt tolerance in rice. *Plant Cell Physiol.* **2019**, *60*, 2220–2230. [CrossRef] [PubMed]
58. Zhao, J.; Wang, S.; Qin, J.; Sun, C.; Liu, F. The lipid transfer protein Os LTPL 159 is involved in cold tolerance at the early seedling stage in rice. *Plant Biotechnol. J.* **2020**, *18*, 756–769. [CrossRef] [PubMed]
59. Wang, X.; Zhou, W.; Lu, Z.; Ouyang, Y.; Yao, J. A lipid transfer protein, OsLTPL36, is essential for seed development and seed quality in rice. *Plant Sci.* **2015**, *239*, 200–208. [CrossRef] [PubMed]
60. Zhang, X.; Li, D.; Zhang, H.; Wang, X.; Zheng, Z.; Song, F. Molecular characterization of rice OsBIANK1, encoding a plasma membrane-anchored ankyrin repeat protein, and its inducible expression in defense responses. *Mol. Biol. Rep.* **2010**, *37*, 653–660. [CrossRef]
61. Pooja, S.; Sweta, K.; Mohanapriya, A.; Sudandiradoss, C.; Siva, R.; Gothandam, K.M.; Babu, S. Homotypic clustering of OsMYB4 binding site motifs in promoters of the rice genome and cellular-level implications on sheath blight disease resistance. *Gene* **2015**, *561*, 209–218. [CrossRef]

62. Takai, T.; Fukuta, Y.; Shiraiwa, T.; Horie, T. Time-related mapping of quantitative trait loci controlling grain-filling in rice (*Oryza sativa* L.). *J. Exp. Bot.* **2005**, *56*, 2107–2118. [CrossRef]
63. Sekhar, S.; Kumar, J.; Mohanty, S.; Mohanty, N.; Panda, R.S.; Das, S.; Shaw, B.P.; Behera, L. Identification of novel QTLs for grain fertility and associated traits to decipher poor grain filling of basal spikelets in dense panicle rice. *Sci. Rep.* **2021**, *11*, 13617. [CrossRef] [PubMed]
64. Song, X.-J.; Huang, W.; Shi, M.; Zhu, M.-Z.; Lin, H.-X. A QTL for rice grain width and weight encodes a previously unknown RING-type E3 ubiquitin ligase. *Nat. Genet.* **2007**, *39*, 623–630. [CrossRef]
65. Zhang, X.; Wang, J.; Huang, J.; Lan, H.; Wang, C.; Yin, C.; Wu, Y.; Tang, H.; Qian, Q.; Li, J. Rare allele of OsPPKL1 associated with grain length causes extra-large grain and a significant yield increase in rice. *Proc. Natl. Acad. Sci. USA* **2012**, *109*, 21534–21539. [CrossRef] [PubMed]
66. Xu, F.; Sun, X.; Chen, Y.; Huang, Y.; Tong, C.; Bao, J. Rapid identification of major QTLs associated with rice grain weight and their utilization. *PLoS ONE* **2015**, *10*, e0122206. [CrossRef]
67. Wei, X.; Jiao, G.; Lin, H.; Sheng, Z.; Shao, G.; Xie, L.; Tang, S.; Xu, Q.; Hu, P. GRAIN INCOMPLETE FILLING 2 regulates grain filling and starch synthesis during rice caryopsis development. *J. Integr. Plant Biol.* **2017**, *59*, 134–153. [CrossRef] [PubMed]
68. Wang, Y.; Xiong, G.; Hu, J.; Jiang, L.; Yu, H.; Xu, J.; Fang, Y.; Zeng, L.; Xu, E.; Xu, J. Copy number variation at the GL7 locus contributes to grain size diversity in rice. *Nat. Genet.* **2015**, *47*, 944–948. [CrossRef]
69. Wang, S.; Wu, K.; Yuan, Q.; Liu, X.; Liu, Z.; Lin, X.; Zeng, R.; Zhu, H.; Dong, G.; Qian, Q. Control of grain size, shape and quality by OsSPL16 in rice. *Nat. Genet.* **2012**, *44*, 950–954. [CrossRef]
70. Sun, L.; Li, X.; Fu, Y.; Zhu, Z.; Tan, L.; Liu, F.; Sun, X.; Sun, X.; Sun, C. GS 6, a member of the GRAS gene family, negatively regulates grain size in rice. *J. Integr. Plant Biol.* **2013**, *55*, 938–949. [CrossRef]
71. Shomura, A.; Izawa, T.; Ebana, K.; Ebitani, T.; Kanegae, H.; Konishi, S.; Yano, M. Deletion in a gene associated with grain size increased yields during rice domestication. *Nat. Genet.* **2008**, *40*, 1023–1028. [CrossRef]
72. Qi, P.; Lin, Y.-S.; Song, X.-J.; Shen, J.-B.; Huang, W.; Shan, J.-X.; Zhu, M.-Z.; Jiang, L.; Gao, J.-P.; Lin, H.-X. The novel quantitative trait locus GL3. 1 controls rice grain size and yield by regulating Cyclin-T1;3. *Cell Res.* **2012**, *22*, 1666–1680. [CrossRef]
73. Mao, H.; Sun, S.; Yao, J.; Wang, C.; Yu, S.; Xu, C.; Li, X.; Zhang, Q. Linking differential domain functions of the GS3 protein to natural variation of grain size in rice. *Proc. Natl. Acad. Sci. USA* **2010**, *107*, 19579–19584. [CrossRef]
74. Li, Y.; Fan, C.; Xing, Y.; Jiang, Y.; Luo, L.; Sun, L.; Shao, D.; Xu, C.; Li, X.; Xiao, J. Natural variation in GS5 plays an important role in regulating grain size and yield in rice. *Nat. Genet.* **2011**, *43*, 1266–1269. [CrossRef] [PubMed]
75. Jiang, H.; Zhang, A.; Liu, X.; Chen, J. Grain size associated genes and the molecular regulatory mechanism in rice. *Int. J. Mol. Sci.* **2022**, *23*, 3169. [CrossRef]
76. Ishimaru, K.; Hirotsu, N.; Madoka, Y.; Murakami, N.; Hara, N.; Onodera, H.; Kashiwagi, T.; Ujiie, K.; Shimizu, B.-I.; Onishi, A. Loss of function of the IAA-glucose hydrolase gene TGW6 enhances rice grain weight and increases yield. *Nat. Genet.* **2013**, *45*, 707–711. [CrossRef]
77. Breviario, D.; Gianì, S.; Morello, L. Multiple tubulins: Evolutionary aspects and biological implications. *Plant J.* **2013**, *75*, 202–218. [CrossRef]
78. Wasteneys, G.O. Microtubule organization in the green kingdom: Chaos or self-order? *J. Cell Sci.* **2002**, *115*, 1345–1354. [CrossRef]
79. Mathur, J.; Hülskamp, M. Microtubules and microfilaments in cell morphogenesis in higher plants. *Curr. Biol.* **2002**, *12*, R669–R676. [CrossRef] [PubMed]
80. Hashimoto, T. Microtubules in plants. *Arab. Book Am. Soc. Plant Biol.* **2015**, *13*, e0179. [CrossRef] [PubMed]
81. Segami, S.; Kono, I.; Ando, T.; Yano, M.; Kitano, H.; Miura, K.; Iwasaki, Y. Small and round seed 5 gene encodes alpha-tubulin regulating seed cell elongation in rice. *Rice* **2012**, *5*, 4. [CrossRef]
82. Yang, G.; Jan, A.; Komatsu, S. Functional analysis of OsTUB8, an anther-specific β-tubulin in rice. *Plant Sci.* **2007**, *172*, 832–838. [CrossRef]
83. Blume, Y.B.; Lloyd, C.W.; Yemets, A.I. Plant tubulin phosphorylation and its role in cell cycle progression. In *The Plant Cytoskeleton: A Key Tool for Agro-Biotechnology*; Springer: Dordrecht, The Netherlands, 2008; pp. 145–159.
84. Chen, J.; Cao, F.; Li, H.; Shan, S.; Tao, Z.; Lei, T.; Liu, Y.; Xiao, Z.; Zou, Y.; Huang, M. Genotypic variation in the grain photosynthetic contribution to grain filling in rice. *J. Plant Physiol.* **2020**, *253*, 153269. [CrossRef] [PubMed]
85. Watson, D.K.; Kitching, R.; Vary, C.; Kola, I.; Seth, A. Isolation of target gene promoter/enhancer sequences by whole genome PCR method. In *Transcription Factor Protocols*; Springer: Berlin/Heidelberg, Germany, 2000; pp. 1–11.
86. Tanaka, W.; Toriba, T.; Hirano, H.Y. Three TOB 1-related YABBY genes are required to maintain proper function of the spikelet and branch meristems in rice. *New Phytol.* **2017**, *215*, 825–839. [CrossRef] [PubMed]
87. Mhamdi, A.; Van Breusegem, F. Reactive oxygen species in plant development. *Development* **2018**, *145*, dev164376. [CrossRef] [PubMed]
88. Jasid, S.; Simontacchi, M.; Bartoli, C.G.; Puntarulo, S. Chloroplasts as a nitric oxide cellular source. Effect of reactive nitrogen species on chloroplastic lipids and proteins. *Plant Physiol.* **2006**, *142*, 1246–1255. [CrossRef] [PubMed]
89. Dietz, K.-J.; Turkan, I.; Krieger-Liszkay, A. Redox-and reactive oxygen species-dependent signaling into and out of the photosynthesizing chloroplast. *Plant Physiol.* **2016**, *171*, 1541–1550. [CrossRef] [PubMed]
90. Lacza, Z.; Pankotai, E.; Busija, D.W. Mitochondrial nitric oxide synthase: Current concepts and controversies. *Front. Biosci.* **2009**, *14*, 4436. [CrossRef] [PubMed]

91. Huang, S.; Van Aken, O.; Schwarzländer, M.; Belt, K.; Millar, A.H. The roles of mitochondrial reactive oxygen species in cellular signaling and stress response in plants. *Plant Physiol.* **2016**, *171*, 1551–1559. [CrossRef] [PubMed]
92. Sandalio, L.; Romero-Puertas, M. Peroxisomes sense and respond to environmental cues by regulating ROS and RNS signalling networks. *Ann. Bot.* **2015**, *116*, 475–485. [CrossRef]
93. Corpas, F.J.; Hayashi, M.; Mano, S.; Nishimura, M.; Barroso, J.B. Peroxisomes are required for in vivo nitric oxide accumulation in the cytosol following salinity stress of Arabidopsis plants. *Plant Physiol.* **2009**, *151*, 2083–2094. [CrossRef]

Disclaimer/Publisher's Note: The statements, opinions and data contained in all publications are solely those of the individual author(s) and contributor(s) and not of MDPI and/or the editor(s). MDPI and/or the editor(s) disclaim responsibility for any injury to people or property resulting from any ideas, methods, instructions or products referred to in the content.

MDPI AG
Grosspeteranlage 5
4052 Basel
Switzerland
Tel.: +41 61 683 77 34

Agronomy Editorial Office
E-mail: agronomy@mdpi.com
www.mdpi.com/journal/agronomy

Disclaimer/Publisher's Note: The title and front matter of this reprint are at the discretion of the Guest Editors. The publisher is not responsible for their content or any associated concerns. The statements, opinions and data contained in all individual articles are solely those of the individual Editors and contributors and not of MDPI. MDPI disclaims responsibility for any injury to people or property resulting from any ideas, methods, instructions or products referred to in the content.

www.ingramcontent.com/pod-product-compliance
Lightning Source LLC
LaVergne TN
LVHW072316090526
838202LV00019B/2297